Second Growth

T0143277

Second Growth

THE PROMISE OF TROPICAL FOREST REGENERATION IN AN AGE OF DEFORESTATION

ROBIN L. CHAZDON

THE UNIVERSITY OF CHICAGO PRESS

Chicago and London

The University of Chicago Press, Chicago 60637
The University of Chicago Press, Ltd., London
© 2014 by The University of Chicago
All rights reserved. Published 2014.
Printed in the United States of America

23 22 21 20 19 18 17 16 2 3 4 5

ISBN-13: 978-0-226-11791-1 (cloth)
ISBN-13: 978-0-226-11807-9 (paper)
ISBN-13: 978-0-226-11810-9 (e-book)
DOI: 10.7208/chicago/9780226118109.001.0001

Library of Congress Cataloging-in-Publication Data

Chazdon, Robin Lee, 1957- author.
Second growth : the promise of tropical forest regeneration in an age of deforestation / Robin L. Chazdon.
pages cm
Includes bibliographical references and index.
ISBN 978-0-226-11791-1 (cloth : alk. paper) — ISBN 978-0-226-11807-9 (pbk. : alk. paper) — ISBN 978-0-226-11810-9 (e-book) 1. Reforestation. 2. Forest ecology. 3. Tropics—Ecology. I. Title.
SD409.C525 2014
577.3—dc23
2013033834

⊗ This paper meets the requirements of ANSI/NISO Z39.48-1992 (Permanence of Paper).

To my mother,
who nurtured my love for
nature and for people

CONTENTS

ILLUSTRATIONS

Plates

Following page 204

Figures

Tables

Boxes

PREFACE

As a child, the closest I got to a jungle was the jungle gym at Jeffrey Park in Chicago's South Side. It wasn't until my second year in college that I had an opportunity to visit a real jungle. During that semester in Costa Rica, keeping tropical forests alive and well became a life-long passion that continued to grow during my graduate work, postdoctoral research, and 25 years as a university professor. I have returned to La Selva Biological Station in Costa Rica (owned and operated by the Organization for Tropical Studies) nearly every year since 1980, at first to study the light environments and ecology of understory palms, then to study responses of shrubs to light variation. And then, I really saw the light. Naturally regenerating forests were springing up inside and outside of La Selva's boundaries, replacing pastures that earlier had replaced old-growth forest. In 1992, together with Julie Denslow, I embarked on the first steps of what has become a life-long journey to understand how tropical rain forests grow back after they have been deforested.

My research on tropical forest regeneration has since become a multidisciplinary exploration, with a strong leaning toward anthropology and geography. As I became better acquainted with the plant and animal life in these young forests, I came to realize that a regenerating forest is a nexus between people and nature. The species that establish in former pastures or cultivated fields arrive from the surrounding landscape, where people live and work the land. Regenerating forests are the collective "backyard" of the local community. Whether their actions help or hinder forest regrowth, people are part of the forest ecosystem. Regenerating and restored forests are also the nexus between conservation and development, and between social and natural scientists. Forest regeneration and reforestation can restore the goods and services of previously forested landscapes, providing resources for people and for a wide array of plant and animal life. Regenerating tropical forests are the foundation of a renewed future for tropical landscapes and rural communities.

The knowledge that tropical forests are malleable is empowering. It is time to use this power to help tropical forests regenerate wherever and whenever it is possible. Forest regeneration can happen in ways that have positive consequences for the billions of people who depend on forests for their livelihoods and well-being. I wrote this book primarily to convey this urgent message. Tropical forest regeneration is a key to sustaining tropical forests, their unique biodiversity, and their ecosystem functions. All earthlings depend on tropical forests in one way or another.

Chapter 1 frames several key issues and introduces themes that are developed in later chapters. I illustrate different perceptions of tropical forests, describe their regeneration, and present information on the extent of regenerating forests across tropical regions. Human activities have shaped forests and landscapes throughout the tropics in ways that we are just beginning to understand. Chapters 2 and 3 explore the legacies of ancient human occupation and land uses in the tropics. Chapters 4 and 5 focus on disturbance regimes in tropical forests and the nature of successional pathways that characterize forest regeneration. Forest dynamics, described in chapter 4, reflect different types of disturbances and analyses at different spatial scales. Chapter 5 provides a broad conceptual overview of successional patterns and phases in tropical forests and discusses different approaches to studying forest succession.

Chapters 6–9 provide a summary of forest regeneration following different types of disturbances, including succession on newly created substrates (chap. 6), following different types of human land use (chap. 7), following hurricanes and fires (chap. 8), and after logging (chap. 9). Chapters 10–12 delve into the details of how regenerating forests change in structure, species composition, and ecosystem properties. The functional traits that influence plant community assembly are described in chapter 10, and chapter 11 focuses on the recovery of key ecosystem processes during succession, nutrient and carbon accumulation, and hydrological functions. Chapter 12 summarizes information on animal diversity and plant-animal interactions during forest regeneration.

The closing chapters of the book look forward to the future of regenerating forests in the tropics. Chapter 13 examines different pathways toward reforestation and restoration of tropical forests, and chapter 14 examines the socioecology of regenerating forests from a landscape perspective. The concluding chapter provides an overall synthesis and returns to the themes developed in chapter 1. Through this journey, my goal is to show that regenerating tropical forests are and have always been an integral component of tropical ecosystems, and that understanding, promoting, and managing forest regeneration are key to sustaining tropical forests worldwide.

ACKNOWLEDGMENTS

One cold December morning I woke up with a vision of this book in my head. I quickly typed an outline into my computer, convinced myself that I was not crazy, and sent a slightly more fleshed-out version to Christie Henry, who provided essential encouragement and feedback. Now that this vision has grown by megabytes, I have many people to thank. First, I want to thank the funding agencies that have supported my research on tropical forest regeneration over the past 22 years: Andrew W. Mellon Foundation, AAUP Travel Awards, Blue Moon Fund, Fulbright Foundation, NASA, National Science Foundation, Organization for Tropical Studies, and University of Connecticut Research Foundation. This funding enabled me to visit tropical forest sites, colleagues, and stakeholders around the world. These funding agencies also supported the work of local paraforesters and parataxonomists who trudge out to research sites in Costa Rica day after day, rain or shine, to document the rise and fall of trees, saplings, and seedlings. Second, I thank the University of Connecticut College of Liberal Arts and Sciences Book Support Committee and the Department of Ecology and Evolutionary Biology for providing generous funding for the color plates. Profound thanks to Jeanette Paniagua, Marcos Molina, Bernal Paniagua, Enrique Salicetti, and Juan Romero—you are the best! A special thanks for Orlando Vargas for all he has done in teaching me, my students, and my staff to identify plants in the field.

My graduate students collaborated in much of this research over the past 20 years: Pablo Arroyo, Vanessa Boukili, Catherine Cardelús, Alexander DeFrancesco, Juan Dupuy, Zbigniew Grabowski, Silvia Iriarte, Susan Letcher, Rebecca Montgomery, Adrienne Nicotra, Manette Sandor, Uzay Sezen, and Amanda Wendt. I also deeply thank my collaborators and colleagues over the years, who have broadened and deepened my research on tropical forest regeneration around the world: Patricia Balvanera, Frans Bongers, Anne Chao, David Clark, Deborah Clark, Robert Colwell, Julie Denslow, Bryan Finegan, Manuel Guariguata, Deborah Lawrence, Miguel Martínez-Ramos, Rita Mesquita, Natalia Norden, Edgar Ortiz, Sassan Saatchi, Nate Swenson, Maria Uriarte, Braulio Vilchez, Bruce Williamson, Mike Willig, and Jess Zimmerman. Many of the ideas generated by our conversations are somewhere in this book.

Portions of this book were written while I was on sabbatical at the Ecosystems Research Center (CIECO) of the National University of México in Morelia and at the Center for Macroecology, Evolution and Climate (CMEC) at the University of Copenhagen in Denmark. CMEC provided financial support during my three-month stay. The snowiest winter in 20 years gave me ample reason to stay inside, look out the window, and write about tropical forests. Special thanks to Carsten Rahbek for his support and friendship. To paraphrase the Carlsberg beer slogan, CMEC is probably the best place in town. I especially want to thank the University of Connecticut Library and the interlibrary loan department for providing prompt and professional service from wherever I was. I couldn't have written this book without access to the Virtual Personal Network (VPN). Echo Valley Ranch provided solitude, space, and scenery, all of which helped me to complete the final manuscript and revisions.

Numerous people reviewed early drafts of chapters and text boxes, helping me to present accurate information in a clear and organized way. Many thanks to Ellen Andresen, Patricia Balvanera, Warren Brockelman, David Clark, Deborah Clark, Patrick Dugan, Giselda Durigan, Ren Hai, Rhett Harrison, Henry Howe, David Lamb, Carlos Peres, Tom Rudel, Michael Swaine, Lawrence Walker, Robert Whittaker, and Bruce Williamson. I especially thank Eduardo van den Berg, Manette Sandor, Susan Letcher, and Vanessa Boukili for constructive reviews of multiple chapters. Several colleagues provided valuable information, papers in press, and photographs. Many thanks to Mitch Aide, Doug Boucher, Mark Bush, Jeffrey Chambers, Charles Clement, David Douterlunge, James Fairhead, John Hoopes, Dennis Knight, Richard Lucas, Akane Nishimura, Stephanie Paladino, Steward Pickett, Dolores Piperno, Michael Poffenberger, Ferry Slik, Michael Swaine, Christopher Uhl, John Vandermeer, Lawrence Walker, Joe Wright, and Jianguo Wu. I am especially grateful to two astute and thorough peer reviewers who read over the entire book manuscript and made numerous valuable suggestions that greatly improved the final version.

I am deeply grateful for the support of my family in so many ways. Carol Chazdon and Bob Amend shared their home with me on many occasions while I was writing and helped me to stay focused during challenging times. My children, Rachel and Charles, were of the right age and academic background to read some chapters and provide feedback and encouragement. Discussing anthropological issues with Rachel was particularly enriching. Rob Colwell, my husband, friend, and colleague, has always been there to exchange ideas, to feed me (intellectually, emotionally, and literally), and to listen patiently to my rants, despite his own pressing deadlines. His belief in me kept me going.

CHAPTER 1

PERCEPTIONS OF TROPICAL FORESTS AND NATURAL REGENERATION

Indigenous knowledge does indeed hold valuable information on the role that species play in ecologically sustainable systems. Such knowledge is of great value for an improved use of natural resources and ecological services, and could provide invaluable insights and clues for how to redirect the behavior of the industrial world towards a path in synergy with the life-support environment on which it depends.—Gadgil et al. (2003, p. 156)

1.1 Viewing Forests as a Cycle

In the holistic worldview of indigenous resource managers, the forest has no end and no beginning—the forest is a cycle that is managed to provide for their needs. The Dayak are forest-dwelling tribes in Borneo with a deep knowledge of forest regeneration and management. For over 4,000 years, their lives have revolved around a system of shifting cultivation based on the regenerative capacities of the tropical forest ecosystem. Their life cycle is completely interwoven with the forest's own regeneration cycle. The Benuaq Dayaks of Datarban in East Kalimantan recognize that many factors affect the rate of forest regeneration, including soil conditions, rainfall, temperature, slope, and aspect. They define five phases of forest regeneration following a short (1–2 yr.) period of cultivation in swidden fields (*ladang*) that are carved out of the primary forest. Within 1–3 years after cultivation ceases, dense young scrub (*kurat uraq*) covers the former swidden field (Poffenberger and McGean 1993). This phase can take 3–5 years if soils are compacted, eroded, or heavily leached, and is dominated by light-demanding grasses, perennial shrubs, herbs, and fast-growing tree species that reach heights of 3–4 meters.

The second phase (*kurat tuha*) follows 2–5 years after fields are fallowed. Trees reach 5 centimeters or more in diameter and heights of about 5–6 meters. The undergrowth is filled with dense shrubs, lianas (woody vines), and herbaceous species that grow rapidly at high light levels. After 3–10 years on good quality soil, the third phase begins (*kurat batang muda*). Pioneer trees now reach 10–15 centimeters in diameter. The upper canopy begins to close, reducing light in the understory. Understory grasses and herbs then start to die out. After 9–16 years, depending on soil quality, the fourth (and longest) phase begins (*kurat batang tuha*). Shade-tolerant tree saplings fill in the

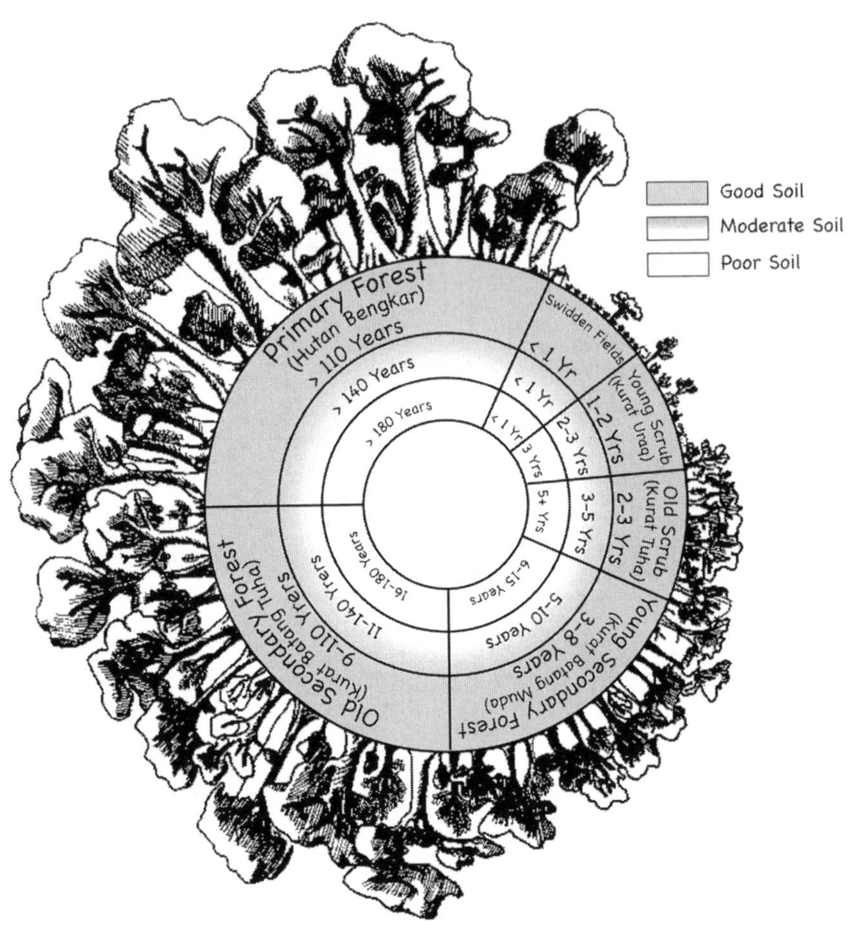

Figure 1.1
Benuaq Dayak phases of forest regeneration. The time required for each phase depends upon soil quality, with good soils favoring more rapid regeneration. Source: Redrawn with permission from Poffenberger and McGean (1993, fig. 6).

spaces in the understory vacated by shrubs, herbs, and grasses. The forest canopy is now virtually closed by trees with diameters well over 10 centimeters. Over the next 100–180 years, the forest changes gradually in structure, composition, and spatial complexity until it returns to the status of primary (old-growth) forest (*hutan bengkar*; fig. 1.1).

The Dayaks' understanding of natural regeneration strikingly resembles that of forest ecologists and foresters. This knowledge has been part of their cultural tradition for as long as there have been Dayaks (Sardjono and Samsoedin 2001; Setyawan 2010). Dayak farmers have careful procedures for determining whether a site is suitable for opening a new swidden field. One test involves cutting a length of stem from a wild ginger plant and burying it in the forest soil. If the stem resprouts within 3–5 days, the conditions of soil moisture and nutrient availability are considered suitable for cultivation. Certain tree species in the Dipterocarpaceae family and certain understory herbs are additional indicators of desirable soil conditions. When the field is cleared, large valuable trees are spared to provide timber and honey in future years. Stumps are left to resprout as coppice regeneration, accelerating regrowth and reducing weed growth.

The Dayaks' sophisticated understanding of forest regeneration is shared by other forest-dwelling peoples who have practiced shifting cultivation and harvested forest products for millennia (Wiersum 1997). Over 16,000 kilometers away, the Yucatec Maya have practiced shifting cultivation for over 3,000 years in the northern Maya lowlands of Mesoamerica. Their way of life also depends on understanding and managing forest regeneration. Their vocabulary includes 6 terms for successional stages of the forest and more than 80 terms for soil characteristics (fig 1.2; Gómez-Pompa 1987; Barrera-Bassols and Toledo 2005). The Soligas of the Western Ghats in India have developed similar traditional knowledge to support their shifting-cultivation lifestyle (Madegowda 2009). Indigenous peoples have learned how forests respond to different types of disturbance and which species of plants and animals appear and

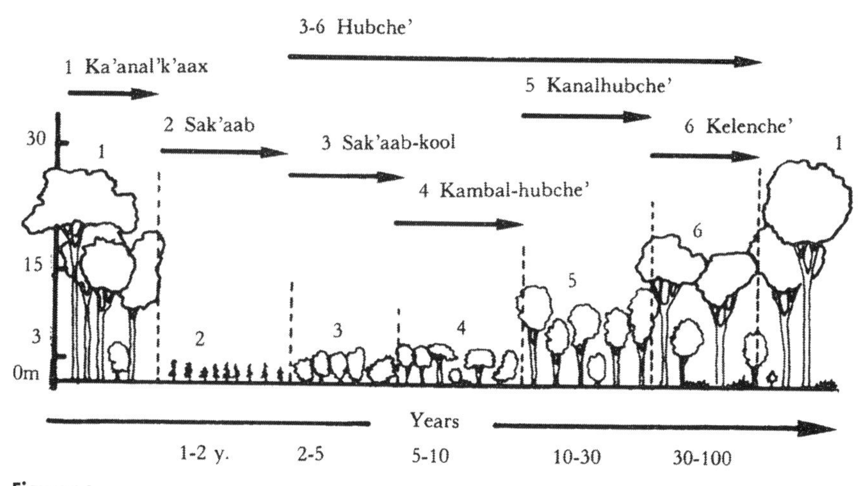

Figure 1.2
Six stages of forest regeneration as defined by the Maya people.
Source: Gómez-Pompa (1987, fig. 1).

proliferate in different phases of forest regeneration. They recognize particular species as indicators of soil conditions. Traditional ecological knowledge enables them to practice adaptive management, adjusting their impact on the forests to sustain regeneration, even over multiple cycles of cultivation.

1.2 The Resilience of Tropical Forests

Traditional shifting cultivators have developed adaptive management practices that sustain the forest regeneration cycle. Their lives and the future of their forest areas depend on this partnership. When the cycle is broken, forests lose their intrinsic capacity to regenerate. There is a limit to the resilience of tropical forests, to their ability to recover from disturbances and reassemble all of their parts. The traditional practices of the Dayaks of Datarban are now no longer possible, because the area of forest available for their use is insufficient to allow long-term rotations (Poffenberger and McGean 1993). In their traditional practice, forests would not be cleared until they reached the fourth stage (*kurat batang tuha*). But increasing population pressures on limited forest lands have forced many farmers to reduce their fallow periods to 5–10 years. Young fallows are being cleared for fields before they have fully recovered soil organic matter and nutrients and before weeds have been controlled by regrowth. Fires are frequent in young scrub, suppressing woody regrowth and favoring weeds and invasive grasses, such as *Imperata cylindrica*. This new regimen reduces the forest's regenerative capacity. The shifting cultivation system that was sustainable for over 4,000 years has now become unsustainable (Coomes et al. 2000; Lawrence et al. 2010).

Tropical forests are often considered to be highly fragile and vulnerable ecosystems because of their complex vertical structure, high species diversity, and intricate networks of species interactions (Chazdon and Arroyo 2013). Immediately following a major disturbance, devastation appears irreversible. However, what makes a forest ecosystem fragile is not its immediate destruction but rather interference with its intrinsic resilience (see figs. 1.1 and 1.2). Tropical forests, like all ecosystems, are naturally subjected to disturbances of variable intensity, frequency, and duration. Forest ecosystems are always in flux. Forces of nature, such as hurricanes, floods, or volcanic eruptions, cause major disturbances. Human activities—such as shifting cultivation, permanent cultivation, grazing, and logging—also cause major disturbances by removing and fragmenting vegetation and degrading soils. Human and natural disturbances often act together to influence forest dynamics. For example, selective logging increases the susceptibility of forests to burning.

Resilience is a feature of a complex adaptive system that is capable of self-reorganization in the aftermath of disturbance—the capacity to return, over time, to a state similar to the predisturbance state (Holling 1973; Levin 1998). This reorganization, in the case of ecosystems, is embodied in the concept of succession and in the forest regeneration cycle (see fig. 1.1; Messier et al. 2013). The gradual reassembly process can take more than a century, especially for ecosystems composed of hundreds of species of long-lived trees. Moreover, each phase in the process can be influenced by factors internal and external to the system (Bengtsson et al. 2003; Chazdon and Arroyo 2013). Tropical forest regeneration is influenced by prior land use, initial colonization, climate, soils, and seed dispersal from forests in the surrounding landscape. When obstacles to regeneration prevail, succession is arrested and a new type of ecosystem develops, such as a grass- or fern-dominated ecosystem (see box 7.1). In these cases, careful human intervention is needed to get succession back on track. Evaluating the resilience of tropical forests involves more than understanding the ecological process of forest regeneration. Humans are active participants in both the disturbance and the recovery of natural systems; the delineation between social and ecological systems is artificial and arbitrary (Berkes and Folke 1998). Resilience is a feature of the combined socioecological system that includes both ecological and human components.

1.3 Forest Regeneration, Succession, and Forest Degradation

Following a small, localized disturbance, spontaneous forest regeneration occurs in disturbed patches within the forest matrix. Gap-phase regeneration is a normal part of forest dynamics (see box 4.1). During forest succession, the spatial unit undergoing regeneration is the entire forest stand (Chokkalingam and de Jong 2001). *Regeneration* is a commonly used term to describe regrowth following forest disturbance at a range of spatial scales, analogous with the regeneration, regrowth, or reconstitution of tissues or organs following damage or loss. Natural regeneration can apply to an individual tree or population, a single tree species, a small forest patch, an entire stand, an assemblage, or an ecosystem, and refers to the regrowth or reestablishment of these units.

Succession is a process linked to the assemblage of species composing a particular ecosystem. Natural regeneration of populations, species, and assemblages occurs throughout all stages of succession. Trees that regenerate during the early phases of succession constitute a different set of species from trees that regenerate during later phases of succession (see box 5.1). Following selective logging disturbances, many trees remain standing in the forest, but they may sustain considerable damage. Successional processes in formerly selectively logged forests drive stand-level changes in structure and composition, but from a different starting point than in former agricultural sites.

The designation *naturally regenerating forest* used in the Forest Resource Assessment (FRA) of the United Nations Food and Agriculture Organization (FAO) can be applied to forests regenerating following selective logging of trees *or* to forests regrowing on former agricultural land that was completely cleared (FAO 2010). A "primary forest" is relatively stable in structure and composition, whereas a "secondary forest" develops after the original forest has been cleared and regenerates spontaneously. The term *secondary forest* also widely refers to selectively logged forests, creating much ambiguity (Chokkalingam and de Jong 2001). Throughout this book, I distinguish between selectively logged forests and second-growth forests on cleared land.

In 2011, the FRA published a working paper entitled "Assessing forest degradation: Towards the development of globally applicable guidelines" (FAO 2011). More than 50 definitions of *forest degradation* have been formulated for different applications (Lund 2009). Forest degradation is caused by humans and is often ascribed to poor or inadequate management or misuse. In 2002, the International Tropical Timber Organization estimated that up to 850 million hectares of tropical forest and forest land were in a degraded state (ITTO 2002). The International Union for the Conservation of Nature (IUCN) and the Global Partnership for Forest Landscape Restoration (GPFLR) have initiated a global effort to restore 150 million hectares of degraded and deforested land by 2020, a fraction of the 2 billion hectares worldwide that offer opportunities for landscape-level restoration (GPFLR 2012). According to ITTO, a degraded forest supplies fewer goods and services and maintains limited biological diversity (ITTO 2002, 2005). The Convention on Biological Diversity (CBD) stated that "a degraded forest is a secondary forest that has lost, through human activities, the structure, function, species composition or productivity normally associated with a natural forest type expected on that site" (Secretariat of the Convention on Biological Diversity 2002, p. 154). If we were grading a forest's condition on a US university scale, a degraded forest would receive a grade less than an A (94%–100%). But it could be anywhere from an A–to an F (failing).

These definitions imply that any forest regenerating on former pastures or cultivated fields is a degraded forest, because the structure, function, composition, productivity, and services are reduced compared to the original, or "natural," forest. Indeed, this inclusive approach is implicit in the CBD's definition. But this definition reflects underlying confusion over the definition of *secondary forest.* Putz and Redford (2010) proposed that secondary forests that develop after complete deforestation should be distinguished from degraded forests, which are derived from old-growth forests and retain some of the original forest structure and composition. Secondary forests are young, second-growth forests that take time to develop the features of a "mature," or old-growth, forest in the same region and climate zone. Yet young forests are labeled *degraded* simply because they are young, regardless of their potential for recovery. They are often viewed as "damaged goods." In fact, the term *secondary* implies that these forests are secondary in quality and in value. For this reason, I prefer to use the term *second-growth*, or *regenerating, forests* in this book.

How do we escape this quagmire of terms? The way forward is to view forests as dynamic entities. The current application of the label *degraded forest* to any forest that has been impacted by human activities (hunting, logging, fragmentation, grazing, or cultivation) reflects a static view of the forest that fails to consider the potential for spontaneous natural regeneration to regain "lost" properties. Rather, forests should be viewed as resilient systems, with intrinsic capacities to reorganize and recover. If all forests affected by humans are lumped into a single category of *degraded forest* by forestry, biofuel industry, and climate mitigation policies, we are overlooking one of the greatest conservation opportunities in human history. In India, ecosystems that experience moderate to heavy disturbance are classified as wastelands (Ravindranath et al. 1996). The way forward is to move beyond the labels of *degradation, deforestation*, and *devastation* to promote forest regeneration, regrowth, and restoration on a massive scale. We can work toward re-creating sustainable forest regeneration cycles and rebuilding resilient ecosystems. We should not simply condemn all forests affected by human activities and auction them off to the highest bidder. If forests can regrow following complete sterilization of the island of Rakata in 1883, our hope can be restored too (see box 6.1).

1.4 The Geographic Extent of Deforestation and Forest Regeneration across the Tropics

Deforestation in the tropics is a major environmental issue of our time, with far-reaching consequences for earth's biodiversity, climate, and life-support systems (MEA 2005). During the last 100 years, old-growth tropical forests have been cleared and replaced with agriculture, pastures, plantations, and young regenerating forests at unprecedented rates. From 1990 to 2000, 8.6 million hectares were deforested across the tropics, including wet and dry zones (see plate 1 for "hot spots" of deforestation; Mayaux et al. 2005). Approximately 42% of the world's tropical forests are seasonally dry forests, which have experienced even higher rates of deforestation than humid tropical forests. Only 16% of the dry forests of South and Southeast Asia remained in 2001, compared to 40% in Latin America (Miles et al. 2006).

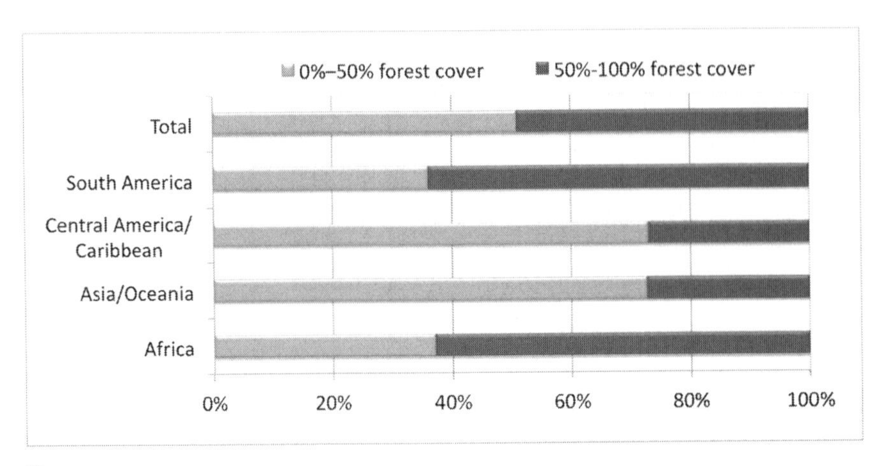

Figure 1.3
Geographic distribution of humid tropical forest areas with 0%–50% forest cover and 50%–100% forest cover in 2005. Source: Asner et al. (2009, table 1).

As of 2005, roughly half of the humid tropical forest biome showed less than 50% cover. South America (northwest Amazon basin) and the eastern Congo basin had the largest extents of intact humid tropical forest, whereas Asia/Oceania and Central America had the lowest extents. Selective logging operations affected 3.9 million square kilometers or 20.3% of global humid tropical forest area (fig. 1.3; Asner et al. 2009).

How much tropical forest is actually undergoing regeneration now? Estimating the extent of regenerating forests in tropical regions is a far more difficult task than estimating rates of deforestation for several reasons (Asner et al. 2009). First, regrowth of vegetation is difficult to monitor using traditional satellite observations, because successional forests cannot be distinguished from old-growth forests beyond the early stages (Lucas et al. 2000). Second, even if areas of young second growth can be identified, these areas may be fallows that are cleared after only a few years. Third, regenerating forests are difficult to distinguish from tree plantations using low-resolution satellite imagery. The best approach to tracking areas of forest regeneration is the sequential analysis of high-resolution satellite imagery, but such studies are limited in spatial and temporal scale (da Conceição Prates-Clark et al. 2009). Forests in late stages of regeneration are the most difficult to distinguish from old-growth, or primary, forests.

Over 30 years ago, Brown and Lugo (1990) estimated that second-growth forests and selectively logged forests combined covered about 26% of the forest area in the tropics. Asner et al. (2009) estimated the geographic extent of deforestation, selective logging, and forest regeneration in humid tropical forests worldwide. Based on 23 regional studies, they concluded that at least 1.2% of the humid tropical forests of the world were undergoing secondary succession for at least a 10-year time span in the year 2000. Wright (2010) corrected these results, using the actual area surveyed rather than the total area of the biome to compute rates of forest regrowth. His calculations, which are thus more accurate, indicated that 11.8% of the tropical forest areas evaluated were undergoing regeneration. This estimate cannot be extrapolated beyond the areas surveyed, however, as the 23 regional studies were not randomly sampled.

In 2010, the FRA began reporting statistics on reforestation as well as on deforestation. But the report points out discrepancies in the quality of data contributed by different countries that have not yet been resolved. Based on the 2010 FRA data, primary forests have been reduced to less than 30% of the tropical forest cover worldwide FAO 2010). Across the tropics, South America claims the highest remaining proportion of forest as primary forests (79%), whereas eastern/southern Africa and the Caribbean claim the lowest proportion (< 6%). The highest proportion of forest cover in tropical regions other than South America consists of naturally regenerating forest (fig. 1.4).

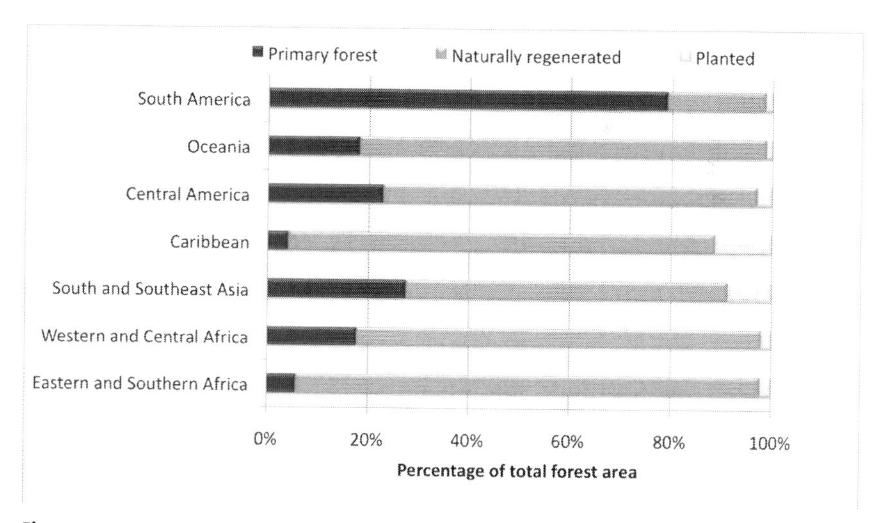

Figure 1.4
The distribution of primary (old-growth) forests, naturally regenerated forests (logged and successional forests), and planted forests (plantations) in different geographic regions based on the 2010 Global Forest Resource Assessment of the Food and Agriculture Organization of the United Nations. Source: FAO (2010, FRA table 7).

This snapshot of forest status is vastly different from the regional studies presented by Wright (2010), which encompass all land uses in the landscape, including agriculture. Moreover, the *naturally regenerated* category in the FRA is dominated by selectively logged forests, with a relatively small contribution of successional forests growing on land previously cleared (FAO 2010).

The fact is we do not have an accurate assessment of the global extent of regenerating tropical forests today, 10 years ago, or 30 years ago. Several issues plague the accurate global assessment of tropical forest cover and its classification into different forms. One major issue is the definition of *forest cover* itself, apart from the challenges of distinguishing successional stages from plantations, old-growth forest, or selectively logged forest across different tropical regions. Systematic attention has not been focused on constructing accurate global knowledge of tropical forest area, biodiversity, and ecosystem services. Sadly, the formal institutions necessary for ensuring quality and consistency in tropical forest assessments at regional and subregional scales have not yet been established (Grainger 2010).

1.5 The Tropical Forests of the Future

Since primary tropical forests are more critical for conserving biodiversity worldwide (Gibson et al. 2011), why should we care about regenerating forests? Foremost, regenerating forests are the predominant form of forest cover in the tropics worldwide. Only 19 of the 106 tropical nations (18%) that supplied data for the 2010 FRA reported more area of primary forest than naturally regenerating forest (FAO 2010). For 51 nations (48%), regenerating forests covered more area than primary forests, and 36 nations (34%) reported that the only natural forest cover remaining was in secondary forests. As crude and inconsistent as these reported statistics are, they highlight the critical importance of regenerating forests for over 80% of tropical nations. Without regenerating forests in tropical landscapes, we stand to lose most of the global carbon stocks and a significant fraction of the world's species and ecosystem services. Much is at stake.

Decades of carefully executed research have revealed much about tropical forest regeneration. This book synthesizes results of hundreds of studies of forest regeneration following different types and intensities of disturbance across different regions and forest types. An understanding of forest regeneration pathways is essential to manage and restore forests and to forecast forest change in a changing climate. Studies of forest regeneration inform how different types and intensities of land use influence successional trajectories of forest structure, species composition, and accumulation of carbon and nutrients in forest ecosystems. Regenerating forests provide habitats for a large proportion of forest-dwelling animal species that disperse seeds, pollinate flowers, and drive trophic interactions. By understanding successional pathways, we can better diagnose the causes of arrested succession and prescribe treatments to overcome barriers to natural regeneration. We can learn from traditional shifting cultivators how to enrich regenerating forests with products for domestic or commercial use. And we can learn how geographic and socioeconomic factors influence deforestation and forest regeneration at multiple spatial and temporal scales.

By understanding forest regeneration, we also learn about the impacts of human activities on forest ecosystems today and in the past. Based on features of successional vegetation, we can identify periods of human disturbance followed by regeneration in tropical forests over the past 2,000–45,000 years. Today's primary forests were once secondary forests. If they are given a second chance, today's secondary forests will become the tropical forests of the future.

CHAPTER 2

ANCIENT HUMAN LEGACIES IN TROPICAL FOREST LANDSCAPES

It is now evident that much so-called "primary" tropical forest has been cultivated at some distant period and that, in fact, the term has usually been used to contrast "secondary" forest composed of the first invaders of cleared land with forest from which these pioneers have gone and which shows no sign of recent disturbance, but which has not necessarily attained its climax condition.—E. W. Jones (1955, p. 564)

2.1 Overview

In the mid-1950s, Eustace W. Jones of the Imperial Forestry Institute of Oxford published a two-part detailed description of the plateau forest of the Okomu Forest Reserve in southwest Nigeria (Jones 1955, 1956). An earlier account by Richards (1939) described the forest under study as primary forest based on several criteria including the absence of cut stumps, relict cultivated plants, or other obvious signs of interference. While sampling soils, however, Jones (1955) noted frequent shards of pottery, charred fragments of oil palm seeds, and charcoal in the soil pits, causing him to question the undisturbed nature of this forest. He observed that dominant emergent tree species of the reserve, such as *Alstonia boonei* and *Lophira alata*, are characteristic of older second-growth forest vegetation and lacked smaller size classes in the forest (Jones 1956). Based on ages of canopy trees, Jones estimated that this was a second-growth forest established at least 200 years earlier on land that had previously been intensively cultivated and densely populated.

Jones was right about the secondary nature of the forest, but he was wrong about its age. White and Oates (1999) retraced Jones's steps and obtained radiocarbon measurements for charcoal and pottery from his study plots and soil pits. Charcoal samples dated to 760 ± 50 cal BP (calibrated calendar years before present, based on radiocarbon-dated years), between AD 1177 and 1378, during the Late Iron Age. White and Oates speculated that 700 years ago the site where Okomu Forest now stands was an oil palm plantation associated with the ancient town of Udo. For reasons that are not well understood, the plantation was abandoned and gave rise to what is now a mahogany forest, dominated by several genera in the Meliaceae family. The oil palms remain only as charred legacies in the soil.

The Okomu Forest history is not unique within Africa (White 2001a). Based on the abundance of oil palm nuts, Fay (1997) proposed that much of northern Congo, southeastern Cameroon, and southwestern Central African Republic was cultivated until around 1,600 years ago. Iron Age remains are abundant in west-central Africa (Oslisly 2001). Throughout Central Africa, former Iron Age settlements can be identified by the presence of economically useful trees (Oslisly and White 2007). In central Gabon, Oslisly and White (2003) discovered 50 iron-smelting furnaces within two hectares of apparently mature rain forest. Legacies of past shifting cultivation 300–400 years ago are clearly evident in the tree species composition and size distribution in lowland rain forests in southern Cameroon (van Gemerden, Olff, et al. 2003). After learning more about the ancient history of rain forests in West Africa, Richards (1963, p. 125) revised his vision. Upon finding fragments of pottery and charcoal in soil in sample plots of "primary forests" of the Bakundu Forest Reserve in southern Cameroon, he wrote, "It seems very unlikely that we have in fact any knowledge of 'virgin' forest in this part of the Cameroons."

Legacies of ancient human impacts extend well beyond the African tropics. William Denevan (1992b, p. 370) wrote, "By 1492 Indian activity throughout the Americas had modified forest extent and composition, created and expanded grasslands, and rearranged microrelief via countless artificial earthworks. Agricultural fields were common, as were houses and towns and roads and trails." Regarding the islands of the Pacific, Kirch (1997, p. 3) wrote, "Indeed, it is becoming clear that hardly any islands in the Pacific—even those that had no human populations at the time of European contact—were without human impact at some point in prehistory." Whether considering the past, present, or future, tropical forest dynamics are intimately linked to human activities.

Although some tropical regions apparently were not colonized by ancient peoples, the work of archaeologists, paleoecologists, paleoclimatologists, and historical ecologists reveals long-term legacies of prehistoric human occupation on the structure, composition, and geographic distribution of tropical forests (Stahl 1996; Willis et al. 2004; Willis et al. 2007). Prehistoric human impacts vary widely within and across tropical regions, qualitatively and quantitatively, because patterns of human settlement, occupation, and land use were subject to many constraints. Most of what we know about prehistoric human impacts on tropical forest has been revealed within only the past 20 years, and new discoveries are surfacing at a rapid pace. Some studies clearly demonstrate cases of repeated patterns of forest clearing and regrowth over millennia within the same location (e.g., on La Yeguada, Panama, see Piperno 2007). In other cases, tropical forest has never reestablished due to continued human occupation and landscape transformation over the

Table 2.1. *Late Quarternary geological periods described in this book*

Period	Approximate time interval (cal BP)
Late Pleistocene	126,000–18,000
Terminal Pleistocene	18,000–11,700
Early Holocene	11,700–7800
Mid-Holocene	7800–4000
Late Holocene	4000–present

Note: Units of time are in calibrated years before present (cal BP), based on radiocarbon dating.

past 10,000 years (e.g., on Kuk swamp, New Guinea, see Denham et al. 2003; Haberle 2003). Several paleoecological studies demonstrate little or no evidence of human occupation in some regions of the South American tropics (e.g., see Piperno and Becker 1996; Bush and Silman 2008; Barlow et al. 2012; McMichael, Piperno, et al. 2012).

The complex history of human impacts on tropical forests can be reconstructed by tracing the history of human colonization and occupation throughout the tropics (Hayashinda 2005; Williams 2008). Here, I synthesize information on the effects of prehistoric human occupations on tropical forest structure and composition around the world, beginning with the late Pleistocene and continuing through the late Holocene (table 2.1). Prior to the spread of agriculture, early hunter-gatherers began to alter forests, disperse key food plants, and hunt certain species to extinction. The origins and spread of agriculture during the Holocene led to increased human populations and to growing demands for food, leading to extensive land clearance and biomass burning in many tropical regions. Human impacts on tropical forests intensified over the past 5,000 years, following extensive landscape transformations. These ancient impacts made indelible marks on the extent and composition of forests that persist in many tropical regions today.

2.2 The Peopling of the Tropics

Tropical forest formations as we know them today did not coalesce until 8,000–11,000 years ago, after humans had dispersed to all tropical regions (Colinvaux and Oliveira 2000; van der Hammen 2001; Piperno 2006). At the beginning of the Holocene, about 10,000 years ago, climates became warmer and wetter throughout the world, creating suitable conditions for forest growth and expansion. During the first 2,000 years of the Holocene, tropical evergreen forests replaced deciduous and semievergreen forest in Petén, Guatemala, much of the Pacific lowlands of Central America, and parts of northern South America (Piperno 2007). In southern peninsular Thailand, the early Holocene was a period of low seasonality and high precipitation, leading to the expansion of arboreal wet forest taxa (Kealhofer 2003). Throughout tropical Africa, evergreen forests expanded under the warm, moist conditions between 10,000 and 7000 cal BP (Livingstone 1975). Maximum forest extension in Cameroon occurred around 9,500 years ago (Maley and Brenac 1998). At 9000 cal BP, tropical forest reached its highest extent in South America and Africa (Servant et al. 1993). The assembly of modern tropical forests was therefore coincident with the development of more complex human societies, migrations, and cultural transformations.

Anatomically modern humans evolved in equatorial Africa about 150,000 cal BP (Mellars 2006). Human populations first expanded within Africa and then migrated out of Africa following a period of intense drought between 135,000 and 70,000 cal BP (fig. 2.1; Cohen et al. 2007; Carto et al. 2009). One group traveled up the Nile corridor or across the mouth of the Red Sea into Arabia and then followed a coastal route through South Asia, later reaching

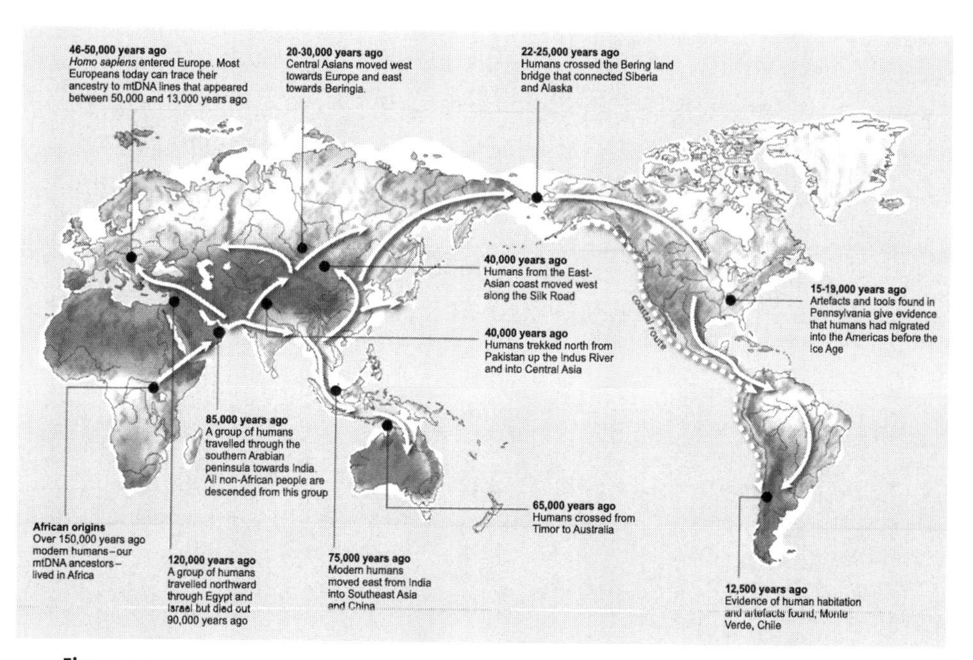

Figure 2.1

Modern human dispersal out of Africa, as reconstructed by Oppenheimer (2009, fig. 1).

the Pleistocene continent of Sahul (New Guinea, Australia, Tasmania) across ice-age land bridges. Modern humans initially migrated into the Indian subcontinent as late as 80,000–50,000 cal BP, but archaeological evidence for the migration of modern humans into southern Asia is lacking (Pope and Terrell 2008; Oppenheimer 2009). Much archaeological information on Pleistocene populations remains buried beneath the ocean in areas that were coastal lands during transcontinental human migrations.

By 40,000 cal BP, modern humans had encountered, crossed, and settled in tropical forests of all types during their expansion across the vast expanses of land bridging Africa and Australia (Mercader 2003; Pope and Terrell 2008). Recent excavations at several locations in the Ivane Valley of New Guinea confirm early human occupations dating to 49,000–43,000 cal BP (Summerhayes et al. 2010). Human populations began occupying tropical forests as hunters and gatherers in Equatorial Guinea, Cameroon, and Zaire between 40,000 and 23,000 cal BP. In southwestern Sri Lanka, hunter-gatherers inhabited rain forests as early as 40,000 years ago (Kourampas et al. 2009). The first record of human presence on the Malay Peninsula dates to 38,000 BP (Kealhofer 2003). The earliest record of human colonization of Australia dates to a similar time (50,000–45,000 cal BP; Bowler et al. 2003). Humans later arrived in South America, reaching Monteverde, Chile, by 13,000 cal BP to complete the peopling of continental tropical forests (Dillehay 1997).

2.3 Impacts of Early Hunter-Gatherer Societies

Terminal Pleistocene populations sustained themselves through hunting, gathering, and harvesting rich coastal resources. They used low-intensity fires to clear brush and create openings in the forest to encourage animals to forage where they could easily be hunted. Abundant archaeological evidence refutes the idea proposed by Bailey et al. (1989) that early hunter-gather societies could not survive within tropical forests (Mercader 2003). Early colonization of lowland tropical rain forests led to the development of forest-based subsistence societies, beginning an unprecedented era of landscape management in regions of Africa, South America, New Guinea, and Southeast Asia (Gnecco and Mora 1997; Roosevelt 1999a, 1999b; Kealhofer 2003; Premathilake 2006; Bayliss-Smith 2007; Rostain 2013).

Across the tropics, forest dwellers increased the concentration, overall abundance, and geographic ranges of plant species used for food and shelter (box 2.1). Archaeological studies at Niah Cave in Sarawak, Malaysia, indicate that the small groups of hunter-gatherers that used the cave and surrounding lowland tropical rain forest about 45,000 years ago also had sophisticated methods for exploiting plant sources of carbohydrates. These food sources included yam tubers (*Dioscorea*), aroid rhizomes (*Colocasia, Alocasia,* and *Cyrtosperma*), breadfruit (*Artocarpus altitis*), jackfruit (*Artocarpus heterophyllus*), *Pangium edule* nuts, and *Eugeissona* spp. Many of these species have high levels of natural toxins and required time-consuming processing before they could be safely consumed (Barker et al. 2007; Barton and Paz 2007).

Haobinhian hunter-gatherers occupied rain forests of the Malay Peninsula during the early Holocene and practiced early cultivation of forest tubers and fruit trees (Bellwood 1997; Kealhofer 2003). The Nong Thalee Song Hong sediment core from southern Thailand provides evidence from fossilized pollen and microscopic silica deposits from plant tissues (phytoliths) for forest disturbance, burning, and cultivation of economic species, particularly tree taxa, during the early Holocene (Kealhofer 2003). These taxa include *Areca* (betel nut palm), *Caryota* (sugar palm), dipterocarp species, *Artocarpus*, and *Garcinia* (Maloney 1999). In India, hunter-gatherer societies persisted in areas of the west coast and Western Ghats until iron was introduced in the region, which was only 3,000 years ago. Many traditions of early hunter-gather societies are still evident today, including the protection of sacred groves, forest management, and the maintenance of *Ficus* trees (Gadgil and Chadran 1988).

Early subsistence strategies in Melanesia generally combined elements of hunting and gathering with arboriculture and wild plant management (Bayliss-Smith 2007). Late Pleistocene colonists in the Ivane Valley in the New Guinea Highlands used waisted stone axes with side notches for attaching a wooden handle. These axes were used for tree felling and possibly for hoeing or digging, enhancing the supply of *Pandanus* seeds, an important staple food (Summerhayes et al. 2010). These early occupations are also associated with increasing frequency of fire in the Ivane basin (Hope 2009; Summerhayes et al. 2009).

In the Neotropics, Paleoindians hunted, fished, and gathered fruits along rich alluvial floodplains, bluffs of major rivers, and coastlines (Denevan 1996; Roosevelt 1999a, 1999b; Oliver 2008; Rostain 2013). Shellfish middens are found in many locations throughout eastern coastal South America (Heckenberger and Neves 2009). Several thousand shell mounds, called *sambaquis*, dot the southern Atlantic coast of Brazil, relicts of prehistoric hunters and gatherers that occupied the Atlantic Forest region about 7000 cal BP (Fairbridge 1976). Hunter-gatherers occupied seasonal foraging camps on the Tapajos River near the current Brazilian cities of Monte Alegre and Santarém, between 11,200 and 9800 cal BP (Roosevelt 2000; Roosevelt et al. 2009). Subsistence depended heavily on turtles, fish, and shellfish, supplemented with tree fruits and seeds (Roosevelt 1998). Human populations extended into

Box 2.1. Arboriculture, Forest Gardening, and Anthropogenic Forests

Prehistoric arboriculture increased the abundance and geographic spread of many useful tree species across tropical forests (Yen 1974; Balée 1989; Peters 2000). Many of these sun-loving and fast-growing species now reach their highest abundance in secondary forests and agricultural fallows. Knowledge of the early utilization and semidomestication of these species enhances understanding of their modern ecology. Although many utilized species were not fully domesticated, their landscapes certainly were (Terrell et al 2003).

Anthropogenic forests are defined as forests dominated by species (or oligarchies of species) that have a clear association with humans. Indigenous peoples of Mesoamerica cultivated trees and managed forest patches for over 3,000 years. When Europeans arrived in Amazonia, 138 species of plants were under cultivation or management, and 68% of these were trees or woody perennials (Clement 1999). The strongest evidence for ancient silvicultural practices comes from studies of the Maya, who planted home gardens and managed forest fallows and forests. High-density aggregations of useful tree species in forests surrounding archeological sites today provide strong evidence of silvicultural management by ancient peoples (Gómez-Pompa 1987; Rico-Gray and García-Franco 1991; Fedick 1995; Campbell et al. 2006; Ford 2008; Ross 2011).

Forest gardens were so widespread during the Mayan Preclassic Period that contemporary forests of southeastern Petén, eastern Guatemala, and western Belize are considered to be anthropogenic in origin (Gómez-Pompa and Kaus 1999; Peters 2000; Campbell et al. 2006; Ford 2008). *Manilkara zapota* and *Brosimum alicastrum*, two of the most abundant and widespread trees in the region, were important sources of food and timber for the ancient Maya. Other tree species important for Mayan subsistence, such as *Protium copal*, *Ceiba pentandra*, *Dialium guianense*, *Haematoxylon campechianum*, and *Swietenia macrophylla*, are dominant elements of the local flora (Peters 2000).

Swidden cultivation fallows or postagriculture secondary forests have a higher abundance of species used by humans compared to mature forests and therefore can be considered anthropogenic forests (Chazdon and Coe 1999; Voeks 2004). Human management of fallows can be direct, as in the case of active silviculture and management, or indirect, as with species associated with Ka'apor indigenous fallow management. In this case, managed fallows are dominated by babaçu palm (*Attalea phalerata*), hog plum (*Spondias mombin*), tucumã palm (*Astrocaryum vulgare*), and inajá palm (*Maximiliana maripa*) due to the spread of seeds discarded in settlements by agoutis rather than by deliberate planting by people (Balée 1993). Many palm forests in Amazonia are anthropogenic (Balée 1988; Balée and Campbell 1990; Erickson and Balée 2006). Based on archaeological evidence, *Acrocomia aculeata* (coyol) was dispersed by humans from South to Central America, whereas *Oenocarpus batua* and *Elaeis oleifera* were widely dispersed within regions of South America (Morcote-Ríos and Bernal 2001). Guix (2009) went so far as to suggest that humans replaced extinct Pleistocene megafauna as primary seed dispersers of many large-seeded fruiting plants in the Amazon basin, including the widespread Brazil nut tree (*Bertholletia excelsa*; Shepard and Ramirez 2011).

A complex arboricultural system is associated with the Austronesian colonization of islands in Oceania from 1600 to 500 BC. On the Mussau Islands, over 20 species of trees were cultivated, including *Canarium indicum*, *Spondias dulcis*, *Pometia pinnata*, *Aleurites molucana* (candlenut), and *Burckella obovata* (Kirch 1989). *Cocos nucifera* (coconut) fruits were widely consumed by humans and pigs. These species have wide distributions in the tropical Pacific and were likely dispersed by humans. These fruits attract flying fox, native chicken, pigeon, swamp hen, and other feral animals, which are hunted for food (Bayliss-Smith et al. 2003;Yen 1974).

Legacies of ancient arboriculture systems are still evident in the current vegetation of the Solomon Islands and throughout Melanesia and Oceania. *Pommetia pinnata* is a common tree species in the Solomon Islands, often occurring as a dominant in mature and secondary forests and in shifting cultivation fallows (Yen 1974). *Canarium* trees form groves in well-developed secondary forests, and inland tracts of *Terminalia brassii* occupy former taro pond fields (Bayliss-Smith et al. 2003).

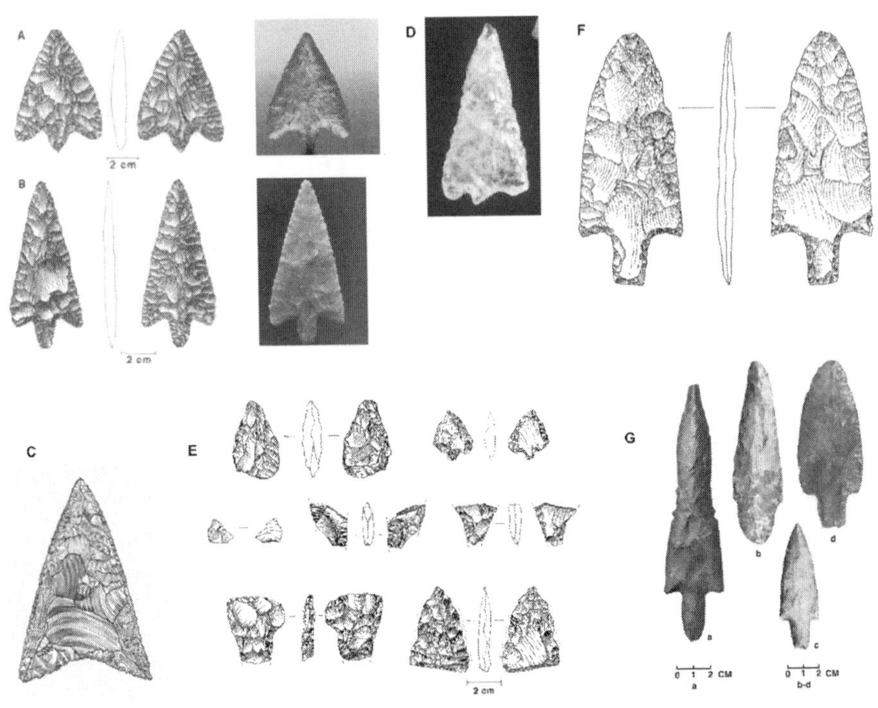

Figure 2.2
Photographs and drawings of projectile points from multiple *terra firme* sites in the Amazon basin: *A*, wide point of hyaline quartz crystal from lower Tapajós River, 6.4 centimeters; *B*, long point of red-brown chalcedony, middle Tapajós River, 8.5 centimeters; *C*, drawing of finely flaked triangular point from Santarém, 12 centimeters; *D*, triangular point from Monte Alegre, 8 centimeters; *E*, drawings of bifacial points from Caverna da Pedra Pintada, Monte Alegre; *F*, stemmed subtriangular point from the Upper Río Negro, Brazil, 15 centimeters; (G) stemmed subtriangular points from Guyana. Source: Roosevelt et al. (2009, fig. 1).

upland areas, where they engaged in broad spectrum hunting and collecting in areas away from streams (Roosevelt 1999b). Archaeological excavations on the middle Caquetá River in Amazonian Colombia reveal a wide diversity of preserved palm seeds and fruits, flaked and ground stone implements, and abundant charcoal dating from 9250 to 8100 cal BP. These early occupants practiced small-scale gardening in addition to hunting and gathering (Oliver 2008); included among the assembled implements were waisted stone tools (fig. 2.2). Archaeological and paleoecological data indicate that humans colonized the Lake Yeguada watershed in Panama by 12,900 BP and largely eliminated forest from the watershed by 7000 BP (Piperno et al. 1990; Bush et al. 1992).

Aboriginal peoples began to occupy rain forests of northeastern Australia 8,000 years ago, soon after these forests expanded during a wetter climate phase. Intense El Niño/La Niña Southern Oscillation (ENSO) events from 2500 to 1700 cal BP coincide with increased Aboriginal activity in forests, as people were forced to occupy rain forests permanently (Cosgrove et al. 2007). In the southeast Cape York region, use of caves and rockshelters increased threefold during the middle to late Holocene, particularly after 4500–3500 cal BP (Haberle and David 2004).

The most irreversible impact of late Pleistocene human populations in tropical forests and woodlands was through hunting. As many Pleistocene megafauna were herbivores in open woodlands and savannas, their extinction likely led to increasing fuel load in fire-prone tropical ecosystems (Rule et al. 2012). Although there is still healthy debate regarding the role of human hunting versus climatic change in driving these extinctions (Wroe et al. 2006), strong evidence shows that heavy extinctions of large-bodied animals on islands and island continents in the late Pleistocene (Australia), terminal Pleistocene (Americas), early to mid-Holocene (West Indies, Mediterranean islands), and the late Holocene (Madagascar, New Zealand, Pacific islands) always coincided with human colonization (box 2.2). Across the tropics, early hunter-gatherer societies left their mark on forests and woodlands, altering the fauna and flora and increasing the use of fire, even before the development of shifting cultivation and intensive agriculture.

2.4 The Development of Agriculture

Along with climate changes, the early to middle Holocene signaled a major shift in human societies, with the independent development of agriculture in at least eight regions of the Old and New World from 10,000 to 5000 cal BP (Piperno 2006). The development and spread of agriculture profoundly impacted the structure and composition of tropical forests precisely at the time when these forests were assembling their modern species composition (Piperno 2007, 2011). Tropical foragers expanded along with the forest and became horticulturalists.

The most important lowland crops in Central and South America were first cultivated and domesticated in seasonal tropical forests (Piperno and Pearsall 1998; Piperno 2006; Piperno et al. 2009). Plant domestication began independently in Mexico and Peru. Maize and cotton (Mexico) and manioc and potatoes (Peru) were probably domesticated only once, whereas squash, common beans, and sunflower were probably domesticated more than once (Iriarte 2007). Genetic evidence suggests that manioc (*Manihot esculenta*) was domesticated in Amazonia, mostly likely in the southwest

Box 2.2. Species Extinctions Linked to Prehistoric Human Colonization

Conclusive cases of species extinctions linked to early human occupations come from tropical islands, where the "triple threat" of predation, biotic introductions, and vegetation alteration left no refuge for vulnerable species (Grayson 2001). Early evidence linking megafaunal extinction with human occupation in New Guinea comes from several sites in the Balim Valley (Fairbairn et al. 2006). The disappearance of a large fruit bat (*Aproteles bulmerae*) during the early Holocene in the New Guinea Highlands at Kiowa is attributed to overhunting by humans (Sutton et al. 2009). In Madagascar and the South Pacific islands, chronologies for extinction events suggest that megafaunas declined following human colonization (Burney and Flannery 2005). In the West Indies, megafaunal extinction followed human colonization and landscape transformation about 5,500–6,000 years ago (MacPhee 2008).

Austronesian colonization in Polynesia about 3,500 years ago (Hurles et al. 2003) had a devastating impact on native birds (Steadman 1995, 1997) and large-bodied herpetofauna (Pregill and Dye 1989; Steadman et al. 2002). In western Polynesia (Mussau, New Caledonia, Tikopia, Anuta, Fiji, Tonga, Futuna, and Samoa), many bird species did not survive the first millennium of human occupation. Extinctions of large frugivores have particularly impacted dispersal and recruitment of large-seeded tree species (McConkey and Drake 2002; Fall and Drezner 2011). Extinctions of seabirds in Polynesia were particularly severe for petrels and shearwaters, whereas losses of landbirds were greatest for rails, pigeons, doves, and parrots. No family of landbirds or seabirds was spared. Prehistoric Polynesians were highly skilled at catching birds, often by hand. They also gathered eggs and frequently visited small, uninhabited islands to gather seabirds and their eggs (Steadman 1997).

After human arrival in the Hawaiian Islands, 1,500–2,000 years ago, 60 endemic species of landbirds became extinct, and 44 endemic species of landbirds were lost in the last millennium in New Zealand. Easter Island appears to have lost more of its indigenous biota than any island its size in Oceania. Of the 30 breeding species of seabirds that nested on the island before human occupation, only one species still nests there today (Steadman 1995).

Intentional or accidental prehistoric introductions of plant and animal species onto islands often had devastating consequences. Human introduction of wild animals from New Guinea into island Melanesia during the Pleistocene had a small but significant impact. The gray cuscus (*Phalanger orientalis*) was introduced to New Ireland during the late Pleistocene. Northern pademelon (*Thylogale browni*) and a large rat (*Rattus praetor*) were early Holocene introductions. Only a single mammal in the New Ireland sequences, *Rattus mordax*, went locally extinct. Introduction of the dingo to Australia 4,000 years ago led to the disappearance of the thylacine (*Thylacinus cynocephalus*) and the Tasmanian devil (*Sarcophilus harrisii*) from mainland Australia (Allen 1997). Mammals introduced prehistorically to Polynesia, particularly dogs, pigs, and rats, preyed on native birds. Rats were by far the major predator of eggs, chicks, and adult native birds. *Rattus exulans*, the Pacific or Polynesian rat, was introduced prehistorically throughout Oceania, intentionally or as a stowaway on voyaging canoes (Steadman 1997).

Colonization of Remote Oceania was devastating for the flora as well as the fauna (Prebble and Wilmshurst 2009). Late Holocene paleoecological records from several islands provide strong evidence of human-mediated decline and extinction of *Pritchardia* palm species, which occupied soils highly suitable for taro cultivation (Prebble and Dowe 2008). *Rattus exulans* is strongly implicated in the decline and loss of palm species in the Hawaiian Islands, and it contributed to (if not caused) the extinction of the palm *Paschalococcus disperta* on Easter Island (Athens et al. 2002; Hunt 2007; for a contrary view, see Mann et al. 2008; Diamond 2007).

In contrast to the many cases of extinction associated with human occupation of tropical islands, there is sparse archaeological evidence that small-scale societies caused extinction on continental regions of the tropics (Grayson 2001). Despite 4,000 years of hunting, forest clearance, and pervasive landscape transformations, no archaeological evidence of late Quaternary plant or animal extinction associated with human occupation has emerged from the Mayan region (Gómez-Pompa and Kaus 1999; Emery 2007).

region (Arroyo-Kalin 2012). Starch grains and phytoliths of *Zea mays* and *Cucurbita* sp. date from 8700 cal BP in the central Balsas River valley, Mexico, where wild progenitors of these species are found (Piperno et al. 2009). By 6,300 years ago maize had dispersed into South America from the Balsas River valley in Mexico (Piperno 2007). Paleoecological studies indicate that resource-rich lake-edge habitats were favored for seasonal cultivation during the dry season when lake margins became exposed, facilitating the use of fire to clear land for planting crops (Ranere et al. 2009). The oldest paleorecord of agriculture from the Amazon basin is from the Ayauch site in lowland eastern Ecuador, where *Zea* pollen was dated to 2850 BP (Bush and Colinvaux 1988).

In Mesoamerica, profiles of pollen, charcoal, and plant phytoliths in lake and swamp sediments from numerous sites in Belize, Costa Rica, El Salvador, Guatemala, Mexico, and Panama show sequences of burning coincident with evidence of crop cultivation (largely maize) and declines of arboreal pollen during the early and middle Holocene (10,000–7000 cal BP; Piperno 2006, 2007). Seasonal variation in rainfall favored land clearance and cultivation, supporting the archaeological and molecular evidence for the primary role of seasonal tropical forests from Mexico to Brazil in the origin and spread of domesticated crops in the New World (Ranere et al. 2009). In the Lake Yeguada watershed in central Panama, slash-and-burn agriculture appeared around 7000 BP (Bush et al. 1992). Pollen and phytolith records show a decline in arboreal forest taxa, an increase in woody second-growth forest taxa, and high levels of charcoal (Piperno 2006).

Coastal agriculture developed independently in Ecuador as early as 10,000 cal BP (Piperno and Stothert 2003). The preceramic Las Vegas people subsisted on marine and estuarine resources and began cultivating both seed plants (*Cucurbita* sp. and *Lagenaria siceraria*, the bottle gourd) and root crops (*Calathea allouid*, leren) in local gardens by 9000 cal BP (Stothert et al. 2003). This same group of early domesticated species is found together with abundant remains of palm fruits at the Peña Roja site located on a terrace of the middle Caquetá River in eastern Colombia dating to 8090 cal BP (Gnecco and Mora 1997; Piperno and Pearsall 1998).

Another independent origin of agriculture occurred in the New Guinea Highlands. Rather than originating in the lowlands, agriculture emerged from practices of plant exploitation in the highlands that enabled permanent occupation of the interior of New Guinea during the late Pleistocene. In the Upper Wahgi valley, the Kuk swamp provides a rich record of early Holocene occupation and forest clearing before 7800 cal BP. Phytoliths, pollen, and starch grains indicate that *Colocasia* (taro) and *Eumusa* bananas were culti-

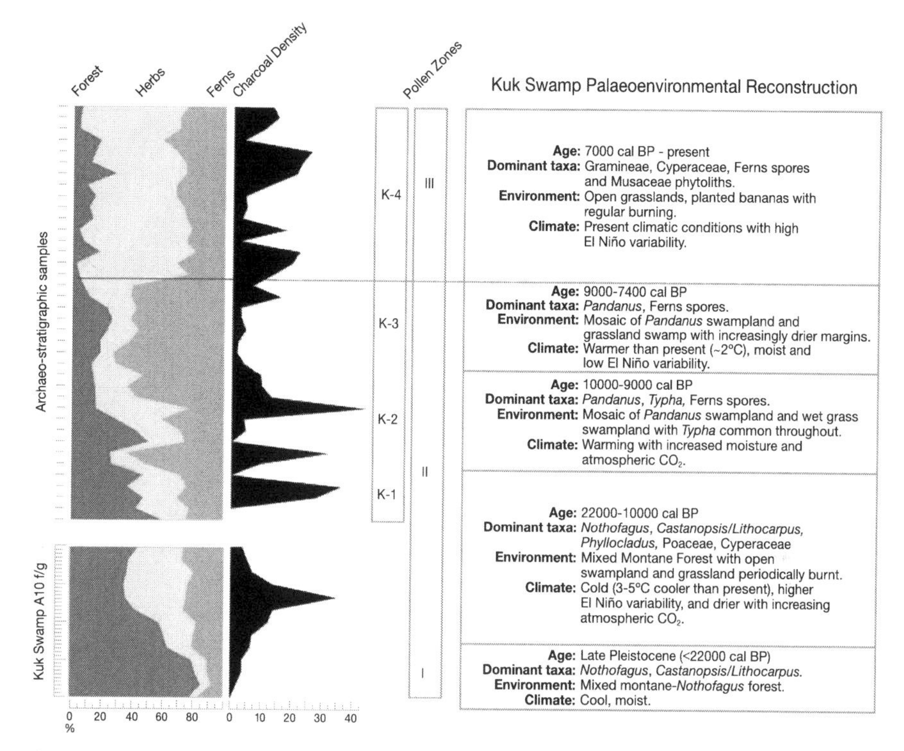

Figure 2.3

Paleoenvironmental reconstruction for Kuk swamp catchment in the Upper Wahgi valley in Papua New Guinea based on a composite pollen diagram. Source: Denham et al. (2004, fig. 6).

vated here, along with *Castanopsis* spp., *Musa* bananas, and *Pandanus* spp. (fig. 2.3). The landscape of the Upper Wahgi valley was a continually changing mosaic of old-growth and second-growth forest, regrowth and grasslands, and variously disturbed riparian and wetland environments during the early Holocene (Denham et al. 2003; Denham et al. 2004).

Based on phytogeographic and genetic affinities, *Colocasia*, *Eumusa* bananas, *Artocarpus altilis* (breadfruit), *Metroxylon sagu* (sago palm), *Saccharum* spp. (sugarcane), and *Dioscorea* spp. (yam) are considered indigenous to New Guinea and Melanesia and were domesticated there (Denham et al 2004). Lowland areas of northern New Guinea were rich in fruit-, nut-, and starch-bearing trees that were heavily utilized by coastal populations during the late Pleistocene and Holocene, such as *Artocarpus altilis*, *Barringtonia* spp., *Canarium* spp., *Cocos nucifera*, *Pandanus* spp., *Terminalia* spp., and *Metroxylon sagu* (Yen 1996). These species figure prominently in later development of arboriculture systems in Southeast Asia, Melanesia, and Oceania (see box 2.1; Yen 1974; Kirch 1989).

In southeastern Asia (Jiangxi Province of China), phytoliths of rice (*Oryza sativa*) recovered from pottery date from 14,000 to 9000 cal BP (Premathilake 2006). *Hordeum* sp. and *Avena* sp. were cultivated in central Sri Lanka from 10,000 to 8700 cal BP. Wild progenitors of these cereal species were growing naturally within this region before the late glacial maximum. But during the early and mid-Holocene, neither cereal nor rice cultivation was suitable for equatorial latitudes in islands of Indonesia, Melanesia, and Oceania. Rather, these regions saw the development of tuber crops (yam, taro), starch-bearing palm species crops (*Metroxylon, Corypha, Arenga, Caryota, Eugeissona*), and fruit trees (*Artocarpus, Pandanus*, rambutan, durian, among others). Wet rice cultivation technologies later spread from northern Vietnam and Thailand around 2,500 years ago to Java, Bali, and Luzon, causing a massive transformation of these islands (Bellwood 1997).

Austronesian agriculturalists emigrated from Taiwan and southern China 5,000–6,000 years ago, colonizing the Philippines and islands of Southeast Asia and Near Oceania (Bellwood 1997; Spriggs 1997). By 1500 BC, the Austronesians had reached the western borders of Melanesia, bringing their agricultural techniques of cereal and tuber cultivation and animal husbandry, as well as chickens, dogs, pigs, and stowaway rats. Austronesian dispersal continued eastward into previously uninhabited islands of Remote Oceania, Polynesia, and Micronesia and westward to Madagascar (Bellwood 1995). Rapid deforestation followed Pacific island colonizations causing environmental degradation, erosion, and extinction of many endemic species (see box 2.2; Kirch 2005).

In Africa, crop cultivation came much later than in other tropical regions, long after the domestication of animals. Cattle were probably domesticated from wild populations of *Bos primigenius* by hunter-gatherers in the eastern Sahara of North Africa 8,000–10,000 years ago (Marshall and Hildebrand 2002; Höhn et al. 2007). Shortly thereafter, sheep and goats were domesticated in the eastern Sahara and Red Sea hills, probably using stock originating from western Asia. Several reasons for the late domestication of African crops include harvesting practices, unpredictable rainfall patterns of the Sahara and its margins, and the mobility of early herders. The first cultivation of plants in Africa occurred in the Nile valley, around 7000 cal BP, using crops from southwest Asia. The domestication of indigenous African crops took place in widely dispersed regions and under a wide range of conditions (Marshall and Hildebrand 2002). Pearl millet (*Pennisetum glaucum*) appeared around 4,000 years ago after Saharan groups migrated into West African grasslands.

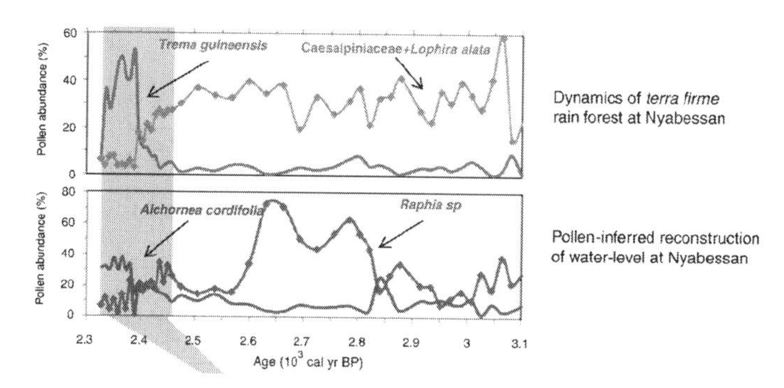

Figure 2.4
Late Holocene pollen data from southern Cameroon, Nyabessan, showing changes in the percentage abundance of pollen from mature evergreen forest taxa (*Caesalpiniaceae*, *Lophira alata*), pioneer taxa (*Trema orientalis*, *Alchornea cordifolia*), and swamp forest taxa (*Raphia*). The shaded area indicates evidence for pearl millet cultivation in archaeological sites. Source: Ngomanda, Neumann, et al. (2009, fig. 5).

Early rain forest farmers combined millet cultivation with harvesting of oil-rich tree fruits (Neumann et al. 2012).

In contrast to rain forest blocks in the New World, New Guinea, and Southeast Asia, there is no evidence of early or mid-Holocene plant cultivation in Central African forests. A period of climate change around 2500 cal BP led to increased seasonality in the Central African rain forest block of southern Cameroon, creating favorable conditions for millet farming (Neumann et al. 2012). Pollen cores from Nyabessan swamp show a corresponding rapid shift in vegetation from swamp and evergreen Caesalpiniaceae forest to dominance by early successional tree species such as *Trema orientalis* and *Alchornea cordifolia* (fig. 2.4; Ngomanda, Neumann, et al. 2009). Further, the opening of this rain forest block provided an opportunity for the expansion of Bantu-speaking agriculturalists and sedentary ceramic-producing populations (Schwartz 1992).

Although distinct crops and agricultural practices developed in different tropical regions and at different times, the land transformations that accompanied the development and spread of agriculture were coincident with periods of climatic change and forest reassembly. The confluence of these factors strongly affected tropical forests in regions where agriculture was practiced, leading to concentrated human settlements and land-use change that can be reconstructed today using pollen and phytolith profiles from lake sediments. The conversion of tropical forests to agricultural land, beginning 8,000–10,000 years ago, created patches of forest disturbance and regeneration within human-occupied landscapes and increased the abundance and geographic distributions of species favored by forest disturbance and human use.

2.5 Holocene Climate Variability, Forest Change, and Agricultural Expansion

Climatic variability in tropical regions during the early to mid-Holocene is strongly linked to changes in the extent of forests and savannas, patterns of human settlements, and cultural practices (Haberle and David 2004). These linkages make it difficult to separate the effects of climatic changes from those of human occupation and land use on tropical forest dynamics and composition. Integrated paleoecological and archaeological data are required to understand the complex causal relationships between climatic variability and human impacts on forest dyamics and composition (Mayle and Iriarte 2013). Forest clearing and regrowth were both heavily influenced by climatic variability, a trend that continues to this day.

2.5.1 *Holocene Forest-Savanna Dynamics*

In tropical regions of Australia, Africa, and South America, dense rain forest vegetation expanded at the end of the Pleistocene and contracted during the early to mid-Holocene (Bush et al. 2007; Servant et al. 1993). Holocene droughts created open forest formations, sclerophyll forest, or savanna vegetation, with major effects on fire regimes across northern South America (Servant et al. 1993; Bush et al. 2000; Mayle and Power 2008). From ca. 8000 to 4000 cal BP, climates in the tropical Andes were significantly drier than present, and cloud forest was replaced by lowland forest taxa (Bush et al. 2004). In the eastern Brazilian Amazon at Carajas, Pará State, reduced precipitation led to the replacement of forest by open savanna, but forest returned to the region during the late Holocene (Mayle and Power 2008). Evergreen moist forests of the Upper Beni region of Bolivia were not replaced by savanna, despite frequent fire during early and mid-Holocene drought periods (Urrego et al. 2012).

Similar shifts occurred in the northern Amazon, leading to the expansion of gallery forests within the Colombian llanos savannas (Behling and Hooghiemstra 2000), and in the campos of southern Brazil, where *Araucaria* forests expanded during the late Holocene (Behling 1997). In the southern Brazilian highlands, warm, humid conditions were associated with the expansion of *Arauraria angustifolia* forest into the highlands about 4,000 years ago (Iriarte and Behling 2007). Another expansion of *Araucaria* occurred 1,000–1,500 years ago in Paraná, Rio Grande do Sul, and Santa Catarina during another humid period. Bitencourt and Kraspenhar (2006) suggested that human management aided the highland dispersal of this dominant species, as *Araucaria* seeds were an important staple in the diet of the indigenous hunter-gatherers that occupied this region. The Amazon rain forest extends further south today than at any time during the past 50,000 years (Taylor et al. 2010).

Holocene aridity was more severe and occurred 4,000 years later in Africa than in South America (Morley 2000). Western equatorial and Central Africa were covered in closed forest in the early and mid-Holocene, as forests expanded following glacial retreat. But an opening of the forest and extension of savanna and pioneer forest formations occurred in many sites throughout Central and West Africa beginning around 4500–4000 cal BP. Evergreen forest formations remained stable in some regions interior to the Congolese forest block until 3000–2500 cal BP, when forest "breakdown" occurred due to decreased precipitation or longer dry seasons. Second-growth forests dominated by oil palm (*Elaeis guineensis*), *Macaranga* species, and *Alchornea cordifolia* followed the widespread establishment of grassland savannas during this period (Brncic et al. 2009; Ngomanda, Chepstow-Lusty, et al. 2009).

Humid conditions returned to Central and West Africa beginning ca. 2000–1400 cal BP, creating conditions favorable for forest expansion. West Cameroon and coastal areas of Gabon are still in an expansion phase (Delegue et al. 2001; Maley 2002; Brncic et al. 2009; Ngomanda, Chepstow-Lusty, et al. 2009). During the last millennium, however, changes in vegetation in lowland forests of the Congo basin and an associated major increase in charcoal in sediments clearly indicate human impacts and extensive biomass burning, despite the return of humid conditions from 1880 to 1345 cal BP (Brncic et al. 2007; 2009). In Lopé, Gabon, savannas that would have been colonized by forest have been maintained through human fires (White and Oates 1999). Pioneer tree species dominate the pollen record of the Congo basin over the past 900 years. Patches of vegetation dominated by the light-demanding herb *Megaphrynium macrostachyum* (Marantaceae) are associated with subsoil charcoal and are likely related to past burning events (Brncic et al. 2009). Hilltop plant associations in Lopé National Park in Gabon include a large number of diagnostic species, such as *Aucoumea klaineana*, *Ceiba pentandra*, and *Pentaclethra macrophylla*, that were probably managed for their economic, sacred, and medicinal value as early as the Late Stone Age over 4,000 years ago (Oslisly and White 2007).

Climate change and Aboriginal burning in wet tropical regions during the Holocene also had local effects on the reexpansion of rain forest vegetation in wet tropical regions in Australia (Head 1989; Hopkins et al. 1993; Bowman

1998; Kershaw et al. 2007). Although climate change was the predominant cause of vegetation change, Aboriginal burning may have hastened the spread of more open sclerophyll vegetation (Bowman 2000; Kershaw et al. 2007). When humans arrived on the scene in Australia, the expansion of *Eucalyptus* forest into areas occupied by *Araucaria*-dominated rain forest vegetation was well underway (Kershaw et al. 2007). There is no evidence that human-caused burning prohibited the expansion of rain forest in the early Holocene (Bowman 1998). Rather, aboriginal burning created small-scale habitat mosaics that favored the abundance of some mammal and tree species.

2.5.2 *Agricultural Expansion and Climate Change*

In the New World and Asian tropics, agriculture expanded greatly during the mid to late Holocene, extending settlements to moist, less seasonal environments. This expansion took place during the Holocene thermal maximum, a 5,000-year period of relatively warm, moist, and stable conditions throughout the New World tropics. Archaeological evidence supports the view that maize-based cultivation enabled settlement in evergreen tropical forests of the Caribbean watershed of Panama about 7,000 years ago (Neff et al. 2006). As populations grew, agricultural intensification was required to meet increased demand for food. Throughout Mesoamerica, abundance of burned phytoliths of Poaceae and *Heliconia* clearly indicate human-set fires in early successional vegetation, evidence of short-fallow shifting cultivation systems (Piperno 2007).

From 4,000 to 6,000 years ago, horticulturalists expanded across Mesoamerica, from the Gulf Coast of Mexico to western El Salvador and Honduras (Neff et al. 2006). By 5,000 years ago, severe forest clearance associated with maize cultivation occurred over a wide area in the Neotropics from the Colombian Amazon to northern Belize (Piperno and Pearsall 1998). Small-scale horticultural populations occupied a large part of what later became the Maya Empire (Ford and Nigh 2009). In the rain forests of Darién, Panama, major forest clearing and burning began over 4,000 years ago, coincident with maize cultivation (Bush and Colinvaux 1994). Charcoal fragments are coincident with the appearance of maize pollen and an abrupt decline in tree pollen in sediment cores in the Kob and Cobweb swamps in Belize and areas to the south (Jones 1994).

Forest clearance and burning for maize cultivation were widespread across virtually the entire Pacific coast of Mesoamerica during the mid to late Holocene (Horn 2007; Dull 2008). In the Maya lowlands of Guatemala, however, fossil pollen studies indicate that forest reduction was associated with a circum-Caribbean drying trend that started 4,500 years ago and lasted for

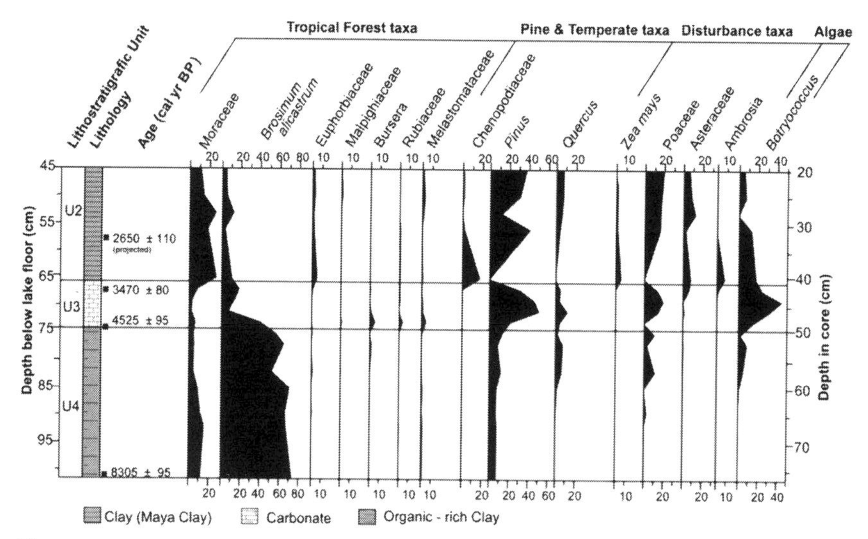

Figure 2.5

Pollen percentage diagram for Lake Petén Itzá, Guatemala, over 8,305 calibrated years. Pollen taxa are grouped as tropical forest taxa, pine and temperate taxa, and disturbance taxa. Corresponding changes in lithostratigraphic composition are shown on the left side of the diagram. Note that decline of tropical forest taxa between 4525 and 3470 cal BP occurred prior to the occurrence of *Zea mays* pollen. Source: Mueller et al. (2009, fig. 6).

about 1,500 years (fig. 2.5; Mueller et al. 2009). The decline in tropical forest taxa and increase in *Pinus, Quercus*, grasses, and second-growth taxa preceded the appearance of *Zea mays* pollen, thus suggesting that drought was a major driver of initial vegetation change in the lowland Mayan region (Ford and Nigh 2009).

Climatic instability during the mid-Holocene was also associated with the spread of agriculture in Africa. Pervasive extreme drying events began around 4,000 years ago, associated with increasing frequency of ENSO events. Holocene droughts and increased seasonality affected forests throughout equatorial zones of Africa from Madagascar to West Africa, beginning abruptly in 4000 BP (Marchant and Hooghiemstra 2004). In Central and West Africa, droughts during this same period caused the shrinkage of rain forests and the expansion of savannas, favoring the migration of Bantu-speaking agriculturalists, as noted above (Ngomanda, Chepstow-Lusty, et al. 2009a). Large increases in charcoal are observed in the Ituri forest soils after 4000 BP (Hart et al. 1996).

Climatic fluctuations, particularly driven by ENSO events, strongly influenced geographic and temporal patterns of human settlement and land use from the late Pleistocene to the mid-Holocene. But humans were not always at the mercy of the environment. Archaeological research, including several very recent studies, provides evidence that feats of engineering and development of technologies enabled human populations to thrive over millennia in the midst of climatic and biotic change.

2.6 Conclusion

Applying knowledge from current ecological studies can be particularly useful in interpreting the role of past human effects on tropical forest dynamics. Given appropriate caveats, prehistoric impacts can be interpreted in light of our current understanding of modern land-use legacies on forest regeneration (D. B. Clark 1996; Foster 2000; Sanford and Horn 2000; Willis et al. 2004; Hayashida 2005; Froyd and Willis 2008; Williams 2008). The same ecological processes that influence contemporary forest regeneration following large-scale forest clearance have molded tropical forest structure, composition, and dynamics over the past 10,000 years. Current studies of forest regeneration can illuminate the nature of past human impacts on tropical forests.

Deforestation, landscape transformation, and forest regeneration have been ongoing processes in tropical regions for over 30,000 years. Although the genetic entities that compose the species pool of tropical forests evolved millions of years ago, modern tropical forests have reassembled and expanded geographically in the presence of one new species in the landscape—*Homo sapiens*—and we cannot ignore that undeniable fact. Forests regenerating during prehistoric times in regions of human settlements bear the obvious and subtle impacts of the activities of humans who practiced arboriculture, forest burning, cultivation, and hunting. Paleobotanical research provides strong evidence that secondary vegetation and forests enriched with useful species predominated in many tropical regions during past millennia, as they do today. Paleoecological studies are complemented by ecological studies that demonstrate long-term legacies of land use and climate variability on forest succession following abandonment (Chazdon 2003). Understanding the short- and long-term impacts of past and contemporary indigenous cultures can provide models for projecting effects of current land-use practices and climate change on future forest structure, composition, and geographic extent (Gómez-Pompa 1971; Roosevelt 1999a, 1999b; Erickson 2003; Heckenberger et al. 2007).

CHAPTER 3

LANDSCAPE TRANSFORMATION AND TROPICAL FOREST REGENERATION THROUGH PREHISTORY

Since the late Pleistocene, forest ecology has been part of human ecology, and forest history, part of human history.—Anna Roosevelt (1999b, p. 373)

3.1 Overview

Ancient humans were not passive occupants of tropical forests. They engineered their environments, developed sedentary agriculture, and established complex societies and urban centers. Despite all of their technological achievements, human societies have always been and will always be inextricably linked to climate change. The late Holocene civilizations of the Maya in Mesoamerica and the Khmer in Cambodia were unable to conquer the forces of extreme drought or floods (Haug et al. 2003; Diamond 2009; Buckley et al. 2010).

In this chapter, I summarize many ways in which human activity—or its cessation—has transformed tropical forest landscapes throughout the world since the end of the Pleistocene. Archaeological studies have revealed widespread landscape transformations in tropical forest regions as agriculture intensified during the late Holocene. These transformations include construction of earthworks to extend crop cultivation into swampy or seasonally flooded habitats and on steep hillsides, planned burning to manage forests and facilitate hunting, and soil modification to enhance fertility and permit intensive agriculture. To a significant extent, today's tropical forest landscapes reflect legacies of past episodes of colonization, exploitation, cultivation, abandonment, and regrowth shaped by human occupations, population growth and decline, cultural change, and climate change.

3.2 Earthworks and Landscape Transformations

The first clear case of anthropogenic landscape transformation in the tropics was identified at a Pleistocene-era site, dating to 49,000 years ago, in the highlands of New Guinea (Summerhayes et al. 2010). Microcharcoal accumulation indicates burning of the upper montane *Nothofagus* and *Eleocarpus* forest before 36,000 cal BP (calibrated calendar years before present, based on carbon dating; Fairbairn et al. 2006). Vegetation disturbance and burning in the Upper Wahgi valley created a unique Holocene agricultural landscape beginning around 20,000 years ago (see fig. 2.3; Haberle 2003).

Human colonization and exploitation practices varied regionally within New Guinea, Near Oceania, and Australia. Swidden agriculture was widespread in New Guinea but did not develop in Australia (Yen 1995; Denham et al. 2009).

Mid-Holocene forest clearance occurred in various locations throughout the New Guinea Highlands, transforming the area into an agricultural landscape with patches of second-growth forest, grasslands on dryland slopes, and disturbed riparian (streamside) and wetland environments (Denham et al. 2004). Early cultivation in the New Guinea Highlands was concentrated in wetland margins, with a later expansion of wetland agriculture during the mid to late Holocene (Haberle 2007). At the Kuk swamp site in the Upper Wahgi valley (fig. 3.1), a network of channels and mounds was created over 6,000 years ago among older paleochannels formed over 9,000 years ago (Denham et al. 2004; Golson 1991; Denham et al. 2003). Archaeological and paleobotanical evidence indicates that crops intolerant of water-logging were grown on the mounds (sugarcane, bananas, yams, and ginger), whereas water-tolerant crops, such as taro (*Colocasia*), were grown between mounds.

Colonization of remote islands in Polynesia and Oceania by seafaring Austronesian peoples between 1,600 and 2,500 years ago caused major extinctions of animal and plant species (see box 2.2) and dramatic environmental changes (Kirch 1996). Following about 1,000 years of shifting cultivation

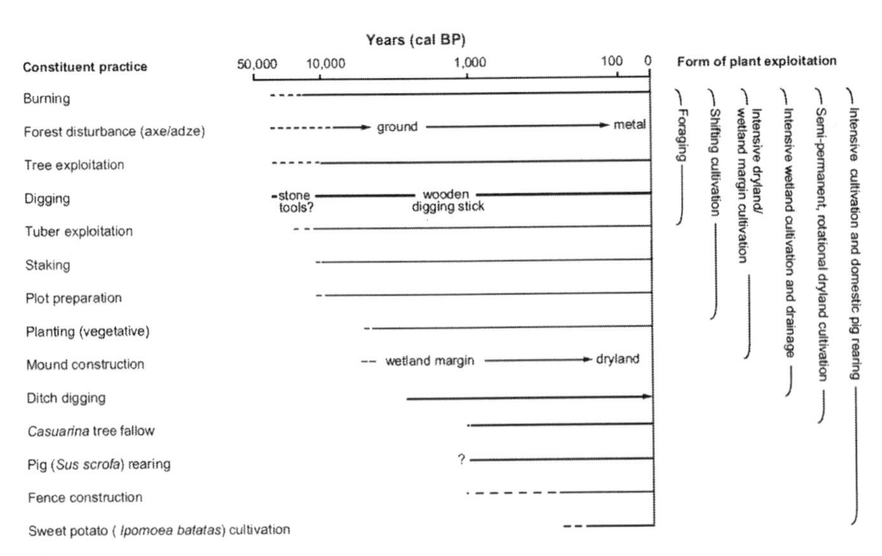

Figure 3.1
Chronology of practices and forms of plant exploitation in the Upper Waghi valley of Papua New Guinea. Source: Denham et al. (2009, fig. 6).

on hillsides that led to heavy erosion in many islands of Polynesia, farmers developed a highly sophisticated system of irrigated taro cultivation, which transformed alluvial valleys into reticulate grids of permanently cultivated pond fields irrigated by streams (Kirch 1997). Intensive agriculture using irrigated taro terraces and long high mounds for dry yam cultivation was widespread on the islands of New Caledonia and across Oceania from 1,000 years ago. Tens of thousands of taro terraces transformed the hilly landscape of La Grande Terre during the millennium prior to European contact, whereas extensive dry-land agriculture covered the flat plains (Sand et al. 2006; Kuhlken 2002).

In Mesoamerica, as populations grew during the late Holocene, agriculture further intensified and extensified (Hammond 1978). Annual or semiannual cultivation was practiced throughout the Maya lowlands region, encompassing southern Mexico and the Yucatán Peninsula, Belize, Guatemala, El Salvador, and northern Honduras. Fallow periods were short, resulting in frequent burning of young successional vegetation (Turner 1978). Tree cropping provided additional subsistence, as shifting cultivation alone could not support the growing population (Rice 1978).

Agricultural intensification transformed both upland and lowland areas of Mesoamerica. During the height of the Classic Period (1,050–1,700 yr. ago), the central Maya lowlands (Yucatán Peninsula and adjacent lowlands in Mexico, Belize, and Guatemala) were transformed from seasonally dry tropical forests in the uplands and seasonal wetlands to an open landscape of orchards, house gardens, and patches of intact forest (Whitmore and Turner 2001). Useful tree species were often spared during land clearing for swidden agriculture; these species were commonly cultivated in household garden plots (see box 2.1). The management of tree cover in Mayan landscapes likely favored the rapid regeneration of forest following the Maya Collapse or, several hundred years later, after the Spanish conquest (see box 3.1).

The wetlands (*bajos*) surrounding a significant area of the Maya lowlands were utilized on a large scale for agriculture during the Classic Period (Beach et al. 2009). Drainage canals occupied an area of over 12,000 square kilometers in northern Belize (Adams et al. 1981). A large number of raised-field complexes may have been constructed in the flat zones of Quintana Roo and the northeastern part of the Petén in Guatemala and in northern Belize (Turner 1978; Pohl and Bloom 1996). Intensive wetland agriculture using the *chinampas* system, raised fields built from canal excavation and dredging, was widely developed by the Aztecs in the basin of Mexico and the Puebla-Tlaxcala basin during the Late and Middle Classic Periods (AD 400–900).

These fields covered an area of about 12,000 hectares around the Aztec capital of Tenochtitlán (Denevan 1992b). A variety of crops were cultivated on the fields and fish were farmed in the canals (Frederick 2007).

Landscape modification moved upslope during the Late Classic Period (AD 600–900), as agricultural production intensified to meet a growing population. Terracing was implemented to prevent soil erosion on slopes, to expand land area under cultivation, to manipulate soil moisture, and to create an adequate soil depth for crops (Turner 1978; Dunning et al. 1997). Terracing was prevalent throughout highland Mesoamerica as well as in the central Andes Mountains (Whitmore and Turner 2001; Denevan 1988; Turner and Butzer 1992). Relic Mayan terraces cover extensive areas of southern Mexico and northern Guatemala; the Rio Brec region displays hundreds of relic terraces and related stoneworks covering an area of 10,000 square kilometers. Abundant stone walls likely represent demarcation of permanent fields, indicating intensive land use (Turner 1978).

In South America, extensive raised-field complexes were constructed in wetland regions to elevate planting surfaces in the seasonally flooded savannas of Bolivia, Surinam, French Guiana, Ecuador, Venezuela, and Colombia (Rostain 2013). Raised fields originated nearly 3,000 years ago in the flooded savannas of Llanos Mojos of Bolivia, where manioc, sweet potatoes, peanuts, beans, squash, and possibly maize were cultivated (fig. 3.2). Here, Amazonian

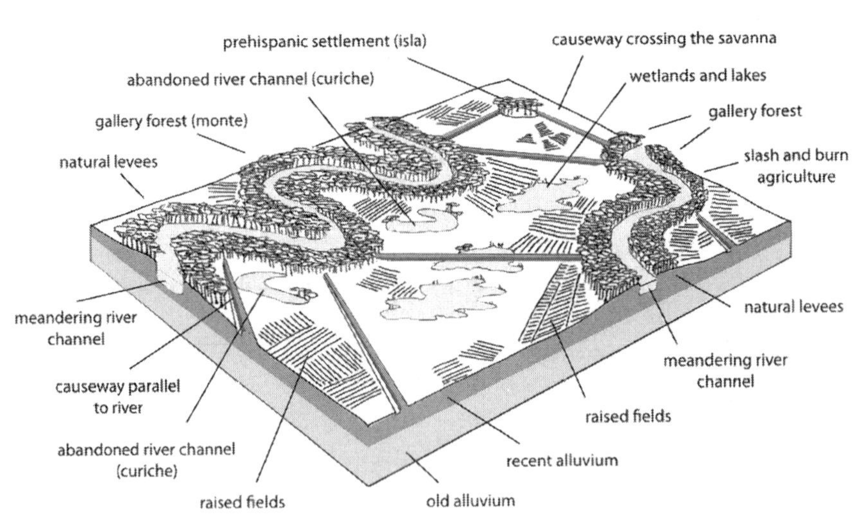

Figure 3.2
Reconstruction of landscape complex of causeways, canals, and raised fields in Llanos de Mojos, Bolivia. Source: Erickson and Walker (2009, fig. 11.5).

peoples constructed a landscape likened to a "huge aquatic farm," with approximately 550 square kilometers of fish weirs to harvest fish from the canals and intensive agriculture on the raised fields (Erickson 2006). This production system covered 50,000 square kilometers of seasonally flooded savanna (Mann 2008). Thousands of forest islands that dot the savanna landscape of the Bolivian Amazon today are thought to be largely anthropogenic in origin (Mayle et al. 2007). Excavations revealed that all intact forest islands contain refuse debris from long-term pre-Columbian settlements (Erickson 2006). The raised fields, fish weirs, causeways, and canal networks that surround forest islands are now covered in dense stands of *Mauritia flexuosa* palm forest. Only areas with extensive pre-Columbian earthworks currently support tall, mature, continuous-canopy forest. In other areas of the Bolivian Amazon, open savannas have been maintained by frequent burning by native peoples as well as current populations and do not currently support forest (Mayle et al. 2007).

Lombardo and Prümers (2010) located hundreds of mounds and forest islands and identified 957 kilometers of canals and causeways within a 4,500-square-kilometer region of the Llanos Mojos, east of the Bolivian city of Trinidad. The canals and causeways enabled movement throughout the region during the entire year, whereas the mounds may represent raised fields designed to allow cultivation during periods of seasonal flooding (Lombardo et al. 2011; Rostain 2013). Over 200 sites with geometrically patterned earthworks have been observed in a region roughly the size of England, extending from northern Bolivia to the eastern part of the state of Acre in Brazil and the southern part of Amazonas State. These "geoglyphs" are located in both upland areas (interfluves) and floodplain areas (fig. 3.3). They were abandoned roughly 500 years ago and became heavily covered with forest regrowth. Recent deforestation reveals their impressive structure and extent, although their function remains unknown. The ditched geometric earthworks in Acre State have been dated as early as 2000 cal BP and may have been used for religious celebrations. Construction of these major earthworks required a massive labor force, indicating substantial populations living within the areas. An estimated workforce of 80 individuals for 100 days was required to build a single geoglyph (Pärssinen et al. 2009; Schaan et al. 2012).

In Amazonia, urban settlements were associated with rivers. Along the Upper Xingu River of Mato Grosso State, Brazil, lies another area with extensive earthworks, where early settlements date to about 1500 cal BP. By 750–350 cal BP, densely packed "urban" clusters of settlements, connected by road networks and surrounded by a matrix of agricultural land, covered an area of perhaps 30,000 square kilometers. The area surrounding abandoned

A

Figure 3.3
A, geoglyphs from Fazenda Colorada site in the Rio Branco region of Acre State, Brazil. *B*, map of 281 earthworks in Acre State, Brazil. Sources: Sanna Saunaluoma (reprinted with permission) and Schaan et al. (2012, fig. 1), respectively.

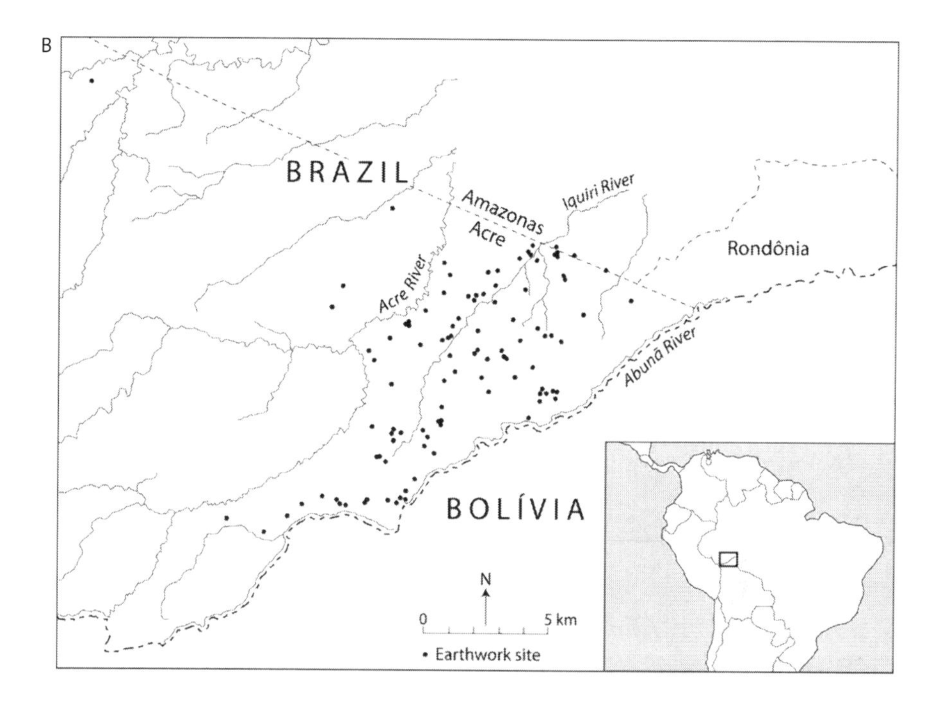

villages is rich in Amazonian dark earth (*terra preta*; see sec. 3.4). Along the Amazon River in Brazil, complex systems of ancient wetland management involving mounds and ponds were practiced, as evidenced by archaeological sites in Marajó Island, Santarém, and the central Amazon. In Marajó, a variety of river products were managed, including fish and riverine palms, such as açai (*Euterpe oleraceae*) and *Euterpe edulis*, which is widely used for palm heart (Heckenberger et al. 2007; Heckenberger et al. 2008; Heckenberger and Neves 2009). Archaeological sites in and around the current city of Santarém suggest a fairly extensive urban settlement with a complex series of mounds, pits, and large middens of black earth (Roosevelt 1999a). Intensive floodplain and upland agriculture supported this broad settlement, which extended 25 kilometers along the Amazon River (Heckenberger and Neves 2009).

Earthworks also characterize ancient human settlements in Southeast Asia. The capital of the Khmer Empire reached its peak population around 912–712 BP (^{14}C yr.) in what is now Cambodia. The city of Angkor was established in the ninth century AD and became the largest preindustrial urban complex in the world. An extensive hydraulic network of canals, embankments, and reservoirs spanning 1,000 square kilometers was constructed by the end of the 12th century to cope with highly seasonal monsoon rainfall.

Human modifications of the natural waterways of the region changed the hydrology of the area in ways that are still evident today (Evans et al. 2007; Kammu 2009). Despite these and other engineering feats, decades-long droughts and intense monsoon rains contributed to the collapse of Angkor in the 15th century AD (Buckley et al. 2010).

Aside from direct impacts on erosion and sedimentation (Beach et al. 2006; Kammu 2009), landscape transformations in the tropics had significant effects on aquatic ecosystems and on soil nutrient-cycling processes in surrounding watersheds. In western Uganda, forest clearance around 1000 cal BP led to substantial changes in the biogeochemistry of Lake Wandakara. Even though forest regrowth occurred 300 years ago, nitrogen concentrations in the lake have remained elevated (Russell et al. 2009). Late Holocene deforestation around Lake Zoncho in southern Costa Rica was associated with altered phosphorous cycling, but this transient effect was reversed following a reduction in agricultural activity after the Spanish conquest (Filippelli et al. 2010).

Sustaining and expanding agricultural productivity across the tropics required more than simply cutting and burning tropical forests. Entire landscapes were transformed, modifying drainage patterns, topographic relief, and soil fertility. These practices left their mark on the forested landscapes of these regions. In some regions, these transformations favored postagricultural forest regeneration, whereas forests have never returned in other regions. The examples presented here illustrate that forest regeneration is a socioecological process reflecting changes in land use and the surrounding landscape over many decades and centuries.

3.3 Prehistoric Fires: Synergies between Natural and Human Causes

Fire has always been a major tool used by humans for environmental change and landscape transformation. Natural processes, such as lightning and volcanic eruptions, also initiate fires (Titiz and Sanford 2007). Forest flammability is increased by prolonged droughts, creating a synergy between droughts and human forest clearing (see box 9.2). Given the tendency for human-initiated fires to escape and spread today, it is likely that accidental burning scenarios also increased rates of biomass burning in the past (Dull et al. 2010).

Although regional-scale climatic forces influenced fire regimes during the Holocene, fire frequency and intensity varied spatially and temporally (Mayle and Power 2008). Novel methods of soil charcoal sampling provide evidence for fire events in western Amazonia, between 4000 and 1000 cal BP, associated with mid-Holocene drought events (McMichael, Correa-Metrio, Bush

2012). Humans were key agents of Holocene forest disturbance in many parts of the Amazon basin particularly in central and eastern Amazonia. Several sites in eastern Amazonia show early to mid-Holocene peaks in *Cecropia* pollen spanning several millennia, suggesting that forests were maintained in early successional states by frequent burning. High spatial heterogeneity in charcoal signals suggests that fires were generally restricted to relatively small spatial scales (Bush et al. 2007; McMichael, Correa-Metrio, Bush 2012). Peak fire activity in the Amazon basin about 1,200 years ago coincided with increased El Niño/La Niña Southern Oscillation (ENSO) activity (Bush et al. 2008). A surge in fire events between 1800 and 1500 cal BP in Amazonia coincided with the spread of agriculture and the formation of *terra preta* soils (see sec. 3.4). The northern Amazon basin (5° N–5° S) shows the highest concentration of fires during the past 2,000 years, whereas the southern Amazon basin indicates two peaks of biomass burning: one around 7500–6000 cal BP and another between 3000 and 500 cal BP (Carcaillot et al. 2002).

Biomass burning in Central America, increased gradually during the early Holocene, reaching high levels from 7000 to 3000 cal BP, which likely corresponds to an extended period of agricultural expansion (see chap. 2, sec. 2.5). Pollen and charcoal records show forest clearance, burning, and agriculture beginning ca. 4000–3000 cal BP or earlier in the Maya lowlands and Central America (Horn 2007). This period was followed by a decline in biomass burning during the last 1,000 years, reflecting land abandonment after the Mayan collapse and later population declines following the Spanish conquest (box 3.1; Carcaillot et al. 2002; Mann 2005; Nevle and Bird 2008; Dull et al. 2010). Coastal savannas of French Guiana show opposing temporal trends in fire frequency. Fires were more frequent during colonial land management following the European conquest than during native raised-field farming prior to the conquest (Iriarte et al. 2012).

Fires caused by natural events or by humans occurred throughout the tropical regions of the world during the late Pleistocene and Holocene (see table 2.1; Goldammer and Seibert 1989). Frequent fires in peatlands of East Kalimantan during the late Holocene (3000–2000 cal BP) are linked to anthropogenic burning during ENSO events (Hope et al. 2004). Charcoal from early to mid-Holocene periods is commonly found beneath soils in tropical wet forests of Costa Rica, the northern Amazon basin, Guyana, and French Guiana (Sanford et al. 1985; Saldarriaga and West 1986; Horn and Sanford 1992; Hammond et al. 2006; Titiz and Sanford 2007). In Ituri Forest in the eastern Democratic Republic of Congo, almost 70% of 416 soil pits contained charcoal with the majority of radiocarbon dates falling within the last 2,000 years (Hart et al. 1996). In Amazonia, a majority of sites where soil charcoal

Box 3.1. Did Postcontact Forest Regrowth Contribute to the Little Ice Age?

Paleoclimatologist William Ruddiman (2003, 2005) put forth the controversial hypothesis that anthropogenic emissions of CO_2 and methane have been affecting global climate for 8,000 years, much earlier than the Industrial Revolution. Ruddiman proposed that these emissions actually reversed a natural decrease in greenhouse gases that otherwise would have initiated a new period of glaciation. Then came the European conquest of the Americas in the late 15th century, followed by the demographic collapse of Amerindian populations. The subsequent abandonment of croplands and the cessation of forest burning led to large-scale regeneration of forests throughout the Americas and high rates of carbon sequestration. Ruddiman further claimed that this sudden carbon sequestration caused a 5–7 parts per million decrease in CO_2 concentrations and a change in the carbon isotope signature ($\delta^{13}C$) of atmospheric CO_2 that contributed to the global thermal anomaly of about–0.1 °C that is known as the Little Ice Age, from AD 1500 to 1750, in the Northern Hemisphere.

These views invited both criticism and support (Brook 2009). The extent to which changes in human populations in the Americas affected forest-burning regimes during prehistory remains highly controversial (Barlow et al. 2012). Marlon et al. (2008) and Power et al. (2013) attested that global-scale climate changes, rather than human population changes, were the main drivers of temporal patterns of biomass burning in Central and tropical South America during the past 2,000 years. Their data showed that the largest decrease in biomass burning in the Neotropics preceded population declines. Power et al. (2008) analyzed global biomass burning changes in sedimentary carbon over the past 21,000 years. Their analysis showed regional coherence in fire regimes and emphasized the role of large-scale climate controls modulated by local changes in vegetation and fuel load. Power et al. (2013) concluded that decreased biomass burning was underway at a global scale before AD 1500, consistent with climate change during the Little Ice Age.

Other interdisciplinary teams of scientists have rallied behind Ruddiman's hypothesis, providing compelling evidence of the human footprint on atmospheric CO_2 levels before the European conquest and of forest regeneration afterward. Faust et al. (2006) claimed that terrestrial carbon sequestration provides the best explanation for both the rapid decline in atmospheric CO_2 concentrations (5–8 ppm within 20–100 yr.) and the increase in $\delta^{13}C$. Nevle and Bird (2008) ruled out sunspot and volcanic activity as drivers of global cooling during the Little Ice Age. Reconstructions of Neotropical fire history based on sedimentary and soil charcoal records from sites throughout Central and South America show a sustained period of reduced biomass burning after ca. 500 cal BP. Proxy data on solar radiation, ENSO events, precipitation, and temperature cannot explain this period of reduced biomass burning (Nevle and Bird 2008; Nevle et al. 2011).

The collapse of human populations in the Americas from 16th- and 17th-century epidemics is well documented (Denevan 1992a; Turner and Butzer 1992; Denevan 2001). Dull et al. (2010) estimated that 19.4 million people were living in the forested lowlands of the Neotropics at the eve of the Columbian Encounter (8.1 million in South America and 11.3 million in Mesoamerica and the Caribbean). In the tropical lowlands of Central and South America approximately 95% of the inhabitants perished primarily due to disease but also to warfare, slavery, and starvation. Dull et al. (2010) claimed that approximately 20% of the earth's human inhabitants were lost in these two centuries, but their calculations do not take into account the high number of Mayan people that died 500–600 years earlier during the Maya Collapse (Gill 2001).

Dull et al. (2010) estimate that the total carbon sequestration potential from postcontact reforestation in the Neotropics ranged from 2 to 5 petagrams of carbon, assuming that 0.9–1.5 hectares per person were abandoned and regenerated. This amount of sequestered carbon represents from 6% to 25% of the terrestrial carbon sequestration required to produce the atmospheric CO_2 decrease of 5 parts per million from AD 1500 to 1750. This is a low estimate, as it does not include land burned through unintentional escape of fires or forest regeneration in North America.

Natural regeneration of Neotropical forests in the aftermath of European contact may have contributed substantially to declining atmospheric CO_2 concentrations and the Little Ice Age of AD 1500–1750. These findings underscore the vast potential for CO_2 sequestration following large-scale forest regeneration in the Neotropics (see also box 11.2).

has been documented in paleoecological studies are in moderately seasonal forests (Bush et al. 2008). In regions supporting moist or wet tropical forests, naturally occurring fires are restricted to periods of high climatic variability (Haberle and Ledru 2001).

Haberle et al. (2001) constructed a long-term cumulative record of microscopic charcoal abundance over the past 20,000 years in Indonesia and Melanesia. Fire frequencies are not clearly related to changes in subsistence patterns in human populations over this time period. Charcoal values are consistently high during the last glacial period (17,000–12,000 BP) and the later half of the Holocene (4000–2000 BP) and are generally low during the early Holocene (9000–5000 BP). The record of biomass burning (based on a normalized comparative index of sedimentary charcoal anomalies called the *charcoal index*) between 8000 and 3000 BP in Indonesia and Papua New Guinea parallels the global atmospheric CO_2 concentration from Antarctica (Carcaillet et al. 2002). This period of biomass burning corresponds with agricultural expansion and with intensification of ENSO, which led to severe drought (Haberle 1996; Haberle and Ledru 2001).

The historic trends in biomass burning described above are based solely on available data from long-term paleoecological studies in selected sites, rather than on systematic sampling of these regions (Barlow et al. 2012). Therefore, it is premature to assume that the patterns examined to date accurately reflect the actual spatial or temporal distribution of prehistorical biomass burning within these regions (Bush and Silman 2007). Lake sediments containing evidence of long-term settlement, burning, and cultivation cannot be assumed to indicate spatially or temporally extensive landscape transformations (McMichael, Bush, et al. 2012). Based on soil charcoal studies in lowland Neotropical forests, lake sediment records do not provide evidence of fire occurring in areas more than five kilometeters away (McMichael, Correa-Metrio, Bush 2012).

3.4 Ancient Soil Modifications

Human activity over hundreds to several thousands of years produced changes in soil properties across the Brazilian Amazon basin and in regions of Columbia, Ecuador, Peru, Venezuela, Bolivia, the Guianas, and West Africa (Eden et al. 1984; Glaser et al. 2001; Graham 2006; Fairhead and Leach 2009). *Terra preta do indio* is an anthropogenic dark-earth soil (ADE), with approximately three times the levels of organic matter, nitrogen, and phosphorus than surrounding ferralsols and acrisols (see plate 2; Glaser and Birk 2012). Characteristically, ADE has numerous broken ceramic pieces, fish bones and scales, shells, and plant remains, as well as ash and a high charcoal content. Basically, it is composed of household waste. ADE typically exhibits elevated levels of nitrogen, phosphorus, potassium, calcium, magnesium, manganese, zinc, and other nutrients compared to surrounding soils (Schmidt and Heckenberger 2009). Over 350 sites of *terra preta*, up to two meters deep, have been located in the Brazilian Amazon basin in sites that range from less than one hectare to several hundred hectares (fig. 3.4; Bush and Silman 2007; Denevan 2007). *Terra preta* is found in all ecoregions and landscapes of Amazonia, with most sites associated with unflooded plateaus (*terra firme*) near white-water rivers (Glaser and Birk 2012).

The association of *terra preta* with human settlements suggests long-term site occupation and intensive cropping systems, rather than shifting cultivation with long fallow periods (Neves et al. 2003). *Terra preta* may have enabled permanent cultivation to support increasing populations over 1,000–2,400 years ago (Glaser et al. 2001). Formation of ADE began around 2,000–2,500 years ago and is clearly associated with the widespread appearance of intensive agriculture in Amazonia. Most *terra preta* sites are between 500 and 2,500 years old (Neves et al. 2003), and they generally occur along river bluffs, in association with settlements (Lima et al. 2002). A second form of anthropogenic soil commonly found in Amazonia, *terra mulata*, is thought to be formed by mulching and burning in semi-intensive or intensive agriculture. *Terra mulata* is dark brown and has elevated levels of organic carbon, but it is less nutrient rich than *terra preta* and has fewer cultural remains (Schmidt and Heckenberger 2009).

How did these soils form? Neither type of anthropogenic soil develops in areas undergoing shifting cultivation (Glaser 2007). Current thinking is that ADE was created as large middens, or organic trash heaps, in zones surrounding stable settlements (Schmidt and Heckenberger 2009). Over time, these "compost piles" became incorporated into the soils and were used for crop cultivation, as they are used today. This theory is consistent with earlier views that intensive, sedentary agriculture, rather than long-fallow shifting cultivation, was practiced during prehistoric times due to the difficulty of clearing dense tropical forest with stone-axe technology (Denevan 1992b). The high level of charcoal found in ADE and the fact that ADE is not found in all areas of long-term human settlement suggest that its formation was intentional (Sombroek et al. 2003; Erickson 2003, 2008; Graham 2006). These soils were often located in a broad band surrounding villages.

In the Republic of Guinea, West Africa, analogs to *terra preta* and *terra mulata* are found in soils of former village sites. Villagers created soil mounds and raised beds, and incorporated kitchen waste, burned and unburned crop residues, and, in some cases, pottery into soils surrounding villages to

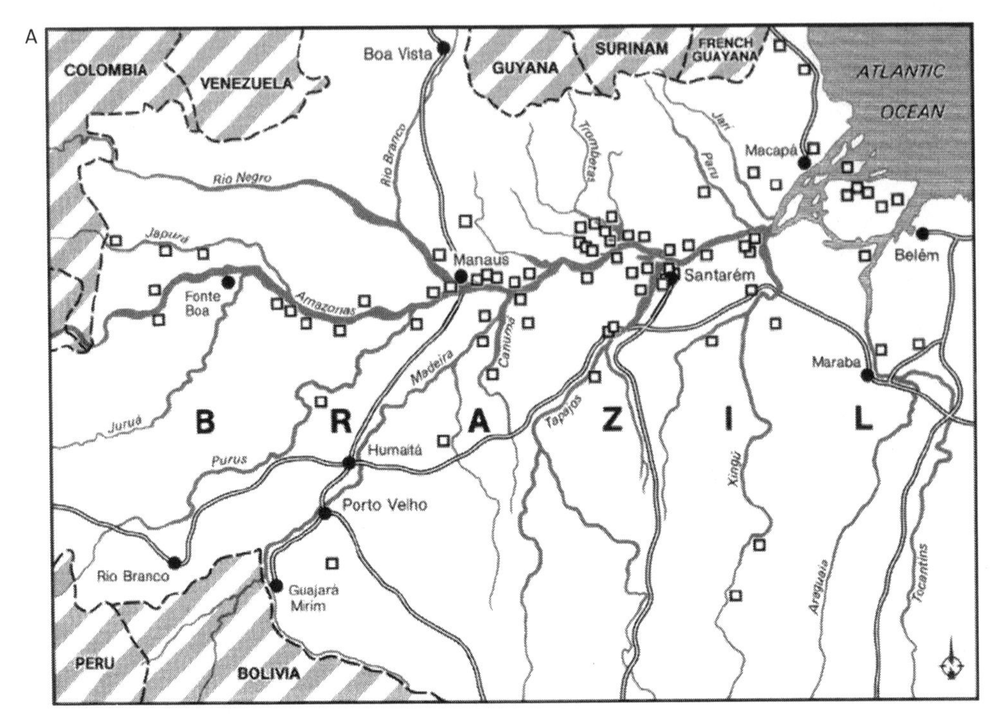

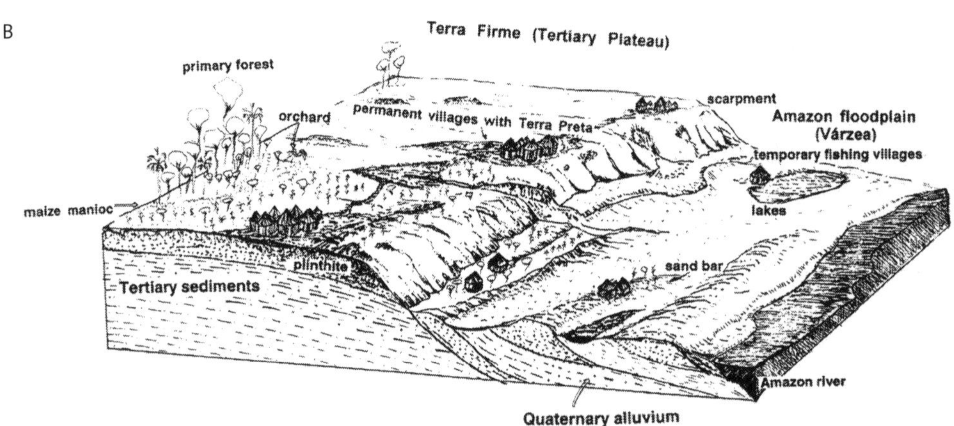

Figure 3.4

A, known occurrence of *terra preta* sites in central Amazonia. *B*, representation of *terra preta* sites and settlements on pre-Columbian river bluffs in Amazonia. The floodplain provided clay for pottery, fish, minerals, and organic material used to enrich soils, but was not suitable for permanent settlement or cultivation. Sources: Glaser and Birk (2012, fig. 2) and Lima et al. (2002, fig. 6), respectively.

improve fertility and moisture retention. These "ripened" soils in ancestral villages are prized for cultivation and are claimed as family lands (Fairhead and Leach 2009). Anthropogenic soils have also been identified in riverside locations of the Malinau district of East Kalimantan that bear some features of ADE (Sheil et al. 2012).

Certain useful plant species show current associations with ADE (Balée 1989; Clement et al. 2003; Fraser et al. 2011). These include locally distributed species as well as semidomesticated species, incipient domesticates, and economically important species. Many species associated with or restricted to ADE are "disturbance specialists" or economically important species, such as brazil nut (*Bertholletia excelsa*), babaçu palm (*Attalea phalerata*), caiaué (*Elaeis oleifera*), and tucumã (*Astrocaryum vulgare*). Clement et al. (2003) proposed that ADE zones currently function as agrobiodiversity reservoirs as a result of the concentration of associated species and genotypes that have been selected over millennia of use in agriculture and arboriculture.

Supporting this hypothesis, Junqueira et al. (2010a) found that density and species richness of domesticated woody species in second-growth vegetation are higher on ADE soils, suggesting that the long-term association of *terra preta* with human cultivation has favored the concentration of agrobiodiversity in these areas (Junqueira et al. 2010a; Junqueira et al. 2010b). In lowland forests of Santa Cruz Bolivia, however, abundances of 17 useful or semidomesticated tree and palm species did not differ significantly among forests growing on *terra preta, terra mulata,* and nonanthropogenic soils. These sites had not been occupied for 380–430 years, so it is possible that initial effects of anthropogenic soils on forest regeneration have been masked by other factors, such as local dispersal or natural disturbances (Paz-Rivera and Putz 2009). The association of *terra preta* soils with permanent settlements provides further evidence that pre-Columbian populations actively managed the forest environment, improving conditions for intensive, short-fallow agriculture. These activities have left long-term legacies on forest composition and settlement patterns across central Amazonia and potentially in other regions.

3.5 The Scale of Prehistoric Human Impacts in the Neotropics

Illuminating the spatial and temporal scale of prehistoric human impacts requires a serious interdisciplinary scientific effort using robust sampling methods and unbiased analysis of data (Bush and Silman 2007; Mayle and Iriarte 2013). Clearly, some tropical regions were heavily impacted during prehistory, whereas others show minimal or no impact (Piperno 2011; McMichael, Bush, et al. 2012; McMichael Correa-Metrio, Bush 2012; McMichael

Piperno, et al. 2012). Based on our current state of knowledge, it would be unwise to assume that forests that appear to be pristine today have truly been unaffected by human activities (Clement and Junqueira 2010). Some formerly impacted areas have recovered to the extent that past disturbances cannot be detected by vegetation surveys. Bush and Silman (2007) and Barlow et al. (2012) rightly cautioned against extrapolating from sites of known human occupation to infer the extent of Amazon-wide levels of human disturbance.

Historical ecologists and anthropologists have highlighted the extent of prehistoric impacts on tropical forests using terms such as *humanized landscapes, built landscapes, cultural parkland,* and *feral forests* (Erickson 2000; Heckenberger et al. 2003; Mann 2005; Campbell et al. 2006). These "anthropogenic" forests include settlements, cultivated areas, managed fallows, and vegetation enriched through arboriculture and forest gardening (see box 2.1). Although these terms are fitting for particular landscapes that have been carefully studied, they do not accurately represent our overall level of knowledge regarding prehistoric human impacts on tropical forests. Balée (1989) estimated that prehistoric peoples used or modified 12% of the Amazonian upland forest, whereas Denevan (1992a) estimated that 22% of Greater Amazonia was deforested prior to the European conquest. Within Amazonia, settlements were concentrated along river basins, which constitute less than 5% of the area of the Amazon basin (Heckenberger and Neves 2009). Areas settled away from major rivers were generally located in seasonal forests in southern Amazonia (Heckenberger et al. 2007). Extensive earthworks in Acre State, Brazil, however, are located in interfluvial regions (fig. 3.3; Schaan et al. 2012).

These claims of pervasive ancient human impacts across Amazonia were flatly refuted by Peres et al. (2010), who maintained that pre-Columbian human influence in interfluvial regions of Amazonia was negligible or undetectable, that less than 15% of Amazonia sustained permanent settlements, and that oligarchic forests could be maintained by localized soil conditions and seed dispersal by seed-caching rodents. McMichael, Bush, et al. (2012) and McMicheal, Piperno, et al. (2012) also challenged the view of widespread human influence in Amazonia. Seasonally dry forests of central and eastern Amazonia burn more easily, are less prone to flooding, and provide far more evidence of fire and *terra preta* soils than do the aseasonal forests of western Amazonia. The upland forest sites they studied in western Amazonia show no evidence for widespread deforestation or settlements but instead suggest occupation by highly localized, small, and shifting populations prior to the European conquest. This conclusion is consistent with work by McKey et al. (2010) and Arroyo-Kalin (2012) that suggested that the sweet (nontoxic) variety of manioc was cultivated along with maize in western Amazonia in small-scale, shifting cultivation settings, whereas the bitter (toxic) variety of manioc was cultivated in *terra preta* soils under more intensive land use along river systems in central and eastern Amazonia, where human populations were greater. The intensity of prehistoric disturbance in interior forests of the Amazon basin declines as the distance to river bluffs increases and as forest seasonality increases (McMichael, Bush, et al. 2012). Confirming these trends, Levis et al. (2012) found a strong negative relationship between the extent of forest manipulation and distance to secondary rivers in interfluvial areas of central Amazonia.

Agricultural systems of the ancient Maya were highly variable spatially and temporally (Dunning and Beach 2000). Human impacts were generally most intense in the lowlands and in highly seasonal forests, which have more fertile soils and are more easily cleared of vegetation. In the southern Yucatán Peninsula, there is little evidence that the bajos (low-lying karstic areas that retain surface water during the wet season) were deforested for cultivation, although tree species were selectively logged. Mayan deforestation was concentrated in the rolling hills of surrounding upland areas. The recruitment sources for forest regeneration in these upland areas may have been the forests in the bajos. The recolonization of upland areas using "stock" from the lowlands may explain the high tree species affinity between the two forest types today (Pérez-Salicrup 2004). The Petén Lake District of Guatemala and the Copan Valley of Honduras were heavily impacted by human forest clearance, soil erosion, and urban settlements (Binford et al. 1987; Beach et al. 2006). But patches of forest vegetation persisted throughout the Mayan landscape, even at the height of agricultural expansion (Wiseman 1978; Ford and Nigh 2009). The continued presence of forest facilitated rapid forest regeneration following population decline during the Mayan collapse and the European conquest (see box 3.1).

There is widespread agreement on one salient feature of prehistoric impacts—that they varied widely in nature and extent. Another point of agreement is that hunting was a pervasive activity across the entire Amazon basin (Junquera and Clement 2012; Barlow et al. 2012). Anthropogenic impacts of prehistoric populations may be far more subtle than direct evidence of forest clearing and burning. Our understanding of the actual extent and nature of anthropogenic impacts and landscape transformations in the American tropics will only be resolved by detailed and systematic sampling and careful extrapolation (McMichael, Piperno, et al. 2012, Piperno 2011).

3.6 Paleoecological Reconstruction of Tropical Forest Regeneration

Reconstructing paleosequences of forest regeneration is challenging for several reasons. First, forest dynamics often more closely reflect Holocene climate variability than the footprint of human occupation and land use. Second, the temporal resolution of pollen profiles in lake or swamp sediments is generally too coarse to document successional changes in species dominance or forest composition. Third, the accurate reconstruction of forest species composition using fossil pollen is impossible for tropical forests, which are dominated by animal-pollinated tree species (Ford and Nigh 2009). Paleoecological studies in the Mayan region rely on fossil pollen from the Moraceae family as a proxy for tropical trees in the landscape (Binford et al. 1987; Islebe et al. 1996). At best, the paleorecord provides a faint sketch of the complex cycles of deforestation and natural regeneration that have recurred throughout history in the tropics (Ford 2008).

3.6.1 *Forest Regeneration in Southeast Asia and West Africa*

Reconstruction of environmental history in mainland Southeast Asia is in its infancy (White et al. 2004). Tropical forest expansion occurred in the terminal Pleistocene and early Holocene, concurrent with habitation by Hoabinian hunter-gatherer societies (Kealhofer 2003). There is no evidence for Neolithic cultures (agricultural production) before 5100 BP (Bellwood 2006). In northeast Thailand, a sediment core from Lake Nong Hang Kumphawapi shows a 2,000-year period of forest burning and disturbance from roughly 6421 to 3812 BP, followed by regrowth of dipterocarp forest by 2900 BP (White et al. 2004). The ancient Cambodian city of Angkor Borei in the Mekong Delta was occupied continuously from about 2000 cal BP, with regular burning, as evidenced by high concentrations of charcoal in sediment cores and abundant grass pollen (Bishop et al. 2003). Charcoal declined abruptly around 1400 cal BP, followed by an expansion of swamp forest and second-growth arboreal taxa such as *Macaranga* and other Euphorbiaceae species.

Around Lake Kassenda in western Uganda, midelevation moist forest was largely replaced by grassland by 1100 BP, coincident with the arrival of iron-using Bantu people from the west. Large, nucleated settlements based on cereal cultivation and cattle grazing appeared. These centers were abandoned around 300 years ago, when settlements became more widely dispersed and populations increasingly relied on pastoralism (Ssemmanda et al. 2005; Taylor et al. 2000; Russell et al. 2009). Lejju et al. (2005) described the recovery of forest concurrent with population declines around 230 years ago in the Munsa archaeological site. Pollen and phytolith evidence show that semideciduous forest taxa, such as *Alchornia*, *Combretum*, *Cyathea*, *Olea*, and *Rapanea*, returned to the site, while nonarboreal pollen sources declined (Lejju 2009). Forest recovery took place at multiple sites within western Uganda, indicating abandonment of large settlements during a period of prolonged drought.

3.6.2 *Forest Regeneration in the Americas*

Paleoreconstructions of forest clearance and regeneration are more numerous in the Neotropics than in other tropical regions. The pollen record of Lake La Yeguada, Panama, provided the earliest evidence of hunter-gatherer societies in Central America, dating to 12,000 cal BP (Piperno et al. 1990; Bush et al. 1992; Ranere and Cooke 2003). La Yeguada was continuously occupied by people throughout the Holocene, with evidence of intensive cultivation and decline of arboreal taxa during the mid-Holocene. Then, around 450 years ago, burning ceased and forest reestablished on the site when Spanish colonists settled the central highlands of Panama.

When Hernán Cortés and his army reached the central Maya lowlands in the early 16th century, dense forests had grown over Mayan ruins. In the Mirador basin of the northern Petén region of Guatemala, sediments of Lago Puerto Arturo reveal a history of forest clearing beginning more than 4,000 years ago, followed by 2,000 years of agricultural activity, then sudden abandonment around 1,100 years ago, at the end of the Late Classic Period (Wahl et al. 2006). This area was never resettled after the Maya Collapse; upland soils now support a dense semideciduous forest. The transition from open forest with agriculture to closed forest occurred in less than 150 years.

A similar history was described for sediments in Lake Petén Itzá by Islebe et al. (1996). Lowland semievergreen forest regrew quickly after over 2,000 years of continuous human impact. Analysis of a short sediment core from Lake Petén Itzá showed that forest regeneration took place from 1112 to 852 cal BP under conditions of increased humidity during the Medieval Warm Period (Mueller et al. 2010). Petén forests recovered rapidly, within 80–260 years, with soil stabilization requiring a slightly longer period of 120–280 years. It is likely, however, that the species composition of the recovered forest was affected by millennia of Mayan land use and management, as well as by changes in climate (Turner 2010b).

A 5,000-year pollen record shows two phases of prehistoric maize cultivation in Sierra de Los Tuxtlas in Veracruz, Mexico, each followed by periods of forest regeneration (Goman and Byrne 1998). Within 300 years of agricultural abandonment, second-growth forests dominated by *Liquidambar* and Urticales (*Trema, Celtis, Ficus,* and other undifferentiated Urticales types) reestablished. At lower elevations in the Sierra de Los Tuxtlas region, analysis of a pollen core from Lago Verde (100 m elevation) showed abandonment of agriculture around 1210 cal BP and regrowth of a diverse lowland tropical forest assemblage within 200 years (Lozano-García et al. 2010). As was observed in Petén by Mueller et al (2010), forest regrowth after 1200 cal BP in the Sierra de Los Tuxtlas region overlaps with a period of increased moisture availability.

In the depths of the Darién forests of Panama, pollen cores studied by Bush and Colinvaux (1994) show nearly 4,000 years of land clearing and maize cultivation in two valleys 15 kilometers apart. These findings are also supported from evidence from phytolith data from the same sites (Piperno 1994). Both sites were abandoned abruptly about 320 years ago, shortly after the Spanish conquest, and have not been resettled. The forests surrounding these areas are now in a late state of secondary succession. As recently as 300 years ago, corn was cultivated adjacent to a swamp at La Selva Biological Station in Costa Rica (Kennedy and Horn 1997). Since canopy trees can live in excess of 300 years, these forests—and many other so-called primary forests in Mesoamerica—are likely still undergoing gradual successional changes in composition and stand structure (Bush and Colinvaux 1994). Current canopy trees may represent only the second generation of trees since forest regeneration began.

Postconquest forest regeneration occurred throughout the Americas following catastrophic population declines (see box 3.1). Despite the massive toll of disease and slavery, small-scale cultivation continued in many areas. In the midelevations of the Coto Brus Valley of southern Costa Rica, Clement and Horn (2001) documented occupation, forest burning, and maize cultivation between 3240 and 460 cal BP. After 460 cal BP, few fires were evident. When Italian immigrants arrived in the 1950s, they encountered a landscape thick with apparently pristine forest, but which was actually 500-year-old regrowth.

Throughout the lowland Mayan region, forest composition during the mid to late Holocene was altered by useful species favored by humans, pioneer species that colonized abandoned areas, and common, generalist species that dominated secondary and old-growth forests (Gómez-Pompa 1971). Consequently, the new forests that developed after agricultural abandonment in the aftermath of the Mayan collapse and the Spanish conquest 900 years later were different in composition from forests that were never cleared or managed. Seed sources for forest regrowth would have originated from surrounding managed forest gardens, which were enriched with useful species (see box 2.1; Ford and Nigh 2009). To this day, species that were heavily utilized by the ancient Mayans and their descendents are highly abundant in these forests (Campbell et al. 2006; Ford 2008; Ross 2011).

Oligarchic forests, characterized by a one or several dominant, economically important species, are widely distributed throughout Amazonia (Peters et al. 1989; Pitman et al. 2001). The linkage of these large-scale geographic patterns of species composition to prehistoric land use and subsequent forest regeneration has been poorly explored. Postconquest wide-scale forest regeneration from local patches of managed forests and fallows could potentially explain the widespread dominance of common tree species in Amazonian upland forests. Most endemic vascular plant species of West African forests have large distribution ranges and exhibit ruderal life-history strategies (Holmgren and Poorter 2007), which may also reflect the long history of human and climatic disturbances that shaped the extent of modern forests.

3.7 Conclusion

Paleoecological studies provide ample evidence for periods of deforestation followed by spontaneous forest regrowth in many tropical regions. Large and small landscape transformations have left obvious and subtle legacies of prehistoric human impacts on tropical forest composition, soils, and geomorphology. Our understanding of these legacies is still at an early stage and will require careful interdisciplinary research. Current tropical forest composition reflects local and regional variation in prehistoric filters linked to both climate change and to human activities, which are inexorably linked. Although not all species survived these filters, the existence of modern forests is testimony to the resilience and regeneration potential of tropical forests, a topic explored in detail in the remaining chapters of this book.

Bush and Silman (2007) expressed concern that assertions regarding the generality of human-generated disturbance in Amazonia can be used to justify further impacts, such as logging and forest clearing. Heckenberger and Neves (2009, p. 260) countered this concern: "Discovering that the region's forested landscapes are not pristine in no way diminishes their relevance in debates on conservation and sustainable development in the Amazon, the poster child of global environmentalism." If we persist in the view that only pristine forest is worth conserving, we might as well forget about tropical forest conservation. As stated by tropical forest ecologist Tim Whitmore (1991, p. 73), "Primeval tropical rain forest, undisturbed and stable 'since the dawn of time' is a myth."

Barlow et al. (2012) claimed that studies of prehistoric human impacts in the tropics provide limited useful information for understanding the resilience of these forests to modern human impacts. I disagree. Although modern impacts are qualitatively distinct from many of the localized forest transformations in the late Pleistocene and early to mid-Holocene, the growing literature on ancient human impacts and land use is highly relevant to our understanding of tropical forest resilience, regeneration, restoration, and conservation today and tomorrow.

CHAPTER 4

TROPICAL FOREST DYNAMICS AND DISTURBANCE REGIMES

Forests are in a continual state of flux, changing all the time and on different spatial scales.—T. C. Whitmore (1991, p. 67)

4.1 Overview

The famous aphorism of Heraclitus "You can't step into the same river twice" applies equally well to forests. Every patch of forest, and its surrounding landscape, has a past, present, and future. Understanding historical human impacts on tropical forests provides a window into the past as well as the present, because the composition of today's forests reflects legacies of forest disturbances that occurred long before scientists began to study tropical forests. Macía (2008) compared forest vegetation in lowland and submontane sites in Madidi National Park in Bolivia. One submontane site showed ruins of an Inca fort dated to be more than 300 years old. The tree size and density at the ruin site was indistinguishable from other forests in the region, but tree species composition was highly distinctive, with a higher abundance of species typical of disturbed forests. Historical legacies are essential components of ecological research on tropical forest dynamics and regeneration (Foster 2000; Chazdon 2003; D. A. Clark 2007).

Rates of temporal change are closely linked to the spatial and temporal scale of analysis. The same forest can be viewed as changing rapidly or as remaining stable, depending on the spatial scale of observation. At a small spatial scale, such as a 10 x 10 m quadrat, individual seedlings, saplings, and trees colonize, recruit, grow, and die at monthly or yearly intervals. Within a larger spatial unit, the number of stems, their size distribution, or even their species composition may remain essentially invariant over many years. Kellner et al. (2009) provided a clear demonstration of the relationship between small-scale dynamics in canopy height (5 m scale) and steady-state equilibrium at large spatial scales in a lowland rain forest in northeastern Costa Rica. Over an 8.5-year interval, increases or decreases in canopy heights were observed in 39% of the forest area. But the overall distribution of canopy height in the forest remained essentially stable.

This chapter focuses on the dynamics of old-growth (or "mature") tropical forests and the legacies (both hidden and obvious) of different types of disturbances on their present structure and floristic composition. Following

D. B. Clark (1996) and Wirth et al. (2009), I use the term *old-growth forest* in this book to describe forests at a late state of succession that are relatively stable with regard to forest dynamics criteria. Disturbance legacies are most pronounced during early stages of forest regeneration, but they may also be observed after hundreds of years as remaining individuals from cohorts of long-lived tree species that established early in stand development. Some forest attributes can reveal past trajectories of change and may be useful in predicting future trends. Research on forest dynamics provides insights into the mechanisms and drivers of forest change beyond a single-time description of structure and composition. From this information, we can also link details about the current structure and composition of forests measured on the ground with characteristics of landscapes and regions revealed by satellite imagery, aerial photography, and other forms of remote sensing (Nelson 1994).

4.2 Disturbance Regimes in Tropical Forest Regions

D. B. Clark (1990, p. 293) defined, based on Pickett and White (1985), disturbance in terrestrial forest ecosystems as "a relatively discrete event causing a change in the physical structure of the environment." Disturbances alter the density, biomass, or spatial distribution of the biota, by affecting the availability and distribution of resources and substrates, or by directly altering the physical environment (Walker and Willig 1999). Periodic natural disturbances such as hurricanes, fires, floods, and landslides vary in intensity and frequency according to other environmental factors such as topography, elevation, seasonality, and rainfall (Whitmore 1991; Whitmore and Burslem 1988). The disturbance regime is the sum of all disturbance events at a particular place and time. Disturbances can arise from within a forest (autogenic disturbance) or from forces acting outside of the forest habitat (allogenic disturbance). Cyclones and volcanic eruptions are allogenic disurbances, whereas the death of a diseased tree and soil disturbance from burrowing animals are considered autogenic disturbances (L. R. Walker 2012).

Disturbances in tropical forests (and elsewhere) are often divided into two broad categories, depending on whether they are caused by forces of nature (natural disturbances) or by human activities (human, or anthropogenic, disturbances). In reality, however, anthropogenic and natural disturbances often occur simultaneously, complicating the task of ascribing vegetation responses to a particular disturbance regime (Chazdon 2003; L. R. Walker 2012). Lentfer and Torrence (2007) described the effects of human disturbances on vegetation recovery following a series of volcanic eruptions on Garua Island,

Papua New Guinea, during the Holocene. Vegetation recovery following these eruptions did not proceed along expected trajectories but was interrupted by intensive burning, land management, and possible cultivation that led to open grassy vegetation rather than to successional forest cover. Second-growth forests on former pastures in Puerto Rico are subject to hurricane disturbances (Zimmerman et al. 1995; Pascarella et al. 2004; Flynn et al. 2010). Forests recovering from human settlement and farming on Kolombangara in the Solomon Islands and Tonga are set back to earlier successional stages by frequent cylone activity (Burslem et al. 2000; Franklin et al. 2004). Vegetation succession following a large landslide in a dry forest region of Nicaragua has been simultaneously affected by human disturbance (Velázquez and Gómez-Sal 2007). Hunting, logging, and fire frequently occur simultaneously (Asner et al. 2009), causing synergistic threats to forest regeneration (see box 9.2; Wright, Stoner, et al. 2007; Dirzo et al. 2007; Laurance and Useche 2009). D. A. Clark (2007) discussed a wide range of anthropogenic and natural disturbances that have affected three active Neotropical research sites, including droughts, fires, floods, harvesting of timber and nontimber products, and soil disturbance by introduced wild pigs.

Disturbance types can be generally characterized and compared in a multidimensional framework that considers four major features: spatial extent, frequency, duration, and severity (Waide and Lugo 1992). Here, I consider above- and belowground damage separately, as these factors are key to forest regeneration following disturbance (fig. 4.1). The framework describes general qualitative trends, reflecting the average or typical pattern for each type of disturbance, based on available literature.

Gaps created by tree and branch falls are the most frequent type of natural disturbance in tropical forests (Hartshorn 1978). Treefall gaps remove relatively little aboveground vegetation and often do not extend completely from the top of the canopy to the forest floor (Connell et al. 1997; Kellner et al. 2009). These are considered autogenic disturbances. Most gaps have a small projected area (< 1000 m^2) and cause minimal soil disturbance, except in uprooting zones. Half of all gaps detected using airborne LiDAR (light detection and ranging; 5 m resolution) in 444 hectares of tropical lowland forest in northeastern Costa Rica were less than 25 square meters (fig. 4.2; Kellner et al. 2009). In montane cloud forest of Costa Rica, most gaps were also small (< 30 m^2) and did not cause localized soil disturbance (Lawton and Putz 1988). Kellner and Asner (2009) compared disturbance regimes and forest structure in five tropical rain forest landscapes in Costa Rica and Hawaii, using data from LiDAR remote sensing to measure canopy height and detect canopy gaps. Modelling of the canopy gap-size frequency distributions in these forests showed strikingly similar fits to a power-law function, despite differences in species composition and modes of disturbance across the sites.

Disturbances caused by large windstorms, hurricanes, or cyclones have a considerably larger spatial extent and greater severity (see fig. 4.1). A major convective storm disturbance in 2005 affected 2,668 hectares of forest and caused the mortality of 300,000–500,000 trees in the Manaus region of

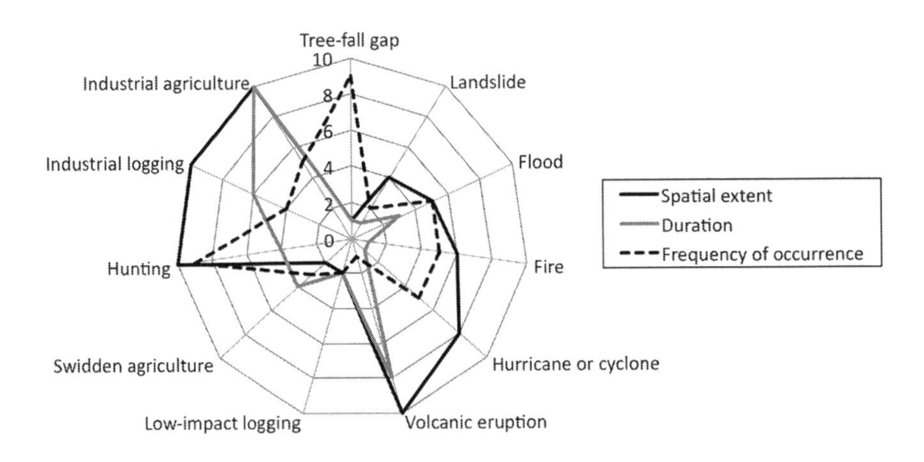

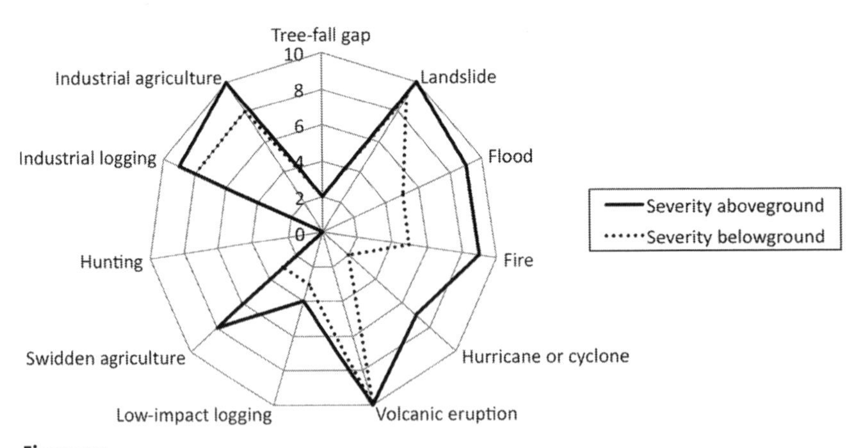

Figure 4.1
A schematic diagram illustrating qualitative trends in five characteristics of 11 types of disturbances that impact tropical forests. The magnitude of each characteristic for each type of disturbance is indicated from 0 (low) to 10 (high).

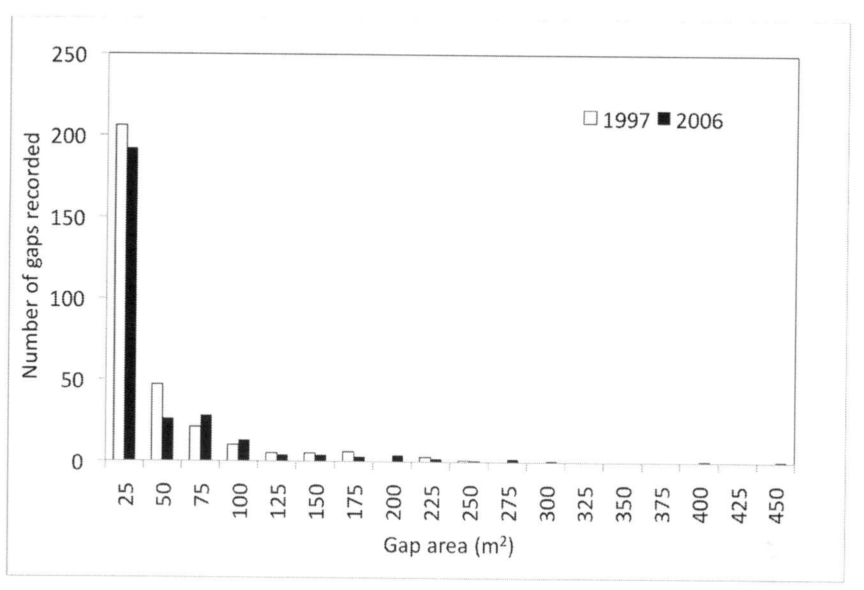

Figure 4.2
Size frequency distribution of canopy gaps in 444 hectares of old-growth forest at La Selva Biological Station. Gap sizes were determined by measuring canopy height at five-meter grid points using LiDAR imagery from 1997 and 2006. Source: Kellner et al. (2009).

central Amazonia (Negrón-Juárez et al. 2010). Catastrophic windstorms have relatively little effect on long-term species composition in tropical forests, as many trees are capable of resprouting (Burslem et al. 2000). Hurricanes cause substantial damage to trees, with little or no soil disturbance. Many damaged trees resprout, leading to relatively rapid reestablishment of the prior species composition.

Fires and floods can have moderate to high spatial extents. During severe droughts, large areas of tropical forests have burned in both wet and dry climatic regions, during the past as well as the present (Goldammer and Seibert 1989). Historical accounts describe fires covering large areas of white-sand forests in the Río Negro basin of Brazil during severe droughts in 1912 and 1926 (Nelson and Irmão 1988). More recent drought events were also associated with large-scale fires in eastern Amazonia and in Southeast Asia (Cochrane and Schulze 1999; Van Nieuwstadt and Sheil 2005). Flood disturbances associated with lateral erosion and channel changes along river systems affect large forest areas throughout the tropics. In the Peruvian Amazon basin, 12% of the lowland forest along rivers is undergoing succession (Salo et al. 1986).

Landslides remove all vegetation and organic soil in steep erosional zones and create downslope depositional zones rich in biotic material (Guariguata 1990). In Mexico and Central America, at least 62 landslide-triggering events occurred between 1888 and 1998; 56 were associated with earthquakes and 6 were associated with major storms. Areas of 17 landslides examined in detail ranged from 300 to 38,300 hectares per event (Restrepo and Alvarez 2006). An estimated 8%–16% of the land surface of Papua New Guinea is disturbed by landslides within a 100-year period (Garwood et al. 1979).

Lava flows from volcanic eruptions cause disturbance on a large spatial scale—completely obliterating vegetation and transforming the soil surface (Turner et al. 1998)—but these occur infrequently (see fig. 4.1). Jago and Boyd (2005) described a series of seven periods of volcanic activity in West New Britain, Papua New Guinea, over the past 2,900 years. Each volcanic eruption caused extensive destruction followed by vegetation recovery. Lava flows can ignite fires during drought conditions in tropical wet forests, as suggested by soil charcoal on the slopes of Volcan Barva in Costa Rica (Titiz and Sanford 2007) and observations in Hawaii Volcanoes National Park (Ainsworth and Kauffman 2009). Mud volcanoes are smaller in extent but can cause severe localized forest disturbances (Ting and Poulsen 2009). Certain areas of the tropics are more prone to particular types of catastrophic or "stand-replacing" natural disturbances. Papua New Guinea has had more than its share of cyclones, earthquakes, volcanic eruptions, and forest fires. In contrast, forests in Malaya, Borneo, and Surinam have been exposed to few catastrophic acts of nature (Johns 1986; Whitmore 1998a).

Differences in disturbance regimes are linked to pronounced differences in forest dynamics, structure, and composition within and across tropical regions. Within Brazil, disturbances are caused by large-scale bamboo mortality in southwestern areas, fires in the seasonal transitional forests, and large blowdowns in the Tefé region. Transitional forests are dominated by fire-adapted babaçu palms (*Attalea phalerata*), and liana forests may also be associated with periodic fire events (Nelson 1994). Baker et al. (2005) reconstructed the disturbance regime of a seasonally dry evergreen forest in western Thailand by studying annual growth rings in a number of the canopy tree species. They uncovered evidence that a catastrophic disturbance on the scale of several hundred hectares impacted this area about 200 years ago. A cohort of *Hopea odorata* trees that colonized the disturbed area continues to dominate the forest canopy today. Since the forest initially began to regenerate, small-scale disturbances in the study plots have provided opportunities for recruitment of a diverse group of tree species, contributing to the high species richness of this forest today.

Human disturbances also vary widely in their extent, frequency, and severity. Hunting is a frequent human disturbance that covers large spatial areas but does not directly cause damage to vegetation or soil (Corlett 2007; Peres and Palacios 2007). Rather, the effects of hunting on vegetation are indirect—that is, through altered abundances of seed predators and seed dispersers in tropical forests (Muller-Landau 2007). Small-scale swidden agriculture strongly impacts aboveground vegetation but creates relatively little soil disturbance. Industrial-scale agriculture and forestry impact large spatial extents of tropical forests over long periods of time; these disturbances eliminate all or most forest cover and significantly modify soils. Mining removes aboveground vegetation, topsoil, and soil organic layers in pit areas (Parrotta et al. 1997; Peterson and Heemskerk 2001).

Many natural disturbances occur quickly, often within a 24-hour period. Postdisturbance forest regeneration processes are initiated with little or no delay. But human disturbances, such as farming or ranching, can persist over decades. For this reason, human disturbances often have more dramatic effects on species composition and on the rate of forest regeneration than do momentary natural disturbances (Zimmerman et al. 1995; Chazdon 2003). But spontaneous forest regeneration after small-scale human and natural disturbances can follow similar stages and pathways. Many local and regional factors ultimately affect the rate and nature of forest regeneration following disturbance.

4.3 Gap Dynamics and the Forest Growth Cycle

Small-scale disturbances are common in tropical forests, mostly resulting from single or multiple treefalls. Spatial variation due to soils, drainage, and microtopography underlie natural disturbance regimes, which have both spatial and temporal components (Whitmore and Burslem 1988). The forest growth cycle has been recognized by plant ecologists for nearly a century. A forest patch (10–100 m in diameter) has its own dynamics defined by localized disturbance and subsequent colonization of a cohort of trees. These patterns of spatial heterogeneity create a complex mosaic of vegetation patches, each with its own internal dynamics, but affected by the dynamics of adjacent units (box 4.1).

Disturbed forest patches can be quite large, such as major blowdowns in central Amazonia that exceed 30 hectares in size (see plate 3; Nelson et al. 1994; Espírito-Santo et al. 2010). These blowdowns occur within a zone traversing the Amazon, from southern Venezuela to northern Bolivia, and are visible on satellite imagery; the largest recorded blowdown was over 2,000 hectares in size (Nelson 1995). Within blowdown areas, blocks of forest undergo phases of disturbance and recovery, creating a landscape mosaic of forest patches

Box 4.1. Aubréville's Forest and the Mosaic Theory of Forest Regeneration

French forester André Marie A. Aubréville (1938) demonstrated that the tree species composition of tropical forests of the Ivory Coast varied spatially. Further, many species represented by large emergent trees were not represented by juveniles, preventing the replacement of those species when they die. Aubréville further noted that second-growth forest had regenerated across many parts of the Ivory Coast over millennia of human disturbance. The presence of light-demanding species characteristic of young "open" forest as large trees in older "closed" forest provided evidence of past human occupation and forest clearing. But Aubréville cautioned, "One must not infer from this that each time one finds these species in closed forest, this is evidence for former felling and burning: in the 'rain forests,' they are able to invade into gaps which occur sporadically due, for example, to the falling of large trees" (1938, p. 46).

Aubréville's findings stimulated much research and gave rise to the "mosaic" theory of forest regeneration later elaborated by Watt (1947) and Richards (1952). Forests are composed of spatially distinct patches, which undergo dynamics at smaller, local scales. These patches may, in some cases, represent distinct successional phases following localized disturbances. Whitmore (1984) called these "gap," "building," and "mature" phases. Deconstructing a forest into a shifting mosaic of patches reduces the spatial scale of regeneration to localized tree replacement dynamics. Differences in tree species composition between individuals in the canopy/emergent layers and those in the undergrowth imply that local forest composition varies temporally as well as spatially (Foster et al. 1996; Burslem and Swaine 2002).

Later studies cast some doubt on the mosaic theory. When larger plots or transects were sampled in a Ghanaian forest, most tree species were represented by juveniles (Hall and Swaine 1976; Swaine and Hall 1988). The extent to which vertical strata of the forest vary in species composition is strongly affected by the spatial scale of sampling, as large spatial scales tend to obscure smaller scale heterogeneity (Newbery and Gartlan 1996). In the case of a forest regrowing after large-scale clearance, however, the pattern is likely to persist even at the largest spatial scales of sampling. The spatial scale at which shade-tolerant canopy species regenerate appears to be considerably smaller than that for light-demanding species (Poorter et al 1996). The frequency and spatial distribution of large gaps (≥ 1,000 m²) strongly influence the colonization and regeneration of light-demanding tree species (Hartshorn 1978; Whitmore and Burslem 1998; Chambers, Robertson, et al. 2009).

On Barro Colorado Island (BCI), Panama, Knight (1975) found that, on average, 40% of the canopy trees in the older (mature) forest stands belonged to species lacking representation in small size classes. Knight concluded that the older forest was still undergoing successional changes. Seven tree species in the old forest at BCI with large trees but few juvenile trees within a 50-hectare plot showed population declines from 1982 to 1995 (Condit et al. 1998). These species are long-lived pioneers that colonize during early stages of succession. Similarly, in mature forest at Korup in Cameroon, 27 of the 197 tree species in the largest diameter classes lacked small trees (Newbery and Gartlan 1996). Populations of these species may also be declining, as conditions are not suitable for new recruitment. Do these results indicate that unstable species composition in these forests is due to exogenous factors, such as environmental or climate change? Or could the "primary" forests at Korup and BCI still be undergoing late stages of succession following disturbances that occurred more than 350 years ago?

Aubréville's emphasis on the spatial ecology of tropical forest trees provided a rich foundation for new theoretical advances in ecology. Complementary species distributions across patches of different successional stages were proposed to be an important mechanism for a peak in species richness in forests subjected to intermediate levels of disturbance (Connell 1978; Bongers et al. 2009). It is no coincidence that Connell's intermediate disturbance hypothesis was based on data from Budongo rain forest in Uganda, which is composed of forest patches of different successional stages (Eggeling 1947; Connell 1978; Sheil 1999; Sheil and Burslem 2003). The spatial discontinuity that Aubreville (1938) observed between juveniles and adults was modeled independently by Connell (1971) and Janzen (1970), who predicted that juveniles dispersed away from adult trees should experience higher rates of survival than those located beneath the adults due to reduced distance-dependent effects of host-specific pathogens and seed predators. These models emphasized the effects of biotic interactions on the abundance and spatial distribution of juveniles relative to adults, apart from effects of forest disturbance regimes.

of different age and successional status (Chambers et al. 2007; Chambers, Robertson, et al. 2009; Chambers et al. 2013). Large forest disturbances were assessed across the entire Amazon basin using Landsat imagery and spectral mixture analysis (Espírito-Santo et al. 2010). This analysis detected 279 disturbed patches ranging in size from 5 to 2,223 hectares, constituting a total disturbed area of 21,031 hectares. Recurrence intervals for these large blowdown disturbances were more frequent (27,000 yr.) in the western Amazon region compared to the eastern Amazon region (90,000 yr.).

Single treefalls cause small, localized disturbances within forests, but regeneration in small gaps often involves different processes than in larger-scale disturbances (Whitmore 1978; Hartshorn 1978; Whitmore and Burslem 1988; Janzen 1990; Chambers, Robertson, et al. 2009). Tree recruitment in single treefall gaps (< 500 m^2) usually favors advance regeneration—saplings or small trees that were present prior to the disturbance. In contrast, large disturbances are usually colonized by light-demanding pioneer species that were not previously established within the disturbed site (Whitmore 1991).

In the absence of forest disturbances, forest structure and composition are predicted to be far more homogeneous and less diverse, dominated by a limited number of disturbance-sensitive species. At the other extreme, in chronically disturbed forests only disturbance-tolerant species are able to persist, also reducing overall species diversity. Where disturbance recurs at a moderate level, however, disturbance-tolerant species can persist, while those more sensitive to disturbance can also colonize and enrich the forest community. Connell (1978) formalized this notion, proposing that intermediate levels of disturbance prevent competitive exclusion, permit more species to coexist, and promote high local species diversity. Under intermediate disturbance regimes and during intermediate stages of succession, larger numbers of ecologically similar species can coexist when periodic local disturbances reduce their competitive interactions. The intermediate disturbance hypothesis (IDH) became the first major nonequilibrium theory to explain the maintenance of high species richness in tropical forests (see box 4.1; Burslem and Swaine 2002).

Tests of the IDH have met with mixed results, in part due to different approaches to measuring forest disturbance (Sheil and Burslem 2003). As presented by Connell (1978), the theory did not account for the strong effect of individual density on measures of species richness (Denslow 1995; Chazdon et al. 1998; Gotelli and Colwell 2001). Furthermore, Connell's theory only focused on large-scale forest disturbances such as windstorms, ignoring small-scale gap disturbances that are considered to be part of the forest growth cycle (Sheil and Burslem 2003). Small treefall gaps (< 50 m^2) are generally insufficient to initiate early successional processes (Hubbell et al. 1999), whereas larger disturbances created by logging are more consistent with IDH predictions (Molino and Sabatier 2001). Bongers et al. (2009) used inventory data from 2,504 one-hectare forest plots in Ghana to test the IDH across a wet-dry climatic gradient. Their results confirmed that species richness reached a maximum at intermediate levels of disturbance, but disturbance explained little of the overall variation in the number of tree species per plot in wet and moist forests.

4.4 Detection of Historical Tropical Forest Disturbance

Tropical forest disturbance throughout paleohistory is detected by the decline or absence of arboreal pollen and a simultaneous increase in pollen of grasses, cultivars, or other disturbance-loving species in sediment cores of lakes. This information does not reveal the extent or the nature of the disturbance, or its true impact on the local vegetation. Much of the evidence supporting widespread anthropogenic deforestation in the Petén region of Guatemala, for example, was based on a thick, seven-meter deposit of Mayan clay dating between 500–1,000 and 4,000 years ago (Binford et al. 1987; Hodell et al. 2008). Ford and Nigh (2009) pointed out that similar clay intrusions are found in lake sediments in the Petén region prior to human occupation in the area. These clay deposits could be the result of extreme precipitation events in the Cariaco basin record from 3,000 to 4,000 years ago. Anselmetti et al. (2007) emphasized that the greatest rates of soil erosion in this basin occurred during initial land clearance, when human population densities were low and exposed soils were highly susceptible to erosion. The coincidence of soil erosion and microfossil evidence suggests a strong link between early periods of soil erosion and land clearance, despite low human populations.

Dendroecological approaches provide major insights into how disturbance history shapes the structure and dynamics of both old-growth and second-growth tropical forests (Rozendaal and Zuidema 2011). The presence of annual growth rings in some tropical tree species has been well established by radiocarbon dating. Age distributions based on tree-rings in one hectare of a semideciduous forest in central Cameroon revealed that the oldest trees (up to 200 yr.) were long-lived pioneers, confirming the late secondary nature of this forest (Worbes et al. 2003). Growth-ring studies of two nonpioneer trees in a Bolivian moist forest showed no evidence of changes in forest dynamics or tree turnover during the past 200–300 years (Rozendaal et al. 2011). Annual growth rings were detected in 37% of the tree species of wet and dry second-growth forests of southern Mexico. Dendroecologial analysis showed that both pioneer and nonpioneer species established early during succession and continued recruiting individuals for many years (Brienen et al. 2009). Moreover, when pioneer trees were still present, tree-ring dating provided accurate estimates of stand age.

Today, various forms of remote sensing provide a real-time assessment of the spatial extent of forest disturbance, and sequential aerial photos or satellite imagery clearly detect different ages of forest regrowth following large-scale disturbances (see plate 4; da Conceição et al. 2009). Moreover, different forms of forest disturbance can often be distinguished from satellite imagery, such as large blowdowns, fire scars, floods, and landslides (Nelson 1994; Restrepo and Alvarez 2006; Delacourt et al. 2009). Asner et al. (2005) used high-resolution satellite imagery to detect effects of logging operations on the number and size of canopy gaps in the Brazilian Amazon region. High-resolution hyperspectral imagery can also be used to distinguish different stages of canopy disturbance following selective logging (Arroyo-Mora et al. 2008). Although advanced stages of forest regrowth (beyond 25 yr. in many cases) cannot be discriminated from old-growth forest using multispectral imagery (Steininger 1996), Kalacska et al. (2007) and Galvão et al. (2009) showed that hyperspectral imagery improved this discrimination. Nevertheless, several forms of anthropogenic disturbance, such as hunting and small-scale gold mining, and the spread of pathogens or invasive species cannot be detected using remote sensing (Peres et al. 2006).

Surveys of canopy height and ground topography conducted with active sensors are used to detect and measure canopy disturbances related to treefall gaps, landslides, and anthropogenic land-use change in tropical regions (Kellner and Asner 2009). Differential synthetic aperture radar (SAR) interferometry was used to observe a large landslide on La Reunion in the Indian Ocean (Delacourt et al. 2009). Polarimetric radar is being used to predict forest structure in tropical forest at scales from 0.25 to 1.0 hectares (Saatchi et al. 2011). Light detection and ranging (LiDAR) can also be used on a fine spatial scale (5 m resolution) to detect changes in canopy height and canopy openings within a large expanse of continuous forest (see fig. 4.2; Kellner et al. 2009), and at coarser scales, to detect forest regrowth based on spatial variation in forest structure (Drake et al. 2002; Dubayah et al. 2010).

Direct measures of localized canopy disturbance also include gap delimitation based on mapping vertical projections of canopy openings (Brokaw 1982) or bases of canopy trees surrounding the gap (Runkle 1982), mapping of canopy height with a range finder (Welden et al. 1991), hemispherical canopy photography (Whitmore et al. 1993; Nicotra et al. 1999), visual assessments of crown illumination using an ordinal scale (D. A. Clark and D. B. Clark 1992), and quantum sensor or red:far-red sensor measurements (Nicotra et al. 1999;

Capers and Chazdon 2004). Each of these methods has advantages and limitations, but proper use of the methods can provide robust comparative measures of local canopy disturbance within a study area. Subcanopy disturbances are not detectable using hemispherical photography or Brokaw (1982) gap delimitation methods, but these disturbances can significantly influence light penetration and tree growth in the forest understory (Palomaki et al. 2006).

Early studies of forest dynamics and regeneration used characteristics of the vegetation itself to detect and measure disturbance. Whitmore (1989b) classified common tree species on Kolombangara Island in the Solomon Islands into four groups according to each species' requirement for canopy disturbance for seedling establishment and tree recruitment. A mean "pioneer index" was then used to assess changes in disturbance conditions and species composition over 21 years. On average, the pioneer index decreased over time, indicating that canopy disturbance was not sufficient to maintain abundances of pioneer tree species in most study plots.

Molino and Sabatier (2001) and Bongers et al. (2009) used a similar approach to rank study plots according to disturbance intensity. In both studies, an index of disturbance was computed as the percentage of trees in each plot that are classified as pioneer species. This bioassay approach integrates responses of vegetation to combined effects of all types of disturbance through recent history, such as windthrow, logging, or fire (Sheil and Burslem 2003; Bongers et al. 2009). Consistent application of this approach requires a detailed knowledge of the regeneration requirements of tree species. In another bioassay approach, Martínez-Ramos et al (1988) used understory palm stem damage to map the distribution and age of small-scale canopy disturbances within a five-hectare plot in a Mexican rain forest (fig. 4.3). This approach indicated that more than 50% of the 5 x 5 m quadrats suffered disturbance over the last 30 years, and 28% showed multiple disturbance events over the last 70 years.

Most methods used to detect forest disturbance are based on measures of forest structure rather than species composition. As a result, these methods cannot be used to distinguish between old second-growth forests (> 200 yr.) and old-growth forest, which tend to have similar vegetation structure but may still vary in species composition (Chazdon 2008b; Fagan and DeFries 2009). The only conclusive way to distinguish old second-growth forests from primary forests (forests that have not been disturbed by humans or catastrophic forces of nature) is through detailed knowledge of species growth and regeneration patterns and careful studies of past stand history (Knight 1975; Baker et al. 2005). Similarly, characterization of plant and animal species

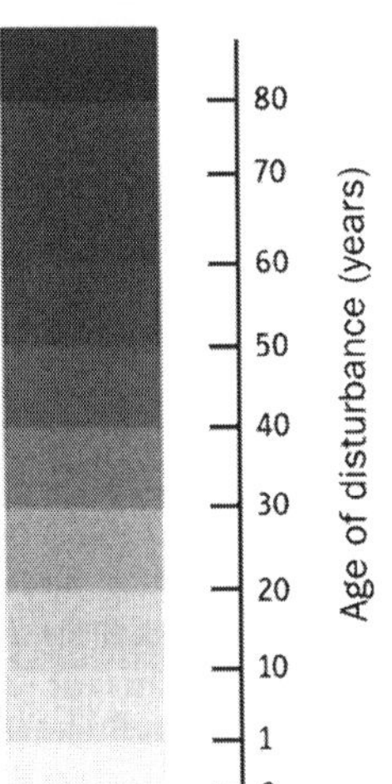

Figure 4.3
The spatial distribution of forest patches exhibiting disturbance at different times within a five-hectare plot in old-growth tropical rain forest at Los Tuxtlas, Veracruz, Mexico. Each square represents a 5 × 5 m quadrat, and shades of gray indicate the number of years since the last treefall was detected in each quadrat. Source: Adapted from Martínez-Ramos et al. (1988, fig. 5).

as "early successional" versus "late successional" should be based on quantitative observations of abundance or occupancy in forests at different successional stages (Sberze et al. 2010; Chazdon et al. 2011).

4.5 Are Old-Growth Tropical Forests Stable?

In 1998, Phillips and his colleagues documented increases in aboveground biomass during the late 20th century in 50 permanent monitoring plots in tropical old-growth forests of South America. These increases were due to lower rates of tree mortality than rates of ingrowth via recruitment and stem growth. These Neotropical forests had no known record of human or natural disturbance. Why, then, would they show increases in tree biomass? Although they lacked specific evidence to explain these responses, Phillips et al. (1998) concluded that biomass gain (or more precisely, increases in plot-level basal area increments) in these forests could be a response to global anthropogenic environmental changes, such as increased atmospheric CO_2 levels, higher temperatures, or nutrient fertilization. Their study raised several important questions that are still actively debated today. Are old-growth tropical forests a sink for carbon? Are old-growth tropical forests at a steady state with respect to the structure and composition of their vegetation? Could these forests still be responding to an "unseen" past disturbance (Wright 2005)?

Several long-term studies in tropical forests have indicated that tropical forest trees may actually be declining in growth rates potentially because of physiological stress from high temperatures or severe drought in some years (D. A. Clark et al. 2003; Feeley, Wright, et al. 2007, D. B. Clark et al. 2010; Wright et al. 2011). In fact, Phillips, Aragão, et al. (2009a) reported no net biomass increase in 55 one-hectare monitoring plots in Amazonian forests during the sampling interval that included the 2005 drought event, reversing a multidecadal carbon sink. Flaws in the methodology used by Phillips et al. (1998) cast doubt on their findings and conclusions (D. A. Clark 2002). D. A. Clark (2004) pointed out that basal area increments are only one component of primary productivity and net aboveground biomass change in forests. Baker et al. (2004) reanalyzed biomass trends, excluding problematic plots, accounting for species differences in wood density, and using different allometric relationships between tree dimensions and biomass. Their reanalysis confirmed a trend of increasing biomass in Amazonian plots but with a slight reduction in magnitude (Lewis et al. 2004).

Another major issue was raised concerning the small size of the plots analyzed in many of the long-term forest dynamics studies (Körner 2003; Chave et al. 2004; D. A. Clark 2004; Fisher et al. 2008). Among the 59 plots analyzed by Baker et al. (2004), 51 were one hectare or less in size. Even within a stable, large old-growth forest area, small patches in a building phase will exhibit increases in biomass, whereas a few patches subjected to recent disturbances will show biomass declines (see box 4.1). Consequently, plot-based studies tend to overestimate regional carbon sequestration, particularly for small plots (Körner 2003). Fisher et al. (2008) and Chambers, Negron-Juarez, et al. (2009) found that spatially clustered disturbances can also lead to underestimation of mortality in small forest plots, even in an undisturbed old-growth forest landscape. Simulations performed by Gloor et al. (2009), based on different assumptions than the model of Fisher et al. (2008), showed that, after accounting for undersampling of large disturbances, net biomass gains were still observed in Amazonian rain forest plots.

What is the appropriate spatial and temporal scale for assessing forest dynamics in old-growth forests? Based on data collected over 15 years for stems greater than or equal to one centimeter in a 50-hectare plot on Barro Colorado Island, Chave et al. (2008) found no evidence for long-term change in aboveground biomass. In this comparison of 10 large forest dynamics plots, 4 plots increased significantly in aboveground biomass, 3 plots showed a nonsignificant trend of increasing biomass, and 3 plots showed decreasing biomass. Forest dynamics in 9 of the 10 plots were consistent with the hypothesis that tropical forests are in advanced stages of recovery from a past disturbance, as biomass increased faster for slow-growing species relative to fast-growing species. In lowland wet forests of Paracou, French Guiana, Rutishauser et al. (2010) reported that increments in forest biomass from 1991 to 2007 were driven by low biomass losses through tree mortality rather than by biomass gains through increased tree growth, suggesting that these forests are responding to a historic disturbance event rather than to increasing atmospheric CO_2 concentration. Plot sizes of 10 hectares or larger are required to detect effects of climate change, increasing CO_2, or nutrient fertilization on biomass dynamics in central Amazonian old-growth forests (Chambers et al. 2013). Landscape-scale estimates of forest biomass need to consider the full continuum of disturbance events, from single treefalls to blowdowns exceeding 1,000 square meters.

Evidence is also mounting that the biomass of lianas, or woody vines, is increasing in forests throughout Amazonia and in Panama (Phillips et al. 2002; Wright, Calderón, et al. 2004; Phillips, Higuchi, et al. 2009b, Ingwell et al. 2010) potentially due to enhancement of photosynthesis by elevated CO_2 levels (Körner 2009). Laurance et al. (2004) found no change in aboveground biomass over 15 years in 18 one-hectare plots in central Amazonia but did observe changes in the generic composition of trees. Genera of faster-growing species increased in abundance, whereas genera of slower-growing species declined, in striking contrast to the findings of Chave et al. (2008).

Increases in aboveground biomass from 1968 to 2007 were observed across 79 plots in African tropical forests. Some of these plots are as small as 0.2 hectares (Lewis et al. 2009). To date, no study has yet demonstrated that increases in atmospheric CO_2 levels lead to increased biomass or stem turnover in tropical forests (D. A. Clark 2007). A study of tree growth over 10–24 years in a tropical wet forest in Costa Rica showed significant effects of annual variation in dry-season rainfall and nighttime temperatures on annual measures of tree growth and wood production but no effects of atmospheric CO_2 (D. B. Clark et al. 2010).

After 15 years of controversy, Phillips, Higuchi, et al. (2009) resolutely maintained that old-growth Amazonian forests are increasing in biomass, have increasing rates of stem turnover, and are changing in species composition. Lewis et al. (2009) proposed that directional changes in African forests are most likely driven by increasing resource availability at a global scale. But these changes may also reflect long-term forest regeneration following past disturbances (Muller-Landau 2009). Increases in aboveground biomass in a 25-hectare forest plot in Ecuador over 6.3 years were caused by a shift toward larger tree size, consistent with expectations for forests that are recovering from relatively recent disturbances (Feeley, Davies, et al. 2007; Valencia et al. 2009). Valencia et al. (2009) concluded that forest structure is rarely stable at decadal scales, even in the absence of human intervention. Short-term responses could simply be a manifestation of the forest growth cycle (see box 4.1). It is important to keep in mind that the mean total observation period for assessing biomass changes in Amazonian forests was 11.3 years (Gloor et al. 2009) and the mean time interval for African and Asian forests was 10.6 years (Lewis et al. 2009). From a forest dynamics perspective, these are very short time intervals indeed.

A medium-resolution (25 m footprint) LiDAR study of the entire forest landscape at La Selva Biological Station in Costa Rica (1,600 ha) mapped short-term changes in estimated forest biomass at a spatial scale of one hectare (see plate 5; Dubayah et al. 2010). This old-growth forest (1,168 ha) is a heterogeneous mix of areas that show no significant change in estimated biomass, with widely scattered source and sink areas, whereas second-growth forest regrowth areas (432 ha) are mostly carbon sinks or neutral. Determining under which conditions mature tropical forests act as carbon sinks or carbon sources is a critically important issue of our time, but according to D. A. Clark (2004, p. 484), "there are no tropical forest sites for which existing plot data are sufficient to determine whether the forest is acting as a carbon sink or source."

4.6 Conclusion

Forests, like any natural system, are dynamic entities. Forest dynamics research in the tropics has revealed responses to a wide range of disturbances at multiple spatial and temporal scales. Natural disturbance regimes interact with anthropogenic disturbance to influence forest composition and structure. Long-term changes in old-growth forests may reveal responses to disturbances that occurred well over 300 years earlier and can also indicate responses to climate change and other environmental factors. Our ability to detect and interpret these responses depends strongly upon the spatial and temporal scale of analysis. Small-scale dynamics are often homogenized and masked by large-scale stability. Rates of recovery of different forest attributes are strongly affected by the spatial and temporal scale of disturbances and by effects on soil properties. Forest regeneration generally proceeds more rapidly following small-scale disturbances that do not disturb the soil, such as treefalls. Human disturbances that mimic small-scale natural disturbances and minimize soil disturbance will also favor rapid rates of forest regeneration (see table 7.1).

Changes in species composition, species functional characteristics, and forest structure are all manifestations of forest dynamics. As these properties usually recover at different rates following disturbances, forest stands may appear to reach stability with respect to structure, while still showing directional change in species turnover. Because old-growth forests are often defined based on forest structure, they may not be stable in species composition. Reconstructing historical trajectories and long-term population and community dynamics long beyond the typical range of 10–25 years will be needed to gain critical insights into how disturbances have shaped today's forests and how forests will change in the future.

CHAPTER 5

SUCCESSIONAL PATHWAYS AND FOREST TRANSFORMATIONS

It is evident that it is no longer useful to speak of tropical succession in the singular. Rather, the tropics encompass a broad array of tropical successions. People are not only the main force today which creates tropical successions, but they thrive in them. It behooves us to appreciate and attempt to understand this variability.—John Ewel (1980, p. 7)

5.1 Overview

Succession is the process of change in an ecological community in a newly formed habitat or following a disturbance that removes existing vegetation. Forest succession involves the gradual replacement of species and populations that establish during initial stages (pioneers) with those that characterize later stages (old-growth species). In forest ecosystems, successional processes are manifest as replacement or turnover of animal, plant, and microbial species (Horn 1974).

Early scholars of succession viewed the successional process as an orderly and progressive sequence of stages that culminate in a stable "climax" stage (Clements 1916; Whittaker 1953; White 1979). This successional paradigm emerged from an equilibrium view of nature, which predicts that natural systems will return to a predictable and stable state following a disturbance (Wu and Loucks 1995). Nonequilibrium theories have displaced equilibrium views, casting new perspectives on the nature of successional change and responses to disturbance (Whitmore and Burslem 1988; Pickett and White 1985). As stated by R. H. Whittaker (1953, p. 59), "No completely rigorous definition of the climax and its distinction from succession has been found, and apparently none need be expected. If the retention of the climax-succession distinction is to be justified, presumably it must be not because the distinction is sharp and invariable, but because the distinction, relative as it is, has some real significance and usefulness." In actual practice, the distinction between a "successional" and a "climax" forest is subjective; there is no magical moment when a forest stops undergoing succession (Chazdon 2008b). Nevertheless, the sharp distinction between successional and climax forests is widely applied today, with major implications for conservation practices and land-use policy. This chapter focuses on successional patterns and processes in tropical forests and how they lead to variation in successional trajectories of vegetation structure and composition following human and natural disturbances. I describe how old-growth forests can be distinguished from second-growth and degraded forests and discuss the approaches used to understand and describe tropical forest succession.

5.2 Variability in Successional Pathways

Even within a single forest type and region, successional changes in vegetation may follow multiple successional pathways (Walker, Wardle, et al. 2010). Divergent successional pathways arise due to local variation in initial conditions, which generates distinct trajectories of species abundance. Variation in initial conditions is caused, in part, by different patterns of agricultural land use (Ferguson et al. 2003; Larkin et al. 2012; Williamson et al. 2012). Mesquita et al. (2001) and Lucas et al. (2002) described distinct successional pathways on former agricultural land in central Amazonia. In sites where land use exceeded four years and fire was used during initial clearance of forest, species of *Vismia* dominated early stages of succession. In contrast, *Cecropia* species dominated in areas where pastures were used for less than two years in both burned and unburned sites (see plate 6). These differences in early species colonization and dominance strongly affected successional trajectories in vegetation structure and composition (Mesquita et al. 2001; Norden et al. 2011; da Conceição Prates-Clark et al. 2010; Williamson et al. 2012).

Pickett et al. (1987) proposed a hierarchical framework of succession, which identifies general causes, ecological processes, and specific factors that determine successional pathways (table 5.1). At the most general level, successional pathways emerge as a consequence of the types and sizes of disturbances, the availability of species to colonize these sites, the life histories and ecophysiological traits of these species, and interactions among colonizing species. This framework may apply to succession within a small, localized disturbance or to succession within a larger, more heterogeneous disturbance.

Based on this framework, it is easy to see how multiple successional outcomes can be observed within a single region, climate zone, or landscape. Successional trajectories are affected by the scale, frequency, and intensity of previous disturbance or land use, by the nature of remnant vegetation, and by conditions following disturbance, including types of management, colonization of invasive species, and dispersal of seeds from surrounding forest areas (Chazdon 2003, 2008b). Although these factors are often affected by conditions in the surrounding landscape or region, vegetation inventories in successional forests are often conducted in plots as small as 0.1 hectares in small patches of regrowth. As vegetation patterns often exhibit small-scale

Table 5.1. *Causes, processes, and specific factors that drive variation in successional pathways within a region or climate zone*

General causes	Processes or conditions	Specific factors
Site availability (chaps. 6, 7)	Large-scale disturbances, topography, drainage	Size, severity, duration, frequency, within-site heterogeneity, resource availability
Species pool (chaps. 6, 7, 12, 14)	Seed dispersal, resprouting, seed bank, seed rain, invasive species	Landscape configuration, dispersal agents, disturbance history, prior land use, remnant vegetation
Species traits (chap. 10)	Ecophysiological, functional, and life-history traits	Germination, establishment, and growth requirements
Intraspecific species interactions (chaps. 10, 12)	Competition, disease	Population size, structure, and dynamics; recruitment; growth; mortality
Interspecific species interactions (chap. 12)	Competition, disease, herbivory, predation, mutualisms (pollination, dispersal, defense, facilitation, mycorrhizae)	Community structure and dynamics, trophic structure, mobile links, facilitation

Source: Adapted from Pickett et al. (1987).

Notes: Species interactions are considered to be general causes of succession. General causes are discussed in more detail in subsequent chapters as indicated. All chapter references are to this book.

variability, the ability to distinguish "signal" from "noise" based on small inventory plots depends strongly on sampling protocols (Chazdon et al. 2007).

Successional trajectories can be modified by subsequent natural and human-generated disturbances (Foster, Knight, Franklin 1998). In Luquillo Forest of northeastern Puerto Rico, land-use history influenced the extent and spatial distribution of hurricane damage and recovery (Zimmerman et al. 1994; Flynn et al. 2010). For example, abundant trees in secondary regrowth following logging and agricultural clearance were more vulnerable to damage by hurricanes, leading to increased canopy openness following Hurricane Hugo in 1989 and Hurricane Georges in 1998 (Uriarte et al. 2004; Comita, Thompson, et al. 2010). Following Cyclone Waka in 2001, early successional forests showed higher levels of tree mortality and damage compared to late successional forests on the Vava'u island group in Tonga (Franklin et al. 2004). These studies clearly illustrate the ways in which successional trajectories can be modified by subsequent natural and human disturbances.

5.3 Successional Stages and Species Classification

Because succession is a continuous process, delimiting successional stages is at best an imprecise science. And because successional pathways and rates vary widely across regions, climate zones, and land uses, standardizing the timing of successional stages poses many challenges. Nevertheless, dividing successional pathways into discrete stages or phases—seres—is a useful approach that enables comparative studies and examination of ecological processes that affect transitions in forest structure, composition, and ecosystem properties during different periods (Chazdon 2008b). Species composition at each stage builds upon previous stages, as the initial floristic composition following disturbance strongly affects successional trajectories (Egler 1954; Swaine and Hall 1983; Finegan 1996; Grau et al. 1997; van Breugel et al. 2007; Chazdon 2008b).

Forest successional stages can be defined based on three major criteria: total aboveground biomass or basal area, age or size structure of tree populations, and species composition (table 5.2). Each of these characteristics varies with spatial scale of measurement; moreover, rates of change vary during succession (Chazdon et al. 2007). Aspects of vegetation structure, such as tree density and size distributions, tend to change more rapidly during succession than does species composition (Chazdon 2008b; Letcher and Chazdon 2009b).

As species abundances change during succession, functional properties of

Table 5.2. *Ecological criteria for characterizing forest successional stages and their ecological determinants*

Criteria for characterizing forest successional stage	Ecological determinants
Aboveground biomass	Tree basal area, rates of tree growth, recruitment, mortality, wood density
Age/size structure of tree populations	History of colonization, demographic rates, seedling establishment requirements
Species composition	Seed availability, long-distance dispersal, community assembly processes

tree species also change, such as wood density, photosynthetic properties, and seed size (Lebrija-Trejos, Pérez-García, et al. 2010). Successional change can also be measured as changes in ecosystem properties, such as buildup of nutrients in soils or accumulation of above- or belowground biomass (Ostertag et al. 2008). Species composition and its variation across size classes can also indicate different successional stages. In early successional stages, the composition of canopy species is usually distinct from the composition of recently established seedlings or saplings (Guariguata et al. 1997; Norden et al. 2009).

5.3.1 *Conceptual Approaches to Characterizing Successional Stages*

The characterization of successional stages in tropical forests has been largely founded on research in wet lowland rain forests in the Neotropics. Based on studies in humid Neotropical forests, Budowski (1965) used 21 different vegetation characteristics to distinguish four seral stages (table 5.3). The "pioneer" stage during the first few years is dominated by a small number of species that colonize rapidly, grow very fast, have low wood density, and are intolerant of shade (table 5.4; box 5.1). The pioneer stage has abundant vines, shrubs, and grasses but few or no epiphytes.

Table 5.3. *Comparison of successional stages of tropical forests under several proposed classification schemes*

Time since disturbance (yr.)	Budowski (1965)	Gómez-Pompa and Vázquez-Yanes (1985)	Finegan (1996)	Oliver and Larson (1996); Chazdon (2008b)
0–1	Pioneer	Herbaceous phase	Herbaceous/ shrub/climber stage	Stand initiation stage
1–3		Shrub stage		
3–15	Early secondary	Pioneer tree stage	Short-lived pioneer stage	
20–50	Late secondary	Secondary tree stage	Long-lived pioneer stage	Stem exclusion stage
30–80		Mature tree stage	Recruitment of shade-tolerant tree species	Understory reinitiation stage
100–200	Climax			
> 200			Old-growth stage	Old-growth stage

Source: Modeled after Franklin et al. (2002).

Table 5.4. *Common pioneer tree species from tropical forests regions of the world*

Stature	Neotropics	Africa	Asia and Melanesia
Small, 2.0–7.9 m	*Aegiphyla* spp. *Carica papaya* *Cordia nitida* *Miconia* spp. *Piper* spp. *Solanum* sp. *Urera* spp. *Vernonia patens* *Vismia* spp.	*Ficus capensis* *Leea guineensis* *Phyllanthus muellerianus* *Rauvolfia vomitoria*	*Commersonia bartramia* *Dillenia suffruticosa* *Glochidon* spp. *Macaranga* spp. *Melastoma* spp. *Phyllanthus* spp. *Pipturus* spp. *Rhodomyrtus tomentosa* *Trichospermum* spp.
Medium, 8–29 m	*Alchornea latifolia* *Cecropia* spp. *Cordia* spp. *Croton* spp. *Guazuma ulmifolia* *Heliocarpus appendiculatus* *Jacaranda copaia* *Muntingia calabura* *Ochroma* spp. *Schefflera morototoni,* *Trema micrantha*	*Anthocleista nobilis* *Cleistopholis patens* *Macaranga* spp. *Maesopsis eminii* *Musanga cecropioides* *Psydrax arnoldiana* *Spathodea campanulata* *Trema* spp. *Vernonia coferta* *Vismia guineensis*	*Acacia mangium* *Adinandra dumosa* *Alphitonia petrei* *Anthocephalus* spp. *Gmelina arborea* *Macaraga* spp. *Morinda elliptica* *Ploiarium alternifolium* *Trema* spp. *Vitex pubescens*
Large, > 30 m	*Apeiba* spp. *Cavanillesia platanifolia* *Cedrela* spp. *Ceiba* spp. *Goethalsia meiantha* *Goupia glabra* *Laetia procera* *Sweitenia macrophylla* *Vochysia* spp.	*Aucoumea klaineana* *Ceiba pentandra* *Entandrophragma cylindricum* *Lophira alata* *Milicia excelsa* *Milicia regia* *Nauclea diderrichii* *Ricinodendron heudelotii* *Terminalia ivorensis* *Terminalia superba*	*Duabanga moluccana* *Eucalyptus deglupta* *Ochreinauclea maingayi* *Octomeles Sumatrana* *Paraserianthes falcataria*

Sources: Adapted from Whitmore (1998a), with additional information from Pompa et al. (1988), Whitmore (1989a), Raich and Khoon (1990), Finegan (1996), Davies and Semui (2006), and Chazdon et al. (2010).

The "early secondary" stage begins at around five years in Budowski's scheme (1965). Pioneer trees have now reached up to 20 meters in height and more than 30 centimeters in diameter, with a second stratum of vegetation below them. A high fraction of pioneers that do not make it to the top of the canopy die during this phase (Chazdon et al. 2005). Rapid tree growth reduces light penetration in the forest understory, preventing establishment of seedlings and saplings of pioneer trees, and causing light-demanding grasses and shrubs to die. Capers et al. (2005) documented high rates of shrub seedling mortality in early second-growth forests of northeastern Costa Rica.

The "late secondary" stage begins at around 20 years after disturbance. A new group of tree species that established during earlier stages now grows into the canopy and replaces the pioneer species. These trees have a longer life span than the early pioneer species. When well illuminated, they grow very fast and some species can reach immense sizes, often becoming canopy emergents in later successional stages. Many are dispersed by wind and have a wide geographic distribution. Even in wet forests, a number of these long-lived pioneer species are deciduous for part of the dry season (Budowski 1970).

Late second-growth forest develops a more complex vertical structure with trees in emergent, canopy, and subcanopy strata. In the understory, shade-tolerant species establish as saplings and recruit as small trees. During this stage, the forest slowly becomes enriched with species, including small-statured subcanopy tree species (LaFrankie et al. 2006; Sheil et al. 2006). Tree sizes become more heterogenous as early and late cohorts mix within the forest. Forests in the late secondary stage are virtually indistinguishable from mature forests using multispectral satellite imagery due to high similarity in overall forest structure (Foody et al. 1994; Steininger 1996; Lucas et al. 2000).

After about 100–200 years, the forest reaches the "climax" stage. Budowski (1965) defined a climax community as relatively stable but not static. The forest loses its even-aged character, as canopy trees die and create gaps that become foci for regeneration of shade-tolerant and shade-intolerant trees (D. B. Clark 1996). Floristic composition is diverse, and the canopy becomes enriched with shade-tolerant species that are capable of regeneration in the understory, thereby stabilizing the species composition. Tree growth is slow, and most trees have dense wood. Species with larger seeds dispersed by small mammals, birds, and gravity become more abundant. Epiphytes increase in abundance and taxonomic diversity during this most advanced stage of succession (Budowski 1965; D. B. Clark 1996; Martin et al. 2004; Howorth and Pendry 2006). Large woody lianas become more common (D. B. Clark 1996; DeWalt et al. 2000), and shade-tolerant shrub species are diverse but low in

Box 5.1. Forest Pioneers

INITIATORS OF SUCCESSION

In ecological terms, a pioneer species colonizes only in canopy gaps, clearings, or large-scale forest disturbances. Van Steenis (1958) called these species "nomads," because successive generations move around forested landscapes colonizing recently created disturbances. The pioneer lifestyle requires fast growth (particularly in the vertical dimension), high fecundity, and widespread seed dispersal (Turner 2001). The seeds of pioneer tree and shrub species are widely dispersed by wind, bats, and birds. Many pioneer trees have small and numerous seeds that accumulate in the soil seed bank (Hopkins and Graham 1983; Whitmore 1983; Saulei and Swaine 1988; Dalling and Denslow 1998; Dalling et al. 1998;).

Pioneer tree and shrub species cannot survive as seedlings beneath a forest canopy. Swaine and Whitmore (1988) first proposed that pioneer species should be defined on the basis of their requirements for seed germination and establishment, but subsequent work has suggested that this definition is too strict (Alvarez-Buylla and Martínez-Ramos 1992; Kennedy and Swaine 1992; Whitmore 1996; Jankowska-Blaszczuk and Grubb 2006). Some pioneer species can germinate in forest understory conditions (Kyereh et al. 1999). Germination requirements of pioneer species in Panama vary with seed size; larger-seeded pioneer species germinate equally well in light and darkness (Pearson et al. 2002).

The dichotomy between pioneer and nonpioneer species based on seed and seedling characteristics does not apply to postseedling stages (Poorter et al. 2005). Some tree species exhibit pioneer features at the seedling establishment stage and nonpioneer features as juvenile trees (Dalling et al. 2001). Furthermore, many tree species that colonize clearings or former fields and pastures are not strictly pioneer species, as they are also capable of regenerating beneath a closed canopy (Chazdon 2008b; Chazdon et al. 2010).

Pioneer trees colonize a wide range of disturbances (Denslow 1987). Within a region, different types of disturbances may be associated with different pioneer species. In tropical dry deciduous forests of Nizanda, Mexico, dominant pioneer species of early secondary forests and fallows are virtually absent from natural regeneration in gaps of mature forest (Lebrija-Trejos et al. 2008). Similarly, pioneer trees that colonize degraded land in Singapore and the Malay Peninsula are not typical components of the rain forest flora and do not colonize forest gaps or clearings (Corlett 1991). In contrast, pioneer species that colonize fallows and former pastures in wet forests of Costa Rica, Mexico, and Panama also colonize large gaps in mature forests (Hartshorn 1980; Alvarez-Buylla and

Martínez-Ramos 1992; Brokaw 1987). Fox (1976) noted that a number of Bornean pioneer tree species also occur within mature forest. The basis for regional differences in the habitat specificity of the pioneer flora is poorly understood.

Across tropical forest regions, pioneer species constitute a small percentage of the total flora and are missing from many families (Whitmore 1991, 1989a). Van Steenis (1958) estimated that 20% of the Malaysian flora consists of "nomads," which are concentrated in relatively few plant families. In northeastern Costa Rica, 23% of the tree species were classified as second-growth specialists (Chazdon et al. 2011). Whitmore (1991) noted that there are fewer pioneer species in America than in Africa and Asia, attributing this difference to diversification within a single large Old-world pioneer genus *Macaranga* with over 250 species. Pioneer species often have widespread geographic distributions, with Asian species showing the most restricted distributions due to ocean barriers (Whitmore 1998b).

Many pioneer trees are short-lived and of relatively small stature (maximum height < 30 m), reaching their peak population density during the stand initiation stage (Swaine and Hall 1983). In early stages of succession, pioneer species commonly tower above a mass of shorter vegetation (see plate 6). But this dominance is short lived. In Asian tropical forests, the canopy of an early successional stand of *Macaranga* trees disintegrates within about 20 years (Kochummen 1966). Long-lived pioneer species are relatively numerous in wet Neotropical forests and West African rain forests but appear to be absent from dry forest succession (Ewel 1980; Lebrija-Trejos et al. 2008). Forests of Asia, Australia, and Melanesia have few species of large-statured pioneer trees (Whitmore 1998b). Pioneer species, such as *Ceiba pentandra, Laetia procera*, and *Cavanillesia platanifolia*, can reach immense sizes, often becoming canopy emergents in old secondary forests (van Steenis 1958; Budowski 1965; Knight 1975).

Long-lived pioneers establish along with short-lived pioneers during the stand initiation phase, but they do not dominate the canopy until the stem exclusion and understory reinitiation phases. Thus, growth and mortality of saplings and small trees of long-lived pioneers are highly sensitive to variation in light availability beneath the canopy of short-lived pioneers (Finegan 1996). Thinning treatments in secondary forests and following logging significantly increased growth rates of long-lived pioneer species (Finegan et al. 1999; Guariguata 1999; Villegas et al. 2009). These experimental treatments are of particular interest to foresters, as long-lived pioneer species across tropical regions have properties of excellent timber trees, such as fast growth and light but strong wood (Whitmore 1998a, Turner 2001). Because of their key role in initiating succession, pioneer species are often planted to establish tree cover in reforestation projects.

abundance. Trees of long-lived pioneer species can still persist through this stage, remaining for centuries as legacies of past disturbance (Budowski 1965; Condit et al. 1998).

In their extensive studies of secondary succession in tropical wet forests of Mexico, Gómez Pompa and Vázquez-Yanes (1981) outlined five stages of succession (see table 5.3). They included an early herbaceous stage, followed by a brief stage of shrub dominance, before the pioneer tree stage. The extent to which herbs and shrubs dominate early phases of succession depends strongly on previous land use or forest disturbance. In some forests that are cleared and not cultivated, pioneer tree regeneration begins immediately, bypassing the herbaceous and shrub phases. An excellent example of "direct regeneration" occurred in Nicaraguan forests heavily damaged by Hurricane Joan in 1988, where abundant regeneration of seedlings and saplings of *Vochysia ferruginea* rapidly restored populations (Boucher et al. 1994; for more on Hurricane Joan, see chap. 8, sec. 8.2). In contrast, areas with intensive land use and a history of frequent burning often show persistent dominance of grasses and ferns that arrest or delay later stages of succession.

Later refinements in delimiting successional stages in tropical forests closely followed Budowski's 1965 framework, although ecologists have disagreed regarding the presence of long-lived pioneers in old-growth or "climax" stages (Wirth et al. 2009). Finegan (1996) emphasized the important distinction between short- and long-lived pioneer species; the latter group has fewer species and dominates in his third successional stage (see table 5.3 and box 5.1). Oliver and Larson (1996) called this the "stem exclusion stage," referring to the competitive exclusion and thinning of shade-intolerant species in the shaded understory (Chazdon 2008b). Shade-tolerant species colonize continuously during succession: before, during, and after the stem exclusion stage. In the framework of Finegan (1996), the recruitment of shade-tolerant trees in the canopy ushers in the fourth stage of succession, also known as the "understory reinitiation stage" in Oliver and Larson's (1996) scheme. By the end of the stem exclusion phase, the understory vegetation has been largely transformed, and species composition resembles understory assemblages found in old-growth forests (Norden et al. 2009).

Reaching the final stage of succession, referred to as "old growth" here, can take centuries (D. B. Clark 1996; Finegan 1996). New species continue to disperse and recruit as seedlings, saplings, and trees, a process that gradually enriches species composition in all forest layers and size classes. Slow-growing tree species that colonize during the understory reinitiation stage may take well over 100 years to reach the subcanopy or canopy (Lieberman et al. 1985; Hubbell and Foster 1991; D. A. Clark and D. B. Clark 2001). Continued seed dispersal is essential for new species arrival, emphasizing the critical importance of the surrounding landscape on the recovery of biodiversity during secondary succession (Chazdon, Peres, et al. 2009).

Oliver and Larsen (1996) conceptualized successional dynamics based on the successive replacement of pioneer and midsuccessional cohorts with late successional species (table 5.5). In their view, the old-growth stage is reached when all trees in the stand have regenerated in the absence of allogenic disturbances; that is, when pioneer and midsuccessional species that colonized following the disturbance have been replaced by late successional shade-tolerant or gap-phase specialists. Wirth et al. (2009) found this conceptualization problematic, because it excludes long-lived pioneer species that co-exist in the canopy with mid and late successional species. According to a strict view of old-growth forest as lacking species that regenerated in prior allogenic disturbances, all of the pioneers, including long-lived pioneers and early colonizing midsuccessional species would have to be replaced by late successional species. The conceptual model of stand development illustrated in figure 5.1 reflects our current understanding of successional transitions in tropical forests with long-lived pioneers. Based on this model, forests with remnant cohorts of long-lived pioneer species should be considered late successional forests rather than old-growth forests.

Successional pathways in dry tropical forests are distinct from wet forests because of the slower rates of canopy closure, the deciduous leaf phenology of trees, and the higher importance of resprouting (Holl 2007). Arroyo-Mora, Sánchez-Azofeifa, Kalacska, Rivard et al. (2005) delimited four successional stages of tropical dry forest succession in northwestern Costa Rica based on vertical and horizontal vegetation structure and leaf-flushing dynamics. Following the pasture stage, early successional stages are composed of sparse patches of woody vegetation, shrubs, and pastures with a single stratum of tree crowns reaching 6–8 meters. Forests at intermediate stages have two vegetation layers, reaching 10–15 meters in height, and up to 80% of the trees in both layers are deciduous during much of the dry season. Late successional stages have three vegetation layers and attain heights of 15–30 meters. Evergreen crowns constitute 50%–90% of the canopy and exhibit only short periods of deciduousness at different times of the year.

5.3.2 *Characterizing Successional Affinities of Species and Functional Groups*

Forests undergoing different stages of succession are dominated by groups of tree species with distinct functional traits (chap. 10, sec. 10.4). The most easily characterized group is the pioneers that first colonize disturbed

Table 5.5. *Vegetation dynamics processes associated with stages of secondary succession in tropical forests*

Successional stage			
Stand initiation	Stem exclusion	Understory reinitiation	Old growth
Germination of seeds in seed bank	Canopy closure	Mortality of canopy trees	Mortality of pioneer cohort in canopy
Resprouting of remnant trees	High mortality of lianas and shrubs	Formation of small canopy gaps	Range of gap sizes
Colonization of short- and long-lived pioneer trees	Recruitment of shade-tolerant seedlings, saplings, and trees	Canopy recruitment and reproductive maturity of early-colonizing species	Recruitment of shade-tolerant and gap-requiring canopy species and emergents
Rapid height and diameter growth of woody species	Growth suppression of shade-intolerant trees in understory and subcanopy	Increased heterogeneity in understory light availability	Spatial heterogeneity in biomass and microtopography
High mortality of herbaceous species	High mortality of short-lived, pioneer trees	Seedling and sapling establishment of shade-tolerant tree species	Large woody debris
High rates of seed predation	Dominance of long-lived pioneer trees	Tree recruitment of early-establishing shade-tolerant species	Maximum diversification of trees and epiphytes
Seedling establishment of shade-tolerant tree species	Development of canopy and understory tree strata		
	Seedling establishment of shade-tolerant tree species		

Source: Adapted from Chazdon (2008b).

Notes: Successional stages based on the framework of Oliver and Larson (1996).

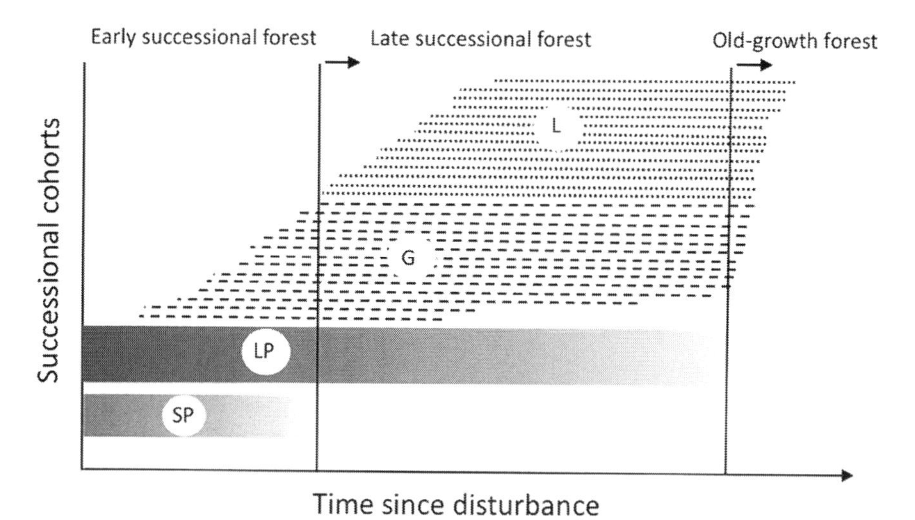

Figure 5.1
A conceptual illustration of criteria for defining early successional forest, late successional forest, and old-growth forest. Cohorts of short-lived (SP) and long-lived (LP) pioneer trees are assumed to initiate early and nearly simultaneously during succession, whereas generalists (G, midsuccessional species) and late successional specialists (L) colonize continuously. The transition from early to late successional forest is marked by the replacement of the short-lived pioneer cohort by late successional species, whereas the transition from late successional to old-growth forest is marked by the decline of the long-lived pioneer cohort, the last legacy of the original allogenic disturbance. Source: Adapted from Wirth et al. (2009, fig. 2.2).

areas (see box 5.1). The studies described in the previous section distinguished other groups of species, according to their dominance in early versus late successional stages and their abundance in old-growth forests. Foresters classify species based on whether they are light-demanding or shade-tolerant, have slow or fast growth rates, and have soft or hard wood. Forest ecologists classify trees into functional groups based on their regeneration requirements in old-growth forests: light-demanding species require large gaps for establishment and regeneration (they are essentially successional pioneers), shade-tolerant species do not require gaps for establishment and recruitment, and gap-requiring species require gaps for recruitment into the canopy but not for seedling establishment (Whitmore 1978; Hartshorn 1980; Denslow 1987).

Most studies of successional vegetation dynamics define species groups based on observations of species abundance in different successional stages (Zhang et al. 2008). Early studies classified tree species into two groups: "pioneer" and "nonpioneer," or "forest," species. "Nonpioneer species" establish slowly and gradually during the very first years of succession, as the high density of short-lived pioneer species declines (Swaine and Hall 1983; Uhl 1987; Myster 2007). In shifting cultivation fallows in the Río Negro region of the Venezuelan Amazon basin, pioneer trees dominated by *Vismia* species colonized quickly and then experienced high rates of mortality after three years (see plate 12; Uhl 1987). Other pioneer species increased in density. Establishment of old-growth forest species (these were called "primary species") was very slow; by year 5 the plot had 3.2 mature forest trees per 100 square meters; 79% of these stems had established from seed and the remainder were sprouts (fig. 5.2). Swaine and Hall (1983) observed similar dynamics during stand initiation following forest clearance for bauxite mining in Ghana. Pioneer trees reached their peak species richness within 2.5 years of clearance, coinciding with a peak in stem density. Mature forest species slowly accumulate in number, density, and size. Through this process, the forest gradually becomes enriched with tree species during succession. After 15 years of regeneration, pioneer trees still dominated the tree assemblage, but nonpioneer species dominated the assemblage of small stems below five centimeters in diameter at breast height (DBH) and composed over 80% of the stems below 1.3 meters tall.

In a study of forest succession on Ilha Grande near Rio de Janeiro in Brazil, de Oliveira (2002) classified tree species in four groups: pioneers, early successional, late successional, and climax (old-growth). Pioneer and early

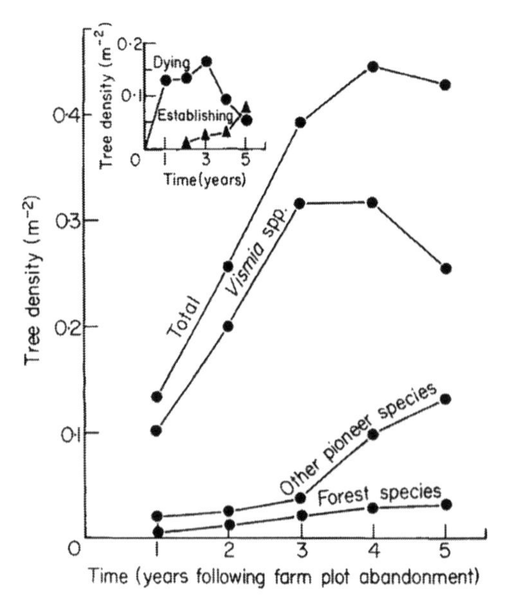

Figure 5.2
Changes in density, establishment, and mortality for trees greater than or equal to two meters tall in a 0.15-hectare study plot during the first five years of succession following shifting cultivation near San Carlos de Río Negro, Venezuela. Source: Uhl (1987, fig. 1).

successional species were most abundant in the 5-year-old regeneration, late successional species dominated in 25- and 50-year-old regenerating forest, and climax species dominated in the old-growth forest sites. A few old-growth tree species were present during early stages of succession, whereas some pioneer species were present in old-growth sites.

Classifications of successional affinities of species based on field observations can be highly circular if certain species are used as indicators of particular successional stages. Furthermore, classifications based on local observations cannot be generalized among different study regions with different species pools. New approaches have applied rigorous statistical criteria to delimit functional groups of species during succession and to classify successional specialists and generalists. Chazdon et al. (2010) classified tree species in lowland wet forests of northwestern Costa Rica into five functional groups based on diameter growth rates in old-growth forest and canopy height. Population densities of fast-growing subcanopy and canopy trees peaked early in succession, whereas the basal area of fast-growing canopy and emergent trees continued to increase beyond 40 years. The relative abundance of understory trees and slow-growing canopy and emergent trees was higher in old-growth than second-growth forests.

Based on the relative abundance of trees in second-growth and old-growth forests in northeastern Costa Rica, Chazdon et al. (2011) developed a multinomial model to classify tree species into successional specialists, old-growth specialists, and generalists. Tree classifications using this model showed strong successional trends in relative abundance of these three groups. The relative abundance of second-growth specialists declined dramatically over 13 years in a young regrowth forest, while the relative abundance of generalists increased steeply. Old-growth specialists increased slowly but steadily in their relative abundance (fig 5.3). Successional generalists have similar relative abundance in second-growth and old-growth forests. These species are widely dispersed by animals and have seedlings that can establish under closed canopy or gap conditions.

Although successional *stages* are defined by characteristics of a forest stand (see tables 5.3 and 5.4), successional *trajectories* are fundamentally determined by rates of recruitment, growth, and mortality of populations of the component tree species. These demographic characteristics have their basis in the life histories and functional traits of different species. The term *successional stage* is apt. Successional pathways can be viewed as an improvisational drama in several acts, with each act featuring the performance of a different set of actors. Some actors perform throughout the drama, but others have cameo appearances in only one act. Although each act sets the stage

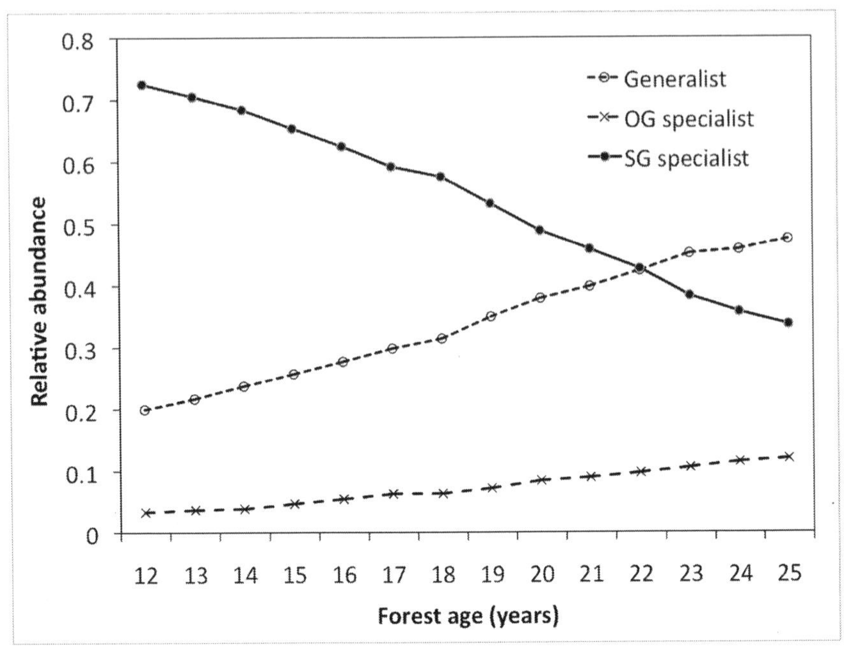

Figure 5.3
Successional trajectories in the relative abundance of trees greater than or equal to five centimeters in DBH classified as second-growth (SG) specialists, old-growth (OG) specialists, and generalists in one hectare of regenerating wet tropical forest in lowland Costa Rica. In two old-growth forests in the same region, old-growth specialists and generalists composed, on average, 43% and 31% of the trees, respectively. Species were classified using the multinomial approach of Chazdon et al. (2011).

for the next, forest regeneration has no director and only a roughly sketched script, creating a high degree of spontaneity, randomness, and uncertainty. Each successional production is unique. Even when people serve as directors during fallow management and active reforestation, their direction serves, at most, to guide forest regeneration into more restricted performances.

5.4 Forest Definitions and Concepts

Even more challenging than delimiting successional stages is defining what is a forest. This definition has major implications for forest policy, global forest monitoring, forest restoration, and climate mitigation efforts (Putz and Redford 2010). Definitions of forest are critically important in distinguishing natural forest cover from plantations and for monitoring land-use and land-cover change. In 2001, the Food and Agriculture Organization

(FAO) of the United Nations developed a definition of forest for its global forest cover monitoring program (Forest Resource Assessment [FRA]). According to this definition, a forest is an area larger than 0.5 hectares with more than 10% tree canopy cover, and a "tree" is a plant capable of growing over five meters tall. Bamboos and palms, but not bananas, are considered trees. "Closed" forest has more than 40% canopy cover, whereas "open" forest has 10%–40% canopy cover (FAO 2001, 2006a).

Within this broad global definition of forest lies the continuum of forest regeneration and degradation. At one extreme of the continuum are cleared areas of forest that give way to second-growth forests, also known as "secondary" or "regrowth" forests, which are simple in structure and have relatively low tree diversity. At the other extreme are old-growth forests with the maximum development of forest height and complex vertical and horizontal structure, diverse age and size composition of trees, dense canopy cover, gap dynamics, and relatively stable species composition. Some widely used terms to describe these forests include *ancient, mature, primary, pristine, climax, virgin, undisturbed, frontier*, and *primeval* (Wirth et al. 2009).

Second-growth forests develop after complete or almost complete forest clearing (Finegan 1992). The International Tropical Timber Organization (ITTO 2005, p. 36) adopted the following definition of second-growth forest: "Woody vegetation regrowing on land that was largely cleared of its original forest cover (*i.e.* carried less than 10% of the original forest cover). Secondary forests commonly develop naturally on fallow land after shifting cultivation, settled agriculture, pasture, or failed tree plantations." Chokkalingham and de Jong (2001, p. 21) offered a broader definition: "Forests regenerating largely through natural processes after significant human and/or natural disturbance of the original forest vegetation at a single point in time or over an extended period, and displaying a major change in forest structure and canopy species composition and/or canopy species composition with respect to nearby primary forests on similar sites."

The term *secondary forest* has been widely applied to forests subjected to logging, particularly in the Asian tropics where logging disturbances are intensive and pervasive (Brown and Lugo 1990; Corlett 1994; Chokkalingam and de Jong 2001). Most ecologists agree, however, that logged mature forests and second-growth forests should remain distinct categories (Sist et al. 1999; Chokkalingam and de Jong 2001; Asner et al. 2009; Putz and Redford 2010). A key aspect of the definition of second-growth forests is a *break in the continuity of forest cover*, which requires that newly arriving seeds originate from sources outside of the disturbed area (Corlett 1994). Forests regenerating on former pastures with isolated remnant trees still satisfy this definition.

Another major issue in defining second-growth forests is whether to distinguish between forests regrowing following human versus natural disturbances. Several reviews have emphasized forests originating from human impacts (Brown and Lugo 1990; Guariguata and Ostertag 2001). Brown and Lugo (1990, p. 3) broadly defined secondary forests as "those formed as a consequence of human impacts on forest lands," excluding landslides, naturally caused fires, and hurricanes. Pathways and rates of forest succession following human disturbances can differ significantly from those following natural disturbances (Chazdon 2003), but second-growth forests can originate from any category of disturbance that causes complete or nearly complete forest clearing. Clarifying the definition of second-growth forests is important for several reasons. First, the dynamics of forests regenerating from cleared or agricultural land are substantially different from the dynamics of forests regenerating following selective or partial logging (chap. 9, sec. 9.2). Second, old second-growth forests reveal legacies of former allogenic disturbances and are still undergoing species turnover of canopy trees. These differences have implications for forest management, biodiversity conservation, and ecosystem processes.

Second-growth forests are often categorized by age, even though age may be a poor indicator of successional development or successional stages (Arroyo-Mora, Sánchez-Azofeifa, Kalacska, Rivard 2005). Measures of forest structure, such as basal area, can also be used to describe the progression of community attributes in forest stands during succession (Lohbeck et al. 2012). In a tropical dry forest in Costa Rica, successional stages (early, intermediate, and late) permitted a far better separation of spectral reflectance classes than did age classes (Arroyo-Mora, Sánchez-Azofeifa, Kalacska, Rivard 2005). The terms *young* and *old* are often applied inconsistently to second-growth forests. The "young" forest on Barro Colorado Island is now 100 years old (Knight 1975). But in Rondônia, Brazil, an "old" second-growth forest has remained standing for 20 years (Helmer et al. 2009). The age structure of second-growth forests varies considerably across landscapes and regions (Chazdon, Peres, et al. 2009), further complicating comparisons of successional pathways and processes.

In 1891, American forester Bernard Fernow coined the term *old-growth forest* to describe a forest composed of old trees (Spies 2009). According to the US Forest Service's National Old-Growth Task Group, "Old growth encompasses the later stages of stand development that typically differ from earlier stages in a variety of characteristics that often include tree size, accumulations of large dead woody material, number of canopy layers, species composition, and ecosystem function" (quoted in Putz and Redford 2010, p. 13). Structure and composition of old-growth forests and their successional counterparts vary widely among climatic, elevational, and edaphic gradients within the tropics (Richards 1996; Primack and Corlett 2007; Ghazoul and Sheil 2010). Structure and composition of old-growth forests also vary considerably at landscape scales (D. A. Clark et al. 1995; D. B. Clark and D. A. Clark 2000).

The notion of primary or virgin forest is distinct from the old-growth concept. The FAO/FRA defines primary forest as "naturally regenerated forest of native species, where there are no clearly visible indications of human activities and the ecological processes are not significantly disturbed" (FAO 2006b, p. 13). Given the problems associated with determining whether or not forests show indications of prior human activities, the term *old-growth* is more appropriate for classifying tropical forests that have reached a late stage in forest succession and are relatively stable with respect to their structure and composition (D. B. Clark 1996). Old-growth forests can develop following human or natural disturbances, given sufficient time.

Disturbances such as wildfires, logging, hunting, and fragmentation cause sudden or gradual loss of ecosystem functions, structure, and composition in old-growth forests. Forest degradation is the reduction of the capacity of a forest to provide goods and services (Simula 2009). Today, forest degradation is mostly measured in terms of the loss of carbon stocks or biodiversity relative to intact old-growth forests (Gibson et al. 2011). The process of forest degradation moves in the opposite direction of succession but can proceed considerably faster.

Any serious effort to understand the causes and consequences of forest disturbances requires distinguishing different forest states. Because forests are dynamic systems, degradation and regeneration processes can be assessed only by comparing forest structure, composition, and ecosystem functions over time. Measuring forest degradation requires looking backward, comparing the current state of a forest to its previously undisturbed state. Measuring forest regeneration, in contrast, requires looking forward, beginning with the period of disturbance, focusing on recovery of carbon stocks or species diversity. Despite the practical applications of defining forest status, in reality all forests lie along the regeneration-degradation continuum and their static conditions may not clearly reveal in which direction they are moving.

5.5 Approaches to Studying Tropical Forest Succession

Forest succession can be investigated using experiments, observations of forests over time, and chronosequences. Each approach has advantages and limitations. A chronosequence is a series of sites that differ in age but otherwise occur on similar soil types and environmental conditions

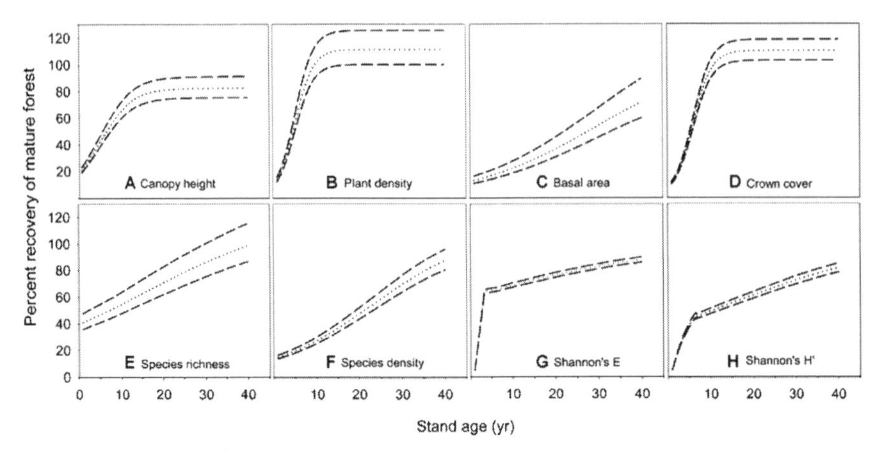

Figure 5.4
Changes in forest structure and composition during succession based on a 40-year dry forest chronosequence in Nizanda, Oaxaca State, Mexico. Source: Lebrija-Trejos et al. (2008, fig. 5).

within the same climatic zone. Usually, each site is studied at a single point of time, yielding a "snapshot" of sites arranged along an axis of time since disturbance. Chronosequence studies substitute different aged sites for time, relying on the critical assumption that fields abandoned at different times in the past experienced the same changes, processes, and conditions during their formation (Pickett and Cadenasso 2005). When sites are arranged in a chronological sequence, changes in various stand attributes, such as canopy height, stem density, basal area, and species richness, can be compared across a successional sequence (fig. 5.4; Lebrija-Trejos et al. 2008). This space-for-time substitution permits the study of a wide range of successional stages and makes it possible to sample considerably more sites than could be accommodated through site-intensive long-term studies. The chronosequence approach is appropriate for studying single linear or cyclic successional pathways, or when convergence occurs early in succession (Walker, Wardle, et al. 2010).

Most of our understanding of forest succession in the tropics derives from chronosequence studies. Estimates of rates of biomass accumulation in tropical forests are based almost entirely on chronosequence data (Brown and Lugo 1990; Marín-Spiotta et al. 2007). Chronosequence approaches have also been used to study primary succession in the tropics following landslides and river channel flooding (Guariguata 1990; Terborgh et al. 1996). Chronosequence studies require knowledge of detailed land-use history and time since disturbance or since abandonment of each stand, a major challenge for areas that are older than the age of local informants or lack other historical information. In some tropical forests, trees with annual growth rings can be used to date forest age (Devall et al. 1995; Worbes et al. 2003; Baker et al. 2005; Brienen et al. 2009). Ages of pioneer trees, such as *Cecropia* species, can also be assessed using the number of internodes on the main stem and branches (Zalamea et al. 2012). In most chronosequence studies, the oldest successional stands are 50 years old (Chazdon 2008b). Few tropical forest chronosequences include sites that are over 80 years old (Saldarriaga et al. 1988; Denslow and Guzman 2000).

It is not always possible to standardize for environmental conditions, land-use history, or landscape surroundings, however, which limits the validity of chronosequence inferences (Foster and Tilman 2000; Johnson and Miyanishi 2008; Walker, Wardle, et al. 2010). Sample plots are usually very small, often less than 0.1 hectares. Well-replicated samples from different forest patches within each age class can compensate for this problem, if there is no systematic bias among age classes in successional trajectories. In tropical landscapes that have recently undergone large-scale deforestation, older second-growth forests may have regenerated following less intensive land use, or may have been embedded in more forested landscapes than recently initiated second-growth forests (Feldpausch et al. 2007).

Observations of forest change over time during succession are far more rare than observations of chronosequences. Several studies document temporal changes in forest structure and composition, providing sufficient data to evaluate the accuracy of chronosequence predictions. Chronosequence data were compared with time-series data in eight successional stands in northeastern Costa Rica. Tree density showed highly variable dynamics over time within stands, and four sites showed little or no change in species density over time, in contrast to the linear increase predicted by the chronosequence. Only basal area showed a consistent trend between the chronosequence and the time series (Chazdon et al. 2007). A similar analysis compared chronosequence data with three-year time-series data for successional forests in Chiapas, Mexico (van Breugel et al. 2006; Chazdon et al. 2007). Fallow age was not a significant predictor of either plot basal area or stem density across the chronosequence. Species density showed a nonlinear chronosequence trend, but most plots showed increases in species density above those predicted by the chronosequence model. Individual sites showed idiosyncratic temporal patterns, driven by initial species composition, site factors, land-use history, and landscape composition.

With few exceptions, successional patterns within individual plots do not closely follow chronosequence predictions. Feldpausch et al. (2007) tested chronosequence predictions in second-growth forests growing on former pastures in central Amazonia. The youngest forests consistently accumulated less biomass per year than the predicted chronosequence trend due to a low initial density of trees. The low stocking rate of forest in the youngest age class was likely the consequence of a longer duration in pasture. These differences in land-use intensity clearly violated the basic assumptions of the chronosequence approach. Even when chronosequences of *Vismia*- and *Cecropia*-dominated sites were evaluated separately, observed successional trajectories in basal area, stem density, and species density often deviated from chronosequence predictions (Williamson et al. 2012). In a dry forest in Jalisco, Mexico, chronosequence data predicted faster rates of change in plant density, species density, and plant cover compared to actual rates of change over three years in young successional forests on former pastures (Maza-Villalobos et al. 2011). In contrast, Lebrija-Trejos, Meave, et al. (2010) observed short-term successional changes in tropical dry forest of Oaxaca, Mexico, that generally matched chronosequence trends, particularly when pioneer and nonpioneer species were considered separately.

Long-term monitoring is costly, slow, and difficult, and may be impossible in some regions where second-growth forests are frequently recleared, leaving no opportunity to study succession beyond the stand initiation stage. Further, long-term studies are usually restricted to a small number of sites, limiting the ability to generalize the results across a landscape or region. Chronosequence studies have a major advantage in allowing greater replication of stand ages (Quesada et al. 2009).

Several studies have used experimental manipulations to study successional processes in tropical forests. Most of these experiments focus on effects of various disturbance treatments on species colonization and growth during initial stages of succession (Uhl et al. 1982). An experimental study by Ganade and Brown (2002) examined effects of neighboring vegetation on seedling establishment of four tree species in early succession in a six-year-old former pasture in central Amazonia. Dupuy and Chazdon (2006) examined responses of seedlings and saplings to manipulated vegetation cover in second-growth forests in northeastern Costa Rica. A subsequent experiment examined the interacting effect of leaf litter and vegetation cover on the establishment of pioneer and shade-tolerant tree species in second-growth forests (Dupuy and Chazdon 2008; Bentos, Nascimento, Williamson 2013).

Experimental studies provide valuable insights into specific mechanisms that affect successional processes and pathways, but most experiments are completed in less than five years, within the time frame of a doctoral research project. Many studies take advantage of "natural experiments" to examine how succession over longer time periods is affected by initial conditions created by different types of human and natural disturbance, and by heterogeneity in these initial conditions. Chronosequences, experiments, and long-term plot-based observations should be viewed as complimentary approaches to studying successional rates and processes. As chronosequence studies do not provide direct information on successional processes or population dynamics, the only way to understand these processes is to monitor changes in forest structure, composition, and function over time within sites of known disturbance history and landscape surroundings.

5.6 Conclusion

The shift in dominance by shade-intolerant species to shade-tolerant species is the most generalizable and predictable feature of successional pathways. Beyond this general trend, successional pathways are highly stochastic, reflecting seemingly random factors that collectively contribute to changes in the structure and composition of "forests under construction." Shade tolerance is not the only driver of successional vegetation change. Initial conditions drive the rate and direction of vegetation change, but changing features of landscapes, climates, and biotic interactions also lead to uncertainty and unpredictability in successional pathways. Successional trajectories reflect the totality of changes in forest structure, population dynamics, and species composition, which all interact in complex ways.

Successional chronosequences are models of a dynamic process based on static data. Because stand dynamics are contingent on past events, it is not surprising that chronosequences will fail to predict individual stand dynamics. New approaches to modeling successional trajectories must incorporate stochastic as well as deterministic factors: "Tropical forest succession is an idiosyncratic process, driven by many factors; the more we understand about how it operates within sites, the more accurately we will be able to generalize about how this complex process operates at large scales" (Chazdon et al. 2007, p. 285).

When viewed as static systems, degraded forests and successional forests share features of structure and composition in comparison to intact old-growth forests. But this superficial similarity belies completely opposing temporal trajectories; successional forests gradually recover structure and composition, whereas forest degradation signals losses in structure and composition. If disturbances cease, a degraded forest can recover lost properties but will follow different regeneration dynamics than those observed during secondary succession on cleared land.

Comparisons of attributes of second-growth forests with attributes of old-growth forests provide limited insight into successional processes and their variability within particular regions. Future successional studies should emphasize the sources of variability among sites in both static and dynamic variables, such as demographic rates, seed dispersal, and changes in species composition. Long-term studies of successional dynamics within plots, in combination with comparisons among sites and regions provide the most valuable insights into the factors that shape successional pathways.

CHAPTER 6

TROPICAL FOREST SUCCESSION ON NEWLY CREATED SUBSTRATES

Volcanoes impact all ecosystems and represent the most intense of nature's forces.—Roger del Moral and Sergeu Yu. Grishin (1999, p. 149)

6.1 Overview

On the slopes of El Reventador volcano in Amazonian Ecuador around 350 meters elevation, a remnant forest fragment stands among pastures. In its structural features, this remnant patch of forest resembles an old-growth forest; tree density, basal area, and mean diameters are similar to those found in other upland forests surveyed in the area. But something is very different about this forest. The number of tree species, genera, and families is greatly reduced, and early successional tree species are nearly twice as abundant. What happened here? Some careful detective work revealed ceramic artifacts and charcoal dating to 520 ± 20 years BP. Clearly, the forest area was populated. But above this layer was a superficial layer of sandy soil 4–38 centimeters thick largely composed of fluvial deposits of volcanic origin. The most likely scenario is that a volcanic eruption or earthquake triggered a catastrophic flood, which was caused by the failure of a natural dam formed on the volcano's slope by an earlier landslide event. The large human population was likely wiped out along with the forest. Fast forward 500 years later. Only 120 of the 239 tree species in nearby upland forests have recolonized and understory trees are notably absent (Pitman et al. 2005). Forest succession on volcanic substrates is a slow process, particularly if undisturbed forest areas are distant. It is likely that other forests like this one remain in the landscape—undetected—as they superficially appear to be "pristine" forests today.

Secondary succession is initiated following a disturbance that removes all or most of the previous vegetation within a local area, leaving the soil relatively intact. Disturbances that initiate secondary succession include blowdowns from high winds, hurricanes, or cyclones, logging clear-cuts, fires, and forest clearances for agriculture or pasture establishment. But some types of disturbance have more dramatic effects than removal of vegetation. Volcanic eruptions cover extensive areas with pyroclastic deposits (see plate 7). During a major landslide, the side of a mountain can wash down into a valley. Fast-flowing rivers flood and erode riverbanks, creating new fluvial bars out of sediment and sand. Bulldozers remove topsoil and compact the mineral soil.

These disturbances initiate primary succession, the gradual development of a plant community on bare soil or rock. The conditions created by such disturbances are inhospitable for the germination and establishment of most plant species. Landslides, volcanic eruptions, and river flooding have one aspect in common—colonists must establish on a newly created substrate. Previous soil has been washed way, eroded, or buried beneath sand, ash, rock, or lava. In this chapter, I show how tropical forests around the world have managed to become established on substrates lacking soil.

The types of disturbances discussed in this chapter originate from natural causes. These natural "disasters," as they have come to be called today, have always been present in the environment of tropical forests. Not surprisingly, many pioneer species that colonize early stages of primary succession also play a major role in early stages of secondary succession following forest clearance by human activities. Many of the functional traits and dispersal syndromes characteristic of early and late successional species are shared between primary and secondary successions. In fact, today's tropical forests would not be as resilient as they are had it not been for the large-scale forest disturbances of past eras leading to the diversification of pioneer species in tropical floras (see box 5.1; van Steenis 1958).

6.2 Biological Legacies and Local Resource Availability

Despite the clear distinction between primary and secondary succession in ecology textbooks, most disturbed sites exhibit a mixture of characteristics of both primary and secondary succession and can be viewed within a single framework. Figure 6.1 illustrates how successional pathways are determined by two major factors: species availability and resource availability (Pickett and Cadenasso 2005). Species availability refers to sources of propagules for regeneration, such as the soil seed bank, local seed rain, resprouts, or remaining trees or patches of vegetation. These "biological legacies" are remnants of vegetation that persist through a catastrophic disturbance (Franklin 1989; Foster, Knight, Franklin 1998). Hurricanes and fires leave behind many biological legacies—for example, damaged trees readily resprout, quickly restoring canopy cover and species composition (see reviews by Bellingham 2008; Lugo 2008; Turton 2008). When fruiting plants are not located nearby, colonization depends upon long-distance dispersal by air, water, or animals, restricting the pool of colonizing species.

Resource availability refers to the supply of soil nutrients, water, or other essential resources for plant growth. Microbial activity also affects soil resource availability, so this is not strictly an "abiotic" axis. Conditions that initiate succession can be located virtually anywhere along these two axes.

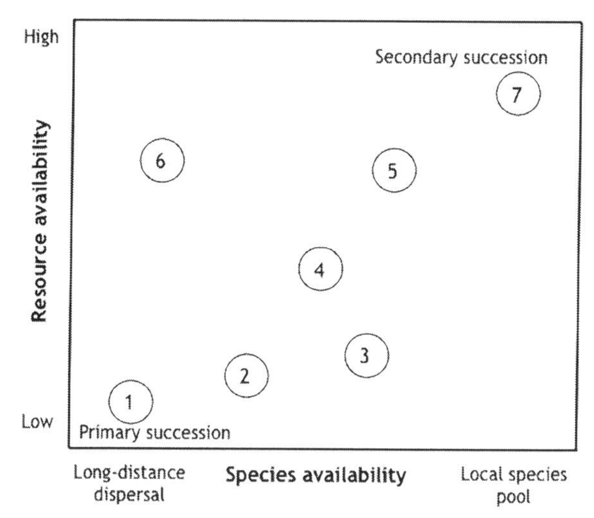

Figure 6.1
A conceptual framework for comparing conditions that initiate successions. Site conditions that initiate primary and secondary succession occur along a two-dimensional gradient of differential species availability and resource availability. Conditions of high local species availability and high resource availability characterize the most rapid secondary successions, whereas conditions of low species availability and low resource availability characterize prolonged primary successions. Increasing numbers reflect increasing rates of successional change in vegetation structure and composition: (1) Krakatau, (2) upper zone of landslide in Puerto Rico, (3) Mt. Lamington, Papua New Guinea, (4) lower zone of landslide in Puerto Rico, (5) pasture succession on fertile soils with remnant trees, (6) pasture succession on fertile soils but no remnant vegetation, (7) swidden fallow embedded in a forested matrix.
Source: Adapted from Pickett and Cadenasso (2005).

This framework illustrates how successional processes are embedded within a broad landscape context, as species availability and resource availability are both affected by features of the surrounding landscape. This view emerges directly from the hierarchical view of succession presented by Pickett et al. (1987) and discussed in chapter 5 (see esp. table 5.1). Martínez-Ramos and García-Orth (2007) presented a similar framework, based on the effects of propagule availability of native species and site quality on regeneration capacity following different types of human land use, although site quality is a broader concept than resource availability.

"Classic" primary successional pathways begin under conditions of low resource availability and low species availability, where there are few local sources of propagules and few or no biological legacies (fig. 6.1[1]). In the most extreme cases, the entire landscape is sterilized by a catastrophic disturbance and dispersal is restricted to long-distance transport. As soil resource availability increases, the trajectory of primary succession becomes more similar to trajectories typical of secondary successions with moderate levels of species availability (fig. 6.1[3, 4, 5]). Secondary successions proceed rapidly where resource availability is relatively high and propagules are available locally for colonization (fig. 6.1[7]). But secondary succession begins under a wide range of conditions and combinations of these two factors. Intensive agricultural practices and large-scale forest clearing often eliminate local propagule sources and exhaust soil nutrients (Aide and Cavalier 1984), reducing species availability and soil resources to the point where colonization and establishment are inhibited and restoration interventions are needed to more rapidly initiate forest regeneration (fig. 6.1[1, 2]). Resource availability and species availability can both vary spatially within the same disturbance location. Following the 1951 eruption of Mt. Lamington in Papua New Guinea, primary succession was highly variable but proceeded rapidly in many areas because of the survival of some trees in the blast areas (fig. 6.1[3]; Taylor 1957). Because of the complex ecological interplay between local resource availability and species availability, processes of vegetation succession and forest regeneration can vary as much within particular types of disturbances as across disturbance regimes.

Facilitation is a key aspect of early colonization during primary succession, as resource availability in soils and related environmental stresses strongly limit initial species colonization. Initial colonization of woody species during primary succession may experience substantial delays due to lack of soil for seedling establishment, insufficient nutrient availability, and extreme dispersal limitation. Facilitation occurs through enrichment of the soil nutrient supply, modifications of soil texture and water-holding capacity, stabilization of soil, or amelioration of environmental stresses. The first colonizers can modify local resource availability and other environmental conditions in ways that favor the establishment of later arriving species. Early colonists create conditions that favor the next successional stage rather than perpetuating their own presence on the site. Alternatively, early colonists can competitively inhibit the colonization of other species, slowing down the rate of species turnover (Walker, Landau, et al. 2010). This notion of *autogenic* change is central to early concepts of vegetation succession (Clements 1916; Tansley 1935). In cases where facilitation or inhibition occurs during succession, vegetation change tends to follow a pattern called *relay floristics*, where colonization of different species takes place in a progressive sequence of seres (successional stages) rather than simultaneously (Egler 1954; Walker and del Moral 2009).

Nitrogen is a major limiting factor for plant growth during primary successions on volcanic substrates in Hawaii, and abundant initial colonizers have the ability to fix nitrogen (Vitousek et al. 1993; Vitousek 1994). The nitrogen-fixing lichen *Stereocaulon virgatum* accounted for 23%–79% of total biomass in young landslides in the mountains of Jamaica (Dalling 1994). Initial colonization of nitrogen-fixing species facilitates later colonization by a more diverse group of species by creating more fertile soil microsites. Following initial colonization, succession proceeds slowly, with gradual enrichment of plant and animal species. Beyond the initial phases of colonization where soil is formed and organic matter accumulates, there is little to distinguish primary from secondary succession, as local species availability becomes a more important determinant of successional pathways than local resource availability.

6.3 Colonization and Succession on Landslides

Landslides are caused by slope failure on steep slopes, which results in the rapid mass downward movement of soil and/or rock. They can be triggered by earthquakes, heavy storms, or human activities such as road construction, deforestation, and urbanization (Restrepo et al. 2009; Walker et al. 1996; Walker et al. 2009). In December 2010, extremely heavy rains and landslides eroded steep slopes in the Panama Canal Watershed, spewing massive amounts of sediment into rivers. The Panama Canal was closed for the first time since 1935. Many factors affect the rate of vegetation recovery in landslides, such as the status of surrounding intact vegetation, recurring landslides, or other disturbances such as hurricanes. Colonizing vegetation can invade from the landslide edge, leading to more rapid regrowth in small versus large landslides (see plate 8; Walker et al. 1996).

Landslides generate a wide range of soil surface conditions, creating heterogeneous patterns of organic matter and exposed mineral soil. The upper edge of a landslide (the erosional zone) is usually steep, eroded, and infertile. Most of the surface soil and organic matter has been removed. In contrast, the deposition zone at the lower edge of a landslide is relatively flat and rich in organic matter and debris deposited from the upper regions (Walker et al. 2009). In eight recent landslides, higher soil fertility and the presence of buried viable seeds in the soil of the lower deposition zones significantly increased the rate of vegetation recovery. A survey of 20 landslides in the upper Luquillo Mountains of Puerto Rico along a 52-year chronosequence showed that stem densities were consistently higher in the lower zones for landslides during the first 25–37 years of regeneration (fig. 6.2; Guariguata 1990).

Several studies emphasize the critical importance of soil resource availability for early plant colonization on landslides (see plate 8). During the first 55 years of landslide succession in Luquillo Experimental Forest, Zarin and

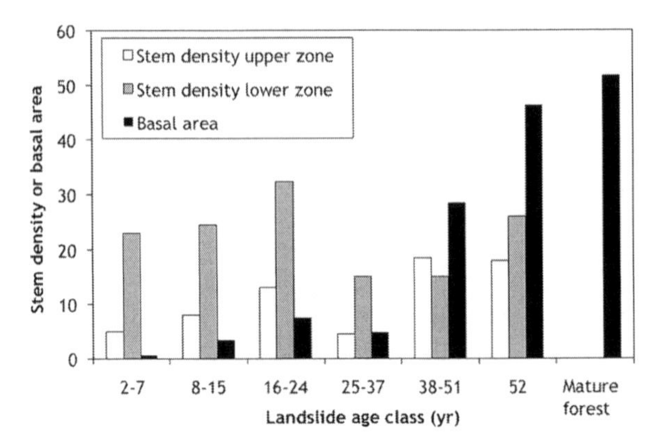

Figure 6.2
Basal area (m^2/ha) and average density (no./100 m^2) of stems greater than one centimeter in diameter at breast height in upper and lower zones of 11 landslides in six age classes and in three mature forest areas in Luquillo Experimental Forest, Puerto Rico. Source: Guariguata (1990, table 6).

Johnson (1995) found that soil nutrient accumulation was controlled by the production and decomposition of soil organic matter. In a more recent study of 30 new landslides in this region, Shiels et al. (2008) found that vegetation cover and plant biomass on volcanic soils was correlated with higher clay content, small participle size, higher total nitrogen content, and high water-holding capacity, but was not correlated with soil organic matter.

Initial landslide colonizers have spores or small seeds that are easily dispersed by wind. Dispersal limitation becomes a stronger determinant of vegetation regrowth and diversity after initial colonization and soil development. Artificial perches significantly increased seed dispersal of bird-dispersed forest species in six landslides in Puerto Rico but did not affect seedling establishment (Shiels and Walker 2003). This finding suggests that postdispersal processes, such as seed predation, seed germination, or seedling establishment ultimately limit seedling recruitment. Dense cover of grasses and ferns can impede germination, establishment, and seedling growth of woody species on landslides (Walker 1994; Russell et al. 1998; Restrepo and Vitousek 2001). Overall, the presence of early successional woody plants and tree ferns increased the species richness of late successional woody plants, facilitating long-term forest development in Puerto Rican landslides (Walker, Laundau, et al. 2010). Dense thickets of tree ferns and scrambling ferns inhibited establishment of late successional plants due to the reduction of light below the fern canopies.

On October 30, 1998, a heavy rainfall event during Hurricane Mitch caused an enormous landslide on the southern slope of the Casita volcano, an inactive stratovolcano in a tropical dry forest region of Nicaragua (see plate 9). Over 2,000 people in the towns of El Porvenir and Rolando Rodríguez were killed by fast-flowing lahar. The landslide was substantially larger (1,120 ha) than other landslides studied in Central America and the Caribbean. Early successional trajectories and rates of revegetation varied strongly within and among the landslide zones. During the first four years of succession, the depositional zone had the fastest rate of regrowth, was dominated by the pioneer tree species *Trema micrantha* (Cannabaceae) and *Muntingia calabura* (Muntigiaceae), and featured the arrival of some tree species found in adjacent forests. In the erosional zones, soil fertility was significantly and positively correlated with species richness, biovolume, and species composition. Variation in successional trajectories within and among zones was strongly influenced by remaining soil, remnant patches of vegetation, and the incidence of clearcutting and burning in the surrounding dry forest landscape (Velázquez and Gómez-Sal 2007, 2008).

Primary succession on exposed bedrock or sandy erosional zones at high elevations may take centuries. Based on studies in the Blue Mountains of Jamaica, Dalling (1994) suggested that plant biomass on landslides could take as long as 500 years to reach levels of mature forests. But studies in Luquillo, Puerto Rico, have suggested much shorter recovery times (Walker et al. 2009). Certainly, elevation and rates of tree growth are important predictive factors, in addition to surrounding vegetation, human disturbances, and soil texture.

6.4 Succession following Volcanic Eruptions

The rate of succession following eruptions varies with impact type, climate, and geographic factors (del Moral and Grishin 1999). As with landslides, vegetation distribution depends strongly on retention of water by fine soil material and on species colonization from adjacent forest areas. Volcanic eruptions can destroy all local and regional vegetation, creating low resource availability and leaving no biological legacies, as occurred on Krakatau (box 6.1). When biological legacies persist as remnant trees or forest patches, colonization proceeds at a more rapid pace.

Taylor (1957) described pioneer vegetation on three volcanoes in Papua, New Guinea. The eruption of Mt. Lamington in 1951 covered a blast area of 200 square kilometers (see fig. 6.1[3]). Within a few years, a wide range of vegetation types was observed within the blast area, reflecting heterogenous patterns of erosion, exposure of topsoil, and depth of ash. At the base of the volcanic dome, areas with surface deposits of coarse gravel and rock supported only scattered moss and ferns, despite an annual rainfall of over 3,600 millimeters. Thick ash layers, consisting predominantly of fine silt, supported a grass community dominated by *Saccharum spontaneum*. The percentage of finer material increased gradually farther down slope, leading to more dense grass cover by *Imperata cylindrica* and *Saccharum*. On sites where only part

Box 6.1. Regrowth of Vegetation on the Krakatau Islands, Indonesia

In 1880, the three volcanic islands of Krakatau in the Sunda Strait were covered with luxuriant vegetation, probably very similar to the current forests of Sumatra and Java, 32 and 41 kilometers away, respectively. In May 1883 the dormant volcano began a pattern of renewed activity that culminated in catastrophic eruptions on August 27 of that year. Pyroclastic flows, volcanic ashes, and tsunamis killed over 36,000 people and destroyed numerous coastal settlements in Sumatra and Java. Only one-third of the largest island, now called Rakata, remained. All three islands (Rakata, Panjang, and Sertung) were effectively sterilized, as the old surfaces were buried by 60 to 80 meters of volcanic products (Whittaker et al. 1989). Yet the islands became covered with vegetation within 50 years. Anak Krakatau, which emerged from the sea in 1927, is still highly volcanically active and is sparsely covered by pioneer vegetation near the shoreline (see plate 7).

The long history of botanical exploration and ecological research in Krakatau provides the best-documented long-term case study of primary succession and species colonization in the tropics. The closest potential source area is the island of Sebesi, 12 kilometers away, but this island was also heavily impacted by the 1883 eruption. Six species of cyanobacteria were noted to be among the first colonizers. They formed a gelatinous layer that is thought to have facilitated the establishment of ferns, which covered the interior of Rakata within a few years. By 1887, a savanna-type grassland covered the interior of Rakata, dominated by *Saccharum spontaneum* and *Imperata cylindrica*, with scattered trees. Early beach vegetation was characterized by the creeper *Ipomoea pes-caprae*, with tree colonization by the strand species *Terminalia catappa*, *Barringtonia asiatica*, *Hibiscus tiliaceus*, *Calophyllum inophyllum*, and *Casuarina equisetifolia* within 14 years of the eruption (Whittaker et al. 1989; Whittaker et al. 1999).

Volcanic ash was slowly transformed into soil through the action of climate, microorganisms, and early colonizing plant life. By 1930, 50 years after the eruption, the forests reached canopy closure, and the successional vegetation assemblages on Rakata resembled those characteristic of secondary successions after forest clearance or farming in other regions of Malaysia (Richards 1996). The *Saccharum* and *Imperata* grasslands in the interior of Rakata were gradually replaced by a mixed second-growth forest dominated by *Ma-*

caranga tanarius and *Ficus* species (Whittaker et al. 1989). Above 550 meters elevation, the forest became dominated by the wind-dispersed early successional tree *Neonauclea calycina*. By the 1950s, the mixed *Macaranga-Ficus* forest at lower elevations in the interior gave way to forest dominated by *Neonauclea calycina*, with a diverse understory and distinct ground and shrub layers. The lowland interior forest remains largely dominated by *N. calycina* to this day (Whittaker et al. 1999). Volcanic disturbances and ash fall from Anak Krakatau since the 1930s on the islands of Sertung and Panjan, however, have set succession back numerous times (Whittaker et al. 1989; Schmitt and Whittaker 1998).

During the 1920s, many species dispersed by birds or bats arrived on Rakata and the other islands, creating a mutually reinforcing effect on frugivores, frugivory, and seed dispersal (Whittaker et al. 1989; Thornton et al. 1996; Whittaker and Jones 1994). As of 1992, 173 plant species found on Krakatau were associated with bat or bird dispersal, representing 42% of the total flowering plant flora (Whittaker and Jones 1994). Fruiting *Ficus* species were first recorded in 1897 (Thornton et al. 1996). Pteropodid bats brought the first wave of *Ficus* species to Krakatau by 1893 (Shilton and Whittaker 2010). Seventeen of the 24 *Ficus* species that have since colonized Rakata were present within 40 years of the eruption (Whittaker and Jones 1994). Thornton et al. (1996) reported 23 species of frugivores on the Krakatau Islands: 16 bird and 7 bat species. The known histories of colonization of these two groups of interacting species are amazingly coincident. Today, *Ficus* species are canopy components within all major forest types on all of the islands. Between 1919 and 1930 the specialist frugivores *Ducula bicolor* (pied imperial pigeon) and *Cynopterus* fruit bats colonized. These frugivores enabled long-distance seed dispersal of 8 *Ficus* species, including some large trees, and contributed importantly to the reforestation of Krakatau (Thornton et al. 1996; Shilton and Whittaker 2010).

At least 300 species of vascular plants have colonized Rakata since the 1883 eruption. The forests of the Krakatau islands remain species poor, dominated by a few species typical of early stages of succession. The islands are well supplied with taxa that have small, light, wind-dispersed propagules, such as ferns and orchids (Whittaker et al. 1989). The specialist bird and mammal species that disperse large-seeded species will not take up residence on the islands until a sufficient number of fruiting trees have established. After 130 years of succession, Rakata remains a forest under construction.

of the topsoil was removed or on sites with thin ash cover, several tree species were recorded, including several *Ficus* species. *Trema* sp. (Cannabanaceae) occurred on steeper slopes where all topsoil had been eroded away.

Regeneration of vegetation on Mt. Lamington was extremely rapid, as numerous species present in the Mt. Lamington blast area survived the eruption. Taylor (1957) further noted that a high proportion of these species reproduced vegetatively by means of underground organs, which favored the persistence of these species. An eruption of similar magnitude to Mt. Lamington occurred around 1870 on Mt. Victory about 100 kilometers to the southeast. Local seed sources were available outside of the blast zone for recolonization, and pioneer tree species rapidly formed a closed woodland community. By 1957, the developing forest had a mixed composition and was rich in species, but it was still far poorer than forest outside the blast zone and will likely require several centuries to reach a mature state.

The 1991 eruption of Mt. Pinatubo in Luzon, Philippines, provided an opportunity to investigate the early stages of vegetation colonization. In contrast to vegetation regrowth on Mt. Lamington and Krakatau, the lowland regions of Luzon were heavily impacted by human settlements. Marler and de Moral (2011) conducted a baseline vegetation study 15 years after the dramatic eruption that spewed five to six square kilometers of ejecta onto the landscape. Along the east flank of the volcano, they sampled vegetation in eight sites along two river gorges from 200 to 750 meters and recorded a total of 58 plant taxa. Among these, 34 (59%) were exotic species that had spread from villages and agricultural areas in the lowlands; cover of exotic species was more than twice as high in lowland sites as in higher elevations. Among the 10 species of woody plants, the most abundant was the native species *Parasponia rugosa* (Cannabaceae), the only nonleguminous species known to fix nitrogen through a symbiosis with rhizobium bacteria (Geurts et al. 2012). *Parasponia* nodules are more primitive than those in legumes and can host a broad range of rhizobium species belonging to four different genera (Op den Camp et al. 2012). The native grass *Saccharum spontaneum* and *Parasponia rugosa* together contributed 86% of the vegetation cover and were inversely related in cover within plots (Marler and del Moral 2011).

Primary succession on volcanic substrates does not always proceed in a sequential progression of plant assemblages. The Hawaiian Islands arose as basaltic shield volcanoes over a hotspot in the Pacific Plate. Evolution in isolation led to a simplified tropical forest flora, lacking fast-growing, short-lived pioneer tree species. All primary successional stages are dominated by *Metrosideros polymorpha* (Myrtaceae), a shade-intolerant pioneer tree species.

Populations of the dominant tree species *Metrosideros* persist indefinitely as "chronosequential monocultures" (Mueller-Dombois 1992). The only significant successional change in vegetation is the decline in tree ferns with substrate age (Crews et al. 1995). Essentially, the forests in Hawaii undergo successional change with minimal floristic change.

6.5 Riverbank Succession

Hydrarch successions are primary successions initiated in a freshwater environment, such as on newly deposited riverine substrates associated with fluvial dynamics. Successional trajectories in river corridors are determined by flooding events, which form new water bodies by channel abandonment (oxbow lakes) and initiate primary succession on newly deposited sediments (Ward et al. 2002). Large rivers originating in the Andes begin to meander when they reach the broad lowland floodplains (Kalliola et al. 1991). As they meander, rivers erode outer (concave) riverbanks and deposit fresh sediment along point bars on the inner (convex) banks. In June 2003, an image showing a 55-kilometer stretch of the Mamore River south of the lowland Bolivian town of Trinidad in the Beni Province was taken from the International Space Station (see plate 10). This stretch (centered at 15.2° S 66° W) was rectified to a 1990 Landsat Thematic Mapper (TM) image. The spatial dynamics illustrated here depict a temporally dynamic system; the river course and its associated floodplain forests are in a constant state of flux.

The western Amazon landscape is dominated by white-water rivers, which drain the geographically young and easily erodible soils of the Andes. Amazonian white-water rivers are rich in nutrients and suspended particles, in contrast to black-water rivers, which drain white-sand forests and are poor in nutrients and suspended particles (Sioli 1950). The floodplain forests associated with these two major river systems have distinctive vegetation composition and dynamics (Prance 1979); forested white-water floodplains are termed *várzea*, whereas forested black-water floodplains are *igapó*. Várzea forest covers approximately 200,000 square kilometers, or 4%–5% of the Amazon basin (Junk 1989). Sediment cores from two wetlands in the lowland western Amazon basin of Ecuador document periods of hydrarch succession during the Holocene beginning around 5800 cal BP (Weng et al. 2002).

Recent alluvial formations occupy approximately 26% of the Amazonian lowlands of Peru, creating a complex landscape mosaic of early and late successional vegetation, swamps, backwaters, and oxbow lakes (Salo et al. 1986; Kalliola et al. 1991; Terborgh and Petren 1991). Salo et al. (1986) estimated that 12% of the Peruvian lowland forest is in successional stages due to the effects of fluvial dynamics. Analysis of a 70-kilometer section of the Río Manú in southwest Peru revealed a mean lateral erosion rate of 12 meters per year over a 13-year period.

From studies of four different rivers in the Peruvian Amazon, Puhakka et al. (1992) estimated that approximately 130 square kilometers of forest are annually eroded and replaced by successional vegetation. Based on analysis of Landsat TM imagery over a 21-year period, Peixoto et al. (2009) found that annual rates of lateral erosion and accretion of new land along the Solimões, Japurá, and Aranupa Rivers in western Brazilian Amazonia were well balanced; pioneer vegetation cover increased by 5.8% of the study area, whereas late successional areas decreased by 5.5%. Over 21 years, pioneer vegetation established on 8,920 hectares of new land formed within the 153,032-hectare study area.

Extreme zonation in successional vegetation is observed on point bars in white-water rivers. Point-bar accretion produces a series of ridges or dunes running parallel to the river, characterized by progressively later successional stages with increasing distance from the river (Salo et al. 1986; Terborgh et al. 1996). At Cocha Cashu Biological Station in Peru's Manú National Park, the Manú River, which has a six-kilometer-wide floodplain, floods every year during the rainy season, temporarily inundating large areas to a depth of 1–2 meters and eroding outer banks at a rate of 25 meters per year at the point of maximum curvature (Ward 2002; Terborgh and Petren 1991). Newly deposited point bars are exposed by low water during the dry season. In one river bend, Terborg and Petren (1991) observed the front of primary successional vegetation extend 115 meters over a 16-year period.

The first colonists are annuals and the woody composite *Tessaria integrifolia*, which can reach heights of two meters before flooding begins again during the next wet season. These stands reach reproductive maturity in 3–4 years, and then become colonized by a stout, rhizomatous cane grass, *Gynerium sagittatum*. The *Tessaria* become overgrown by *Gynerium*, which persists for 15–20 years (Terborgh and Petren 1991). Although the canebrakes are still subject to annual flooding, the dense stands resist the current better than *Tessaria* and provide a more suitable environment for germination of tree seedlings. Even-aged stands of *Cecropia membranacea* then grow up through the cane, reaching heights of over 20 meters.

Over time, short-lived *Cecropia* trees are replaced with a mixed stand composed of *Guarea* (Meliaceae), *Sapium* (Euphorbiaceae), *Guatteria* (Anonaceae), *Inga* (Leguminosae), *Cedrela odorata* (Meliaceae), and *Ficus insipida* (Moraceae), as well as other species. The latter two species grow

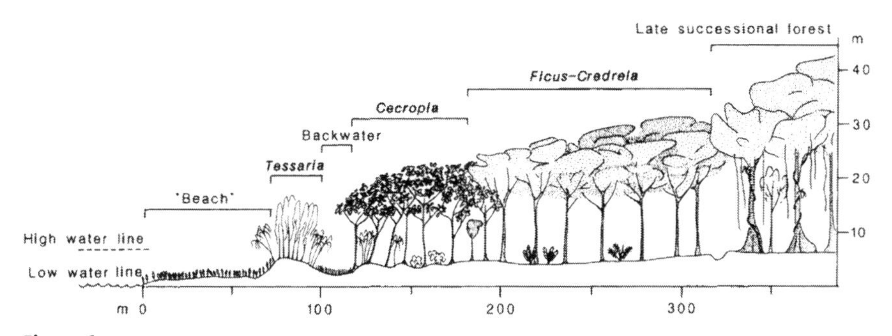

Figure 6.3
Primary succession along a simplified meander loop transect at Cocha Cashu Biological Station in Peru. Source: Salo et al. (1986, fig. 3).

slowly but eventually overtop all of the others, establishing a fairly uniform upper canopy layer at 35–40 meters. This association persists for a century or more. During this long transitional stage, dozens of tree species become established. But two or three more tree generations rotate through the forest before mature phase floodplain forest is formed. The entire primary successional pathway may take 300 years or longer to be completed (fig. 6.3; Salo et al. 1986; Terborgh and Petren 1991).

On point bars in central Amazonia, newly deposited sandbanks with coarse-grained sediments are initially colonized by grasses and sedges, followed by the woody pioneers *Salix martiana* and *Alchornea castaneifolia* (Parolin et al. 2004). These species are highly tolerant of sedimentation and erosion due to the production of new lateral roots near the soil surface (Junk 1989). Their presence favors the establishment of seedlings of *Cecropia latiloba*. Flooded forests on point bars in southeastern Brazil are dominated by *Salix humboldiana* and *Inga vera* (Oliveira-Filho et al. 1994). Parallel to the successional gradient, species show well-defined distributions along a flood-level gradient, separating low-várzea associations from high-várzea associations (Wittmann et al. 2004). The primary succession of várzea forests is linked to the deposition of fine-grained sediment, which is favored by the slowing of river current by dunes and tree stems. Older successional stages (20–60 yr. old) are dominated by *Pseudobombax munguba*. Many other tree species become established during this later stage, slowly increasing the species diversity and structural complexity of the forest over a 300–400 year period (Worbes et al. 1992; Parolin et al. 2004; Wittmann et al. 2004). Richards (1996) described a similar pattern of hydrarch succession on sandbar islands in the Zaïre River in the Congo basin of Africa.

6.6 Conclusion

We live on a restless earth. In forested tropical regions, volcanic eruptions, landslides, and floods cause major destruction to both forests and human settlements, while creating new substrates that slowly recolonize and develop into new forests. From studies of succession following these disturbances, we have learned that early colonists require nitrogen-fixing capabilities that permit growth on nutrient-poor substrates. Enrichment of soil by these early colonists facilitates colonization by more nutrient-demanding species. We have also learned that disturbances that initiate primary successions create environmental heterogeneity, providing diverse conditions for the entire gradient of species adaptations, from disturbance-tolerant to disturbance-sensitive species. Many of the disturbance-tolerant species can be employed to restore degraded soils and accelerate succession.

Forest succession on newly created substrates is slower than the secondary successional pathways described in chapter 5 and may require hundreds of years to recover species composition. But forest structure can recover rapidly, even in isolated areas completely covered by volcanic deposits and ash, as observed on Rakata, Krakatau. In areas such as Mt. Pinatubo, however, where human settlement and agricultural land uses have transformed local forest cover, trajectories of vegetation recovery can be impeded or diverted by invasive exotic species and by restricted dispersal of native species. Understanding the synergistic effects of human activities and "natural disasters" on the destruction and regeneration of tropical forests is an urgent challenge of the 21st century.

CHAPTER 7

FOREST REGENERATION FOLLOWING AGRICULTURAL LAND USES

The fragility of tropical forests has been overemphasized while their resiliency has been underemphasized.—Ariel Lugo (1995, p. 957)

7.1 Overview

The rate and quality of forest regeneration vary widely in the aftermath of human disturbances. Forest disturbances created by human activities range from relatively benign to severe. In contrast to immediate initiation of succession following pulsed natural disturbances that occur during a brief period, long-term agricultural land use can delay or inhibit regeneration after abandonment. Long-term, intensive land use can compromise the resilience of tropical forest ecosystems, leading to alternative stable states (Scheffer et al. 2012; Chazdon and Arroyo 2013). Succession can become arrested or deflected due to a variety of factors, including colonization of invasive exotic species (see plate 11).

In disturbed areas with local seed sources and relatively intact soils colonization of vegetation following human disturbances can proceed rapidly with little or no delay (table 7.1; Chazdon 2003). Species diversity and biomass accumulate rapidly when initial stand disturbances are small, land use is of short duration, and clearings are embedded in a forest matrix, as is common in traditional shifting cultivation systems with long fallow periods (Kassi and Decocq 2008; Piotto et al. 2009). Remnant vegetation, in the form of forest patches, trees, hedgerows, living fences, or resprouts, is the basis of "ecological memory," biological legacies that are internal or external to a regenerating patch, and that represent species previously present within the local landscape (Bengtsson et al. 2003; Chazdon and Arroyo 2013).

This chapter focuses on successional pathways in the aftermath of different types of human disturbances, with an emphasis on agricultural land use. (I will examine regeneration following fires, hurricanes, and logging in chapters 8 and 9.) As is the case with primary succession, the availability of propagules (seeds and resprouting stems) and soil resources critically determine initial colonization during forest succession (see fig. 6.1; Pickett and Cadenaso 2005). The intensity, extent, severity, and duration of land use directly influence the composition of the soil seed bank and newly arriving

Table 7.1. *Ten local and landscape conditions that favor rapid and diverse colonization of former cultivated fields or pastures in tropical regions by promoting the availability of propagules and resources for regeneration of both pioneer and nonpioneer species*

1. Retention of topsoil
2. Proximity to forest fragments
3. Resprouting from tree roots or stems
4. Intact soil seed bank
5. Presence of seeds of early and late successional woody species in the seed rain
6. Continuous colonization of native species from surrounding areas
7. Weed suppression by pioneer shrub or tree colonization
8. Animal and microbial diversity (insects, vertebrates, soil fungi)
9. Protection from recurrent fire
10. Protection from hunting and excessive harvesting of litter and forest products

seeds, soil nutrient availability, remnant vegetation, and conditions for early seedling establishment (Martínez-Ramos and García-Orth 2007). Abiotic factors, surrounding landscape, and land-use history all influence successional pathways in tropical old fields (Holl 2007). The quality and quantity of initial species colonization following land abandonment critically determine successional pathways for decades, if not centuries (Chazdon 2008b).

Successional pathways are determined by abiotic conditions, biotic interactions, and historical contingencies. These broad sets of factors determine where the species that colonize regenerating forests come from and what conditions they encounter following their arrival. If functionally and taxonomically diverse old-growth forests are present within the landscape, and seeds can be dispersed into cleared areas and recruit as seedlings, saplings, and trees, species composition in second-growth forests will more rapidly approach that of old-growth forests over time (Chazdon, Peres, et al. 2009; Dent and Wright 2010). Source populations and dispersal processes link the local process of forest regeneration with features of the surrounding landscape. A key factor influencing successional pathways is the extent to which plant and animal species characteristic of old-growth forests colonize during early and intermediate stages of succession. Continued deforestation and human pressure on tropical forests present the greatest obstacles to forest regeneration and to the conservation of tropical forest species in human-modified landscapes (Melo et al. 2013).

7.2 Effects of Land Use and Biological Legacies on Propagule Availability and Modes of Regeneration

When a treefall gap occurs in an intact forest, new tree recruits come from four major sources: the soil seed bank, newly dispersed seed, resprouting damaged stems, and the local pool of existing seedlings and saplings that survived the disturbance known as advance regeneration (Uhl at al. 1990; Vieira and Proctor 2007). During forest clearance for shifting cultivation, the first colonizers of a field are primarily drawn from the pool of species already present, either as established plants, resprouts, or dormant seeds in the seed bank (de Rouw 1993; Quintana-Ascencio et al. 1996). These species often persist as early successional colonizers if a field is abandoned after one crop and little weeding is done. The vigorous growth of sprouts confers a competitive advantage during early stages of forest succession. In shifting cultivation fallows of Bragantina, Brazil, in eastern Amazonia, 81%–86% of the individuals and 68%–81% of the species of trees above five centimeters in diameter at breast height (DBH) originated from resprouts (Vieira and Proctor 2007). In San Carlos de Río Negro in Venezuela, 54% of the trees above five centimeters in DBH in three-year-old fallows were resprouts (Uhl et al. 1982). Resprouts constituted 83% of the total stem basal area five years after abandonment of shifting cultivation on Hainan Island, south China (Ding and Zong 2005). Resprouting is a more common mode of regeneration in seasonally dry tropical forests, as water limitation favors increasing energy investments in roots (Ewel, 1977; Vieira and Scariot 2006; Holl 2007). By persisting as an ecological memory of past vegetation, resprouts enhance continuity between former forests and early stages of regeneration following shifting cultivation, fire, logging, and wind disturbance.

Even after forest clearing and burning, many pioneer and nonpioneer species are capable of resprouting from roots and/or stems (Kammesheidt 1999). In 7- to 10-year-old shifting cultivation fallows in a dry forest area of Laos, up to 30% of the individuals of nonpioneer species were resprouts from surviving root systems (McNamara et al. 2012). The importance of resprouting can vary substantially, depending on land use. In 3- to 8-year-old *Vismia*-dominated stands in central Amazonia, every *Vismia* stem examined originated from a resprout. Resprouting of *Vismia* is favored by frequent burning of pastures prior to abandonment. In contrast, tree recruits in *Cecropia*-dominated stands of similar age (including *Vismia* stems) originated mostly from seed (Wieland et al. 2011). The importance of resprouting in shifting cultivation fallows was demonstrated in a comparison of former rice and poppy fields in Chiang Mai Province in northern Thailand. Tree stumps were left in swidden rice fields, but trees and shrubs were uprooted and burned in preparation for poppy cultivation. Species diversity and aboveground biomass was higher in 30- to 49-year-old second-growth forest stands growing after rice cultivation compared to similar aged stands following poppy cultivation (Fukushima et al. 2008). Resprouting is also an important mechanism for tree regeneration following logging and cyclones, which will be discussed in chapters 8 and 9.

The soil seed bank is also an important source of recruits during old-field succession. The importance of the soil seed bank for early recruits of both herbaceous and woody seedlings increases with the size of clearings. Many pioneer trees with small seeds in the seed bank are photoblastic, requiring direct exposures of red light for germination (Vázques-Yanes and Orozco-Segovia 1993). Several pioneer tree species, such as *Ochroma lagopus* and *Heliocarpus donnell-smithii*, require variation in soil temperatures to germinate. Seeds of pioneer tree species can persist in the soil seed bank for decades. Soil cores from Barro Colorado Island, Panama, revealed that *Zanthoxylum ekmanii* (Rutaceae) seeds can germinate after 18 years in the soil, *Trema micrantha* (Celtidaceae) after 31 years, and *Croton billbergianus* (Euphorbiaceae) after 38 years (Dalling and Brown 2009). In contrast, seeds of late successional species are rarely found within the soil seed bank; these species have relatively short seed longevity, cannot withstand drying, and germinate shortly after dispersal (Vázquez-Yanes and Orozco-Segovia 1993). In pastures of Chiapas, Mexico, the seed bank was more important than the seed rain for recruitment of early successional tree species, but the opposite was true for late successional species (Benítez-Malvido et al. 2001). Repeated burning can eliminate soil seed banks as well as "sprout banks," leaving seed rain as the only source of new tree recruitment during initial stages of succession.

Intensity of pasture management in eastern Amazonia strongly affected early plant colonization following abandonment (table 7.2). Some former pastures persisted as grass- and shrub-dominated old fields for many years. Pastures that were abandoned within one year of formation showed rapid rates of forest regrowth and colonization by tree species from old-growth forest areas, but these pastures accounted for only 20% of the former pastureland in the late 1980s (Uhl, Buschbacker, and Serrão et al. 1988; Nepstad et al. 1991). Species availability is a major limiting factor for regeneration in heavily-used pasture sites, as seed rain is the only potential source of new colonists, and rates of seed predation and seedling mortality are high.

Studies throughout the world's tropical regions have demonstrated that most seeds are dispersed short distances, with seed rain declining with increasing distance from forest vegetation. A higher proportion of tree species are dispersed by wind in tropical dry forests compared to rain forests (Chazdon

Table 7.2. *Effects of intensity of pasture management on woody regeneration eight years after pasture abandonment in Paragominas, Pará, State, Brazil*

	Pasture use (yr.)	Pasture use (activity)	Aboveground biomass (Mg/ha)	Number of tree species (/100 m^2)	Sources of tree regeneration
Mature forest	0	n/a	285–328	23–29	Advance regeneration, seed rain, seed bank, resprouts
Light use	≤1	No weeding, light grazing	90	21–25	Resprouts, seed bank, seed rain
Moderate use	6–12	Weeding, burning every 1–3 yr.	33	16–19	Resprouts, seed bank, seed rain
Heavy use	6–13	Bulldozed, disked, mechanical mowing	5	0	Seed rain

Source: Based on Uhl et al. (1988).

et al. 2003; Vieira and Scariot 2006). In a dry forest region of India, seed rain of wind-dispersed seeds in former agricultural clearings declined steeply with distance from forest, but vertebrate-dispersed species showed no spatial pattern. Although wind-dispersed seeds strongly dominated seed rain in clearings, seedlings and saplings of vertebrate-dispersed species were three times more abundant in regenerating vegetation (Teegalapalli et al. 2010).

Distance from the forest edge can reduce the diversity of colonizing vegetation well beyond the initial stages of succession (Günter et al. 2007). Proximity of forest vegetation is particularly important for the arrival of larger-seeded, late successional species (Norden et al. 2009). Abundance of late successional species increases in the seed rain during succession when there are nearby forest fragments to serve as seed sources (del Castillo and Pérez Ríos 2008). Floristic development in old fields in Veracruz, Mexico, was positively correlated with the perimeter of forest remnants in the surrounding landscape (Purata 1986). The extent and distribution of forest cover within the landscape, the external ecological memory, is therefore a key factor determining the gradual recruitment of old-growth forest species during secondary succession on a local scale (Chazdon, Peres, et al. 2009).

Following land uses that remove the soil seed bank and tree roots and stems, seed rain is the only source of new colonization. Such is the case following gold-mining activities where forest is cleared, superficial soil layers are removed, and soils are extracted from deep pits. Rodrigues et al. (2004) monitored early stages of tree and shrub regeneration following gold-mining activities in Mato Grosso, Brazil, after pits were filled with pebbles and sand and the remaining soil was deposited over the area. The number of individuals, the number of species, and the species diversity were highest adjacent to a remnant forest fragment (fig. 7.1). Species typical of later successional stages were found only in the samples adjacent to the forest fragment. These results clearly emphasize the importance of remnant forest patches for forest regeneration in degraded areas; even small forest remnants provide critical seed sources and refuge for seed-dispersing animals.

Species availability can also limit early colonization in shifting cultivation fallows (Delang and Li 2013). During repeated cycles of shifting cultivation in Kalimantan, Indonesia, species diversity of both early and late successional species declined with increasing numbers of cycles (fig. 7.2). These declines

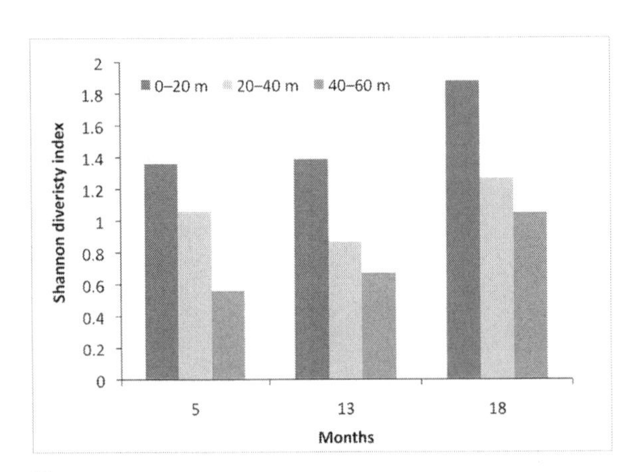

Figure 7.1
The diversity of shrubs and trees greater than or equal to 0.50 meters in height in 20 × 160 m plots parallel to a remnant forest patch following forest clearance and gold mining in Mato Grosso, Brazil. Shrub and tree diversity declined with increasing distance from the remnant forest patch (0–20 meters and 40–60 meters). Data were collected 5, 13, and 18 months after the degraded area was leveled, pits were refilled with pebbles and sand, and the remaining soil was distributed over the area. Source: Rodrigues et al. (2004, table 2).

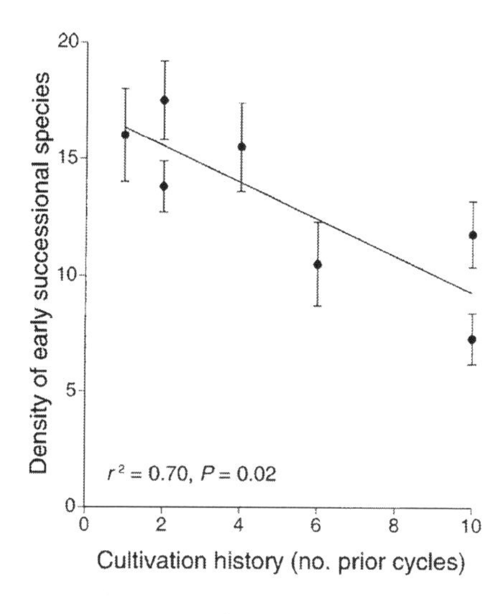

Figure 7.2
Decline in the density of early successional trees (no./300 m²) greater than or equal to 5 centimeters in DBH as a function of the number of prior cycles of shifting cultivation in Kalimantan, Indonesia. Data are means for four plots per site plus/minus one standard error. Source: Lawrence (2004, fig. 8).

reflected changes in the surrounding landscape that affected local species availability, rather than changes in soil nutrient status. Seed rain declined significantly in areas greater than or equal to 300 meters away from old-growth forest. Changes in landscape structure also play a role in the decline of late successional species, as distance to primary forest increased with the number of fallow cycles (Lawrence 2004; Lawrence and Schlessinger 2001; Robiglio and Sinclair 2011).

Two studies in Laos show contrasting effects of shifting cultivation on species availability in young fallows. In shifting cultivation fallows in the lowlands of Laos, species richness, stem density, and basal area declined after the third fallow cycle compared to the first cycle; the reduction in basal area was particularly high (72%; Sovu et al. 2009). Progressive degradation after four to six fallow cycles can lead to arrested succession in some shifting cultivation systems (box 7.1). Another study of shifting-cultivation fallows in Laos, however, showed a high potential for regeneration of nonpioneer species (McNamara et al. 2012). Fallows (7–10 yr. old) that were cultivated only once were compared with similar aged fallows cultivated three to five times over the past 30 years. Species richness of saplings of nonpioneer trees did not differ significantly among the two types of fallows and remnant primary forests, and the composition of nonpioneer trees did not differ significantly between the two types of fallows. The finding that more intensive shifting cultivation did not cause a decline in the regeneration of nonpioneer species reflects the substantial contribution of resprouts from surviving root systems

Box 7.1. Arrested and Deflected Successions

LEGACIES OF INTENSIVE LAND USE

Abandonment of agriculture following intensive land use can lead to the dominance of forms of vegetation that arrest succession, precluding the recovery of forest structure, composition, and ecosystem services in the absence of targeted interventions. Intensive land use and multiple disturbances—such as logging plus fire—create conditions that favor the persistent dominance of a single species capable of outcompeting native pioneer species and preventing the normal progression of successional vegetation.

Arrested succession, usually in the form of persistent cover by invasive grasses and ferns, is highly associated with the reduction in length of fallow cycles in shifting cultivation systems and the frequent use of fire following logging in vast areas throughout the tropics (Cohen et al. 1995; Ashton et al. 1997; Ramakrishnan 1998). A positive feedback cycle between environmental conditions and invasive vegetation, often (but not always) involving fire, prevents successional vegetation dynamics (D'Antonio and Vitousek 1992). Thirty years after tea plantations were abandoned in montane regions of Sri Lanka, grasslands continued to dominate due to the spread of *Cymbopogon nardus*, which was planted to stabilize terrace rises during tea cultivation (Gunaratne et al. 2010).

An estimated 64 million hectares of previously forested lands in Indonesia are dominated by *Imperata cylindrica* (alang-alang or cogongrass), a rhizomatous, perennial pantropical grass (Otsamo et al. 1995; Kuusipalo et al. 1995). The largest *Imperata* grasslands are in the interior of Borneo, where they occupy former dipterocarp forest soils subjected to logging followed by short-fallow shifting cultivation with frequent fire (Kuusipalo et al. 1995; Garrity et al. 1997; Yassir et al. 2010). *Imperata* is able to grow on a wide range of soils. Once established, *Imperata* grasslands are highly flammable. *Imperata* arrests secondary succession by outcompeting early successional vegetation and through allelopathic effects on tree seeds (Kuusipalo et al. 1995). In the absence of frequent burning, *Imperata* grasslands in East Kalimantan will undergo secondary succession; within nine years, cover of *Imperata* was reduced to 18%, whereas cover of shrubs and young trees increased to 44% (Yasir et al. 2010).

Single species dominate fallows in the shifting cultivation system of the Betsimisaraka people in a rain forest region of eastern Madagascar. Early fallow cycles are dominated by pioneer trees; later cycles are dominated by shrubs, and ultimately by ferns and grasses (Styger et al. 2007). Due to reduction in fallow periods, land degradation now occurs within 20–40 years of initial deforestation. Restrictions on forest clearing throughout many tropical regions pressure farmers to reduce fallow periods, promoting land degradation and reducing forest resilience in shifting cultivation landscapes (Schmook 2010; Robligio and Sinclair 2011).

In Panama, the exotic invasive grass *Saccharum spontaneum*, native to Asia, invades former cattle pastures that previously supported tropical rain forest. *Saccharum* is adapted to drought and frequent burning and has deep rhizomes, contributing to its persistence. Early stages of secondary succession are impeded in *Saccharum*-dominated grasslands subjected to fire, requiring intensive restoration to enhance tree regeneration (Hooper et al. 2004, 2005; Jones et al. 2004; Kim et al. 2008).

Intensification of land use has led to the widespread invasion of bracken fern (*Pteridium aquilinum*) in Chiapas and southern Yucatán, Mexico (Douterlungne et al. 2008; Schneider and Fernando 2010). Bracken fern thrives in open areas with frequent burning, preventing seedling establishment, depleting the soil seed bank, and impeding secondary succession. The fern *Dicranopteris pectinata* forms thickets after logging, agriculture, burning, and subsequent soil erosion. These thickets have inhibited natural succession in the Dominican Republic (Slocum et al. 2006).

Invasive trees and shrubs can also arrest succession following abandonment of cultivated areas. The aggressive Neotropical woody shrub *Chromolaena odorata* invades early fallows in West Africa, Indonesia, and Asia, forming a dense canopy and suppressing other plant growth, including other weeds (Koutika and Rainey 2010). In 1998, a 25-hectare tea plantation in the Sri Lankan highlands was abandoned as part of a restoration initiative. Within six months the invasive Neotropical shrub *Austroeupatorium inulifolium* became established and now dominates vegetation cover along with the invasive fast-growing Australian tree *Acacia decurrens* (Pethiyagoda and Nanayakkara 2011). The invasion of these species was likely promoted by heavy soil erosion within the tea plantation and by lack of second-growth forest vegetation as a colonization source in the area. Forest succession is arrested in dense stands dominated by arborescent bamboos (*Guadua* spp.) that cover 180,000 square kilometers in southwestern Amazonia (Nelson 1994; Griscom and Ashton 2003). High bamboo dominance alters vegetation structure and reduces the species richness of trees (Lima et al. 2007). In mixed deciduous forests of

Box 7.1 (*continued*)

Thailand, proliferation of bamboo following forest disturbances reduces tree seedling abundance and species richness (Larpkern et al. 2011).

Interactions between open disturbances, vegetation, and large herbivores can also lead to plant formations that impede succession (Struhsaker et al. 1996; Kasenene 2001). In Kibale National Park, Uganda, large canopy gaps created by selective logging were quickly colonized by an aggressive and persistent herb layer, dominated by the native subwoody shrub *Acanthus pubescens* (Chapman et al. 1999). These gaps showed little forest recovery 30 years after selective logging (Chapman and Chapman 2004; Paul et al. 2004; Lawes and Chapman 2006). The interaction between elephants and open vegetation patches creates a positive feedback that favors persistence of *A. pubescens* in disturbed forest areas, such as logging gaps.

There are limits to the resilience of tropical forest ecosystems. Arrested successions are examples of alternative steady states following intensive land use and multiple disturbances (Scheffer et al. 2012). Reforestation of these lands will require active and potentially costly interventions (Lamb 2011).

and the close proximity to primary forest remnants in the area. In more highly populated areas of Laos, fallow periods are shorter and regeneration is compromised (Sovu et al. 2009).

Seeds, resprouts, and remnant vegetation collectively compose the internal and external ecological memory that is an essential component of forest resilience (Bengtsson et al. 2003; Chazdon and Arroyo 2013). Remnant trees in cultivated fields or pastures promote seed dispersal following abandonment and potentially well beyond the early stages of succession. If the sources of propagules colonizing old fields derive from species characteristic of old-growth forests, the species composition of regenerating forests will increasingly resemble old-growth forests over time. Otherwise, species composition will be dominated by second-growth specialists for many decades (Howe and Miriti 2004; Martínez-Garza et al 2009). Regardless of the sources, propagule availability is highly stochastic spatially and temporally, contributing to localized variation in initial colonization and successional trajectories.

7.3 Effects of Land Use on Site Quality and Resource Availability

Species availability and resource availability, together, determine the rate and quality of forest regeneration (see fig. 6.1). These axes are often interdependent, as remnant vegetation and soil quality are both altered by the extent and intensity of former land use. Remnant trees left in pastures affect both seed dispersal and local environmental conditions for seed germination following pasture abandonment (box 7.2). The availability of soil water, soil nutrients, and safe sites for seed germination and establishment ultimately determine survival of newly colonizing seedlings.

Once a seed is dispersed into an old field, or emerges from the seed bank, it faces new risks of seed predation, seedling predation, drought, and root competition (see chap. 12, sec. 12.4). Many tropical regions have a pronounced dry season, which strongly impacts establishment of small-seeded pioneer species (Nepstad et al. 1991; Nepstad et al. 1996). Competition from herbaceous vegetation present at the time of abandonment also strongly influences the establishment of early successional vegetation (Chazdon 2003; Holz et al. 2009). On degraded cattle pastures on steep slopes, severe soil erosion reduces the abundance and diversity of the spores of mycorrhizal fungi required by most tree species to assimilate soil nutrients (Carpenter et al. 2001). Early colonization of trees and shrubs requiring mycorrhizal inoculation can be substantially delayed if spores are not present in the soil.

Following agricultural abandonment, crop species and related weeds can persist for several years or more in early successional vegetation, significantly inhibiting early woody regeneration (Myster 2004). In former fields of *Saccharum officinarum* (sugarcane), *Musa* sp. (banana), and *Setaria sphacelata* (pasture grass) in Ecuador, these crop species continued to dominate for the first five years of succession (Myster 2007). Pasture grasses introduced from Africa often form dense swards up to three meters tall in former pastures, creating a dense barrier to establishment of woody seedlings (see box 7.1; Aide et al. 1995; Holl et al. 2000). Establishment of woody seedlings and saplings was impeded by high grass coverage in former pastures in Missiones, Argentina. Species composition in second-growth forests less than 20 years old was strongly affected by previous land use for pastures, *Pinus* plantations, annual crops, or *Ilex paraguariensis* (yerba maté) plantations, but differences in structure and composition were not detected in forests older than 20 years (Holz et al. 2009). Based on multiple measures, forest regeneration in Petén, Guatemala, was significantly faster in shifting cultivation fallows and former traditional agroforestry sites than in former pastures or intensive sesame and corn monocultures, where grasses and weedy *Bidens* sp. (Asteraceae) suppressed the establishment of woody regeneration (Ferguson et al. 2003). Disturbance intensity was a more important factor than edaphic variation in explaining the variation in the structure and composition of regenerating forests in Hainan Island, South China (Ding, Zang, Liu, et al. 2012).

Box 7.2. Remnant Trees and Nucleation of Postagricultural Land

The presence of remnant trees in large pastures can significantly enhance seed rain and seedling establishment, enriching early and intermediate stages of secondary succession. When birds and bats are attracted to perching sites or fruiting vegetation within pastures and fallows, the rate of succession accelerates through increased quantity and diversity of seed dispersal and enhanced recruitment of woody species, including species characteristic of old-growth forests.

This nucleation processes (Yarranton and Morrison 1974) can be initiated by isolated remnant trees that were spared cutting during prior forest clearance for pasture or swidden cultivation (Guevara et al. 1986; Guevara et al. 1992), early-establishing saplings (Campbell et al. 1990), early-establishing shrubs (Vieira et al. 1994), or trees that regenerated from seeds or sprouts in pastures or swidden fields prior to abandonment (Toh et al. 1999). Nucleating trees ameliorate microclimates and soil conditions below their crowns, provide perch sites for frugivorous vertebrates, and produce seeds. Seed collectors beneath crowns of *Solanum crinitum* treelets in a former pasture in the Paragominas region of Brazil yielded 400 times more seed than collectors in the open grass/shrub matrix; 18 tree species were deposited under treelets compared to only 2 in the open areas (Nepstad et al. 1991).

In former pastures in a tropical dry forest region of Costa Rica, secondary succession is accelerated by nucleating trees, primarily due to their attractiveness to fruit-dispersing vertebrates (Zahawi and Augspurger 2006). In the absence of nucleating trees, wind-dispersed species are the only colonizing trees in former pastures. In contrast, a more diverse pool of vertebrate-dispersed trees and treelets accumulates beneath the canopy of nucleating trees, creating a positive cycle of increased fruit production, increased frugivore visitation, and increased seed rain. Over time, small patches associated with nucleating trees enlarge and coalesce, transforming a pasture into a young forest (Janzen 1988).

Studies throughout the tropics have demonstrated enhanced seedling establishment of vertebrate-dispersed species beneath remnant trees compared to areas away from remnants within the same site. In open pasture in the Los Tuxtlas rain forest region of Veracruz, Mexico, woody species and vertebrate-dispersed species were more than twice as abundant below remnant tree canopies than in perimeter and open pasture sites, and tree species were three to four times more abundant below remnant trees (Guevara et al. 1992). Over 50% of the species found in sites below remnant trees were zoochorous (animal-dispersed) species compared to 38%–39% in other sites. Further studies in this region confirmed the contribution of bats and birds to seed rain beneath remnant trees in the genus *Ficus* (Galindo-González et al. 2000). Most seeds were from tree and shrub species; 89% were of zoochorous species, and 24.6% were characteristic of late successional stages. Remnant *Ficus* trees frequently attract birds and bats carrying small seeds from parent trees over 75 meters away (Laborde et al. 2008). Seed rain was higher beneath remnant tree species with fleshy versus dry fruits in pastures of Veracruz, Mexico (Guevara et al. 1986) and northeastern Costa Rica (Slocum and Horvitz 2000).

In traditional swidden agriculture, isolated trees are often spared during the clearing of fields. In southern Cameroon, Carrière et al. (2002) compared regeneration beneath remnant trees in 3- to 20-year-old fallows. The relative abundance of tree regeneration and the proportion of vertebrate-dispersed species were both significantly higher beneath remnant trees than away from crowns. Away from crowns, lower seed rain and more exposed conditions favored the establishment of clonal, light-demanding herbaceous monocots.

Aside from nucleating succession, remnant trees in pastures can also facilitate regeneration in at least three other ways. First, isolated trees in pastures serve as sources of seed for tree regeneration beneath other isolated remnant trees (Laborde et al. 2008) and as pollen and seed sources for regenerating trees in forest fragments and second-growth forests (Aldrich and Hamrick 1998; White et al. 2002; Sezen et al. 2007).

Second, isolated pasture trees store seed of forest tree species in canopy soil that accumulates within crowns. In montane forest of Costa Rica, seed banks from the canopy soil of isolated pasture trees were similar to those from intact forest canopy soil (Nadkarni and Haber 2009). Nearly half of the species that emerged from canopy mats of isolated pasture trees were trees, and more than 40% of these were old-growth forest species.

Third, isolated remnant trees in pastures continue to influence tree recruitment beneath crowns beyond the initial stages of succession, even after canopy closure. In montane forests in Costa Rica, density and species richness of recruits were significantly higher beneath remnant trees 3 and 14 years after pasture abandonment (Murray et al. 2008). Remnant trees also influenced the composition of regeneration; late successional species were more abundant

Box 7.2 (*continued*)
than early successional species only beneath remnant trees. In the Caribbean lowlands of Costa Rica, tree and sapling density decreased with increasing distance from remnant trees 23 years after pasture abandonment (Schlawin and Zahawi 2008). Remnant trees also favored the recruitment of trees with seeds greater than or equal to one centimeter in diameter, which are dispersed by birds and bats. The spatial distribution of vertebrate-dispersed seedlings, saplings, and trees in second-growth forests may reflect a long-term legacy of remnant trees and their interactions with vertebrate frugivores.

Variation in soil fertility can strongly affect the composition and rate of forest regeneration (Guariguata and Ostertag 2001). Secondary succession generally proceeds more rapidly on young, fertile, volcanic soils compared to weathered, nutrient-poor soils. Rates of forest regrowth across five regions of the Amazon basin, as measured by stand height, were highest in the region with the highest soil fertility (Moran et al. 2000). In shifting cultivation fallows of white-sand vegetation in the Venezuelan Amazon, saplings of old-growth forest tree species took 60 years to dominate in abundance over early successional species (Saldarriaga et al. 1988). In contrast, sapling assemblages in 15- to 20-year-old second-growth forests on former pastures on young volcanic soils in northeastern Costa Rica were dominated by old-growth forest species. The most abundant canopy tree seedlings in these second-growth forests were of tree species that are also abundant in old-growth forests of the region (Guariguata et al. 1997; Norden et al. 2009).

Rates of forest regrowth in Amazonia are also strongly affected by settlement history and type of land use. In regions of eastern Amazonia with a prolonged history of land use, rates of forest regrowth were slow and species composition of regrowth was impoverished (Tucker et al. 1998). In general, traditional shifting cultivation fallows showed higher rates of total biomass accumulation than more intensive land uses such as pastures or mechanized fields (Moran et al. 2000). Rates of biomass accumulation during forest regeneration across the Amazon basin were strongly affected by the number of previous burns, which often serves as a proxy for the number of shifting cultivation cycles (Zarin et al. 2005). On average, stands with five or more previous burns showed a reduction in biomass accumulation of more than 50% compared to stands that burned only once or twice.

Sustainable shifting cultivation relies upon replenishment of soil nutrients and elimination of weeds during fallow periods (Aweto 2013, Delang and Li 2013). In a tropical forest region of Madagascar, tillage practices, cropping duration, and fallow age were all shown to have significant impacts on the diversity and structure of woody vegetation during fallow ages up to 29 years (Randriamalala et al. 2012). Heavy tillage and long cropping durations (5–15 yr.) led to a slow recovery of woody regrowth and favored dominance of wind-dispersed herbaceous species in fallows. High-tilling regimes significantly reduced species richness, woody basal area, vegetation height, and the percentage of animal-dispersed species. In contrast, no tillage with short cropping durations (1–2 yr.) and long fallow ages favored the highest species diversity and woody biomass accumulation. Longer cropping duration increased herbaceous species and reduced vegetation height in intermediate and older fallow age classes. Repeated burning and plowing removed most woody stumps and roots and favored establishment of wind-dispersed herbaceous species, significantly delaying woody vegetation succession.

Repeated cycles of shifting cultivation can also reduce levels of available soil nutrients during fallow periods, affecting initial colonization (Aweto 2013). Phosphorus frequently limits tropical forest growth. In tropical dry forests of the Yucatán Peninsula in Mexico, after three cultivation-fallow cycles, levels of available phosphorus declined 44%, creating a negative feedback cycle of increasing phosphorus limitation in regenerating vegetation (Lawrence et al. 2007). Long fallow periods are needed to replenish lost phosphorus in these dry forest ecosystems. In dry forests of the Yucatán, aboveground biomass accumulation and levels of coarse woody debris were significantly reduced in regrowth following three to four cultivation-fallow cycles compared to following a single cycle (Eaton and Lawrence 2009). The shortening of fallow periods increased the abundance of herbaceous weedy species and reduced the abundance of pioneer trees and vertebrate-dispersed species in the dry forest region of Quintana Roo, Mexico (Dalle and de Blois 2006). Moreover, shortening the fallow cycle compromises the sustainability of ecosystem services and forest products, such as firewood, that the local Yucatec people rely upon (see box 7.1).

Early stages of forest succession are particularly sensitive to severe soil disturbance or soil degradation. Activities such as bulldozing, creating skid trails in logging operations, and subsurface mining remove topsoil, organic matter, and soil fungi and microbes, as well as destroy the soil seed bank (see plate 12). Without soil structure, organic matter, or mycorrhizal inoculum, few species can colonize. Active reforestation interventions can restore soil quality and reestablish the capability for forest regeneration on these sites (Chazdon 2008a).

7.4 Conclusion

What takes a few moments to destroy by machete, bulldozer, or chain saw can take decades or centuries to reconstruct through the process of secondary succession. Early stages of secondary succession are highly sensitive to prior land use in tropical regions. When land use is not prolonged or highly intensive, secondary succession often proceeds rapidly, providing new habitats for forest plant and animal species and providing essential ecosystem services. Land uses that promote the availability of propagules and resources for regeneration of both pioneer and nonpioneer species will maximize the potential for forest regeneration within human-modified landscapes (see list in table 7.1). The rate and quality of forest regeneration is determined more by the effects of the disturbance on propagule availability and resource availability than by whether the disturbance originated from human agency or natural causes.

The structure and biomass of tropical forests are surprisingly resilient to both human and natural disturbances, although the recovery of species composition is a long-term process that plays out over decades or centuries. Forest regeneration following agricultural land use can be impeded or altered by subsequent fires, repeated logging, or large-scale natural disturbances. Recovery of ecosystem processes following human disturbance can be relatively rapid following low-intensity land use.

Clearly, there are limits to the regenerative capacity of tropical forest ecosystems, and these limits vary with climate, soil type, and landscape structure. Restoration efforts can hasten the recovery of forest cover, soil quality, and species composition in many tropical ecosystems. These efforts focus on increasing both species availability and resource availability, establishing the conditions that promote long-term forest succession (see fig 6.1).

Although the ecological conditions produced by former land use strongly affect successional pathways, the main obstacles to forest regeneration in the tropics emerge from socioeconomic, cultural, and political factors that govern patterns of land occupation and land use. These factors and their interactions are discussed in chapter 14. Ultimately, long-term commitments to reforestation require reduction of human pressure on tropical forests in the form of logging, cultivation, hunting, and mining, so that spontaneous regeneration, assisted natural regeneration, or reforestation can proceed on cleared forest land in close proximity to old-growth forest remnants.

CHAPTER 8

FOREST REGENERATION FOLLOWING HURRICANES AND FIRES

The crux of the fire problem in tropical rainforests is not so much the introduction of fire into these ecosystems but the frequency with which they are being burned.—Mark Cochrane (2003, p. 914)

8.1 Overview

Tropical cyclones (also known as hurricanes and typhoons) and fires are the most frequent types of large-scale natural disturbances affecting tropical forests. Following landslides, flooding, and volcanic eruptions, trees are uprooted or completely destroyed, and soils are eroded or covered with thick layers of ash or lava. In contrast, hurricanes and typhoons damage tree crowns and may cause some uprooting, but forest soils remain intact and most adult and juvenile trees survive, often through resprouting (Boucher 1990; Bellingham et al. 1992; van Bloem et al. 2003; Lugo 2008). Similarly, forest fires do not cause complete tree mortality; they produce spatial mosaics of forest patches with varying degrees of damage (Cleary and Priadjati 2005).

Large and infrequent disturbances, such as hurricanes and forest fires, leave behind a patchy and heterogeneous legacy, reflecting a complex interplay of conditions and events that modulate the intensity, quality, and extent of disturbance within a region (Baker et al. 2008; Turton 2008). The impacts of these disturbances may vary between island and mainland ecosystems as well as along gradients of rainfall, seasonality, elevation, topography, soil texture, and soil nutrient availability. Following hurricanes, fires, and selective logging, significant ecological memory, both internal and external, is retained. Forest succession following hurricanes and fires therefore differs from old-field succession following agricultural abandonment, because some trees remain and produce seeds and resprouts, promoting more rapid recovery of vegetation structure and composition (Chazdon 2003).

Species or populations persist following a large-scale disturbance through high rates of survival (persistence) or high rates of turnover (replacement) among individuals. Species or populations with low mortality following major disturbances are considered "resistant," whereas species or populations with high rates of both mortality and recruitment are considered "resilient." Both of these mechanisms promote reassembly of forest structure and composition. In addition to these two major types of species, assemblages of resistant and resilient species become enriched with a third group of species, "usurpers," that was previously rare or absent but proliferates following hurricanes and fires. A fourth category of species has high rates of damage and low rates of recruitment or growth following large-scale disturbances; these species undergo long-term declines in relative abundance. Regeneration of these "susceptible" species may require particular conditions or species interactions that are disrupted, at least temporarily, by the disturbance. This framework has been used to describe tree population responses to hurricanes and other large-scale disturbances (table 8.1; Boucher et al. 1994; Bellingham et al. 1995; Ostertag et al. 2005). A similar framework developed by Bond and van Wilgen (1996) compared species responses to fire. Resistance and resilience are major themes of disturbance ecology that can be applied to individuals, populations, assemblages, and entire ecosystems (Holling 1973; Harrison 1979; Halpern 1988).

This chapter focuses on patterns of regeneration following hurricanes and fires in tropical forests. In seasonally dry tropical forests and wooded savannas, fires are a regular feature of the ecology and vegetation dynamics, and dominant tree species are adapted to fire regimes (Baker et al. 2008). They are essentially "fire-resistant" species. Fires are rare in tropical wet forests and occur only following severe droughts, which can also directly cause tree mortality (Chazdon 2003; van Nieuwstadt and Sheil 2005). Accordingly, few tree species in tropical wet or moist forests have characteristics that confer fire resistance. In these cases, more trees die following fires, and the regeneration of forest vegetation depends heavily on new seedling recruitment, as during early stages of succession on cleared land. In areas prone to hurricanes and cyclones, forests are subjected to repeated wind damage, and species have adapted to this disturbance regime (de Gouvenain and Silander 2003). Successional pathways following cyclones and fires therefore strongly depend

Table 8.1. *Classification of tree species based on level of damage sustained during a hurricane and posthurricane responsiveness, as measured by population growth rates*

	Posthurricane growth rate	
Level of damage	Low	High
Low	Resistant	Usurpers
High	Susceptible	Resilient

Source: Based on Bellingham et al. (1995).

on the extent to which forests have been historically exposed to these disturbance regimes and are dominated by "resistant" species.

The excessive fuel load created by hurricanes can increase the risk of fires in tropical wet forests, particularly during droughts. During the dry season of 1989, three months after Hurricane Joan struck the southeast coast of Nicaragua (box 8.1), fires burned throughout many of the wind-damaged forests. This was an unusually dry year due to the strong El Niño Southern Oscillation (ENSO). Trees and limbs downed by the hurricane provided extra fuel for the fires. Prior to Hurricane Joan and the subsequent fire, Laguna Negra was surrounded by mature swamp forest vegetation (Urquhart 2009). After these events, grasses, sedges, cattails (*Typha* sp.), and the fern *Blechnum serrulatum* colonized the burned and blown-down areas.

History repeated itself at Laguna Negra. Pollen and sediment cores revealed that prehistoric Hurricane Elisenda struck Laguna Negra between 3830 and 2820 BP (Urquhart 2009). After this event, a series of fires occurred over approximately 200 years, followed by regeneration of swamp forest. After about 75 years of regeneration, fires began again, reducing tree pollen and increasing grass and fern pollen. After another 75 years, swamp forest trees increased and charcoal levels declined around Laguna Negra. Thereafter, a mature swamp forest developed on the site. Regeneration of this swamp forest took 400–500 years due to repeated setbacks from fires. Forests are dynamic entities that upon closer inspection reveal cyclical histories of disturbance and recovery. The long history of disturbance from hurricanes, droughts, and fires in southeastern Nicaragua provides essential background for understanding the rapid regeneration of forests in this region following Hurricane Joan.

8.2 Hurricane Damage and Regeneration

Cyclones do not occur uniformly across the globe. Close to 90% of all tropical cyclones form in a region between 10 and 20 degrees north and south of the equator. There are six major "hurricane belt" regions in the world: the southwest Indian Ocean off the east coast of Madagascar, the north Indian Ocean off the east coast of India, the southwest Pacific Ocean off the east coast of Australia, the northeast Pacific Ocean off the east coast of the Philippines, the Caribbean, and the west coast of Mexico (see plate 13). Cyclonic windstorms take on different names in different oceanic regions of the world. Typhoons occur in the northwest Pacific Ocean; hurricanes occur in the North Atlantic, northeast Pacific, or South Pacific Oceans; and cyclones occur in the Indian Ocean. Average yearly frequencies of recorded tropical storms from 1972 to 2000 were highest in the northwest Pacific Ocean (including the Philippines, Taiwan, coastal Indochina, and Peninsular Malaysia),

Box 8.1. Forest Regeneration following Hurricane Joan in Southern Nicaragua

On October 22, 1998, Hurricane Joan struck southeastern Nicaragua, the largest area of rain forest in Central America. Winds at speeds in excess of 250 kilometers per hour toppled or snapped 80% of the trees in 500,000 hectares of lowland rain forest. The area affected by Hurricane Joan represented more than 15% of the forested area of Nicaragua. Three weeks after the storm, only 27% of the trees remained standing and only 18% had leaves (Yih et al. 1991; Mascaro et al. 2005). But three months after the hurricane, 77% of the trees had leaves and 77 of 79 tree species observed had some resprouting individuals. Only a single pioneer species, *Croton killipianus*, was observed as seedlings in the forest areas inventoried; none of the typical pioneer species that colonize cleared areas were observed. Yih et al. (1991) concluded that forest regeneration in this area occurred directly through the resprouting and recruitment of tree species already present in the forest rather than through the new colonization of pioneer species typical of early states of secondary succession.

Later studies in long-term observation areas established within this region confirmed rapid posthurricane recovery dominated by resprouting damaged trees and saplings of individuals present before the hurricane (Vandermeer et al. 1995; Vandermeer et al. 1996; Vandermeer et al. 1997; Vandermeer et al. 1998; Vandermeer et al. 2001). Forest regeneration followed two phases. The first six years, constituted an initial "building phase" characterized by resprouting trees and saplings and rapid growth of preexisting seedlings that formed a dense 8- to 10-meter canopy. The intense competition within this dense canopy led to a "thinning" phase characterized by an overall reduction in density and a vertical separation between dominant and suppressed individuals (Vandermeer et al. 1998; Vandermeer et al. 2001). These phases roughly correspond to the stand initiation and stem exclusion stages described by Oliver and Larson (1996; see table 5.5). Canopy thinning was pronounced between 2000 and 2003, with surviving trees significantly taller than trees that died. Trees above 10 meters in height grew rapidly, whereas those below 10 meters showed reduced or no growth and were more susceptible to mortality. Throughout both phases, damaged adult trees remained standing and formed a sparse upper canopy. Pioneer species occurred only in low abundance in the thinning canopy (Vandermeer et al. 2001; Vandermeer and Granzow et la Cerda 2004).

On average, 76% of the aboveground biomass of these forests was lost immediately following Hurricane Joan. Over time, aboveground biomass increased steadily, with an average accumulation rate of 5.36 megagrams per hectare per year. These rates are lower than those observed in forests damaged by Hurricane Hugo in Puerto Rico in 1983 and in cases of agricultural abandonment in many regions. Several factors may explain the low rate of biomass accumulation following Hurricane Joan. First, the persistence and resprouting of trees and saplings allowed the forest composition to quickly return to prehurricane conditions, dominated by slow-growing tree species with limited establishment of fast-growing pioneer species. Second, the large-scale damage of forests in this topographically flat region left few seed trees, reducing seed dispersal and limiting colonization by pioneer species. Third, the degree of biomass loss following the hurricane was far more severe in Nicaragua than in Puerto Rico (Mascaro et al. 2005; Scatena et al. 1996).

The limited recruitment of pioneer species in the long-term study sites had another major consequence for forest regeneration: the species richness of trees increased dramatically during the 10 years after the hurricane. Species richness of stems greater than or equal to 3.2 centimeters in diameter at breast height (DBH) in hurricane-affected forests was two- to threefold higher than in forest areas to the north unaffected by Hurricane Joan (Vandermeer et al. 2000). Despite the catastrophic impact of Hurricane Joan on the forests of southeastern Nicaragua, forest regeneration was rapid and diverse, reflecting a fundamentally different mode of succession than that observed in former agricultural areas of the region (Boucher et al. 2001). Trajectories of species composition in six long-term plots over 12 years did not show a tendency for increasing similarity, however, suggesting that species composition of these forests does not converge on a single equilibrium state (Vandermeer et al. 2004).

which experienced an average of 6.3 hurricanes per year (de Gouvenain and Silander 2003).

8.2.1 *Forest Responses to Hurricane Frequency and Intensity*

The extent and nature of tree damage and regeneration following a hurricane event vary widely within the same area (Webb 1958; Weaver 1986; Bellingham 1991; Heartsill Scalley et al. 2010). In hilly terrain, valleys and leeward slopes are generally less damaged than ridgetops and windward slopes (Laurance and Curran 2008). After Hurricane Gilbert struck the montane forests of the Blue Mountains in Jamaica in September 1988, breakage of stems and the extent of crown loss, defoliation, and resprouting were significantly higher on the forest slopes with southern exposures and on the ridgetops compared to slopes with northern exposures, reflecting the prevailing wind direction (Bellingham 1991). Major physiographic features may buffer some forests from hurricane wind forces (Whitmore 1974). Hurricane Hugo struck Puerto Rico in September 1989, causing a 50% reduction in biomass in Luquillo Experimental Forest (Scatena et al. 1993). After 15 years of regeneration, prehurricane differences in species composition among slope, valley, and ridge locations were still not reestablished (Heartsill Scalley et al. 2010).

Hurricane impacts on forest structure, composition, and regeneration pathways also vary with the frequency and history of hurricane events. In regions frequently affected by hurricanes, a distinct pattern of tropical forest structure is observed: an even and smooth canopy of short stature dominated by species that are highly resistant to hurricane damage (Lugo 2000; de Gouvenain and Silander 2003). Short return times create forests in a perpetual state of disturbance and regrowth. The short canopy and high rate of basal sprouting of trees in the semideciduous Guánica Forest along the southwest coast of Puerto Rico is considered to be the result of frequent hurricane disturbance (Van Bloem et al. 2003, 2007). Beard (1955) referred to palm-dominated vegetation on steep slopes of Caribbean islands as a "disturbance climax" that remains in a perpetual stage of succession, never reaching a "true climax." Longer return times, in contrast, provide an opportunity for forest structure and biomass to attain old-growth conditions. When hurricanes occur every 100–200 years, tree species that are highly susceptible to major damage can accumulate in the forest canopy, leading to high rates of damage in these species when a hurricane does eventually strike (Canham et al. 2010).

Hurricane frequencies vary widely among tropical regions. During the 20th century, 30 major hurricanes passed over the Caribbean coast of Mexico (Sánchez Sánchez and Iselbe 1999). Hurricanes strike Luquillo Experimental Forest in Puerto Rico every 50–60 years (Canham et al. 2010) and impact the island of Tonga in the southwest Pacific every 10–40 years (Franklin et al. 2004). The return time of hurricanes in southern Nicaragua is about 100 years (Boucher 1990). Individual hurricanes have unique and spatially variable effects. Hurricane Georges hit Luquillo Experimental Forest in 1998, only 9 years after Hurricane Hugo. Canham et al. (2010) found a striking lack of correspondence between tree damage from the two hurricanes. In a nearby rehabilitated forest area, however, Ostertag et al. (2005) found that trees damaged during Hurricane Hugo were more likely to be damaged again during Hurricane Georges.

In the aftermath of a major hurricane, regeneration comes from three major sources: (1) resprouting of damaged or undamaged stems, (2) advance regeneration (growth of existing undamaged seedlings and saplings), and (3) new seedling recruitment from the seed bank or newly dispersed seeds. New species enter the forest only through new seedling recruitment. The relative importance of these three different recruitment sources depends upon many factors, including anthropogenic disturbance history, time since the last major hurricane, availability of nearby seed sources, availability of seed-dispersing vertebrates, and the species composition of trees prior to the hurricane event. In southeastern Nicaragua, forest regeneration following Hurricane Joan in 1988 quickly reestablished the tree species composition before the hurricane due to massive resprouting of adult and juvenile trees (see box 8.1)

Early regeneration of forests in Luquillo Experimental Forest in Puerto Rico after Hurricane Hugo was dominated by the growth and recruitment of pioneer species, such as *Cecropia schreberiana*, that established after the disturbance. Increased light availability in the understory following tree damage leads to higher species richness as predicted by the intermediate disturbance hypothesis (Scatena et al. 1996; Drew et al. 2009; Vandermeer et al. 2000). Where regeneration is dominated by resprouting, however, the tree species composition returns quickly to prehurricane condition. After Hurricane Georges passed over Guánica dry forest in Puerto Rico, basal sprouting was observed on 48% of stems that were structurally damaged, on 32% of defoliated stems, and on 29% of stems with no visible damage (van Bloem et al. 2003).

Tree basal area, aboveground biomass, and composition of dominant tree species tend to recover quickly after most hurricane events. Permanent forest monitoring plots established in 1943 in Luquillo Experimental Forest have shown a continuous increase in basal area and stand biomass despite two major hurricane events (Drew et al. 2009). The two dominant tree species in 1943, *Dacryodes excelsa* (Burseraceae) and *Manilkara bidentata* (Sapotaceae), were still the two dominant tree species in 2005. Similarly, the relative ranking of density and basal area of 12 common tree species in lowland rain forests of Kolombangara, Solomon Islands, did not change significantly over 30 years despite major cyclone activity (Burslem et al. 2000). And in montane forests of Jamaica, the rank of species abundance was strongly correlated after a 30-year span including a major hurricane (Tanner and Bellingham 2006).

Hurricane effects are strongly mediated by successional status, tree size distribution, and forest diversity at the time of hurricane impact. In Kolombangara, Solomon Islands, the forest types with the highest rates of canopy disturbance and postcyclone population turnover had the highest precyclone species richness and were composed of fast-growing tree species with low-density timber (Burslem and Whitmore 1999). In Puerto Rico, prior agriculture use strongly affected the nature of disturbance and regeneration following Hurricane Georges; large trees in older sites suffered more damage than trees in early successional forests (Flynn et al. 2010). After the hurricane, seedlings of early successional species were generally more abundant in areas formerly used for agriculture (Comita, Thompson, et al. 2010). Following Cyclone Waka in Tonga in 1991, the opposite trend was observed; early successional forests and species suffered more damage than late successional forests and species (Franklin et al. 2004). In montane forests of Jamaica, less diverse forests on less fertile soils were more resistant to the effects of Hurricane Gilbert than were more diverse stands on fertile soils (Tanner and Bellingham 2006). The more diverse forests had higher rates of mortality and recruitment and less resprouting following the hurricane. A complex set of factors, including forest age, composition, and soil time, determines rates of damage and recovery following hurricanes.

The rate and composition of forest regeneration following a hurricane also strongly depends on seed dispersal by vertebrate frugivores. Visitation of frugivorous birds, bats, and terrestrial mammals is reduced immediately after a hurricane due to declining food supplies, thus restricting the dispersal of seeds from nearby areas of forest (Boucher 1990; Hjerpe et al. 2001). Indirect effects of a hurricane on food supplies are greater than direct effects of storm winds and rain. Four months after Hurricane Gilbert in Jamaica, populations of 83% of the fruit- and seed-eating birds declined in montane forests compared to 50% of resident insectivorous species (Wunderle et al. 1992; Shilton et al. 2008). In Tongan forests, the density of fruiting trees and the population of flying foxes (*Pteropus tonganus*) were reduced to 15% and 20%, respectively, of precyclone levels 6 months after Cyclone Waka (McConkey et al 2004). Following Hurricane Iris in Belize in 2001, population density of black howler monkeys (*Aloutta pigra*) initially dropped by 42% and continued to decline over 29 months (Pavelka and Chapman 2005). Common Phyllostomid bat species at Luquillo Experimental Forest showed striking declines immediately following Hurricane Hugo but differed in population recovery. *Artibeus jamaicensis* and *Monophyllus redmani* populations rebounded within two years, whereas populations of *Stenoderma rufum* did not recover within three years (Gannon and Willig 1994). Most of these studies examine immediate or short-term effects of hurricanes on frugivore populations; long-term effects are poorly understood.

8.2.2 *Characteristics of Resistant and Resilient Tree Species*

The features of dominant tree species are major determinants of forest response to hurricane impacts. Susceptibility to damage and breakage varies widely among tree species. Following Hurricane Joan in southeast Nicaragua, *Vochysia ferruginea* (Vochysiaceae) suffered complete mortality of adult trees, but rapid growth and high survivorship of seedlings and sprouting saplings quickly restored population sizes. In contrast, all of the trees of *Qualea paraensis* (Vochysiaceae) initially survived the hurricane, although mortality increased in later years; populations were maintained by ingrowth rather than by new recruitment. *Vochysia* was described as a resilient species, whereas *Qualea* was described as a resistant species (Boucher et al. 1994).

Bellingham et al. (1995) classified 20 common tree species into four different groups based on whether trees in montane Jamaican forests sustain low or high damage and whether posthurricane growth rates are low or high (see table 8.1). Only a single species was classified as resilient—*Hedyosmum arborescens* (Chloranthaceae)—with high rates of mortality and high rates of posthurricane recruitment. The largest group (11 species) was classified as resistant, with low rates of damage and low posthurricane recruitment. If many common species are resistant species, forest structure and composition recovers rapidly following a hurricane.

How do functional characteristics of tree species influence trees' susceptibility to damage during hurricanes and rates of posthurricane growth? A general consensus emerging from many studies is that slow-growing trees suffer less damage from hurricanes than fast-growing trees (Zimmerman et al. 1994; Franklin et al. 2004; Ostertag et al. 2005; Canham et al. 2010). Wind-resistant species have,on the one hand, low growth rates, low resprouting capacities, and low height-to-diameter ratios, but, on the other hand, high wood densities, high moduli of elasticity of wood, and high specific leaf areas (Laurance and Curran 2008). Following Hurricane Iniki in Kauai, Hawaii, in 1992, resistance to stem snapping was correlated with a low elastic modulus of wood tissue but not to wood density (Asner and Goldstein 1997). Among six species in rain forests on the Atherton Tablelands in tropical north Queensland, resistance to major structural damage following Cyclone Larry in March 2006 was significantly correlated with stem wood density (Curran et al. 2008). For the same group of species, rates of biomass accumulation were significantly negatively correlated with wood density, suggesting a functional trade-off between resistance to damage and rapid biomass accumulation through fast growth. Early successional species tend to have fast growth rates, low wood densities, and high susceptibilities to physical damage during hurricanes (see chap. 10, sec. 10.4; Canham et al. 2010).

Hurricanes can accelerate invasion of exotic species, as occurred with *Pittosporum undulatum* (Pittosporaceae) following Hurricane Gilbert in montane forests of Jamaica (Bellingham et al. 2005). On the island of Mauritius in the Indian Ocean, cyclone disturbance of evergreen forests promoted invasion of weedy, exotic plant species (Lorence and Sussman 1986). But nonnative species may be less adapted than native species to hurricanes and can therefore suffer greater damage when hurricanes do occur. Vine tangles of nonnative species can negatively influence forest regeneration by competing with native vine species, strangling native tree seedlings, and causing more trees to tip up following subsequent wind storms (Putz and Holbrook 1991; Horvitz et al. 1998).

In places where cyclones strike frequently, the persistence of forests requires that dominant species are resistant to damage. Successional pathways never have an opportunity to reach a relatively stable old-growth stage. Resistance of tree species to hurricane damage is conferred by wood characteristics that tend to favor bending over breaking and that enable resprouting in trees that sustain stem and crown damage. But when cyclones occur with far lower frequency, tree species that colonize rapidly, grow fast, and have lower wood density will become more dominant, even if these species suffer more damage during the next storm that comes along.

8.3 Tropical Forest Regeneration after Single and Recurrent Fires

Forest fires burn when sources of ignition conspire with water deficits and the accumulation of sufficient flammable material. In seasonally dry forests dominated by deciduous tree species, leaf litter accumulates from annual leaf fall, creating a high fuel load; fires occur with relatively high frequency in these forests. In moist and wet tropical forests dominated by evergreen tree species, however, high fuel loads are not typical of most years. Deep root systems tap underground sources of water available throughout the year, redistribute water from lower soil layers to surface layers, and create humid microenvironmental conditions in the forest understory (Nepstad et al. 1994). But even these forests can become susceptible to burning when the forest understory desiccates and dead organic matter accumulates following severe drought (Cochrane 2003), hurricanes, or logging. As discussed in this section, most wet tropical forests are resilient to a single fire event, but repeated and frequent burning can lead to dramatic and permanent changes in forest structure and nutrient cycling.

8.3.1 Fire History and Fire Regimes in Tropical Forests

Fire is no stranger to tropical forests. Tropical evergreen forests show evidence of burning during the Holocene, but in most places fires occurred infrequently, usually at intervals of 500–1,000 years (Sanford et al. 1985; Piperno and Becker 1996; Hammond and ter Steege 1998; Turcq et al. 1998). Severe droughts associated with "mega-Niño" events in Amazonia roughly 400, 700, 1,000, and 1,500 years ago caused widespread fires in evergreen forests that normally do not burn (Meggars 1994; K. E. Clark and Uhl 1987; Uhl 1998). In the Guianas of northeastern South America, forests on well-drained sandy soils have been more susceptible to fire during ENSO periods over the last century, and charcoal is common in soils under well-developed forests (Charles-Dominique et al. 1998; Hammond and ter Steege 1998; Hammond et al. 2006). Along an elevational transect from 50 to 2,600 meters in Costa Rica, soil charcoal was found consistently at every elevation sampled beneath mature rain forests that receive more than 4,000 millimeters of annual rainfall (Titiz and Sanford 2007).

Given the evidence for past fires in forests that otherwise resemble undisturbed forests, the prospects for long-term recovery of forest structure and composition appear to be good for forests that do not suffer from recurrent fires or other large-scale disturbances. But to date, there are no detailed published accounts of long-term recovery of tropical forests after fire. An extensive tract of forest in Kelantan, northeast Malaya, was damaged by a major windstorm followed by a fire in November 1880. Over 70 years later, in 1953, the legacy of this disturbance was still evident in the low diversity of upper canopy species and the concentration of trees in small size classes (Wyatt-Smith 1954).

The fire history of tropical forests can be documented by charcoal evidence in soils, the presence of fire-tolerant vegetation, and fire scars on living trees. Fire-scarred trees of the endemic pine (*Pinus occidentalis*) in the Dominican Republic showed evidence for 41 fire years between AD 1727 and 2002, with fire frequencies ranging from 9.8 to 31.5 years (Martin and Fahey 2006). In Roraima State, Brazil, a recent investigation revealed fire scars in a forest previously thought to be undisturbed by fire (Barlow et al. 2010). Obvious physical deformities caused by fire were observed on 8.2% of the trees of the Maracá Ecological Station. It is not known precisely when this forest burned. Unlocking the fire history of more tropical forest sites will provide critical insights into long-term effects of fires on forest structure, composition, and ecosystem processes.

In seasonal tropical monsoon climates, dry season fires have shaped forest composition, phenology, and productivity (Goldammer 2007). Low-intensity fires burn through forests every few years in strongly seasonal regions of continental Southeast Asia, with little impact on most trees (Baker et al. 2008; Baker and Bunyavejchewin 2009). More than 50% of the tree species in eastern Amazonia in regions that receive 1,700 millimeters of annual rainfall are capable of vegetative resprouting after burning. The high resprouting capacity of tree species in seasonally dry forests may be an adaptive response to more frequent fire regimes than those experienced in less seasonal rain forests (Kauffman 1991; Kammesheidt 1999). In a dry forest in western Nicaragua, all but two species of saplings resprouted in response to an experimental burn treatment and 46.5% of saplings survived (Otterstrom et al. 2006). Tree species that dominate in savannas are distinct from those that dominate in forests, and these species possess suites of plant traits uniquely adapted to savanna conditions, including frequent fire (Bowman 2000; Hoffmann et al. 2012). (Postfire regeneration of woody vegetation in tropical savannas is outside the scope of this book.)

Recent changes in fire regimes in tropical regions present major implications for the future of tropical forests and for the recovery of forests that have burned previously (Laurance 2006; Shlisky et al. 2009). In eastern Amazonia, over half of the forests in two study areas experience fire every 5–10 years (Cochrane 2001). Satellite imagery detected 44,734 separate fires in Amazonia during a four-month period following an extreme drought in 1997 (Schwartzman 1997). Throughout the tropics, fires following extreme El Niño droughts since 1982 were triggered by forest fragmentation, extensive logging activities, and widespread use of fire for land clearance and pasture maintenance (Cochrane and Laurance 2002). These changes in fire regimes—induced by the combination of human impacts and climate changes—have increased the spatial extent of fires, reduced the time interval between successive fire events, and increased the vulnerability of tropical forests to repeated burning (see box 9.2; Barlow and Peres 2004). The increasing frequency of fires in humid tropical forests poses a major threat to forest regeneration, reforestation, and conservation of biodiversity and ecosystem services in existing intact forests and forest fragments.

8.3.2 *Tree Mortality and Postfire Regeneration*

Prior to 1988, little was known about the effects of fire on tropical forest flora and fauna (Uhl, Kaufmann, and Cummings 1988). Early studies focused on initial stages of forest regeneration following shifting cultivation (Uhl et al. 1981). Most of our knowledge about forest regeneration after fires is limited to the initial 10–15 years. Moreover, most studies of the impacts of tropical forest fires have been conducted in only two major regions, Amazonia and Borneo. Rates of fire-related tree mortality vary considerably among forests with different historical fire and rainfall regimes. Trees in seasonal evergreen forests at the margins of the Amazon basin in Roraima, Brazil, and Bolivia show significantly lower rates of postfire mortality compared to interior Amazonian forests (Barlow and Peres 2006a, 2006b; Balch et al. 2011). Few tree species in the Brazilian Amazon are resistant to fire. Most have thin bark and buttresses, characteristics associated with a high sensitivity to fire (Uhl and Kauffman 1990; Pinard and Huffman 1997; Barlow et al. 2003).

Most fires that affect tropical evergreen forests are slow-moving, low-intensity surface fires, with a mean char height below two meters (Barlow and Peres 2004). Small trees and understory vegetation are killed, whereas most large canopy trees survive. But following an extreme drought, large trees become highly susceptible to mortality (Nepstad et al. 2007). In central Amazonia, postfire mortality for trees ranges from 36% to 64%. Tree mortality is lower (8%–23%) following fires in transitional forests along the southwestern edge of Amazonia (Barlow and Peres 2006a; Balch et al. 2011). Mortality of large trees can be delayed for several years, however, leading to considerably increased rates of mortality and biomass loss compared to values determined within one year of burning (Barlow et al. 2003). Tree mortality is strongly size dependent following initial burns, with trees below 30 centimeters in DBH showing higher rates of mortality than larger stems (Woods 1989; Holdsworth and Uhl 1997; Cochrane and Schulze 1999; Haugaasen et al. 2003; Slik and Eichhorm 2003). Bark thickness increases with stem size, conferring greater fire resistance to larger trees (Uhl and Kauffman 1990; Pinard and Huffman 1997; van Nieuwstadt and Sheil 2005; Baker et al. 2008; Balch et al. 2011).

Overall rates of tree mortality in eastern and central Brazilian Amazonia and Borneo increase dramatically following recurrent fires, reaching 80% or higher (Barlow and Peres 2006a). Furthermore, after recurrent fires, all stem sizes become equally vulnerable to mortality from catastrophic heat damage (Cochrane and Schultz 1999; Barlow and Peres 2004). Extreme droughts can also cause significant tree mortality, particularly for large trees, creating highly flammable conditions. In East Kalimantan (Indonesian Borneo), van Nieuwstadt and Sheil (2005) observed increased mortality rates of large trees in unburned forest after the 1997–1998 drought (see plate 14). Drought effects alone accounted for an estimated 30% of large tree deaths in burned forests, leading to a 57% reduction in aboveground biomass. After 21 months, overall mortality following drought and fire averaged 64.2%.

Correcting for drought-induced mortality, species-specific rates of mortality for canopy trees in Sungai Wain in East Kalimantan varied from 5% to 67% 21 months after the 1997–1998 fires (van Nieuwstadt and Sheil 2005). For all species combined, the size-specific mortality rate attributed to fire declined linearly with average bark thickness of trees in that class, and bark thickness increased linearly with DBH across species (fig. 8.1). But bark thickness did not explain across-species variation in tree survival. Rather, species-specific mortality rates were negatively correlated with wood density. Species represented primarily by large trees, such as many Dipterocarpaceae, had low rates of mortality, whereas species represented by smaller stems suffered high rates of mortality. Palm mortality was only 3% after drought and 10% after burning.

As in hurricane-damaged forests, fire-damaged forests regenerate by resprouting, germination of seeds in the seed bank, and germination of newly dispersed seeds. But seedlings and saplings in evergreen forests are usually completely killed by fires, requiring that the understory be entirely reassembled (Slik et al. 2002). Regeneration postfire can therefore be considerably slower than regeneration following hurricanes or logging (Cochrane and

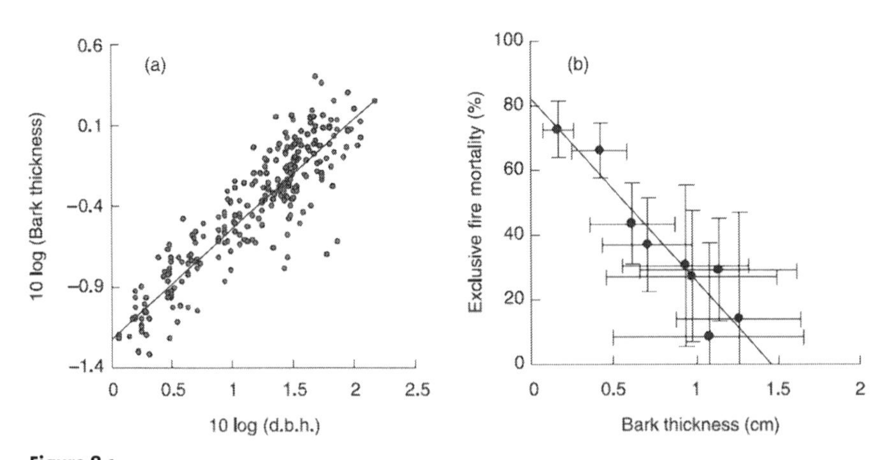

Figure 8.1
Effects of tree size on bark thickness and effects of bark thickness on fire-induced tree mortality in Sungai Wain forest. Source: van Nieuwstadt and Sheil (2005, fig. 7).

Shultze 1999). Early recruitment of seedlings is favored by wind dispersal, which is more predominant among species of dry forests than wet forests (Chazdon et al. 2003; Otterstrom et al. 2006). Although small trees are more susceptible to fire, they are also more likely to resprout shortly after fire damage (Baker et al. 2008). Following the 1997–1998 low-intensity surface fires in East Kalimantan, less than 1% of trees below 8 centimeters in DBH survived, but about 45% of trees above 30 centimeters in DBH survived, ranging from 20% to 95% across species (van Nieuwstadt et al. 2001). Sprouting frequency of small trees that were killed aboveground was about 17% compared to less than 10% for trees above 10 centimeters in DBH. Two years after the fire, 25% of the sprouts had outgrown competing ferns and shrubs and contributed to persistence of many primary forest species. Although seeds in surface soils were killed, seed density in soil layers below 1.5 centimeters of soil was not reduced by fire, producing a dense carpet of pioneer seedlings within four months after the fire in areas with high densities of pioneer trees prior to burning (box 8.2).

A major source of postfire regeneration of mature forest tree species is through seed production by surviving trees in patches of unburned forest or by surviving individuals in burned areas. In central Amazonia, total fruiting tree abundance in once- and twice-burned forest was reduced by 83% and 39%, respectively, compared to unburned forest (Barlow and Peres 2006a). In East Kalimantan and other areas of Southeast Asia, many canopy tree species are mast fruiting and do not produce seeds every year. El Niño droughts are a trigger for mast fruiting of dipterocarps (Curran et al. 1999; Kettle et al. 2011). Many species in Sungai Wain Protection Forest in East Kalimantan fruited in 1997 during the prefire drought and did not fruit again during the first seven years after the fire (van Nieuwstadt et al. 2001; Slik et al. 2011). The lack of a mast fruiting during early postfire regeneration in Sungai Wain reduced levels of regeneration of late successional species in the burned forest.

In a semideciduous mesophytic forest of southeastern Brazil, recovery of floristic richness in a burned forest fragment was rapid, but species composition remained distinct (Rodrigues et al. 2005). In contrast, black-water floodplain (igapó) forests along the Río Negro in northwestern Brazil are neither resistant nor resilient to fires. These low-nutrient forests on white-sand soils have a highly developed root mat and become highly flammable following severe droughts (Flores et al. 2013). Trees exhibit high rates of mortality (75%–100%), and postfire succession occurs at a rate slower than in swidden cultivation fallows in the same type of vegetation. Reasons for the low rate of postfire regeneration in these forests include seed dispersal limitation, fire damage to

Box 8.2. Drought, Fire, and Postfire Regeneration of Rain Forests in East Kalimantan, Indonesia

Severe droughts and forest fires have occurred repeatedly in Borneo in modern as well as prehistoric times (Goldammer and Siebert 1989). The Norwegian zoologist and explorer Bock (1882) noted that about a third of the trees in the Kutai district of East Kalimantan, Indonesia, died as a result of an extended drought in 1878. In Sabah, Malaysia, forest fires in 1914–1915 burned 80,000 hectares of rain forest in an area that now forms the Sook Plain grassland (Cockburn 1974; Goldammer 2007). Numerous fires in Borneo during the 20th century have been associated with periodic droughts linked to El Niño Southern Oscillation (ENSO) events (Leighton 1984; Goldammer 2007; Langner and Siegert 2009).

During the extreme ENSO event in 1982–1983, forests in East Kalimantan received only 35% of the region's average annual precipitation. Extreme drought in combination with land-clearing activities caused fires to burn out of control across the entire island of Borneo. Within the Mahakam basin of East Kalimantan, 2.7 million hectares of tropical rain forests were affected by the 1982–1983 fires, and previously logged forests were most strongly affected (Goldammer 2007). But the fires following the 1997–1998 ENSO event were even more devastating. In East Kalimantan, 2.6 million hectares of tropical forests burned. The impact of these fires was greatest on recently logged forests and peat forests. Moreover, many areas had been previously burned in the 1982–1983 fires and during recurrent fires in the 1990s (Siegert et al. 2001). Ten years of satellite data since 1997 have recorded more than 320,000 active fires on the island (Langner and Siegert 2009). In addition to the expanding presence of human ignition sources, burning coal seams and peat deposits below the soil surface create permanent ignition sources for forest fires whenever a severe drought occurs (Goldammer 2007).

Few long-term studies have examined postfire forest regeneration in Borneo. Five years after the 1982–1983 fires swept through Bukit Soeharto Education Forest, monitoring of forest regeneration was initiated in three areas of lowland dipterocarp forest that were subjected to selective logging prior to 1978 (Toma et al. 2000). In lightly and moderately disturbed plots, the dipterocarp trees that survived the fire still contributed most of the aboveground biomass in 1997, despite rapid growth of pioneer tree species, such as *Macaranga gigantea*, *M. triloba*, *M. hypoleuca*, and *Omalanthus populneus*

(Toma et al. 2000; Toma et al. 2005). Highly disturbed sites showed low rates of aboveground biomass accumulation, as few trees remained after the fire. Based on the observed rates of aboveground biomass change from 1988 to 1997, Toma et al. (2005) estimated that it would take more than 100 years for the lightly disturbed site to reach the level of aboveground biomass present in primary forests of the region (450–500 Mg/ha). Biomass recovery in these forests requires mature forest species to colonize the forest, as pioneer species are short-lived and have low-density wood (Slik et al. 2002; Toma et al. 2005; Hiratsuka et al. 2006).

Drought and fire revisited these forests in 1997–1998. All of the *Macaranga* trees were killed, but vigorous seedling regeneration emerged shortly after the fire. Aboveground biomass accumulation from 2000 to 2003 was strongly dependent on the dominant pioneer tree species *Macaranga gigantea* and *M. hypoleuca* (Hiratsuka et al. 2006). Forest regeneration after two devastating fires resembled secondary succession on bare ground.

Slik et al. (2008) studied forest regeneration in an area about 40 kilometers southwest of Bukit Soeharto Education Forest in a matrix of unburned, once-burned, and twice-burned forests (see plate 14). Species composition of trees changed dramatically after fire, shifting from heavy- to light-wooded tree species. Burned forests showed no significant recovery of composition or aboveground biomass within 7 years. The smallest stems showed strong increases in species diversity and average wood density, indicating rapid recruitment of species characteristic of unburned mature forests. Canopy openness, leaf area index, herb cover, and stem density quickly returned to prefire conditions (Slik et al. 2002; Slik et al. 2008; Slik et al. 2011). Twice-burned forests showed a higher density of early successional shrub and tree recruits than once-burned forests, potentially reflecting the accumulation of early successional species in the soil seed bank over 15 years.

The Borneo forest fires also had long-term effects on wildlife. Three years after the 1997–1998 fires in East Kalimantan, the number of bird species and bird abundance were similar between unburned, once-burned, and twice-burned forests (Slik and van Balen 2006). Species diversity and turnover among sites were lower in burned forests, and species characteristic of closed forest were poorly represented. Fruit availability declined dramatically after the 1997–1998 fires, affecting food sources for obligate and facultative frugivores, such as the sun bear. Fruit availability also declined due to drought-induced tree mortality in unburned forests (Fredriksson et al. 2006).

roots and mycorrhizal associations in the root mat, annual flooding regimes that reduce rates of tree growth and seedling establishment, and repeated burning of low, open patches of vegetation.

Changes in the species composition of birds provide another indicator of postfire forest recovery. The composition of bird species remained distinct for over 25 years in areas where black-water floodplain forests burned (Ritter et al. 2012). Following low-intensity forest fires in central Amazonia, bird species richness recovered within 1 year, but the species composition in burned areas remained highly distinct for over 10 years (Mestre et al. 2013). Burned areas favored nectar feeders, understory frugivores, and omnivores, whereas mixed-flock specialists and ant-following birds were more abundant in unburned areas.

Frequent recurring fires alter the balance between dominance of woody versus herbaceous vegetation. In central Amazonia, the mean density and species richness of herbaceous plants was higher in burned *Vismia* stands compared to nonburned *Cecropia* stands (Ribeiro et al. 2010). Herbaceous species with belowground storage organs, such as *Heliconia acuminata* (Heliconiaceae) and *Calathea altissima* (Marantaceae), are common in both burned and unburned sites. In a lowland dipterocarp forest in southwest Sabah, Borneo, herb and fern cover dominated for the first two years after a fire in 1982–1983. Over time, however, biomass of grasses and herbs lost their competitive advantage and declined, and after eight years, total aboveground biomass reached 24% of the prefire forest biomass. The site was dominated by woody pioneer species, such as *Macaranga* species, along with a few primary forest species that seeded in from nearby mature forest areas (Nykvist 1996). If a fire recurs before the seed bank and seed rain are reestablished, seed limitation can severely limit forest regeneration, leading to arrested succession (see box 7.1). Postfire recovery depends upon secondary forest trees shading out aggressive herbaceous growth (Woods 1989). Wind-dispersed species, such as the grass *Imperata cylindrica* (alang-alang), bracken fern (*Pteridium caudatum*), and bamboo species, can invade the understory and outcompete woody seedlings. These herbaceous species also increase flammability, creating a positive feedback between dominance by herbaceous vegetation and fire (van Nieuwstadt et al. 2001). Frequent burning stimulates the growth of *Imperata*, promoting the conversion of forest to grassland throughout Borneo. Forests that are burned three times within a two-year period can become converted to *Imperata* grassland (Kartawinata 1993). A similar positive feedback has been described for Amazonian forests in seasonally dry climates (see box 9.2).

Similar to regeneration following other forms of disturbance discussed in chapters 6 and 7, postfire regeneration is determined by the availability of seeds and resprouts and by conditions for seedling colonization and tree growth. Changes in climate, seasonal rainfall, and the presence of ignition factors also tip the balance between frequent and infrequent fire events, with critical consequences for the long-term dynamics, composition, and biodiversity of tropical forests.

8.4 Conclusion

Despite elevated tree mortality and massive structural damage, tropical forests are highly resilient to hurricane and fire damage. Following these disturbances, forests rapidly begin to recover through resprouting and recruitment of new seeds. In many cases, recovery of the original species composition is more rapid than during postagricultural succession due to the high incidence of resprouting. But if fires or hurricanes recur frequently, forest structure and composition do not have ample time to recover completely. Legacies of past fires and wind disturbances can be evident for more than a century.

Fires and hurricanes create favorable conditions for colonization of exotic invasive species, which can alter fire regimes and arrest succession. Furthermore, these conditions are exacerbated by logging and by climate change. These synergisms between human and natural disturbances pose enormous challenges for management and for predicting future effects of climate change on tropical forests.

CHAPTER 9

FOREST REGENERATION FOLLOWING SELECTIVE LOGGING AND LAND-USE SYNERGISMS

The more that comes off (or more disturbance) the further back in the process of succession is the stand placed.—J. E. D. Fox (1976, p. 41)

9.1 Overview

Globally, selective logging is a much more pervasive form of forest disturbance than hurricanes or fires. Selective-logging operations occur within an estimated 20.3% of humid tropical forests across the world (Asner et al. 2009). Over 4 million square kilometers (403 million ha) of tropical forests are officially designated for timber production (Blaser et al. 2011). In Indonesia alone, 41.1 million hectares were designated as production forests in late 2000, composing 33.9% of the total forest area (Kartawinata et al. 2001). In Central Africa, 30% of the forest area is under logging concessions (Laporte et al. 2007). Across five timber-producing states in the Brazilian Amazon, 12,075 to 19,823 square kilometers of forest was logged each year between 1999 and 2002, accounting for 60%–120% more forest area than deforestation (Asner et al. 2005).

Forest logging often sets the stage for a host of other interventions and synergisms that can lead to complete deforestation, forest degradation, or conversion of forest to other land uses (fig. 9.1; Kartawinata et al. 2001; Putz and Redford 2010). Nowhere is this more evident than in Malaysia and Indonesia, where logged forests are frequently converted into oil palm plantations (Edwards et al. 2011). In the Brazilian Amazon, from 1999 to 2004, the probability of clearing previous logged forest was two to four times greater than for intact forest at distances 5–25 kilometers from main roads; overall, the mean conversion rate of logged forests was 32.7% over four years (Asner et al 2006). Commonly, 40%–50% of canopy cover is removed during logging, dramatically altering the forest microclimate and increasing flammability (Holdsworth and Uhl 1997; Uhl and Vieira 1989; Cochrane et al. 1999; Siegert et al. 2001). Of the estimated one million hectares of forest in Sabah, Malaysia, affected by fire during 1982–1983 (see box 8.2), 85% were selectively logged forests (Dennis et al. 2001). The impacts of hunting are also higher in selectively logged forests than in unlogged forests in protected areas (Bennett and Gumal 2001; Sethi and Howe 2009).

If logged forests escape burning or conversion to other land uses, prelog-

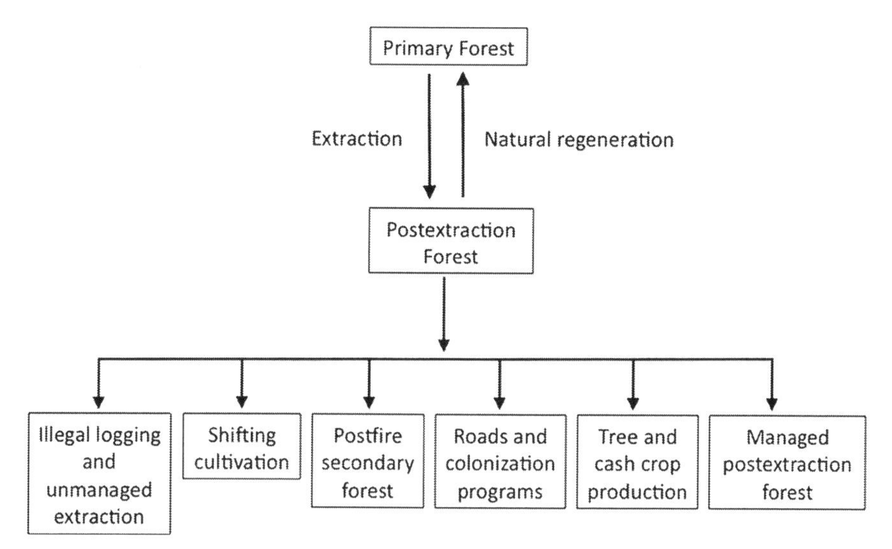

Figure 9.1
Postextraction forests throughout the tropics are subjected to many potential fates that prevent natural regeneration. The most common pathways in Indonesia are shown here. Illegal logging leads to degradation of postextraction forests. Pressures on postextraction forests lead to conversion to shifting cultivation or to tree and oil palm plantations. The largest proportion of postextraction forest is maintained as part of the permanent forest estate. Source: Redrawn from Kartawinata et al. (2001, fig. 1).

ging forest structure and composition gradually reestablish, as also occurs after hurricanes and fires. The good news is that a high fraction of forest-dependent species persists in logged forests. Moreover, natural regeneration processes can restore populations of both commercial and noncommercial tree species given sufficient time (Putz et al. 2012; Edwards et al. 2012). In this chapter, I summarize how tropical forests regenerate after logging and describe interactions between logging operations and other disturbances that can seriously derail natural regeneration processes. Particularly in tropical Asia, partially logged-over forests are described as secondary forests because of the successional processes that are initiated following these disturbances. I emphasize here that forests regenerating after selective logging follow successional trajectories distinct from those associated with shifting cultivation or pasture abandonment. To reduce ambiguity and confusion, the term *secondary forests* should only be applied to those forests regenerating following complete or nearly complete clearance of the original forest (Corlett 1994; Chokkalingam and de Jong 2001).

9.2 Harvesting Intensity, Forest Disturbance, and Postlogging Forest Regeneration

Logging intensity and logging practices vary widely, leading to enormous variation in the extent of tree and soil damage. At the low end of the spectrum, only one to three species of trees are selectively harvested in remote areas of the Amazon basin at an average logging intensity of 0.3 cubic meters per hectare (Putz et al. 2000). At the high end of the spectrum, logging intensities reach 150 cubic meters per hectare in Sabah, Malaysia (Pinard and Putz 1996). Logging intensities are highest in Asian dipterocarp forests, as a large number of species have commercial value (Whitmore 1985). In a typical selective logging operation in Sabah, 8–15 trees are felled per hectare, producing 50–120 cubic meters of timber. Mean logging intensities in Amazonia are 23 cubic meters per hectare, whereas in French Guiana they are considerably lower at 14 cubic meters per hectare (Blanc et al. 2009).

Logging operations, particularly those using bulldozers, lead to extensive damage, including soil compaction along skid tracks and log landings, forest clearance for logging roads, and damage to residual trees during felling (box 9.1 and fig. 9.2). In a typical commercial logging operation, 40%–70% of the residual trees are damaged (Pinard and Putz 1996). Skidding, hauling, and log yarding can disturb as much as 30% of the ground surface. Implementation of reduced-impact logging (RIL) guidelines reduced damage to residual trees less than 60 centimeters in diameter at breast height (DBH) from 41% to 15%, substantially reducing biomass loss and hastening recovery of stand-level productivity (Pinard and Putz 1996). For these reasons, RIL can increase carbon retention during logging (Putz et al. 2008). Sist et al. (2003) showed that tree damage was reduced by 40% using RIL techniques compared to conventional logging practices, particularly skid track damage. Four years after logging in East Malaysia, RIL treatments showed higher density and species richness of regenerating saplings compared to conventionally logged areas (Pinard et al. 2000). Even following RIL, however, trees show an elevated rate of postlogging mortality attributed to tree damage (Mazzei et al. 2010).

As noted by Johns (1988), the term *selective logging* is misleading for logging operations in lowland dipterocarp forests at Sungai Tekam in West Malaysia. Although only 3.3% of the trees were felled as timber, 50.9% of the trees were destroyed. Further, logging operations effectively removed trees at random and did not substantially alter the relative proportions of different tree families sampled before and after logging.

Regeneration of forests following high-intensity conventional logging essentially begins at the stand initiation stage, which is characterized by the colonization and establishment of fast-growing pioneer species. In contrast,

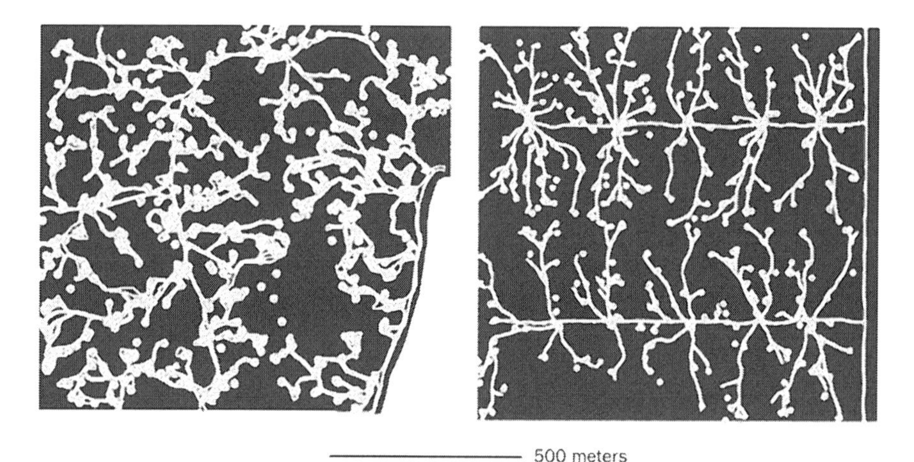

Figure 9.2
Maps of forest areas subjected to conventional (CL, *left*) and reduced-impact logging (RIL, *right*) at Fazenda Cauaxi, Pará, Brazil. Mean gap size after conventional logging was 473 square meters compared to 277 square meters following RIL. Dark gray areas represent residual forest and white areas disturbances by logging gaps, roads, log decks, and skid trails. Source: Schulze and Zweede (2006, fig. 1).

regeneration following low-intensity logging resembles successional dynamics in large treefall gaps within intact forests, which are characterized by accelerated growth of previously established trees and saplings and relatively little colonization by pioneer species. Mazzei et al. (2010) simulated aboveground biomass recovery under three logging scenarios in eastern Amazonia and predicted that prelogging levels would be reached within 15 years following a logging rate of three trees harvested per hectare, within 51 years at six trees per hectare, and within 88 years at nine trees per hectare. In East Kalimantan, recovery of logged forests to conditions similar to unlogged forests is projected to take at least 150 years (Riswan et al. 1985) compared to 45 years in French Guiana (Blanc et al. 2009). On nutrient-poor soils, such as heath forest on white-sand soil, postlogging forest regeneration will be even slower than on nutrient-rich soils (Riswan and Kartaniwata 1988). Variations in logging disturbances, along with other constraints due to seed dispersal and other forms of disturbance, underlie the wide range of recovery times observed and projected for selectively logged tropical forests.

Selective logging contributes substantially to long-term losses of carbon from tropical forest ecosystems. From 1999 to 2002, 15%–19% more carbon was emitted due to selective logging than from deforestation in the Brazilian Amazon (Huang and Asner 2010). In the Brazilian Amazon, Huang and Asner

Box 9.1. Vegetation Succession on Logging Roads and Skid Trails

Logging roads, skid trails, trail edges, logging gaps, and log landings create distinct microhabitats and conditions that differentially affect establishment of plant species. Bulldozer paths typically cover 15%–40% of the ground area in logging operations in eastern Malaysia. Reduced-impact logging guidelines can reduce this damage to 6% (Pinard et al. 2000). In areas with high densities of harvested stems, logging roads traverse more than 20% of the total ground surface and the canopy is completely eliminated, while the forest is left unchanged in other areas (see fig. 9.2; Uhl and Vieira 1989). These heterogeneous conditions create localized variation in disturbance regimes and successional pathways, with long-term effects on tree regeneration and forest structure and composition.

Four years after both conventional and reduced-impact logging in East Malaysia, stem density and species density of woody saplings were significantly lower in skid trail tracks than along track edges or in residual forest (Pinard et al. 2000). Following conventional logging, pioneer tree saplings were most abundant in microsites along the edge of skid tracks, whereas dipterocarp saplings were most abundant in residual forest. Significant effects of skid trails on tree regeneration were still evident 6 and 18 years after logging.

Similar effects of logging roads were observed in tropical wet forests of Costa Rica 12–17 years after abandonment (Guariguata and Dupuy 1997). Logging road tracks supported the lowest densities of stems and species of tree saplings, whereas edge plots supported the highest stem and species densities. For larger trees, the basal area and stem density in logging track plots were 25% of values in edge or forest plots. Stem and species densities of woody species were significantly lower in plots on logging roads compared to adjacent upslope or downslope edges 25 years after logging on Hainan Island, China (Ding and Zang 2009). Four pioneer species showed a significant association with logging tracks. Reduced mean tree diameter and canopy height in logging track plots were attributed to growth under less favorable soil conditions, such as high soil compaction and reduced organic matter.

Environmental conditions, rather than seed dispersal, appear to be the strongest factor limiting woody species recruitment on logging roads. High compaction and reduced soil fertility impede plant establishment, particularly for large-seeded species (Ding and Zang 2009). Seeds and seedlings easily wash away from logging roads and skid tracks following heavy rains. Along the edges of roads, however, tree regeneration can be enhanced by deposition of scraped topsoil and accumulation of seeds from the seed bank.

Pioneer tree species thrive in the localized soil and canopy disturbances created by logging roads and skid trails. Seedling density of the commercial species *Schizolobium amazonicum* in logging gaps was 10 times higher in soils scarified by logging skidders than in unscarified areas (Fredericksen and Pariona 2002). Regeneration of *Ficus boliviana*, *Terminalia oblonga*, and *Ampelocera ruizii* was enhanced along former logging roads in Bolivia compared to forest areas (Nabe-Nielsen et al. 2007). In a selectively logged semideciduous forest in Mexico, soil disturbance by skidders enhanced regeneration of shade-intolerant tree species in logging gaps (Dickinson et al. 2000). Reduced competition from established vegetation and disturbance of soil and litter likely favored establishment of these species.

Over half of Ghana's 214 forest reserves have been selectively logged (Hawthorne et al. 2012). Following selective logging in Ghana, densities of seedling recruits were higher in felling gaps and skid trails than in unlogged areas (Swaine and Agyeman 2008). Mean species density of seedlings was higher in felling gaps and skid trails than in unlogged forest largely due to the influx of pioneer timber species that are uncommon or absent in unlogged areas. After commercial logging operations in Ghana, basal area remained lower in areas directly affected by tree harvesting (apparent extraction network) compared to nonimpacted areas, and pioneer tree species continued to dominate even 30 years after logging (Hawthorne et al. 2012). The current 40-year felling cycle will not permit sustainable harvest.

The differential responses of plant species to soil and canopy disturbance during logging operations in different tropical forests have critical implications for forest management and sustainable timber yields. Sustainable management and silvicultural practices must be tailored to local biological constraints. Unfortunately, most tropical forests are currently being logged at rates two to three times higher than rates that will support sustainable timber production (Zimmerman and Kormos 2012). Initiatives to retain forest carbon stocks, reduce forest degradation, promote sustainable legal extraction of timber, and transfer management rights to empowered local communities are critically needed (Putz et al. 2012).

(2010) estimated a 29% loss in live biomass and a 56% loss of soil carbon following logging. They estimated a mean recovery time of 94 years for wood carbon stocks and a mean recovery time of 51 years for net primary productivity, assuming that forests are not subjected to further disturbance. A meta-analysis of 22 studies revealed that tropical forests retain on average 74% of their aboveground live carbon stocks after they are first logged (Putz et al. 2012). This figure varied from 47% to 97%, depending on harvesting intensity and logging practices.

Most studies of postlogging tree regeneration have focused on commercial tree species from a forest management perspective and have been relatively short-term, usually lasting less than five years. I focus here on long-term studies that track changes in forest structure, diversity, and species composition after selective logging within forest plots or logging treatments. Few studies have followed changes in vegetation structure and species composition within the same stands over time. More often, regeneration patterns emerge from comparisons of different stands at different periods after the logging disturbance, similar to chronosequence studies on unused agricultural lands (for more on chronosequences, see chap. 5, sec. 5.5). These studies suffer from similar limitations, such as differences in initial conditions (including logging intensity), changes in conditions during forest regrowth, and effects of local environmental heterogeneity. Another major problem with studies of logging effects is the lack of proper replication of treatment effects. Among 77 studies of the effects of logging on biodiversity, 68% were clearly pseudo-replicated, lacking replication of treatments across spatially independent study areas (Ramage et al. 2013).

The intensity of logging strongly affects short- and long-term rates of recovery of tree composition. Bonnell et al. (2011) examined forest dynamics over a 37-year period following high, moderate, and light timber extraction in Kibale National Park, Uganda. In 1989, 20 years after logging, early successional species showed increased abundance in the moderately and heavily logged compartments compared to lightly logged and unlogged areas. These compositional differences persisted over the next 17 years, while early successional species disappeared from lightly logged areas. Basal area of lightly logged compartments was not significantly different from unlogged compartments during any census period.

Based on temporal changes in tree basal area, Bonnell et al. (2011) estimated that prelogging conditions would be achieved in 95 and 112 years, respectively, for the moderately and heavily logged compartments. Lightly logged compartments had similar species richness when controlling for stem density, whereas moderately and heavily logged compartments had lower species richness. The severity of logging also influenced the projected time for recovery of abundance and size of food trees for primates. Whereas light logging had no significant effects on primate food sources, recovery of primate food sources was projected to require from 50 to 100 years in moderately logged compartments and up to 158 years in heavily logged compartments. Frugivorous primate species may be more affected by logging than are folivorous species due to the longer recovery time for preferred primate fruit trees.

Despite reductions in stem density and species density (number of species per 0.1 ha plot), conventional logging operations in West Kalimantan, Indonesia, did not cause long-term decline of tree species richness (Cannon et al. 1998). When changes in stem density are taken into account, rarefied tree species richness (number of tree species in a standardized sample of individuals) was actually higher in 8-year-old logged forest than in unlogged forest. These results suggest that logging-induced mortality was concentrated in relatively common species, reducing the total number of stems but eliminating few species. Family-level composition was similar between unlogged and 8-year logged forest. Bischoff et al. (2005) sampled forests 8 and 13 years after logging in Sabah, Malaysia, and found similar species density and species diversity (Shannon index) of trees among logged and unlogged areas. Species composition, however, was affected by logging. Pioneer species contributed substantially higher relative basal area in logged forests compared to unlogged forests, leading to higher mortality and growth rates.

Low-intensity logging treatments in Pará State, Brazil, did not significantly affect the stem density and species density of trees 11 years later, whereas moderate-intensity logging significantly reduced both stem and species density of large trees (Parotta et al. 2002). In contrast, moderate levels of logging in a Bolivian forest had no effect on species richness after 8 years, but new tree recruitment shifted the average values of functional traits (Carreño-Rocabado et al. 2012). In central Amazonia, experimentally logged sites had higher density and species richness of tree saplings 7–8 years after logging compared to unlogged control sites (Magnusson et al. 1999). Species composition was significantly affected by logging and colonization by pioneer species but also varied substantially among control forests. Herb community composition was not significantly affected by logging in these experimental plots but was strongly related to localized disturbances in logging gaps and skidder tracks (Costa and Magnusson 2002).

Because of increased heterogeneity in logging intensity and localized disturbance, logged forests are expected to vary more in tree species composition across a landscape than unlogged forests. Comparisons of tree species composition between logged and unlogged sites are therefore highly dependent on the spatial scale of analysis (Imai et al. 2012). At the local scale (alpha diversity), species richness and diversity of trees did not differ significantly between forests logged 18 years previously and unlogged forests in Sabah, Malaysia (Berry et al. 2008). But variation in species composition at the landscape scale (beta diversity) was significantly greater among logged forests compared to unlogged forests, leading to higher total species richness of trees (gamma diversity) across the entire landscape in logged versus unlogged forests. Landscape-level variation in species composition in logged sites reflected 10-fold variation in logging intensity across sites. Differences in species composition between logged and unlogged forests at the landscape scale were still evident after 18 years; logged forests showed increased abundance of dipterocarps among small trees and pioneer species among large trees compared to unlogged forests. (Berry et al. 2008).

Logging favors the regeneration of timber species and lianas that require high light conditions for establishment. Selectively logged forests in the Ituri forest in the Congo basin in Africa had at least 10 times more seedlings of four species of African mahogany (*Khaya anthoteca, Entandrophragma angolense, E. cylindricum,* and *E. utile*) than unlogged forests (Makana and Thomas 2006). Seedling regeneration of Neotropical mahogany (*Sweitenia macrophylla*) also requires large-scale disturbances (Snook and Negreros-Castillo 2004; Grogan and Galvão 2006). Lianas (woody vines) show rapid recruitment into logging gaps, often resprouting from stems (Pinard and Putz 1994; Schnitzer et al. 2004). Rapid liana colonization can impede tree regeneration (Fox 1976; Schnitzer et al. 2000). Cutting of lianas prior to logging has been shown to significantly reduce liana recruitment following logging (Gerwing and Vidal 2002; Schnitzer et al. 2004). Reduced-impact logging in the eastern Amazon did not sufficiently enhance light availability to promote the regeneration of commercial species, suggesting that additional silvicultural interventions will be needed (Schwartz et al. 2012).

Resprouting is a common means of regeneration of tree species following logging, particularly in seasonally dry forests. In a semideciduous forest in Venezuela logged 5–19 years previously, resprouts were observed in 43% of all species, whereas only 14% of species in unlogged forest had resprouts (Kammesheidt 1999). In a Bolivian dry forest, 45% of canopy tree saplings originated from root or stem sprouts. Root and stem sprouts initially grew faster than seedlings, suggesting that vegetative reproduction is an important mode of tree regeneration following logging disturbances in tropical dry forests (Mostacedo et al. 2009). Of 122 species investigated in semideciduous forest in the Budongo Forest Reserve in Uganda, 119 resprouted in response to cutting or harvesting, predominantly by producing stem sprouts (Mwavu and Witkowski 2008).

These studies clearly show that the effects of logging on forest composition and the rate of postlogging regeneration are directly related to the amount of timber removed and to levels of soil and canopy damage. Patterns of regeneration following intensive logging operations are similar to early stages of succession on cleared land, whereas low-intensity logging leads to regeneration patterns similar to those observed in large treefall gaps within intact forests. Even in cases where local species richness of trees is not significantly different between logged and unlogged forest areas, landscape-level variation in tree species richness, and the composition of vegetation in terms of species and life-forms, can be strongly impacted by logging operations. These impacts have direct implications for ecosystem processes and biodiversity at local and landscape scales.

9.3 Effects of Logging on Animal Abundance and Diversity

When half of the trees in a tropical forest are harvested or die shortly thereafter from logging-related damage, food supply and habitats for animals are drastically reduced. Spider monkeys in Bolivian logging concessions spent 47% of their time feeding in timber tree species and obtained half of their total intake of macronutrients from timber trees. Loss of these critical food resources during logging operations is likely the cause of a 75% decline in spider monkey population density one to two years after logging compared to unlogged forests (Felton et al. 2010). A conventionally logged forest is unlikely to support all of the vertebrate species present before logging but may sustain similar levels of species richness, particularly if logged forests remain connected to larger expanses of forest (Johns 1985). Some primate species compensate for reduction in fruit supply by eating leaves. Following logging in hill dipterocarp forest, lar gibbon (*Hylobates lar*) and banded leaf monkeys (*Presbytis melalophos*) shifted their diet from fruit consumption toward leaf consumption.

Populations of dietary specialists are more likely to be negatively affected by logging than are generalists. The bird species most affected by logging are specialist insectivores, whereas species that thrive in forest-edge or disturbed habitats within the forest increase in abundance following logging, such as generalist feeders and disturbance-tolerant bird species. The number of bird species observed in selectively logged forest areas of West Malaysia increased gradually during the first six years of forest regeneration. In dipterocarp forests of Asia, many pioneer tree species do not produce fruit consumed by birds or mammals. The recruitment of canopy tree species harvested or damaged is therefore required to support the full assemblage of vertebrate frugivores prior to logging (Johns 1985, 1988).

Dunn (2004a) reviewed the effects of tropical forest logging on several animal taxa based on more than 30 different studies. Recent logging had no effect on the standardized diversity of ants, birds, or Lepidoptera (butterflies and moths), nor did the diversity change over time since abandonment. Reduced-impact logging in eastern Amazonia did not affect the abundance of common fruit- and nectar-feeding bats (Castro-Arellano et al. 2007; Presley et al. 2008). A meta-analysis of 109 studies of logging effects on biodiversity revealed modest impacts on species richness for several major taxa (Putz et al. 2012). Species richness of birds in logged forests was, on average, 84% of unlogged forest values, whereas invertebrates, mammals, and plants retained 92%–101% of species richness. These results support the findings of Gibson et al. (2011), which were also based on an extensive meta-analysis, that logged forests have high conservation value, even if species composition is altered.

Similar species diversity or richness between logged and unlogged forests does not imply similar species composition. Felton et al. (2008) sampled bird communities in areas logged 1–4 years previously and unlogged areas in Bolivia. Bird species richness was identical in logged versus unlogged areas, but 20% of the species were exclusive to or significantly more abundant in the unlogged areas. Insectivorous or frugivorous birds were more likely to be significantly associated with unlogged areas, suggesting that they have a higher sensitivity to logging disturbance. The bird fauna in selectively logged forests in Kibale National Park remained impacted 23 years after logging (Dranzoa 1998). Most birds in the logged forest were generalists or forest-edge species, but 84% of the forest interior specialist birds that occurred in old-growth forest recolonized or persisted in the logged forest. Seven understory specialist birds did not recolonize the logged forest area even after 23 years. Avian diversity did not fully recover even 30 years after logging in Peninsular Malaysia. Selectively logged forests contained only 73%–75% of the bird species found in unlogged forests in the region (Peh et al. 2005).

Few studies have compared species composition of different animal taxa before and after logging in the same sites. Azevedo-Ramos et al. (2006) sampled ants, spiders, birds, and mammals prior to logging and six months after reduced-impact logging in eastern Amazonia. Logging had only a minor effect on these taxa; spiders were the only taxon that showed a significant effect of logging on species composition. Mammals showed no change in richness, abundance, or composition. The low impact of logging on these taxa may be due to the connectivity of logged and unlogged sites, which promoted rapid recolonization of logged areas.

Moth species richness was not significantly impacted by logging in the Danum Valley of Sabah, Malaysia, and moth assemblages showed a high similarity between logged and unlogged areas (Kitching et al. 2012). But compositional similarity between assemblages decreased with increasing intersite distances only for unlogged forest areas, suggesting that moth assemblages in logged areas were strikingly homogeneous throughout the region. This study emphasizes the importance of evaluating both local and landscape-level effects of logging on species richness and composition.

Several factors influence the sensitivity of birds and mammals to logging and forest fragmentation. Three variables (number of species per genus, number of subspecies per species, and number of islands in Southeast Asia on which each species occurs) predicted with 79% accuracy the sensitivity of mammal species to logging in Indonesian Borneo (Meijaard et al. 2005). For birds and mammals combined, species negatively affected by logging are often endemic to Borneo or Sundaland, and 69% are frugivorous, insectivorous, or carnivorous. Species positively affected by logging tend to be understory species and Sundaland endemics. Mammals intolerant to logging generally have narrow ecological niches, and many are strictly frugivorous, carnivorous, or insectivorous.

Phylogenetic age is the best predictor of sensitivity of mammal species from Borneo to logging (Meijaard et al. 2008). Species that decline following logging are generally species that began evolving during the Miocene or early Pleiocene. They show little geographic variation in morphology, are usually rare on small islands, and tend to occupy narrow ecological niches (as strict frugivores, carnivores, or insectivores). Species tolerant of logging, in contrast, are younger in origin, dating from the late Pliocene or Pleistocene, are common on small islands and across Southeast Asia, and are herbivorous or omnivorous. Although relatively few mammal species show reduced abundances after logging, management within logging concessions should be modified to provide the maximum value for wildlife conservation (Meijaard and Sheil 2008).

Bird species richness and diversity were significantly lower in logged forests of Sabah, Malaysia, compared to unlogged forests, even after 19 years of natural regeneration. Insectivorous species were consistently less abundant in previously logged forests. Rehabilitation through enrichment planting and silvicultural treatments after logging increased the species richness and diversity of bird species to levels found in unlogged forests. Frugivores were less abundant in rehabilitated forests, and overall bird abundance was reduced compared to unlogged forests, an effect attributed to liana cutting (Edwards et al. 2009; Ansell et al. 2011).

Twenty years after logging in moist, semideciduous forest in Ghana assemblages, the abundance and species composition of leaf litter frogs was indistinguishable from unlogged forests (Adum et al. 2013). This recovery occurred in two phases: disturbance-tolerant species that initially dominated following logging declined in abundance, followed by increasing abundance of forest-dependent species. The relatively low intensity of logging in this study area (3 trees/ha) may have allowed the persistence of forest-dependent species at substantially reduced abundances. For forest-dependent species with low mobility, the persistence of some individuals following logging increases the potential for later resurgence of populations and hastens the recovery of assemblage composition.

Plant and animal taxa vary widely in their long-term responses to logging. In the same study area, Berry et al. (2010) sampled individuals from 11 taxonomic groups and compared species richness between unlogged forests and logged forests that had undergone 19 years of natural regeneration. Herbs, trees, mammals, and dung beetles had higher rarefied species richness in previously logged forest, whereas species richness of birds, ants, termites, and butterflies declined in response to logging (fig. 9.3). With the exception of termites, declines in species richness were within 10%. No consistent patterns of abundance changes in these taxa were observed in response to logging. Over 90% of the species recorded in unlogged forest were also present in logged forest, including species of concern for conservation.

Will the conservation value of logged forests hold up after repeated logging? Repeated logging is thought to have an even greater impact on regeneration than initial harvesting (Whitmore 1990), but few studies exist. In the same logging concession where Berry et al. (2008), Berry et al. (2010), and

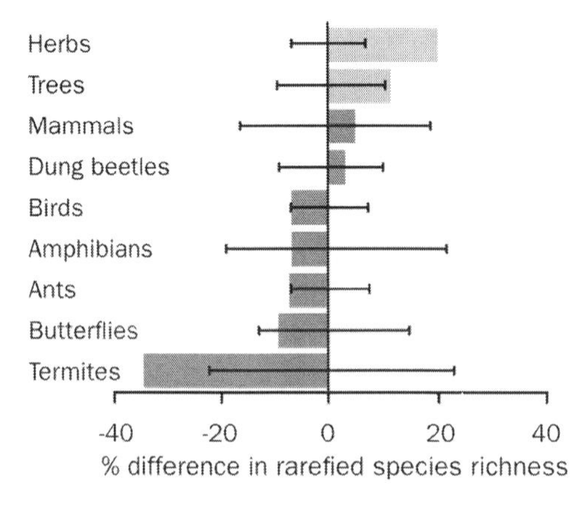

Figure 9.3
The average percent difference in species richness between logged and unlogged forests in Sabah, Malaysia, after 18 years of natural regeneration. A positive difference means more species were found in logged forest. Light bars indicate plant taxa, whereas dark bars indicate animal taxa. Error bars indicate the 95% confidence intervals based on 999 randomizations of individuals assigned to logged or unlogged forests at random. Source: Berry et al. (2010, fig. 2).

Edwards et al. (2009) conducted their comparisons of logging impacts on tree and animal diversity and composition, Edwards et al. (2011) examined effects of a second rotation of harvesting on birds and dung beetles. Twice-logged sites were harvested between 2001 and 2007 and were sampled more than one year after logging. Neither understory birds nor all birds pooled differed in species richness among unlogged, once-logged, and twice-logged forests. Species richness of dung beetles declined by 18% in once-logged forests and by 12% in twice-logged forests. Despite these declines, more than 75% of the bird and dung beetle species from unlogged forest persisted in twice-logged forest. For birds, the second rotation affected species composition more than the first rotation, causing the decline of a number of bird species endemic to Borneo.

Edwards et al. (2012) compared the impacts of a second rotation using reduced-impact logging techniques and conventional logging in the same region of Sabah, Malaysian Borneo, one to eight years after logging. Although 67%–86% of the species of birds, ants, and dung beetles in unlogged forests were found in both types of logged forests after the second rotation, logging led to significant shifts in species composition. Effects of logging on biodiversity of these taxa were similar, suggesting that a second round of logging using reduced-impact techniques does not retain more species than conventional logging. Potential long-term benefits of reduced-impact logging on biodiversity remain to be investigated.

Studies to date suggest that logged forests provide suitable habitats for a high fraction of animal species in tropical regions. Many of these studies are focused on short-term effects and lack suitable levels of replication of treatments. Moreover, few compare species composition before and after logging in the same sites. Clearly, certain types of bird and mammal species are more vulnerable to the effects of logging, such as endemic species, understory specialists, and strictly frugivorous, carnivorous, or insectivorous species. Species with broad ecological tolerances and unspecialized diets are likely to be equally or more abundant in formerly logged forests and regenerating secondary forests than in unlogged forests. Further research is needed to evaluate how local variation of species composition following logging affects landscape-level patterns of species turnover.

9.4 Consequences of Land-Use Synergisms for Forest Regeneration

The indirect effects of selective logging can be even more serious and far-reaching than the direct effects (Uhl and Buschbacher 1985; Putz et al. 2000). Logging roads provide access to formerly remote or inaccessible forest areas, opening them up to colonization, land clearance, and increased fire frequency (see fig. 9.1). This process can also lead to forest fragmentation as blocks of forest become isolated. Logging roads also serve as conduits for the harvest and transport of bushmeat to growing populations of consumers and as pathways for the introduction of invasive plant species.

Selective logging opens up forest canopy, decreasing relative humidity near the forest floor and increasing susceptibility to fire (Uhl and Kauffman 1990; Cochrane and Schulze 1999). In Matto Grosso State (southwestern Brazil), the total amount of forest disturbed by selective logging and fire increased from 5.4% in 1992 to 40.1% in 2004 (Matricardi et al. 2010). Based on the detectability of disturbed forests in satellite imagery, observations show that recovery in logged areas that burned took longer (3–10 yr.) compared to logged forests that did not burn (3–5 yr.). The rate of fuel drying in eastern Amazonian rain forests was affected by canopy openness, time since logging, and logging techniques. A large logging gap (> 700 m^2) became susceptible to burning after only six days (Holdsworth and Uhl 1997). After four years of postlogging forest regrowth, fuel moisture conditions were similar to those observed in old-growth forest, suggesting that fire susceptibility declines strongly over time after logging. Reducing the size of logging gaps through low-impact logging can lower the risk of postlogging fires.

Hunting is commonly associated with timber extraction in community forests and logging concessions (Bennett and Gumal 2001). Despite the widespread recognition that logging indirectly leads to increased hunting in tropical forests (Peres et al. 2006), few studies have quantified this relationship (Fimbel et al. 2001). The main protein source for workers in logging camps in Sarawak is wild meat; one camp consumed yearly approximately 29,000 kilograms of wild meat, mostly bearded pig (Bennett and Gumal 2001). During 1996 in Sarawak, Malaysia, more than 1,000 tons of bushmeat were transported out of the forest primarily using logging roads (Robinson et al. 1999). Commercial logging operations attract large numbers of immigrant workers into the forest or into new urbanized areas near forest concessions. After the civil war in Congo in 1997, the commercial logging sector underwent a massive expansion. Timber concessions in the northern Republic of Congo adopted low-intensity, selective, reduced-impact logging techniques on 12,000 square kilometers (Poulsen et al. 2009). Populations in five logging towns grew by 69.6% from 2000 to 2006, and the corresponding biomass of bushmeat consumed increased by 64%. Most of the bushmeat was hunted by immigrants who settled in the area to work in the logging concessions.

Hunting selectively removes large birds and mammals that are important seed dispersers and seed predators in tropical forests, with direct consequences for tree regeneration and forest dynamics (Wright, Stoner, et al.

2007; Stoner, Vulinec, et al. 2007; Terborgh et al. 2008). Assemblages of frugivorous vertebrates varied significantly across a gradient of forest disturbance in Uganda, with a reduction in specialist frugivores in the most heavily logged and fragmented forest area (Babweteera and Brown 2009). These changes affect the quality of seed dispersal and dispersal distances. The loss of large-bodied specialist frugivores is likely to affect recruitment of many tree species that rely on dispersal of seeds away from parent trees.

The loss of large frugivores in logged and hunted forests in northeastern India negatively impacted seedling recruitment and regeneration of large-seeded tree species, (Sethi and Howe 2009; Velho and Krishnadas 2011). The abundance of large-bodied avian frugivores decreased in logged forests, whereas abundance of small-bodied avian frugivores was unaffected or increased slightly following logging (Velho et al. 2012). Frugivorous birds had lower visitation rates to large-seeded fruiting trees in logged forests than in unlogged forests. Among biotically dispered tree species, large-seeded species had fewer recruits (stems < 8 m in height) in the logged areas, but small-seeded species showed no significant logging effect (fig. 9.4). Reduced recruitment of large-seeded trees likely reflects the compounding effects of hunting and selective logging.

Logging also opens the door to invasive exotic species. The strawberry guava, *Psidium cattleianum* Sabine, was introduced to Madagascar from South America in 1806 and spread throughout southeastern Madagascar where it proliferates in logged forests and other disturbed areas (Brown and Gurevitch 2004). Invasive species, including *P. cattleianum* (Myrtaceae), *Clidemia hirta* (Melastomataceae), *Eucalyptus robusta* (Myrtaceae), and *Lan-*

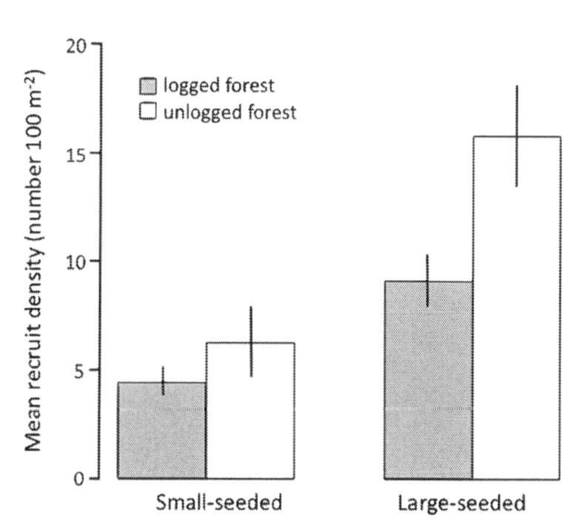

Figure 9.4
The mean density (± 1 standard error) of recruits (seedlings and saplings < 8 m in height) in two logged and two unlogged forests in northeastern India. Unlogged forests were protected from hunting. Small-seeded species have seeds below 10 millimeters in diameter; large-seeded species have seeds greater than 10 millimeters in diameter. Source: Velho et al. (2012, fig. 7).

tana camara (Verbenaceae), constitute from 52% to 83% of the trees in five logged forest stands. Despite 50–150 years of abandonment, species richness has not recovered in these logged stands, but it has quickly recovered in the three years following a cyclone that blew down a forest that had no exotic invasive species. The high concentration of invasive species in these logged forests is likely inhibiting recovery of forest composition and species richness.

The invasive shrub *Chromolaena odorata* (Asteraceae) was introduced to tropical Asia, West Africa, and parts of Australia, and is a major weed in selectively logged forests and swidden fallows (see box 7.1). In a tropical dry forest in Thailand, *C. odorata* dominates the understory of forests subjected to intensive logging, providing increased supplies of floral nectar to butterflies (Ghazoul 2004). As a consequence of the abundance of this invasive species, *Dipterocarpus obtusifolius* receives substantially lower levels of pollination, reducing its reproductive output. A similar scenario may explain declining reproductive output of trees in southern Asian forests dominated by the invasive shrub *Lantana camara* (Verbenaceae), which is native to the Neotropics.

Commercial logging operations are credited with the introduction of the little red fire ant (*Wasmannia auropunctata*) into the interior of forests in Gabon, where it is causing blindness in a wide range of wildlife species (Walsh 2004). The ants were absent from Lope Forest Reserve until logging roads were cleared in the 1970s. They likely hitched a ride on a logging vehicle. Logging vehicles also transport seeds of several invasive grass species that proliferate on log landings and logging roads in eastern lowland Bolivia (Veldman and Putz 2010). The most abundant invasive grass species in this region is *Urochloa maxima*, and several other exotic grass species are dispersed distances of over 30 kilometers by flatbed logging trucks that move among logging areas. Over 60% of log landings in this forest area contain alien grasses, greatly increasing fire risk (fig. 9.5 and box 9.2).

The effects of selective logging and associated silvicultural practices on vertebrate-mediated seed dispersal, seed predation, and seedling recruitment of trees are poorly understood (Jansen and Zuidema 2001). As 70% of woody plants in tropical wet forests and 35%–70% in tropical dry forests produce seeds dispersed by vertebrates, it is likely that indirect effects of logging on hunting will have a major impact on regeneration of many tree populations and will alter the structure and community composition of tropical forests.

9.5 Conclusion

In contrast to conversion of tropical forests to agriculture, logged forests retain much of the biodiversity and biomass of the original forests.

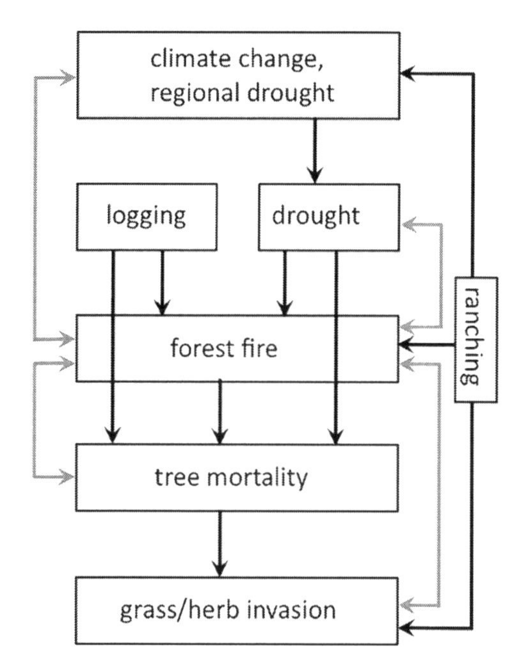

Figure 9.5
Interactions among climate change, logging, drought, and forest fire that lead to positive feedbacks and Amazon forest dieback. Black arrows indicate direct effects, whereas gray arrows indicate positive feedback loops. Source: Redrawn from Nepstad et al. (2008, fig. 2).

Results from studies across the world's tropics show that if these forests are allowed to regenerate without further disturbance, a high fraction of local forest-requiring species will return, and biomass and stored carbon will recover their prelogging levels. Forests in wetter regions (> 1,500 mm annual rainfall) will recover faster than in drier regions (Huang and Asner 2010). But many logged forests will not be given this option; they will be logged again, burned, fragmented, hunted, or converted to farms or plantations. The popular view is that once a forest is "degraded," its value is reduced and it becomes a low priority for conservation or protection from hunting. In many tropical regions, loss of logged forests means not only the loss of virtually all remaining forest areas, but also the loss of forest-dwelling wildlife, and the loss of the diverse ecosystem services that logged forests would continue to provide.

Logged forests are worth much more than their standing timber value. They have high conservation value, particularly when compared to plantations of exotic tree species or oil palms. They are farther along in their successional trajectories than shifting cultivation fallows or former pastures and are more diverse and similar in structure and composition to old-growth forests. As active logging concessions currently extend over 200,000 square

Box 9.2. Multiple Disturbances, Positive Feedbacks, and Forest Dieback in Amazonia

In Amazonia, frequent fires and forest degradation through logging and fragmentation are driving a process referred to as forest dieback. The process begins with tree mortality induced by selective logging, drought, and fire (Barlow and Peres 2008; Nepstad et al. 2008). Selective logging opens up the forest canopy, produces slash on the forest floor, decreases the relative humidity in the understory, and favors the colonization of invasive grasses and herbs. These conditions transform a normally fire-resistant ecosystem into a fire-prone ecosystem that is readily ignited during a severe dry season drought. Ignition sources are plentiful, as many forest areas are adjacent to pastures that are burned annually. In highly fragmented tropical landscapes of Paragominas and Tailândia in eastern Brazilian Amazonia, edge effects significantly increase susceptibility to fire; over 90% of forest areas affected by fire were contacted by one or more forest edges, and most burned forests occurred within 500 meters of a forest edge adjacent to a cattle ranch (Cochrane 2001; Zarin et al. 2005). Within the Paragonominas study area, one to four fires burned every 20 years.

Synergistic effects of logging, drought, forest fragmentation, and fire increase the flammability of forests, exposing them to recurrent burning in a positive feedback loop leading to increasing tree mortality, greater canopy openness, increased penetration of light to the forest floor, invasion by fire-promoting grasses and herbs, and repeated burning (see fig. 9.5; Cochrane and Schulze 1999; Nepstad et al. 2008). The end result is a rapid conversion of a forest ecosystem to a fire-prone grass- or scrub-dominated ecosystem. Within Amazonia, some areas of northeastern Mato Grosso, southeastern and eastern Pará State, and near Santarem indicate that this transformation is already occurring (Nepstad et al. 2008).

The flammability of native and exotic grass species creates positive feedbacks between fire and grass invasion in seasonally dry tropical forests. Balch et al. (2009) described a positive feedback loop that promotes grass invasion in burned forest edges in transitional forest 30 kilometers north of the forest-cerrado boundary in Mato Grosso State, Brazil. The extent and probability of invasion of grass species increased with recurrent fires. Once fire initiates grass invasion, increased grass fuel loads and drier understory conditions create a positive feedback that perpetuates grasses along forest edges.

In a selectively logged Bolivian tropical dry forest, cover of the exotic invasive grass *Urochloa maxima* was six times higher in an experimentally burned forest compared to unburned forest (Veldman et al. 2009). The native fire-adapted bamboo species *Guadua paniculata* increased in abundance in burned areas of the Chiquitano dry forest region in Bolivia, forming dense stands and aggressively resprouting in repeatedly burned areas (Veldman 2008). Repeated fires shift the competitive balance away from tree regeneration toward bamboo dominance (see box 7.1)

In less seasonal Amazonian forests, recurrent fires can transform closed-canopy forests into more open forests dominated by species typical of young secondary forests, a processed termed *secondarization* by Barlow and Peres (2008). In the Arapiuns River basin in central Amazonia, species composition in postfire regeneration was affected by the number of burns, reducing the representation of mature-forest tree species after increasing cycles of burning. Barlow and Peres (2008) proposed that recurrent fires in this region can lead to a shift from closed-canopy mature forests to more open forests dominated by short-lived pioneer species. The abundance and diversity of shade-tolerant, slow-growing species typical of unburned forests are projected to decline due to this changing fire regime.

Several recent climate change scenarios for the Amazon basin emphasize the vulnerability of forests to increasing drought stress and fire risk, potentially causing large-scale forest dieback, particularly in forest areas adjacent to savanna bioclimatic zones (Nobre and Borma 2009; Malhi et al. 2009). Based on satellite imagery between 1996 and 2002, approximately 28% of Brazilian Amazonia faces incipient fire pressure, which is defined as being within 10 kilometers of an ignition source (Barreto et al. 2006). Direct interventions to reduce fire risk and forest fragmentation and to reduce global greenhouse-gas emissions can steer Amazonian forests away from this ominous tipping point (Malhi et al. 2009).

kilometers of Borneo, conservation requires careful attention to the species living in logged forests (Meijaard and Sheil 2007; Edwards et al. 2011). As stated by Johns (1985, p. 371), "Undoubtedly logged forest is less pleasing aesthetically than primary forest, and may be of little use as a recreational amenity, but it is of great potential value in the long-term conservation of rain-forest animal species."

CHAPTER 10

FUNCTIONAL TRAITS AND COMMUNITY ASSEMBLY DURING SECONDARY SUCCESSION

Knowledge of seed biology is essential to understanding community processes like plant establishment, succession, and natural regeneration.—C. Vázquez-Yanes and A. Orozco-Segovia (1993, p. 69)

10.1 Overview

Life-history theory predicts that species adapted to ephemeral, high-resource environments should exhibit fast growth, small size at reproduction, and rapid completion of their life cycles (MacArthur and Wilson 1967). The turnover of woody species during tropical forest succession follows a general theme of replacement of shade-intolerant, fast-growing species with shade-tolerant, slower-growing species (Zhang et al. 2008). Much of this replacement takes place relatively early in succession beneath the developing canopy of short- and long-lived pioneer trees, which themselves become replaced by shade-tolerant and gap-requiring tree species during later stages of succession. Changes in species composition and functional characteristics in the forest understory that are often observed during relatively early successional stages foreshadow subsequent changes in the canopy that occur over decades or centuries (Chazdon 2008b; Norden et al. 2009; Chai and Tanner 2011).

"Pioneer" and "shade-tolerant" trees define the extremes of a continuous life-history spectrum (Swaine and Whitmore 1988; see box 5.1). The contrasting lifestyles of fast-paced pioneers and slow-paced shade-tolerant species illustrate a fundamental ecological trade-off between growth rates in high light and survival rates in low light based on physiological and morphological constraints that underlie specialization (box 10.1). The evolution of the pioneer lifestyle has occurred multiple times within different lineages, giving rise to a diversity of solutions to the ecological challenges of colonizing open fields or cleared land. Similarly, the challenges of seedling establishment, growth, and survival beneath a dense forest canopy require particular functional traits that have evolved within most lineages of higher plants. During the long successional transition from old fields to old-growth forest, changes in species composition reflect the inevitable tension between the filtering effect of ecological specialization across multiple lineages and the diversifying effect of adaptive radiation within lineages. Because many functional traits that influence plant performance are conserved within plant genera or

Box 10.1. The Growth-Survival Trade-Off in Tropical Forests

Among the mechanisms that promote coexistence of large numbers of tree species in tropical forests, the trade-off between species maximum growth rate and mortality rates has received major attention (Grubb 1977; Rees et al. 2001; Wright et al. 2003). This fundamental axis of species variation is thought to underlie patterns of niche differentiation when coupled with spatial heterogeneity in resource availability (Hubbell 1998; Kitajima and Poorter 2008). Successional turnover of species can also be driven by similar patterns of niche differentiation, as resource availability changes dramatically over time (Chazdon 2008b; Chazdon et al. 2010).

Demonstrating clear growth-mortality trade-offs among tropical tree species is challenging, as both growth and mortality rates are highly dependent on plant size and on local resource availability. The growth-survival trade-off is most apparent when considering early life stages. Kitajima (1994) first demonstrated growth-mortality trade-offs among tree seedlings on Barro Colorado Island, Panama. Species with high rates of early seedling mortality in a shaded greenhouse had the highest rates of early seedling growth. The growth-mortality trade-off was observed among seedlings of pioneer trees on Barro Colorado Island (Dalling et al. 1998) but was not evident among species of shade-tolerant trees in North Queensland, Australia (Bloor and Grubb 2003). Gilbert et al. (2006) extended Kitajima's findings to naturally occurring seedlings and small saplings of trees and lianas on Barro Colorado Island.

Tree saplings on Barro Colorado Island also showed a significant trade-off between diameter growth and mortality across 73 species (Wright et al. 2003). In a lowland moist forest of Bolivia, mean rates of height growth were negatively correlated with survival rates among 53 species of saplings (Poorter and Bongers 2006). In lowland mixed dipterocarp forest of Sarawak, Malaysia, rates of diameter growth of 11 *Macaranga* (Euphorbiaceae) species in high light were positively correlated with rates of low-light mortality (Davies 2001).

Species-specific variations in functional traits underlie the observed demographic trade-off between growth and survival of seedlings and saplings. Variation in leaf traits explained variation among species in rates of height growth in gaps and survival in shaded understory (Poorter and Bongers 2006; Sterck et al. 2006). Species differences in the density and structure of wood tissue are also expected to mediate the growth-mortality trade-off (King et al 2006; Kitajima and Poorter 2010; Poorter et al. 2010; Wright et al. 2010). In lowland tropical forests of Brunei, Borneo, wood density was significantly negatively related to both tree mortality and diameter increments. But the direct relationship between diameter growth and annual mortality was insignificant, in part because of a strong effect of adult stature (maximum species height) on diameter growth rates (King et al. 2005; Osunkoya et al. 2007).

Growth-survival trade-offs among tree species are more difficult to demonstrate. When data from five Neotropical forests were combined, a weak but statistically significant relationship between growth and mortality was obtained, but within forests the relationship was significant for only one site (Poorter et al. 2008). Wright et al. (2010) found a highly significant relationship for saplings and trees on Barro Colorado Island using the 95th percentile for relative diameter growth rate and the mortality rate of 25% of the slowest-growing individuals of 103 species. The growth-survival relationship was weak among large trees, with a maximum coefficient of determination of 10%.

Three factors may explain the weak growth-survival trade-off in trees, particularly during tropical forest succession. First, wood density, stiffness, and safety margins *decrease* with increasing adult stature among shade-tolerant species but *increase* with increasing adult stature among pioneer species in tropical rain forests. Large, long-lived pioneer species must be more strongly built than smaller, short-lived pioneer species. In contrast, small, shade-tolerant tree species are confined to the understory for their entire life and must be more strongly built than tall species to withstand damage by falling debris (van Gelder et al. 2006).

Second, high growth rates of pioneer species are rarely sustained beyond the stand initiation stage, as neighboring trees simultaneously increase in both size and density, increasing competition for both above- and below-ground resources (Chazdon et al. 2007). Strong effects of local neighborhood competition can overwhelm species-specific effects of individual tree size on diameter increments.

Third, during secondary succession there is a temporal—as opposed to spatial—separation of conditions under which species exhibit maximum rates of growth and mortality. Short-lived pioneer species exhibit maximum growth rates during the stand initiation phase and maximum mortality rates during the stem exclusion (thinning) phase a decade or more later (Chazdon et al. 2010). Mortality of long-lived pioneers during the stem exclusion phase is concentrated in small size classes that exhibit poor or no growth under suppression (Chazdon et al. 2005; Palokami et al. 2006; Chazdon et al. 2010).

Box 10.1 (*continued*)

The growth-survival trade-off among tree species is a major driver of successional changes in species composition, restricting the recruitment of pioneer species to early stages of succession and favoring the recruitment of shade-tolerant species during stand initiation, stem exclusion, and understory reinitiation phases. Beyond the seedling and sapling phases, however, trade-offs between species-specific functional traits that influence rates of tree growth and survival are strongly mediated by effects of tree stature and local neighborhood structure on resource availability.

families, different phylogenetic lineages tend to dominate during early versus late stages of forest succession (Moreno et al 2009; Letcher 2010; Norden et al. 2012; Letcher et al. 2012).

This chapter examines the life-forms, functional traits, and functional groups of plants that predominate during different successional stages and describes patterns and mechanisms of species turnover during succession. Species functional traits, as expressed under prevailing environmental conditions, influence rates of growth and survival during seedling and postseedling stages. These demographic rates drive changes in the structure and composition of species assemblages during succession. Tree longevity varies among species, affecting rates of change in the composition of assemblages during succession. Assemblages change rapidly initially in succession because of the early establishment and dominance of short-lived pioneer species. Changes in assemblage composition during succession strongly influence ecosystem functioning and are associated with changes in species interactions during tropical forest succession.

10.2 Environmental Gradients during Succession

Dramatic changes in environmental conditions and resource availability characterize different stages of tropical forest succession and impose strong filters on species establishment, growth, and recruitment. High levels of solar radiation, photosynthetic photon flux density, high air temperatures, and high evaporative demand characterize old fields and large clearings (Bazzaz and Pickett 1980; Chazdon and Fetcher 1984; Chazdon et al. 1996). These conditions favor the establishment and rapid growth of pioneer species, but they usually prevail for only for a brief period following disturbances or agricultural abandonment.

During fallow succession near San Carlos de Río Negro, Venezuela, daily photosynthetic photon flux density declined from 35 to below 5 moles of photons per square meter per day after 9 years (Ellsworth and Reich 1996). Within 15 years of pasture abandonment, transmittance of diffuse photosynthetic flux density in the understory was, on average, less than 1% of levels above the canopy in tropical wet forests of northeast Costa Rica, which was similar to levels in old-growth forests of the region. Due to more homogeneous, even-aged forest cover, spatial variation in light availability is low in second-growth forests, as canopy gaps are absent and understory vegetation is uniformly dense (Nicotra et al. 1999; Montgomery and Chazdon 2001).

In tropical dry forests, decreases in air temperature, soil temperature, and relative humidity accompany reductions in light availability during succession (fig. 10.1; Lebrija-Trejos et al. 2011). In dry tropical second-growth forests of Oaxaca, Mexico, the fraction of canopy photon flux density reaching the understory declined from 74% to 10% during the first 10 years of succession.

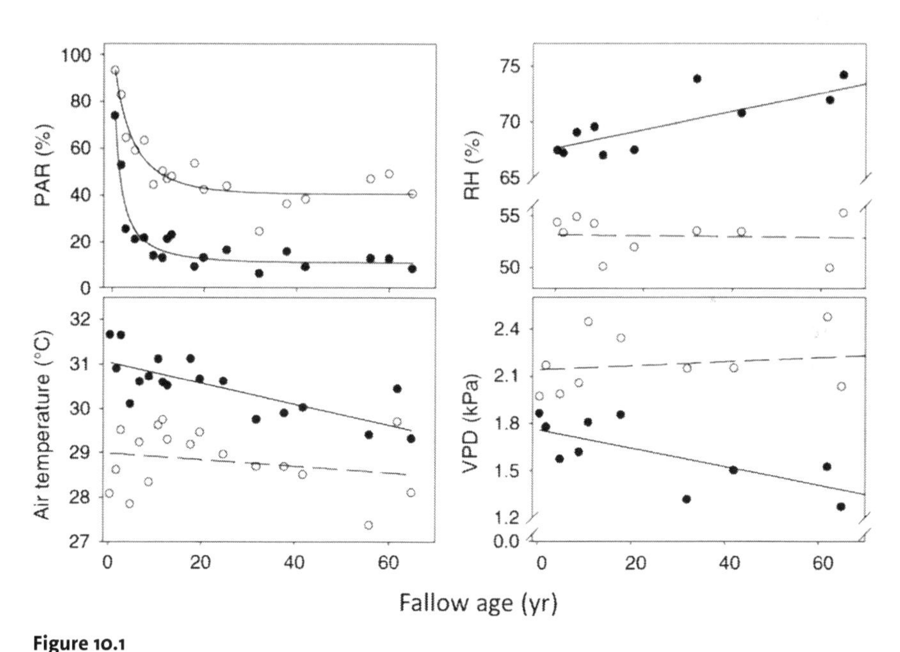

Figure 10.1

Changes in environmental conditions during a 60-year chronosequence of secondary succession in 17 dry forest sites following shifting cultivation. Environmental variables were measured during the dry season (open circles) and the wet season (closed circles). PAR is photosynthetically active radiation, RH is relative humidity, and VPD is vapor pressure deficit. Regression lines were fit using best-fit models; dashed lines indicate nonsignificant relationships. Source: Adapted from Lebrija-Trejos et al. (2011, fig. 1).

Air and soil temperatures declined linearly with increases in tree basal area along a 60-year chronosequence. Environmental changes are more pronounced across wet seasons, corresponding with the period of more active plant growth.

In areas where land use and poor management have caused soil erosion, spontaneous natural regeneration improves soil conditions. On degraded soils in Guangdong, China, regeneration was associated with increasing soil organic matter, increasing nitrogen availability, and declining soil bulk density (Duan et al. 2008). Pioneer species must be able to tolerate harsh abiotic conditions, particularly during seasonal droughts. Improvement of soil texture, water-holding capacity, and fertility facilitate the colonization of species less tolerant of belowground stresses and more tolerant of aboveground shading.

10.3 Successional Changes in Life-Form Composition

Although trees form the structural matrix of tropical forests, a large component of plant diversity consists of other life-forms. During the stand initiation phase, light-demanding species varying in life-form and stature colonize, including short-lived pioneer trees, light-demanding shrubs, grasses, large-leaved herbaceous species, herbaceous vines, woody lianas, and long-lived pioneer trees (Ewel and Bigelow 1996). As succession proceeds and the understory becomes more moist and shaded, shade-intolerant species decline in relative abundance in the understory, while shade-tolerant species in a wide range of life-forms increase in relative abundance (Chazdon 2008b; Muñiz-Castro et al. 2012).

The large-leaved herbs, ferns, grasses, and palms that dominate following large-scale disturbances strongly affect microenvironmental conditions for seedling establishment of woody species (Denslow 1978, 1996; Ewel 1983; Dupuy and Chazdon 2006). Removal of understory vegetation (including lianas, shrubs, ferns, and large herbs) significantly reduced mortality of naturally recruiting tree seedlings in experimental gaps within young second-growth forests of northeast Costa Rica (Dupuy and Chazdon 2008). In the Atlantic Forest of Brazil, the early successional palm *Attalea oleifera* negatively affected seedling density and species richness beneath its crown (Aguiar and Tabarelli 2010). In dry forest regions, shrub establishment may be impeded by dense growth of grasses and herbs. Grasses and herbs dominated the understory in young shifting cultivation fallows in a Bolivian dry forest, but their cover declined dramatically across a 50-year chronosequence (Kennard 2002).

Along with pioneer trees, woody shrubs rapidly colonized old fields in lowland wet forests of Costa Rica. The relative abundance of shrubs was significantly higher in young second-growth forests (10–15 yr. postabandonment) than in old-growth forests, whereas the relative abundance of understory and canopy palms was low and increased during succession (Guariguata et al. 1997). These trends were also confirmed within long-term study plots over time in regenerating forests (Capers et al. 2005). A more extensive chronosequence study in this region showed that large-leaved herbs were common only in second-growth forests less than 20 years old, whereas the relative abundance of canopy palms, understory palms, and canopy tree species increased with forest age class (Letcher and Chazdon 2009b). In shifting agriculture fallows of eastern Madagascar, shrubs and herbs dominated during the first years after abandonment but declined in both abundance and species richness as tree seedlings, saplings, and adults increased in abundance and species richness (Klanderud et al. 2010).

Lianas (woody vines) are most abundant during early stages of forest succession in the tropics. Lianas can negatively affect the development of tree species during forest succession (Uhl, Buschbacker, and Serrão et al. 1988; Zahawi and Augspurger 1999; Schnitzer et al. 2000; Paul and Yavitt 2011). Although liana abundance declined with stand age in a chronosequence in a moist forest of Panama, the basal area of individual lianas increased with stand age (DeWalt et al. 2000). Relative abundance of lianas declined across a Costa Rican chronosequence, but total liana biomass increased, reaching its highest level in old-growth forests (Letcher and Chazdon 2009a, 2009b). Lianas composed 31% of all stems greater than two millimeters in diameter at breast height (DBH) in young fallow vegetation compared to 23% in 12- to 25-year-old regrowth in central Amazonia (Gehring, Denich, and Vlek 2005). In a tropical dry forest in southeastern Brazil, liana density declined from intermediate to later stages of succession (Madeira et al. 2009). The abundance of lianas in regenerating forests is strongly affected by previous land use. In second-growth forests dominated by *Cecropia* species after clear-cutting, liana density was higher than in *Vismia*-dominated second-growth forests subjected to repeated burning in central Amazonia (Roeder et al. 2010).

The structural and taxonomic diversity of vegetation increases during succession. Life-form diversity was higher in old-growth than in 40-year-old secondary riparian forests of the Cordillera Central in the Dominican Republic (Martin et al 2004). Arborescent ferns, canopy palms, vascular epiphytes, and bryophytes were significantly more abundant in old-growth forests, whereas vines and woody lianas were significantly more abundant in second-growth forests. Midstory and understory trees (treelets) had significantly higher densities in the understory of old-growth forests compared to second-growth stands. Old-growth forests in the wet lowlands of Costa Rica had a higher abundance of understory tree species than second-growth forests (Chazdon et al. 2010).

Species richness of vascular epiphytes during succession is positively associated with the diversity of trees and microclimatic conditions that occur during forest regeneration (Barthlott et al. 2001; Cascante-Marín et al. 2006; Benavides et al. 2006; Woods and DeWalt 2013). In a lowland Amazonian forest of Amacayacu National Park in Colombia, the species richness of holo-epiphytes (species not rooted in forest soil) and hemi-epiphytes (species rooted in forest soil) increased significantly across a chronosequence of shifting agriculture fallows 2–30 years old, developing the highest species richness in old-growth forest (Benavides et al. 2006). The density and species richness of vascular epiphytes increased gradually during succession in tropical moist lowland forest in Panama (fig. 10.2; Woods and DeWalt 2013). In 115-year-old second-growth stands, species richness was equivalent to old-growth stands, but density was far lower and only reached 49% of old-growth levels. Epiphyte density is slow to recover and reflects the high degree of dispersal limitation and substrate specificity in epiphytic vascular plants. Similarity of species composition to old-growth assemblages increased over time, reaching 75% in 115 years. In a Venezuelan Andean forest, the composition of vascular epiphyte assemblages differed markedly between old-growth forest and 23-year-old second-growth forest, with fewer orchid species but more Bromeliaceae species in the second-growth forest (Barthlott et al. 2001).

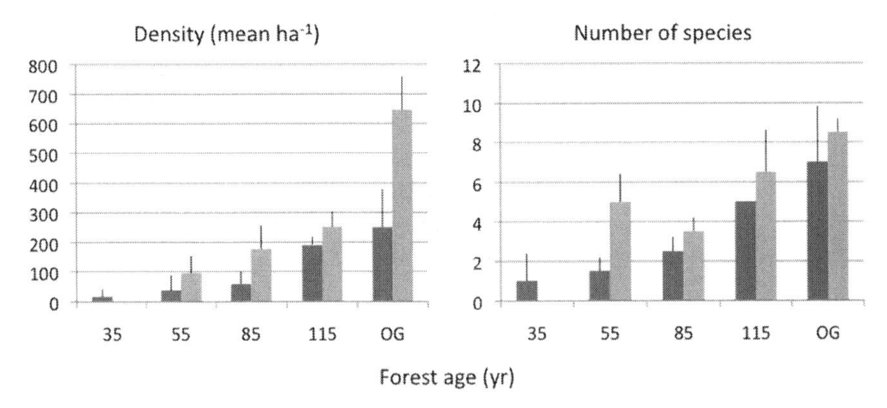

Figure 10.2
Density (mean per ha ± standard error) and species richness (mean raw counts ± standard error) of holo-epiphytes (dark gray bars) and hemi-epiphytes (light gray bars) along a successional chronosequence in the Barro Colorado Nature Monument in central Panama. Holo-epiphytes are not rooted in forest soil, whereas hemi-epiphytes are. Species richness increases more rapidly than density during succession. Source: Woods and DeWalt (2013).

10.4 Functional Traits of Early and Late Successional Species

The life-history traits and functional attributes of pioneer species enable them to colonize and establish in the transitory conditions of high light availability following disturbances, but these characteristics reduce their survival and growth under shaded conditions (see box 5.1; Rees et al. 2001). Here I explore the physiological and morphological traits that distinguish early successional specialists from species arriving later in succession. Functional traits of seedlings, saplings, and trees can be strong determinants of demographic rates that drive changes in species composition during forest regeneration.

10.4.1 Seed and Seedling Traits of Woody Species

Because of the premium on monopolizing space and resources as quickly as possible, seed and seedling characteristics are major components of the pioneer regeneration strategy. Fast-growing early successional species show significantly greater photosynthetic capacity, stomatal conductance, leaf nitrogen content, and specific leaf area compared to late successional species (Bonal et al. 2007). Shade-tolerant seedlings of late successional species, in contrast, make their living by carefully budgeting their limited resources and reducing their risk of mortality. They persist by means of stored reserves, sturdy construction, physical and chemical defense of leaf and stem tissues, and efficient use of light (table 10.1).

Bazzaz (1991) listed 12 physiological characteristics of pioneer species; 4 of these pertain specifically to seed and seeding stages: (1) seeds are often present and abundant in the soil seed bank, (2) seeds have dormancy, (3) seed germination is photoblastic (triggered by exposures to red light) or enhanced by temperature fluctuations or higher nutrient concentrations, and (4) seedlings have aboveground (epigeal) germination and photosynthetic cotyledons. These adaptations ensure successful germination and rapid seedling growth following disturbances, enabling pioneer species to exploit unpredictable and highly transitory conditions of high light availability. With dormant seeds present in the soil seed bank, pioneer species are ready to become seedlings as soon as conditions become appropriate. Shade-tolerant species, in contrast, persist for relatively long periods in the understory, where they may show little or no apparent growth (Kitajima 1992).

Pioneer species typically have small seeds with aboveground (epigeal) photosynthetic cotyledons, whereas belowground (hypogeal) storage cotyledons are generally associated with large seed size and with higher rates of survivorship in shade (Miquel 1987; Hladik and Miquel 1990; Garwood 1996). Seed reserves provide important resources for seedling growth and survival

Table 10.1. *Seed and seedling traits of pioneer and shade-tolerant tree species in tropical forests*

Attribute	Pioneer	Shade-tolerant
Seed size	Small	Large
Initial seedling size	Small	Large
Seed dormancy	Capable	Little or no capacity
Seed longevity	Relatively long-lived	Short-lived
Seed germination physiology	Often require red light	No red light requirement
Seedling germination	Mostly epigeal (aboveground)	Mostly hypogeal (belowground)
Cotyledon morphology	Photosynthetic cotyledons	Nonphotosynthetic, reserve cotyledons
Leaf and stem tissue density	Low	High
Cotyledon/leaf toughness	Low	High
Photosynthetic capacity	High	Low
Specific leaf area	High	Low
Seedling survival rate	Low	High
Seedling growth rate	High	Low
Herbivore resistance	Low	High

Sources: Based on Bazzaz (1991), Kitajima (1996), and Alvarez-Clare and Kitajima (2007).

under shaded conditions. Kitajima (1992) established a strong negative relationship between seed mass and photosynthetic rates of cotyledons for 74 woody species from Barro Colorado Island, Panama. A similar relationship was observed among 53 woody species from Kibale National Park, Uganda (Zanne et al. 2005). Among tropical forest woody species, the relative growth rate of young seedlings is negatively correlated with survivorship in shaded understory (Kitajima 1994; Baraloto et al. 2005). The trade-off between seedling survival and seed size appears to be mediated by cotyledon functional morphology (Kitajima 1996). The higher seedling survival and slower growth of larger-seeded species in French Guiana were better explained by cotyledon type than by seed mass (Baraloto and Forget 2007).

Woody species have been classified into five seedling types, based on cotyledon exposure, position, and morphology (foliaceous or reserve type; Miquel 1987; Hladik and Miquel 1990). Cotyledons that emerge from the seed coat and become fully exposed are phanerocotylar, whereas those that remain enclosed within the seed coat are cryptocotylar. Epigeal cotyledons are positioned above the ground, whereas hypogeal cotyledons are deployed below the ground. The most common type of seedling class in tropical forests, PEF (phanerocotylar-epigeal-foliaceous), accounts for 33% to 56% of woody species in tropical floras. Woody pioneer species are predominantly the PEF type, and the prevalence of PEF species is highest among species with small seed mass (Garwood 1996). In a sample of 209 Malaysian tree species, 78% of the pioneer species had PEF-type seedlings compared to only 29% of nonpioneer species (Ng 1978).

In tropical forests of Mexico and Brazil, initial morphology of seedlings was associated with seed size, seed dispersal mode, and successional status (Ibarra-Manríquez et al. 2001; Ressel et al. 2004). At Los Tuxtlas Biological Station in Mexico, species with PEF seedlings had significantly smaller seed mass than species with CHR (cryptocotylar-hypogeal-reserve) seedlings. PEF species were predominantly nonanimal dispersed and were overrepresented among pioneer species, whereas CHR species were predominantly animal dispersed and were overrepresented among nonpioneer "persistent" species. At Panga Ecological Station in Minas Gerais, Brazil, 75% of the pioneer species had PEF seedlings, compared to only 29% of shade-tolerant species. In contrast, only 3% of pioneer species had CHR seedlings compared to 52% of shade-tolerant species. For seeds less than 0.1 grams in mass, 73% were PEF, whereas 86% of seeds above 1.5 grams were of the CHR type. Among CHR species, 96% were dispersed by animals, compared to 60% of the PEF species (Ressel et al. 2004).

Seed size is positively correlated with seedling size, and larger seedlings tend to have higher rates of survival and slower rates of growth. Among eight species from tropical forests of Paracou, French Guiana, larger-seeded species were more likely to survive to five years of age. Smaller-seeded species, however, have a numerical advantage, as larger numbers of seeds are produced and dispersed more widely (Baraloto et al. 2005). Seedling survival in the understory ultimately depends upon resource allocation patterns that place a high priority on storage reserves and on defense from physical damage and herbivores (Kitajima 1996). Species that establish and survive well in the shaded understory have stem and leaf tissue with a higher modulus of elasticity (stiffness), higher fracture toughness (resistance to tearing), and higher tissue density (Alvarez-Clare and Kitajima 2007). The strongest predictors of seedling survival in the understory are stem and leaf tissue density.

The trade-off between seed number and seed size has important consequences for dispersal and seedling establishment success (Coomes and

Grubb 2003). Seed mass of pioneer species varies over at least four orders of magnitude in central Panama (Dalling and Hubbell 2002) and 5,000-fold in central Amazonia (Bentos, Mesquita, et al. 2013). Smaller seeded pioneers are more abundant in the seed bank but less abundant as seedlings in forest gaps (Dalling et al. 1998). The seed mass of pioneer species is positively correlated with seedling emergence success from the seed bank and with initial seedling survival rates (Dalling and Hubbell 2002).

Life-history traits of liana species sampled across a replicated chronosequence have revealed associations between successional abundance, seedling growth habitat, and seed size (Letcher and Chazdon 2012). The relative abundance of small-seeded lianas declined with forest age class, whereas large-seeded lianas increased. Liana seedling growth form also shifted in importance during succession; freestanding seedlings increased in relative abundance and climbing seedlings declined with stand age. Freestanding and animal-dispersed liana species were significantly overrepresented in larger seed sizes, whereas climbing and abiotically dispersed species were significantly overrepresented in lower seed sizes.

Seed size is a key functional trait that influences seedling establishment, growth, and survival during changing understory conditions during succession. Seed size is strongly linked with dispersal mode, seed number, resource allocation, and seedling growth rate. Seedling types are highly phylogenetically conserved at the level of genus, family, and higher-order clades (Ibarra-Manríquez et al. 2001). Because of these functional and phylogenetic associations, patterns of variation in seed size and associated functional traits strongly reflect larger-scale patterns of evolutionary diversification among tropical forest plant taxa.

10.4.2 Functional Traits of Saplings and Trees

Species-specific growth responses to light availability are not always consistent across ontogenetic stages, adding to the complexity of defining functional groups that respond to environmental conditions in consistent and predictable ways during their entire life span (D. A. Clark and D. B. Clark 1992; Poorter et al. 2005; Chazdon et al. 2010). The classic distinction between pioneer and nonpioneer species proposed by Swaine and Whitmore (1988) is based on seedling and seed responses that do not clearly apply to postseedling stages (see box 5.1). Nevertheless, the distinction between pioneer and shade-tolerant species provides an essential framework for understanding successional dynamics and their mechanistic basis in functional plant traits. As stated by Whitmore (1996, p. 8–9), "The Swaine and Whitmore dichotomy is a necessary, but not a sufficient description of the autecology found in nature."

Pioneer and shade-tolerant species in tropical forests exhibit contrasting values of many functional traits that are important determinants of plant growth and survival (table 10.2; Bazzaz 1979; Bazzaz and Pickett 1980). In addition to leaf and wood traits, pioneer and shade-tolerant trees are also distinguished by species-specific differences in canopy and whole-plant architecture—features that are associated with light interception, tree growth rates, and vertical position within the forest canopy (Sterck and Bongers 2001). Leaf functional traits of dominant tree species in early and

Table 10.2. *Leaf and wood traits of postseedling stages of pioneer and shade-tolerant tree species*

Attribute	Pioneer	Shade-tolerant
Specific wood gravity	Low	High
Radial gradients in specific wood gravity	Present	Absent
Light-saturated photosynthetic rate (area and mass based)	High	Low
Leaf nitrogen content (area and mass based)	High	Low
LMA (leaf mass per unit area)	Low	High
SLA (leaf area per mass)	High	Low
Leaf density	Low	High
Leaf toughness	Low	High
Photosynthetic nitrogen use efficiency	High	Low
Photosynthetic phosphorus use efficiency	High	Low
Leaf life span	Short	Long
Transpiration rate	High	Low
Maximum stomatal conductance	High	Low
Maximum growth rate	High	Low
Vulnerability to cavitation	High	Low
Leaf-specific hydraulic conductance	High	Low
Stem hydraulic conductance	High	Low
Height growth	Rapid	Slow
Herbivore resistance	Low	High

Sources: Trait information was obtained from multiple sources (e.g., Whitmore 1990; Bazzaz 1991; Jurbandt et al. 2004; Hölscher et al. 2006; Markesteijn, Iraipi, et al. 2011; Markesteijn, Poorter, et al. 2011).

intermediate successional stages of seasonally dry tropical forests of Mexico were clearly separated from those of dominant trees in late successional forests (Alvarez-Añorve et al. 2012). In early and intermediate successional tree species, leaf functional traits enabled maximization of photosynthetic rates, photoprotection, and heat dissipation, whereas traits of late successional species enhanced light acquisition.

Most woody species are positioned at intermediate positions on the spectrum from pioneer to shade tolerant (Wright et al. 2003; Wright et al. 2010). This ecological variation is most apparent when comparing species within large, diverse tropical forest genera. Within the diverse pioneer genus *Macaranga* (Euphorbiaceae), for example, a diversity of ecophysiological and life-history traits are expressed among species (Davies 1998). Among nine *Macaranga* species grown under three experimental light levels, maximum photosynthetic rates on a leaf mass basis were positively correlated with the crown illumination index of trees growing in forests of Sarawak, Malaysia, and were negatively correlated with seed mass. Shade-tolerant species (lower crown illumination index) have higher values of leaf mass per area. The genus *Piper* (Piperaceae) shows a similar range of life-history and ecophysiological traits within a Mexican tropical rain forest and associated early successional habitats (Gómez-Pompa 1971; Vázquez-Yanes 1976; Chazdon and Field 1987; Fredeen and Field 1996).

Leaf traits strongly associated with plant growth rates include photosynthetic capacity, stomatal conductance, life span, nutrient concentrations, thickness, mass per area, tissue density, and toughness. These leaf traits vary in concert, forming what is now called the "leaf economics spectrum" (Wright, Reich, et al. 2004). Species with a high photosynthetic capacity on a leaf mass basis tend to also have high leaf nitrogen and phosphorous content, low leaf mass per unit area, short leaf life span, and high dark respiration rates. The 30-fold variation in leaf life span across 23 tree species in the Venezuelan Amazon was associated with species successional status, and with several physiological and morphological traits correlated with photosynthetic rates (Reich et al. 1991). Leaf spectral reflectance characteristics can also provide useful indicators of plant photosynthetic performance and photochemical properties (Alvarez-Añorve et al. 2012).

Differences in leaf functional traits among species that dominate during different stages of succession reflect intrinsic physiological capacities as well as physiological and morphological plasticity in response to changing local microenvironments. Both of these factors were found to be important in determining variation in photosynthesis during the first 10 years in a successional sere in Venezuela. Early successional pioneer species had higher photosynthetic rates per leaf mass than species that occur later in succession. Rates of photosynthesis also declined within species with increasing time since abandonment of shifting agriculture (Ellsworth and Reich 1996).

These findings support the hypothesis that early successional species should exhibit greater environmentally induced variation in leaf photosynthetic characteristics compared to shade-tolerant species due to a high level of resource variability in early successional habitats (Bazzaz and Pickett 1980; Strauss-Debenedetti and Bazzaz 1991). The intrinsically high photosynthetic capacity of light-demanding species essentially provides a "built-in" capacity for plasticity through down-regulation of photosynthetic processes in leaves that become shaded. In contrast, species adapted to low-light conditions have limited potential for photosynthetic acclimation to high light (Chazdon 1992; Chazdon et al 1996; Strauss-Debenedetti and Bazzaz 1991; Valladares et al. 2000; Portes 2010).

Poorter et al. (2004) ranked 15 species from lowland moist forest in Bolivia according to their successional position, based on abundance patterns during succession following shifting cultivation (Peña-Claros 2003). Leaf traits were then compared among saplings of those species that occurred in similar intermediate-light conditions in a mature forest in the same region. Specific leaf area, leaf water content, leaf nitrogen content, and leaf phosphorous content all declined from early to late successional position, whereas leaf carbon-to-nitrogen ratios and lignin content increased (Poorter et al. 2004). Leaf life span increased from early to late successional position, whereas herbivory rates declined (fig. 10.3).

Inter- and intraspecific variation in functional traits of leaves and stems must be interpreted in the context of whole-plant structure and resource allocation. Short-lived pioneer species maintain their dominance early in succession by virtue of their high leaf-level photosynthetic capacity and high instantaneous light interception per unit of leaf mass at the crown level (Ackerly 1996; Selaya et al. 2008). The continuous deployment of leaves at levels of full illumination requires high rates of height growth and high rates of turnover of leaves, which reduce leaf life span and increase whole-plant costs of light interception.

Short-lived pioneer species can reduce their support costs by producing low-density wood and reducing branching, but this growth strategy ultimately limits crown size and total leaf area and reduces competitiveness as the density and height of neighboring trees increase. Short-lived pioneers and late successional tree species showed similar values of total light intercepted per unit of aboveground biomass in a young shifting agriculture fallow in the Bolivian Amazon (Selaya and Anten 2008). Although later successional

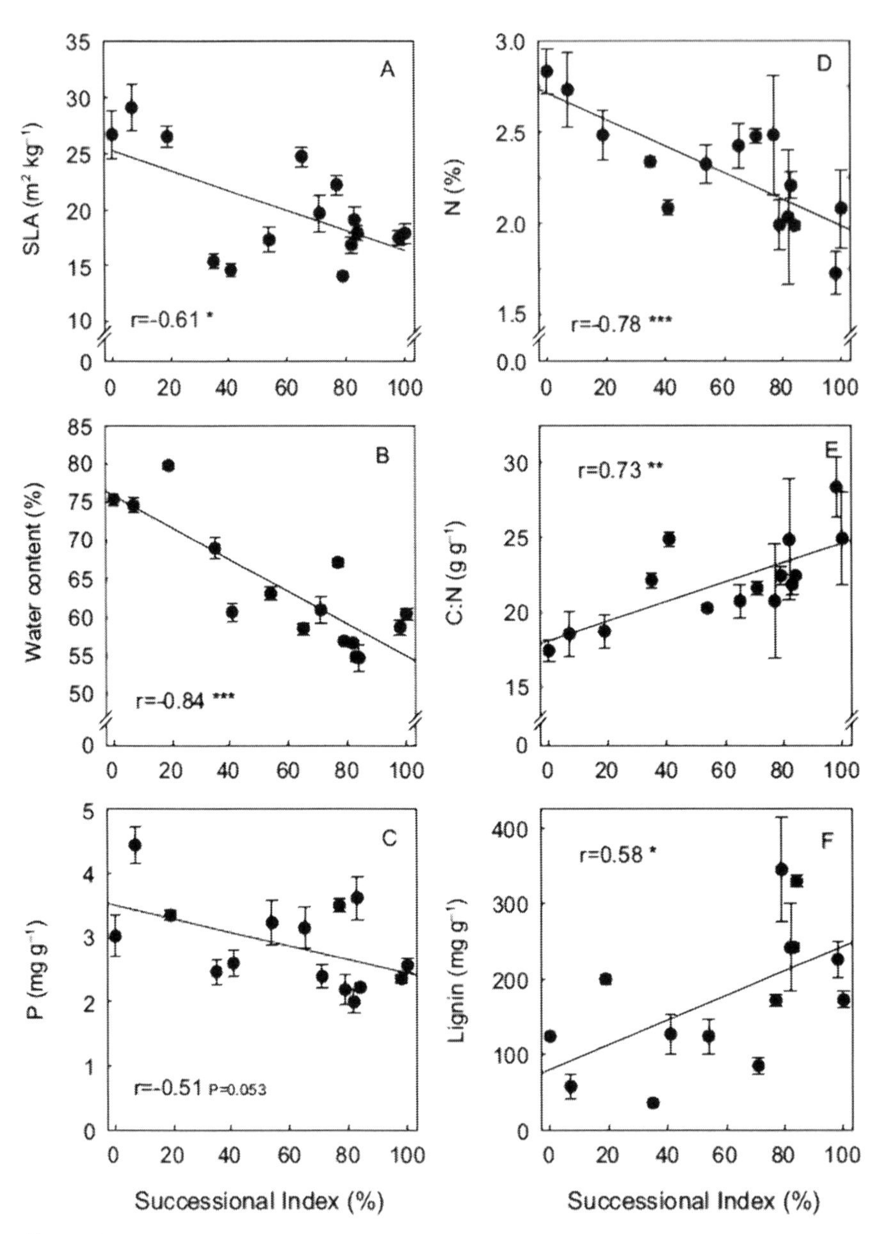

Figure 10.3

Variation in leaf traits across 15 rain forest tree species in eastern Bolivia in relation to a species successional index. The sucessional index was based on the species score on the first axis of a correspondence analysis of species composition including successional and mature forest plots. Sources: Peña-Claros 2003; Poorter et al. (2004, fig. 1).

species have lower leaf-level rates of photosynthesis on a mass basis, their substantially higher leaf longevity led to values of lifetime carbon gain that were similar to short-lived pioneers in fallow vegetation (Selaya and Anten 2010).

Just as leaf traits show consistent patterns of covariation among species along a fast–slow continuum of growth rates, so do wood traits. Chave et al. (2009) described a "wood economics spectrum" of wood traits that show similar trends of variation across a wide range of woody species. These traits include anatomical traits (density and diameter of vessels), hydraulic traits (hydraulic conductivity and resistance to cavitation), mechanical properties (modulus of rupture and Young's modulus of elasticity), and chemical properties of wood tissue (lignin, cellulose, and mineral content). Diameter growth rates of tree species in a Bolivian rain forest were negatively related to wood density and vessel density but were positively correlated with vessel diameter and potential hydraulic conductance (Poorter et al. 2010).

Sustaining high rates of photosynthesis in hot tropical climates requires high rates of transpirational water loss, which demand rapid and efficient transport of water from roots to leaves (Huc et al 1994; Juhrbandt et al. 2004; Hölscher et al. 2006). But efficient water transport through xylem (high hydraulic conductance) poses a risk of xylem cavitation under conditions of reduced water supply. In pioneer tree saplings in a tropical dry forest in Bolivia, wood was significantly more vulnerable to cavitation than in shade-tolerant species (Markesteijn, Poorter, et al. 2011). Vulnerability to cavitation was positively correlated with stem and leaf hydraulic conductance and with wood density. These considerations lead to a functional trade-off between safety (reduced risk of cavitation) and efficiency (high hydraulic conductance) of water transport among plant species (Sobrado 2003; Markesteijn, Iraipi, et al. 2011). High wood density and reinforcement of xylem conduits reduce the risk of implosion of xylem vessels during drought stress (Hacke et al. 2001). Leaf traits and wood traits are therefore tightly linked with regard to hydraulic functioning in tropical forests (Sobrado 2003; Santiago et al. 2004; Meinzer et al. 2008; McCulloh et al. 2011). Among 20 species of canopy trees in two lowland forests in Panama, leaf-specific hydraulic conductivity of upper branches was correlated with photosynthetic capacity and stomatal conductance and negatively correlated with branch wood density (Santiago et al. 2004).

Although few species have been studied in detail, saplings and trees of pioneer species had higher stem hydraulic conductances and leaf-specific hydraulic conductances than those observed in shade-tolerant or late successional species in two Panamanian forests. Rates of sap flux in branches were also significantly higher in pioneer species, indicating that pioneer species are capable of moving greater quantities of water through vascular conduits for a given tension in the water column. The higher leaf-specific hydraulic capacity in the pioneer species examined was correlated with larger xylem diameters and lower wood density in branch tissue rather than to differences in wood density or xylem characteristics in stem tissue (Tyree et al. 1998; McCulloh et al. 2011)

These relationships suggest that leaf and wood functional traits are closely integrated with the growth and regeneration strategies of tropical tree species, but the few species that have been studied may not be representative of forest assemblages. Challenging the view of tight physiological integration, a study of functional traits of 668 rain forest tree species in French Guiana suggested that leaf traits vary independently of wood traits (Baraloto et al. 2010). Moreover, in a comparison of five Neotropical forests, Poorter et al. (2008) found that wood traits are better predictors of relative growth rates and juvenile mortality rates of tropical trees than leaf and seed traits. The generality of these findings remains to be tested among different tropical forest regions based on more extensive sampling of species and successional stages.

A number of fast-growing pioneer species exhibit age-related increases in wood density with increasing radial distance from the pith. Producing wood of higher density at the periphery of stems increases structural rigidity at minimal cost, while also decreasing the risk of xylem cavitation in sapwood. Some pioneer species show percentage increases in wood density as high as 300% from pith to bark (Wiemann and Williamson 1988; Nock et al. 2009; Williamson and Wiemann 2010; Hietz et al. 2013; Schüller et al. 2013). Radial gradients in wood density represent a promising functional trait linked to successional distributions and patterns of tree growth during succession.

The high capacity of photosynthesis and height growth of seedlings, saplings, and adults of tree species that colonize early in succession enables these species to compete effectively for high levels of resource availability and to exploit transient early successional habitats (see tables 10.1 and 10.2). But species with these "fast" traits lose their competitive edge later in succession, when establishment and survival depend more upon "slow" traits that reduce intrinsic rates of growth and increase long-term persistence in shade and resistance against damage by herbivores and falling debris. Shifts in functional traits during succession therefore reflect dramatic changes in resource availability and species interactions.

10.5 Environmental Filtering, Functional Diversity, and Community Assembly during Succession

Secondary succession is described as "community assembly in action" (Lebrija-Trejos, Pérez-García, et al. 2010, p. 387). Species assemblages undergo continual change during succession, reflecting the differential ability of species to arrive, establish, and survive under the prevailing environmental conditions and competitive regimes at different stages. Rates of species turnover are determined by three distinct processes: recruitment of new species, mortality of species present during earlier periods, and persistence of species populations through time. Stem turnover rates are not always good predictors of species turnover rates (Chazdon et al. 2007).

The functional characteristics of species determine their qualifications for membership in a species assemblage associated with particular environmental characteristics. Individuals must pass through several ecological filters to become recruited as seedlings, and later as saplings or trees. Dispersal is the first filter; seeds may be present in the soil seed bank or may be transported by the wind, water, or biotic dispersal agents. Once dispersed, seeds must survive attack by seed predators or fungal pathogens. Germinated seedlings must then possess leaf, stem, and root adaptations that permit growth and persistence in their particular microenvironments. These functional characteristics define niche dimensions that strongly shape the regeneration requirements of plant species (Grubb 1977). Pioneer species do not establish or grow well in shaded forest understory microsites, and slow-growing, shade-tolerant tree species are poor competitors for light, water, and soil nutrients in former agricultural fields and young shifting cultivation fallows (van Breugel et al. 2007).

The notion that species have clearly defined successional niches is pervasive in tropical forest ecology, having originated in an early distinction between "secondary" and "primary" forest species, or early versus late successional species (Whitmore 1984). Although individual species can be assigned to the two extremes of this gradient, the dichotomy between pioneer and shade tolerant is poorly supported by assemblage-wide data. When species are ranked according to their position along a growth–survival gradient, species are not clustered in groups along this gradient, but rather are distributed all along the gradient (Wright et al. 2010).

Following canopy closure, species turnover during succession occurs in vertical waves, beginning in the understory and culminating in the replacement of canopy species. Effects of environmental filtering and community assembly are therefore more evident in understory than in canopy layers

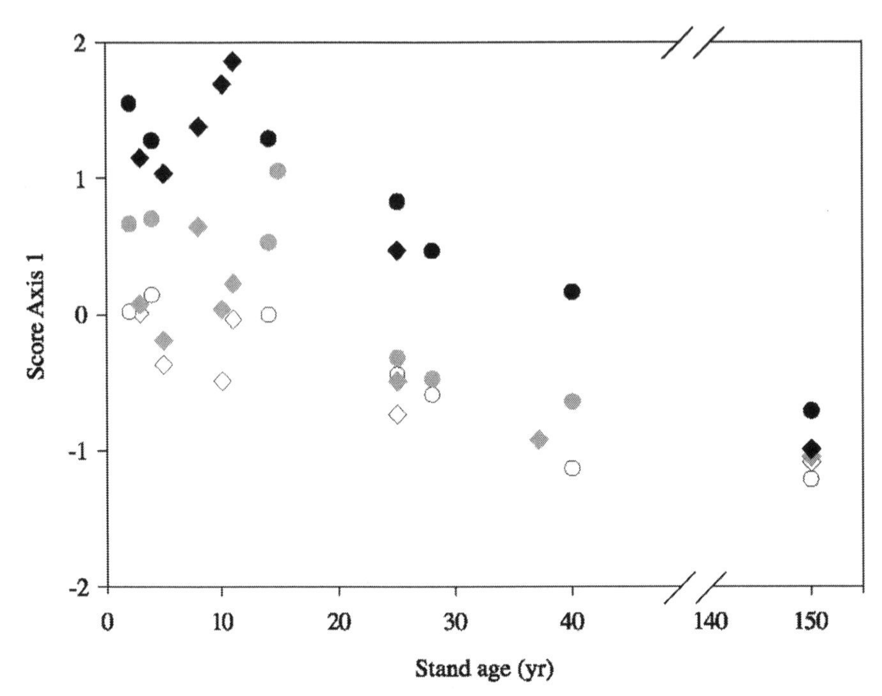

Figure 10.4
Species composition of different canopy strata across a successional chronosequence in two study areas of the Bolivian Amazon. Composition is summarized by an ordination score, with a lower score reflecting later successional assemblages. Open symbols are understory layer, gray symbols are subcanopy layer, and filled symbols are canopy layer. Diamonds indicate data from the Reserve El Tigre, and circles are data from El Turi. Source: Peña Claros (2003, fig. 4).

(Peña Claros 2003; Lozada et al. 2007; Ochoa-Gaona et al. 2007). The species composition of understory layers approaches that of old-growth forests more rapidly than canopy layers (fig. 10.4; Peña Claros 2003; Norden et al. 2009). This pattern reflects the recruitment of shade-tolerant species beneath a canopy of long-lived pioneer species (Guariguata et al. 1997; Finegan 1996). A similar pattern has been observed in lowland wet forests of Costa Rica, where species composition of seedlings and saplings was similar between 12- and 29-year-old second-growth forests and old-growth forests, despite differences in species composition of trees. The shade-tolerant species that colonize young forests tend to be species that are abundant in nearly old-growth forests, have high rates of dispersal, and fruit year-round (Norden et al. 2009).

Ding, Zang, Letcher, et al. (2012) assessed the effects of logging and shifting cultivation on the distribution of functional traits in regenerating vegetation in lowland and montane forests of Hainan Island, China. Formerly heavily disturbed sites showed strong evidence of environmental filtering, where assemblages were dominated by closely related species with disturbance-adapted functional traits. In contrast, multiple processes drive community assembly in lightly disturbed and old-growth forests, where biotic interactions at fine spatial scales play a more important role.

The chronosequence study by Lebrija-Trejos and his colleagues (Lebrija-Trejos, Meave, et al 2010; Lebrija-Trejos, Pérez-García, et al. 2010; Lebrija-Trejos et al. 2011) of tropical dry forests of Nizanda in Oaxaca, Mexico, provides an excellent example of community assembly and environmental filtering during succession. Changes in abundance-weighted functional traits during succession reflect two interacting processes: changes in species composition and species-specific variation in functional traits. Sunny, hot, and dry early successional sites were dominated by species with traits that reduce incident radiation, favor convective cooling over transpirational cooling, and maintain leaf structure under conditions of severe water stress. In contrast, seedling assemblages in relatively cool and moist conditions later in succession were dominated by species with simple, large leaves and long petioles that maximize light interception per leaf area. The abundance-weighted averages of trait values of seedling, juvenile, and adult plant assemblages showed convergent values as stands increased in basal area during succession, with changes occurring more rapidly in smaller size classes (fig. 10.5). Their study clearly demonstrated that, at least in this dry forest area, environmental filtering is a predictable and fundamental mechanism of community assembly.

In landscapes containing remnants of old-growth forests, woody seedling and sapling assemblages of second-growth forests consist of a mixture of species representing both second-growth and old-growth forests (box 10.2). In lowland wet forests of Costa Rica, composition of seedling and sapling assemblages became more similar to tree assemblages of old-growth stands over time. The successful establishment of abundant old-growth tree species as seedlings and saplings in regenerating forests is largely determined by the dispersal of vertebrate-dispersed tree species that are abundant in adjacent old-growth forests (Sezen et al. 2007; Norden et al. 2009).

Environmental filtering acts more strongly to exclude pioneer species from establishing in shaded understory than to exclude nonpioneer species from establishing early in succession. Shade-tolerant species can establish during the first few years of succession, supporting the importance of initial floristic composition (Egler 1954). Ecological studies in Amazonia, Bolivia, Costa Rica, and Mexico found that species from a wide range of functional groups established early in succession and continued to recruit

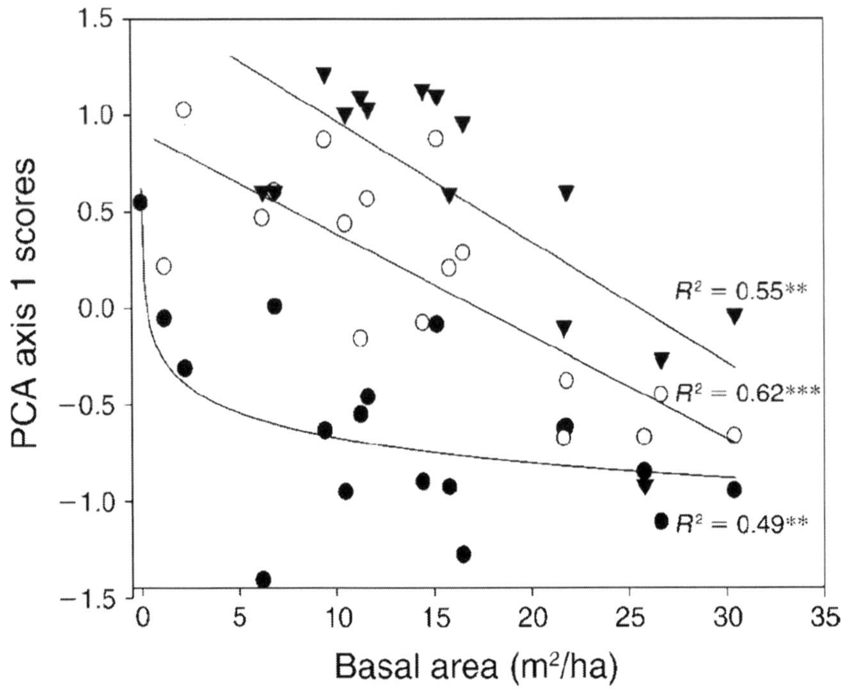

Figure 10.5
Community-level abundance-weighted mean of 23 functional traits (principal components analysis [PCA] axis 1 scores) during forest succession in Nizanda, Oaxaca, Mexico, of assemblages of three size classes: seedling regeneration (solid circles), juvenile (open circles), and adult (solid triangles). Second-growth forests were studied 1–60 years following abandonment after shifting cultivation along with a mature forest (basal area of 31 m^2/ha). Continuous lines represent modeled changes for each community stratum. Source: Lebrija-Trejos, Perez-Garcia, et al. (2010, fig. 5).

following canopy closure (Uhl, Buschbacker, and Serrão 1988; Peña-Claros 2003; van Breugel et al. 2007; Chazdon et al. 2010). As Egler (1954) proposed, changes in species composition during succession reflect a combination of initial floristic composition and relay floristics, in which species colonize sequentially in response to changing forest conditions (Finegan 1996). In Neotropical late successional forests, among the last species to establish are slow-growing, shade-tolerant, small-statured, understory tree species (Pitman et al. 2005; Chazdon et al. 2010). The slow arrival of these species may reflect dispersal limitation and low levels of species abundance in old-growth forests. The gradual reassembly of tropical forest assemblages depends critically on the availability of diverse seed sources in surrounding forest areas and

Box 10.2. Establishment of Endemic Tree Species and Old-Growth Specialists during Succession

An important metric of the conservation value of regenerating forests is the extent to which endemic species and species typical of old-growth forests colonize and establish (Chazdon, Peres, et al. 2009). Former coffee plantations in the Blue and Port Royal Mountains of Jamaica have regrown into forest over the past 150–170 years, providing an unprecedented opportunity to examine long-term recovery of species composition and endemism of trees during secondary succession. The montane rain forest in the Blue and Port Royal Mountains of Jamaica has a rate of tree species endemism of 41% (Tanner 1986). Forests on former plantations did not differ significantly from adjacent old-growth forests in basal area, the overall number of species, or the number of endemic species (Chai and Tanner 2011). But species composition has not recovered to the same extent. Despite 150 years of regrowth, the abundance of endemic species was lower in the second-growth forests. Over half (54%) of the species present in nearby old-growth forest were present in the regrowing plantations, however, suggesting that old-growth tree species are gradually establishing in these forests.

Shifting cultivation fallows in Laos were found to contain a high proportion of old-growth forest species as juvenile stems (84%) after 7–10 years of regeneration (McNamara et al. 2012). Compositional differences between low-use and high-use fallows and old-growth forest fragments were attributed to changes in the relative abundance of species rather than to the absence of old-growth forest species from regenerating assemblages. Of 49 old-growth species surveyed, at least three individuals were recorded in all of the fallow sites. This study highlights the fact that fallow forests that are successfully regenerating over time can contain high levels of tree diversity and should not be classified as degraded forest land.

In Central and West African forests, 80% of the plant species are endemic to the Lower Guinea region (Sayer et al. 1992). Van Gemerden, Shu, and Olff (2003) evaluated the long-term recovery of vegetation in shifting cultivation fallows 10–60 years after abandonment, considering both woody and non-woody species. The proportion of species endemic to Lower Guinea increased with fallow age, whereas the proportion of widespread species decreased. But even after 60 years, the proportion of Lower Guinea endemics in shifting cultivation fallows (about 10%) was significantly lower than in old-growth forest

Box 10.2. (*continued*)

(about 16%). The species richness of endemic understory plants and trees was lower in 15-year-old secondary forests compared to "near-primary" forests in the southwestern lowlands of Cameroon but remained substantially higher than in agroforests and cultivated fields (Waltert et al. 2011).

Endemic tree species are not always highly vulnerable to land-use change. Most of the endemic species within Upper Guinean forests (in Senegal, Guinea-Bissau, Guinea, Sierra Leone, Liberia, Ivory Coast, Ghana, and Togo) have widespread distributions and exhibit ruderal life-history strategies, including shade intolerance, drought tolerance, and wind dispersal (Holmgren and Poorter 2007). These characteristics may be a legacy of past climatic disturbances and period forest expansions and contractions in West Africa (Maley 2001; White 2001a). The paleoenvironmental filter may predispose many of these endemic species to colonize successional forest habitats.

In the southern Atlantic Forest region of Brazil, only 3.1% of the original forest cover remains (see box 13.3). During the past decade, forest cover has increased threefold, following the abandonment of agricultural fields and increased landscape conservation efforts (Mata Atlantica SOS 2002; Piotto et al. 2009). Piotto et al. (2009) inventoried trees in 12 secondary forests in three age classes (10, 25, and 40 yr.). The percent of known local endemics and Atlantic forest endemics increased with age across these stands, whereas widespread species decreased. The similarity of species composition between secondary forests and old-growth forests in the region increased with stand age, due to the gradual establishment of old-growth tree species.

their dispersal into regrowing forests. In nine young secondary dry forests in Mexico, the recruitment rate and species gain of small stems increased with the amount of forest in the surrounding matrix (Maza-Villalobos et al. 2011).

Changes in tree assemblages during succession reflect change in functional diversity as well as species diversity (Brown and Lugo 1990; Guariguata and Ostertag 2001; Chazdon et al. 2007; Lebrija-Trejos, Meave, et al. 2010). Across a chronosequence on former cornfields in Chiapas, Mexico, functional diversity and species diversity increased asymptotically with stand basal area (Lohbeck et al. 2012). Species richness and species diversity were strong predictors of functional richness and functional diversity, respectively. These findings suggest little or no functional redundancy during early stages of successional development. As new species establish during succession, their sets of func-

Using a multinomial classification model based on estimated species relative abundance, Chazdon et al. (2011) classified tree species in northeast Costa Rica as second-growth specialists, old-growth specialists, generalists, or species too rare to classify. These results were then used to assess changes in the relative abundance of old-growth specialists during forest regeneration on former pastures over 13 years. In all sites, old-growth specialists gradually increased their relative abundance in tree assemblages, reaching 35%–40% of trees in 30- to 40-year-old plots.

Across a chronosequence in tropical rain forests of lowland northeast Costa Rica, species composition and species richness of stems greater than or equal to 2.5 centimeters in DBH did not differ significantly between older secondary forests (30–42 yr. old) and old-growth forests (Letcher and Chazdon 2009b). The proportion of old-growth species found in second-growth forest areas increased linearly with age since abandonment, suggesting that the establishment of old-growth species is largely a function of time. In seedling assemblages of 15- to 25-year-old secondary forests, the proportion of old-growth species increased to 59%–75%.

Higher proportions of old-growth specialists are found in the seedling assemblages of shifting cultivation fallows compared to larger stem classes (Ochoa-Goana et al. 2007; Williams-Linera et al. 2011). After several decades, if these fallows develop into forests, these species will become more abundant as trees. The establishment of slow-growing old-growth specialists during succession moves at a slow but steady pace.

tional traits do not overlap with those of existing species. Functional redundancy likely increases during later stages of succession. Katovai et al. (2012) compared functional diversity of plant species across different land-use types on Kolombangara in the Solomon Islands based on three plant traits (growth form, dispersal mechanism, and clonality). Second-growth and old-growth forests had similar values of functional diversity within elevational bands, despite lower species richness in second-growth forests. Second-growth forests here also appear to have lower functional redundancy than old-growth forests.

Changes in community structure during succession are ultimately driven by variation in rates of recruitment and mortality across species. Species-specific variations in functional traits are strong determinants of these demographic rates. At a local scale, community assembly is driven by seed dispersal and variation in environmental conditions, including intra- and interspecific competition. At a larger spatial scale, community assembly depends on the species pool in the surrounding area and the proximity of forested areas that provide seeds. Under optimal conditions, successional forests increase in species diversity and functional diversity over time, as forest ecosystems become more functionally and structurally complex.

10.6 A General Scheme for Community Assembly during Secondary Succession

Tropical old-growth forests are the most species-rich assemblages on earth. How species composition changes over long time scales during succession remains poorly understood, as few studies have followed such changes for more than a few years. Understanding the ecological processes that determine community assembly requires information on vegetation dynamics within individual sites over time. Repeated censuses of marked stems within one or multiple sites provides information on stem turnover and species turnover between time intervals, revealing actual successional pathways and their variability among sites and over time (van Breugel et al. 2007; Chazdon et al. 2007; Norden et al. 2011).

Figure 10.6 illustrates how stem and species turnover influence characteristics of successional assemblages, such as species composition, functional diversity, and phylogenetic diversity. The scheme can also apply to vegetation change following forest degradation, logging, or fragmentation, or to vegetation change across environmental gradients. By examining which species, functional traits, and clades are gained or lost during successive time intervals, we can gain insights into the ecological processes that drive community assembly during succession. For example, if the species that are gained possess different functional traits than the species that are lost, and these functional traits increase competitive ability or are favored by environmental filtering, we will observe directional shifts in functional trait distributions over time (Lebria-Trejos, Pérez-García et al. 2010; Mayfield and Levine 2010). If the species gained derive from phylogenetic lineages distinct from those of existing species, species turnover also leads to increasing phylogenetic diversity (Swenson 2011).

A chronosequence study by Letcher (2010) in Costa Rica revealed changes in the phylogenetic structure of plant assemblages during succession. Species present within a particular site are a nonrandom subset of the taxa that could potentially occupy the site, but as succession proceeds, the taxa present become more distantly related than predicted by chance. This pattern

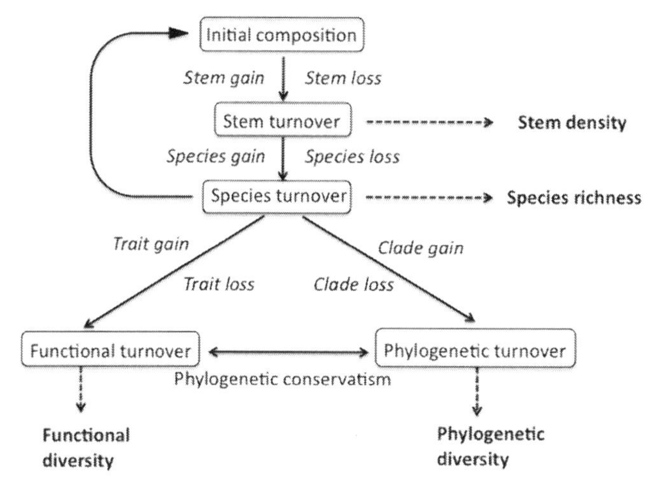

Figure 10.6
A general scheme for community assembly during succession within a single tropical forest area. Following initial colonization of postagricultural or disturbed land, new stems are gained (recruitment), whereas some existing stems remain (persistence), and others are lost (mortality). Stem and species turnover determine changes in species richness and abundance distribution. Rates of stem and species turnover change iteratively during succession. Stem and species turnover affect the gain and loss of functional traits, determining functional turnover and functional diversity. Changes in stem and species turnover also have implications for successional change in higher phylogenetic units, resulting in the gain or loss of evolutionary clades. Phylogenetic turnover determines phylogenetic diversity and phylogenetic community structure. The relationships between functional diversity, species diversity, and phylogenetic diversity are influenced by phylogenetic conservatism of functional traits and by rates of clade gain and loss during community assembly.

of increasing phylogenetic evenness is most pronounced for the smallest stem size class. In the youngest successional stands, trees are more closely related than predicted by chance, suggesting that a species' response to disturbance has a strong phylogenetic component. Early successional pioneer species require traits conferring high dispersal abilities and high growth rates in disturbed environments. Two traits characteristic of pioneer species, seed size and wood density, show strong phylogenetic conservatism (Moles et al. 2005; Chave et al. 2006; Swenson and Enquist 2007). As succession proceeds, recruitment favors species that are more distantly related than predicted by random colonization from the regional species pool. Evaluating these trends over time, Norden et al. (2012) confirmed that, over a relatively short time frame of 12 years, tree assemblages become more phylogenetically even due to recruitment and mortality processes. Stem loss (mortality) is concentrated among closely related taxa, while stem gain (recruitment) is associated with net species gain and net clade gain. The colonization of propagules belonging to a wide array of shade-tolerant lineages increases phylogenetic evenness more rapidly in seedlings than in trees.

Successional studies in Mexico, Brazil, China, and New Guinea have confirmed the trend of phylogenetic clustering in tree assemblages in early successional stands and phylogenetic evenness in late succession and old-growth stands (Letcher et al. 2012; Ding, Zang, Letcher et al. 2012; Whitfeld et al. 2011). A comparison of herbaceous and woody vegetation in second-growth and old-growth transects in tropical dry forests of Veracruz, Mexico, revealed a lower average taxonomic distinctness in 12- to 20-year-old second-growth areas, reflecting lower phylogenetic diversity, despite the higher mean species richness in these areas (Castillo-Campos et al. 2008; Moreno et al. 2009). In 12 0.25-hectare plots regenerating after abandonment of subsistence agriculture and 7 old-growth plots in the New Guinea lowlands, mean phylogenetic distance among trees increased with total basal area per plot (Whitfield et al. 2011). Significant phylogenetic clustering was observed in 5 of the second-growth plots, whereas significant phylogenetic evenness was detected in all of the old-growth plots. These trends are consistent with the hypothesis that tree recruitment during later stages of succession favors species that are more phylogenetically distinct than during earlier stages of succession as biotic interactions become increasingly important drivers of community assembly, particularly at small spatial scales (Chazdon 2008b; Comita, Muller-Landau, et al. 2010).

10.7 Conclusion

The functional traits of plant species are important determinants of establishment, growth, competition, and survival during different stages of succession. But most studies of plant functional traits and their relationship to vital rates have been conducted in mature forests (Poorter et al. 2008; Wright et al. 2010). Future research needs to bridge the gap between inventories of vegetation dynamics and descriptive surveys of functional traits and evolutionary relationships among species, which will provide major insights into the mechanisms of community assembly in tropical forests.

Despite high rates of stem turnover in successional forests, the process of reconstituting the structure, composition, function, and rich evolutionary heritage of old-growth forests requires long time scales, often more than a century, largely because of the long life spans of many tropical tree species. Stem turnover can lead to changes in species richness, functional traits, and phylogenetic diversity over time, directing our focus to the factors that influence the nature of these transitions. The functional and demographic attributes of clades are now recognized as major drivers of community assembly in tropical forests. During forest regeneration, the surrounding landscape also changes, exerting strong influences on surrounding vegetation patches and the composition of the regional species pool. Synthetic approaches combining chronosequence data across sites of different age with long-term monitoring data within sites and their surrounding landscapes may be the only way to gain insight into the factors that control the rate and nature of community assembly during forest regeneration.

While we observe the long-term process of community assembly in regenerating forests, old-growth forests in the surrounding landscapes continue to be transformed. Second-growth forests will not develop into diverse, fully functional forests in the absence of old-growth forests. Only by slowing down rates of deforestation and increasing rates of reforestation in tropical landscapes can we can hasten the natural regeneration of functionally and taxonomically diverse tropical forest ecosystems.

CHAPTER 11

RECOVERY OF ECOSYSTEM FUNCTIONS DURING FOREST REGENERATION

Scientific debate about the role of tropical forests in the global carbon cycle transcends the obvious difficulties caused by poor data bases. Instead, it is a debate on the perception of what tropical forests are and how they function.
—Ariel Lugo and Sandra Brown (1992, p. 240)

11.1 Overview

When a tropical forest is cleared, burned, cultivated, or grazed, a substantial fraction of the living biomass is lost from the ecosystem, exported to the air or leached by the rain. Approximately 48% of the living aboveground biomass in tropical forests is composed of carbon, which is released to the atmosphere as CO_2 upon burning. Deforestation and forest degradation of tropical forests contribute between 10% and 15% of global carbon emissions (van der Werf et al. 2009; Achard et al. 2010; Asner et al. 2010). Nitrogen, easily volatilized by burning, is released to the atmosphere or mineralized into nitrate in the soil, which is easily leached. Mineral nutrients that were contained in the biomass, such as potassium, magnesium, and calcium, are less likely to be volatilized during low-temperature burning and accumulate in ash but can be lost through leaching. In addition to losses of carbon and nutrients, deforestation reduces evapotranspiration, increasing stream-flow and flood frequency, and reducing local rainfall (Giambelluca 2002).

Can the reestablishment of forest on these lands restore the functions and nutrient stocks lost from the original forest ecosystem? The short answer is yes, but some processes take longer than others. Thousands of years ago shifting cultivators recognized that soils that previously supported tropical forests sustain cultivation on nutrient subsidies from residual biomass for only a few years. A fallow period is essential for restoring soil fertility so that the land can again be cleared and cultivated. In fact, early studies of nutrient cycling in tropical shifting cultivation systems by Nye and Greenland (1960) led to our current understanding of nutrient cycling and ecosystem functioning in tropical forest ecosystems (Denslow and Chazdon 2002). Knowledge of ecosystem function during primary and secondary succession has direct application to restoration projects as well as to understanding the dynamics of intact forest ecosystems (Walker, Walker, and Hobbs 2007; Walker and del Moral 2009).

The fate of carbon and nutrient stocks following cycles of forest clearance, cultivation, and regeneration after land abandonment depends on many factors including soil fertility, soil texture, rainfall and seasonality, frequency of burning, initial species colonization, temperature, and site productivity. Productivity and nutrient cycling are strongly linked, because biomass is composed of carbon and mineral elements. The greater the aboveground productivity of recolonizing vegetation, the faster is the replenishment of total carbon and nutrient stocks in the developing ecosystem (Brown and Lugo 1990). Changes in species composition, vegetation structure, and biomass accumulation following land abandonment or large-scale natural disturbances are intricately linked; these changes govern the recovery of carbon and nutrient stocks above and below the ground during secondary succession (fig. 11.1). Leaf traits that underlie species differences in growth rates during succession, such as leaf nitrogen content, specific leaf area, and leaf toughness, significantly affect rates of leaf decomposition, nutrient resorption in senescent leaves, and nutrient concentration in leaf litter (Bakker et al. 2011; Wood et al. 2011).

As woody vegetation increases in diversity and stature, forest biomass and

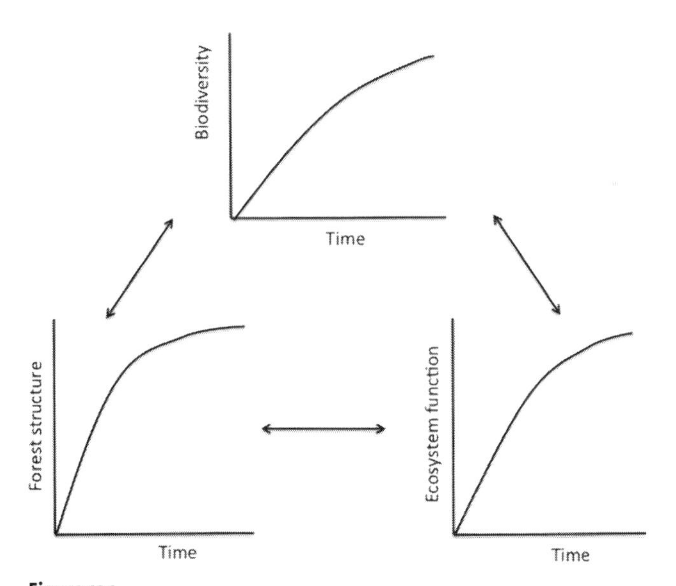

Figure 11.1
Coordinated changes in biodiversity, forest structure, and ecosystem functions occur during forest regeneration. The rates of change of these measures of forest integrity and their complex interrelationships are poorly understood in tropical forest ecosystems. Changes may not increase continuously or linearly over time.

carbon storage increase, and surface soils become enriched with organic matter from decomposing leaf litter (Sang et al. 2012). Changing conditions in soils, nutrient stocks, and forest microenvironments feed back to influence species colonization and vegetation growth rates. All three aspects of forest integrity—biodiversity, structure, and ecosystem functions—change in synchrony during forest regeneration. These changes may not all occur at the same rate, however, as structure and ecosystem functions appear to recover more quickly than does species composition (Chazdon et al. 2007). Although these coordinated changes are poorly understood, our current knowledge suggests an increase in functional redundancy of tree species during later stages of succession in tropical forests (Lohbeck et al. 2012; Chazdon and Arroyo 2013).

11.2 Loss of Nutrients and Carbon during Conversion of Forest to Agriculture

Most of the world's tropical forest grows on deeply weathered, nutrient-poor soils, including soils in the central and eastern Amazon basin and white-sand (heath) forests of South America and Indonesia. The luxuriant development of these forests is attributed to nutrient-conserving mechanisms including development of a 15- to 40-centimeter-thick root mat, where nutrients are directly absorbed by mycorrhizal fungi and translocated to below- and aboveground plant tissues. Acidic conditions in the root mat inhibit denitrifying bacteria, preventing losses of highly mobile nitrate ions. Long-lived, evergreen, sturdy leaves and slow growth rates also minimize nutrient loss from vegetation. When these forests and their associated humus layer are destroyed for agricultural use, nutrients are quickly lost, as the mineral soil has little capacity to retain them (box 11.1; Stark and Jordan 1978; Jordan 1989; Jordan and Herrera 1981).

Tropical forests account for about 55% of the carbon stored in the world's forests, according to recent estimates (Pan et al. 2011). This figure is an underestimate, as terrestrial carbon stocks are generally assessed only for the uppermost soil layer, only 0.3 to 1.0 meters deep. Substantial stocks of carbon at depths below 1.0 meters have been reported in eastern Amazonia and the Caribbean lowlands of Costa Rica in deeply weathered soils (Nepstad et al. 1994; Veldkamp et al. 2003). Nearly one-third of the total global soil carbon stock resides in the top 3.0 meters of soil beneath tropical evergreen and tropical deciduous forests (Jobbágy and Jackson 2000). Estimated carbon emissions from soils in the tropics total 0.2 gigatons per year, accounting for 10% to 30% of total carbon emissions from deforestation (Houghton 1999; Achard et al. 2010). Among tropical forests, total ecosystem carbon stocks and the distribution of carbon between aboveground and belowground pools vary widely. Ecosystem carbon stocks in Neotropical old-growth forests vary fourfold, ranging from 141 to 571 megagrams of carbon per hectare (D. A. Clark et al. 2001; Kauffman et al. 2009). Carbon stocks in moist forests range from 249 to 488 megagrams of carbon per hectare, whereas dry forests range from 141 to 344 megagrams of carbon per hectare.

Overall, 32% of the estimated tropical forest carbon stock is stored in the top meter of soil (box 11.2; Pan et al. 2011). Soil carbon stocks store 39% to 77% of the total ecosystem carbon pool in Neotropical forests that burn infrequently. In seasonal dry forests of the northern Yucatán Peninsula in Mexico, 51.4% of the total ecosystem carbon pool resides in the shallow soils (Vargas et al. 2008), whereas in seasonal dry forests in Jalisco, Mexico, in deeper soils, soil carbon composes 54% of total ecosystem carbon pools (Jaramillo et al. 2003).

After forest cutting and burning, the aboveground pools of carbon and nutrients are more depleted than the belowground pools. Forests with high levels of belowground carbon storage lose less total carbon after clearing, at least in the top meter of soil. When tropical forests are converted to pasture, total ecosystem nitrogen pools decline more than carbon pools (Kauffman et al. 1998; Jaramillo et al. 2003). In moist forests of the Brazilian Amazon, the burning of slash after cutting old-growth forests consumed from 29% to 57% of the total aboveground biomass, whereas slash fires in Mexican dry forests consumed even higher percentages (62%–80%) of the original forest aboveground biomass (Kauffman et al. 2009). In the humid tropical region of Los Tuxtlas in Mexico, conversion of forests to pastures or cornfields led to the loss of 95% of the aboveground pool of carbon, 91% of the aboveground pool of nitrogen, and 83% of the aboveground pool of phosphorus (fig. 11.2; Hughes, Kauffman, and Jaramillo 2000). Approximately half of the combined aboveground and soil pools of carbon were lost as a result of deforestation and land use. In contrast to aboveground pools, soil carbon and nitrogen pools were not strongly affected by conversion of forest to agriculture in the Los Tuxtlas region.

The fate of carbon and nutrients in the soil following forest clearing varies widely depending on the type and intensity of land use, as well as soil texture and clay content. According to a global meta-analysis of 385 studies, on average, conversion of tropical old-growth forests to cropland resulted in a 25% loss of soil organic carbon, twice as much as when old-growth forests are converted to pasture (fig. 11.3; Don et al. 2011). These losses are underestimates, as

Box 11.1. A Pioneering Study of Nutrient and Biomass Dynamics following Shifting Cultivation in the Río Negro Region of Venezuela

The infertile, highly weathered, clay-rich Oxisols of the Río Negro region of Venezuela are prevalent throughout the Amazon basin (Markewitz et al. 2004). In 1975, studies were initiated on nutrient fluxes, tree growth, and litter production in a one-hectare plot in old-growth forest. One year later, an adjacent area of old-growth forest was cut and burned after several months of drying (no wood was harvested). Regrowth in this plot was then followed for five years with no further manipulation (Uhl and Jordan 1984). Another section of the cut-and-burned plot was cultivated with manioc, pineapple, plantain, and cashew trees for three years using traditional methods (Uhl et al. 1982; Jordan et al. 1983; Uhl 1987). The cultivated plot was then abandoned and successional patterns were followed using the same approaches as Uhl and Jordan (1984).

After five years, the canopy of the cut-and-burned (uncultivated) plot had grown 14 meters tall (see plate 15). Total aboveground living biomass reached 16% of the biomass present in the forest control site (Uhl and Jordan 1984). Dead wood from the former forest contributed more biomass than living biomass. Leaves and roots, which are more nutrient-rich tissues than wood, composed 27% of the total live biomass in young regrowth compared to only 18% in the intact control forest. In addition, tissue nutrient concentrations were higher in the pioneer species dominating the regrowth. For these reasons, the percentage of forest stocks of phosphorus (23%), potassium (39%), calcium (48%), and magnesium (45%) prior to burning were higher than for biomass (16%).

Nitrogen stocks were slower to recover than biomass. In the intact forest control, nutrient stocks were concentrated in stems and roots, whereas in the

they do not take into account the loss of carbon stocks in subsoils at a one- to three-meter depth from deeply weathered soils after conversion to pasture (Veldkamp et al. 2003).

Pastures are generally not tilled or cultivated, and little carbon is lost in the top meter of soil (McGrath et al. 2001). Conversion of forest to active pasture in northeastern Costa Rica did not lead to significant loss of soil carbon (Reiners et al. 1994; Powers 2004; Powers and Veldkamp 2005). In some cases, young regrowth forest, most of the nutrient stocks were contained within the dead wood fraction. Although concentrations of potassium, magnesium, and nitrate initially increased in leached soil water after cutting and burning, nutrient concentrations declined to levels similar to the uncut forest after two years.

When the cultivated plot was abandoned in late 1979, 20% of the original forest stock of nitrogen had been lost from the ecosystem. Leaves contained about 3% of the nitrogen stocks in the intact control forest; the nitrogen in dried leaves left as slash after cutting was likely all volatilized during the burn. The bulk of the decrease in nitrogen was due to the decomposition of incompletely burned organic matter (primary trunks and roots) over the course of three years. Nitrogen uptake by the crops and successional vegetation constituted a small proportion of the total nitrogen stock. The amount of nitrogen in the soil did not decrease after cutting, burning, and cultivation for three years. Even so, the accumulation of aboveground biomass after five years was 16% lower than in the cut-and-burned site. After five years of succession, soil organic matter stocks were 35%–40% of uncut forest levels (Jordan et al. 1983; Uhl 1987).

These results demonstrated for the first time that, even following forest cutting, burning, and traditional shifting cultivation on nutrient-poor Oxisols, nutrients and carbon accumulate rapidly during forest regrowth. It is unlikely that nitrogen levels limited forest regrowth in either the cultivated or the uncultivated plot. The Río Negro studies clearly showed that the key to ecosystem recovery after deforestation lies in the gradual accumulation of biomass and nutrients in the tissues of recolonizing vegetation. Further, they showed that intensive land uses that exhaust soil nutrients or remove soil organic matter inhibit vegetation regrowth and significantly slow down the accumulation of biomass and nutrients during succession (Uhl 1987).

soil carbon increases following conversion of forest to pasture, as pasture grasses maintain cover, reduce soil temperatures, and add organic matter to soils (Brown and Lugo 1990; Neil et al. 1997; Guo and Gifford 2002). When pastures are formed directly after forest clearing, pasture soils tend to accumulate carbon (Neill et al. 1997). The surface stock of carbon in forest prior to conversion to pasture is the strongest predictor of soil carbon changes after conversion (Neill and Davidson 2000). The clay content of soils is also an

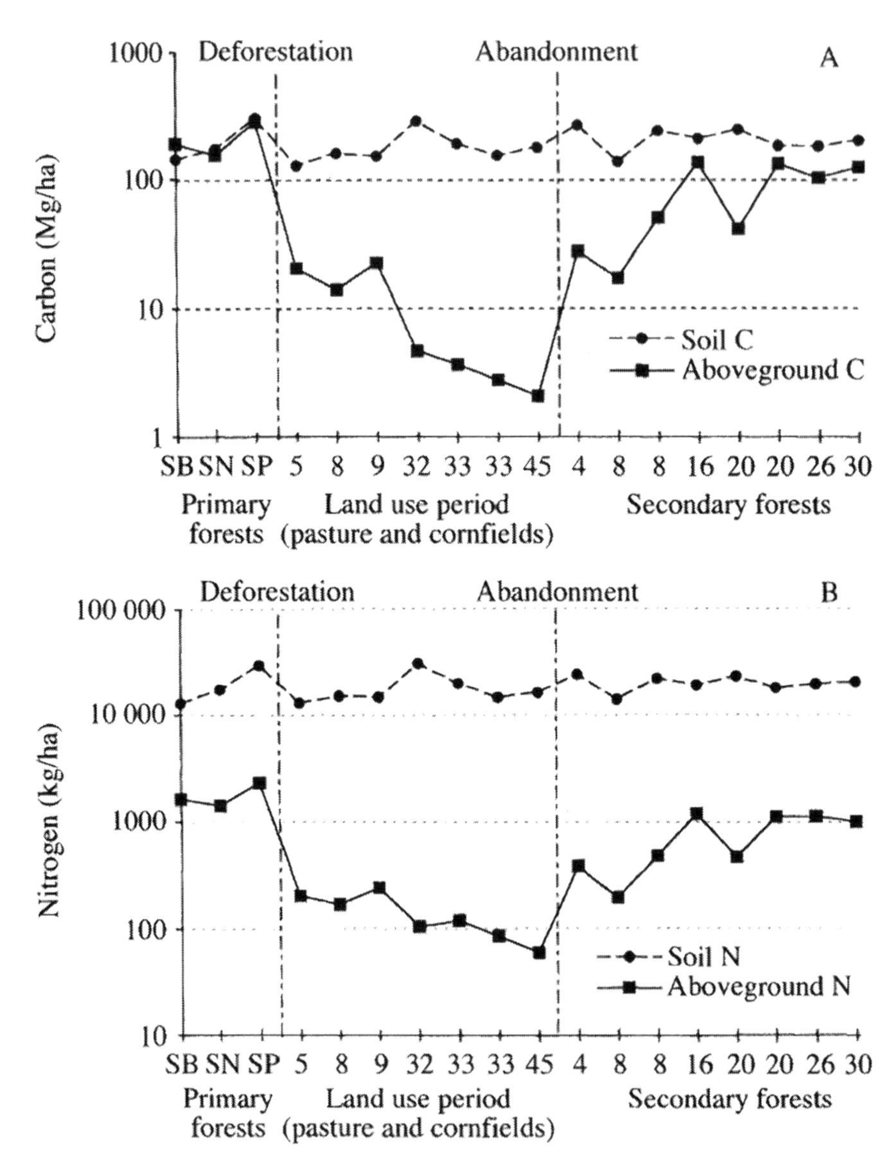

Figure 11.2

Dynamics of carbon (*A*) and nitrogen (*B*) in aboveground biomass and mineral soil pools in intact old-growth forests, pastures, cornfields, and second-growth forests in the Los Tuxtlas Region of Veracruz, Mexico. Values are on a log scale. Values on the *x*-axis represent the number of years of land use or number of years following abandonment. SB, SN, and SP are three different old-growth forest sites. Sources: Hughes et al. (1999, fig. 3). Data for old-growth forests, pastures, and cornfields are from Hughes, Kauffman, and Jaramillo 2000; data for second-growth forests are from Hughes et al. 1999.

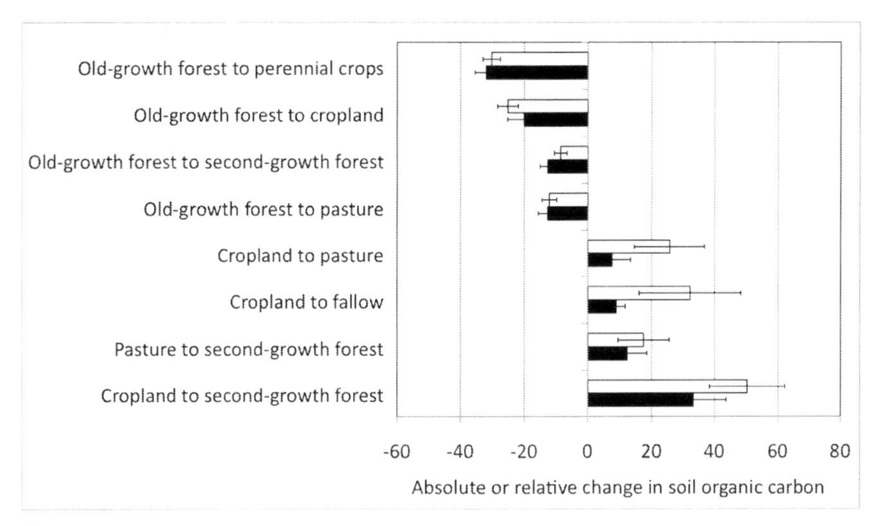

Figure 11.3
Mean absolute (black bars) and relative change (white bars) in soil organic carbon stocks during different types of land-use changes. Error bars are standard deviations. Absolute change is measured in megagrams per hectare, whereas relative change is measured in percent. Data are based on 385 studies on the influence of land-use change on soil organic carbon (0–30 cm depth) in 39 tropical countries. Source: Don et al. (2011, table 1).

important factor affecting accumulation of carbon and nitrogen in pasture soils (Hughes, Kauffman, and Cummings 2002; López-Ulloa et al. 2005; Paul et al. 2008). Ecosystem carbon pools in pastures converted from evergreen forests ranged from 96 to 178 megagrams of carbon per hectare (Hughes, Kauffman, and Cummings 2000; Hughes, Kauffman, and Cummings 2002). In tropical dry forest regions, ecosystem carbon pools of pastures were 77% of old-growth forest levels (Jaramillo et al. 2003).

In contrast to pastures, tillage and cultivation of soils with annual and perennial crops, as in shifting cultivation, causes significant depletion of soil carbon and nutrients (Guo and Gifford 2002; Don et al. 2011). Further, rates of nitrogen mineralization are elevated shortly after cutting and burning (Matson et al. 1987). Exposure of topsoil increases soil temperature and rates of microbial decomposition and mineralization of nitrogen and phosphorus. Ashing of organic matter produces carbonates that increase the pH. Erosion also causes nutrient losses in exposed soils. The legacy of depleted stocks of carbon and nitrogen in soils during shifting cultivation persists for many years in forests regenerating from former croplands (McGrath et al. 2001). The effects of land use and frequent burning of pastures were still detectable in leaf nutrient concentrations of dominant tree species more than two decades

Box 11.2. Carbon Sequestration in Tropical Regrowth Forests

During the early 1980s, as part of efforts to understand changes in global carbon balance, ecologists, geophysicists, and atmospheric scientists began to quantify the losses of CO_2 to the atmosphere from deforestation, burning, logging, and other forms of forest degradation in the tropics, as well as to track the uptake of atmospheric carbon and storage in vegetation biomass through forest succession (natural regeneration), afforestation (plantations, reforestation), agroforestry, and increasing stocks of organic carbon in soil. Early models of global carbon balance overlooked the fact that natural disturbances are common in old-growth forests and affect carbon stocks in biomass and soil organic matter (Houghton et al. 1983). Carbon flux in old-growth tropical forests was assumed to be in a steady-state equilibrium between carbon uptake through photosynthesis and carbon emissions through respiration.

Lugo and Brown (1992) challenged the assumption of old-growth equilibrium. They proposed that both second-growth and old-growth forests functioned as carbon sinks, and estimated that in 1980 between 1.5 and 3.2 petagrams of carbon were sequestered—removed from the atmosphere—in tropical lands across the globe. The carbon sequestered was approximately equal to the carbon released from deforestation that year. The assumption of steady-state carbon flux in old-growth forests has now been disproven by several studies based on forest inventory data and flux tower studies of carbon balance at the ecosystem scale (see box 4.1). In old-growth forests, carbon accumulates as coarse woody debris (0–1 Mg C ha^{-1} y^{-1}), increased tree biomass (1–2 Mg C ha^{-1} y^{-1}), and as soil organic carbon (0.02–0.03 Mg C ha^{-1} y^{-1}). Aboveground carbon accumulation rates are higher in rapidly growing plantations (15 Mg C ha^{-1} y^{-1}) and in secondary forests less than 20 years old (2.0–3.5 Mg C ha^{-1} y^{-1}). These figures support the conclusion that many of the world's old-growth tropical forests are still recovering from past human or natural disturbances.

Although our ability to account for carbon sources and sinks due to land-use change in tropical regions has improved greatly, estimations are associated with many sources of error (Houghton 2010). Our knowledge is limited regarding carbon pools, rates and patterns of biomass loss following different types of land-cover change, and quantification of the carbon storage potential in secondary forests (Kauffman et al. 2009). Uncertainty regarding rates of deforestation accounts for over half of the range of estimations in global carbon flux (Houghton and Goodale 2004). Houghton (2010) estimated that carbon stored in soils and vegetation of secondary tropical forests between 1990 and 2005 was 1.5 petagrams of carbon per year, whereas Shevliakova et al. (2009)

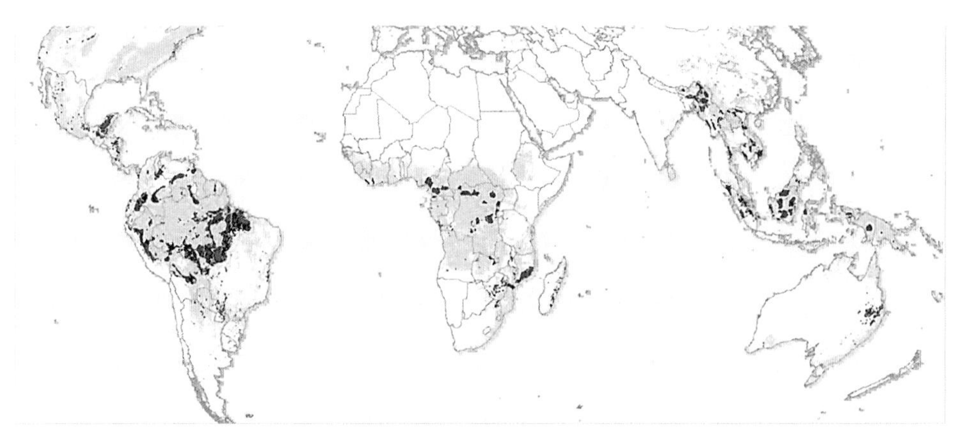

Plate 1
Main tropical deforestation fronts in the 1980s and 1990s based on deforestation hot spots in the tropics identified by three different forest-cover analyses. Red indicates areas of high rates of deforestation, whereas green indicates existing areas of tropical forest. Source: Mayaux et al. (2005, fig. 3).

A

B

Plate 2
Natural and modified soils from Amazonia. *A*, an oxisol, the most common upland soil in Amazonia. *B*, a *terra preta* soil, modified with the addition of ash and organic waste. Source: Bruno Glazer, reprinted with permission.

A

B

Plate 3

A, a view of a blowdown area north of Manaus, Amazonas, Brazil, taken from a helicopter about two years after the event.

B, Giuliano Guimarães assessing vegetation dynamics in a blowdown plot. Source: Jeffrey Chambers, reprinted with permission.

Plate 4
Age classes of second-growth forests regrowing following agricultural abandonment north of Manaus, Amazonas, Brazil, based on analysis of sequential satellite imagery between 1973 and 2003. Ages in years correspond to colors indicated. Dark green is old-growth forest. In 2003, the majority of second-growth sites were less than 30 years old. The edge of the city of Manaus is at the southern end of the image. Source: da Conceição et al. (2009, fig. 4a).

Plate 5

Map of forest areas serving as carbon sources (decrease), sinks (increase), or neutral areas (no change) for La Selva Biological Station. Each square is one hectare. Changes in estimated aboveground biomass from 1998 to 2005 were estimated based on LiDAR data. Second-growth forests and plantations appear uniformly as sinks or neutral areas, whereas old-growth and selectively logged areas are mostly neutral with scattered source and sink areas. Source: Adapted from Dubayah et al. (2010, fig. 11).

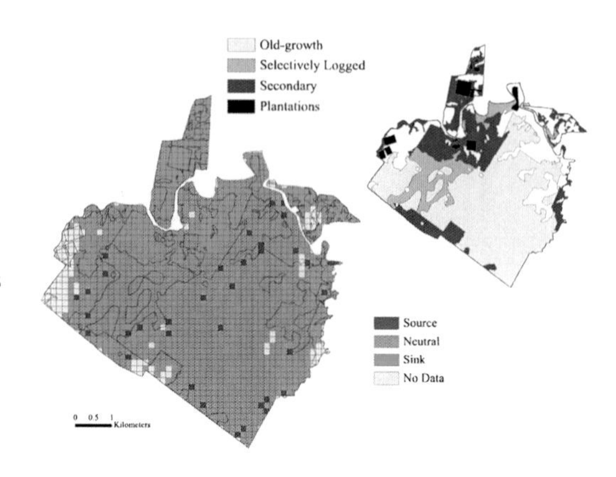

Plate 6

Second-growth forests dominated by several *Vismia* species (*A*) and by *Cecropia sciadophylla* in canopy (*B*) in central Amazonia. Source: Rita G. C. Mesquita, reprinted with permission.

A

B

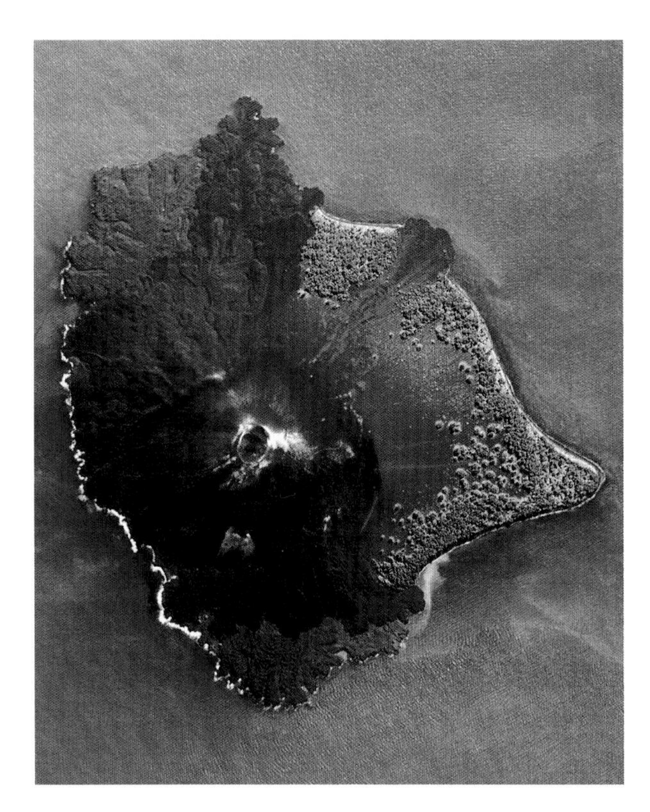

Plate 7
Aerial view of Anak Krakatau volcano in 2005. Primary successional vegetation is visible along the low-lying coastal zones. See box 6.1 for further details. Source: Image courtesy of NASA Earth Observatory.

Plate 8
Time series of a landslide at Luquillo Experimental Forest, Puerto. In 1991 (*middle panel, bottom right*) colonization was more rapid on the residual forest soil. After five years the dominant woody plants were *Cecropia schreberiana* and the tree fern *Cyathea arborea*. Source: Lawrence Walker, reprinted with permission. Walker et al. (1996, fig. 1).

Plate 9
The scar of the landslide and mudslide on Casita volcano in western Nicaragua caused by heavy rains during Hurricane Mitch in October 1998. Source: Photo courtesy of the US Geological Survey.

Plate 10
An image of the Mamore River in lowland Bolivia at the foot of the Andes taken in 2003 by the International Space Station. The Mamore River meanders through a floodplain with numerous contorted channel traces indicating former positions of the river. The darker areas are riverine forest, whereas the lighter areas are tropical savanna. Source: Image courtesy of NASA Earth Observatory.

A

B

Plate 11 Examples of arrested succession: *A*, bracken fern invasion in fallows in Chiapas, Mexico; *B*, former pastures in Soberania National Park, Panama, now covered by three-foot tall *Saccharum spontaneum* grassland. Source: Robin Chazdon.

A

B

C

Plate 12
Early forest regeneration on cleared land in Atewa Range Forest Reserve in Ghana. *A*, study site in February 1975 when the initial transect was laid out. Bauxite ore was removed in the flattened areas with heavy bulldozer use, causing high soil compaction. *B*, 18 months after clearance pioneer trees have established but not on the bulldozed soil. *C*, 4 years after clearance, less bare ground is visible. Pioneer trees dominated by *Musanga cecropioides* are beginning to senesce, while mature forest species increase in density and species numbers. *D*, 15 years later, the site has regrown substantially, and nonpioneer species now dominate. Source: Michael Swaine, reprinted with permission. Swaine and Hall (1983).

D

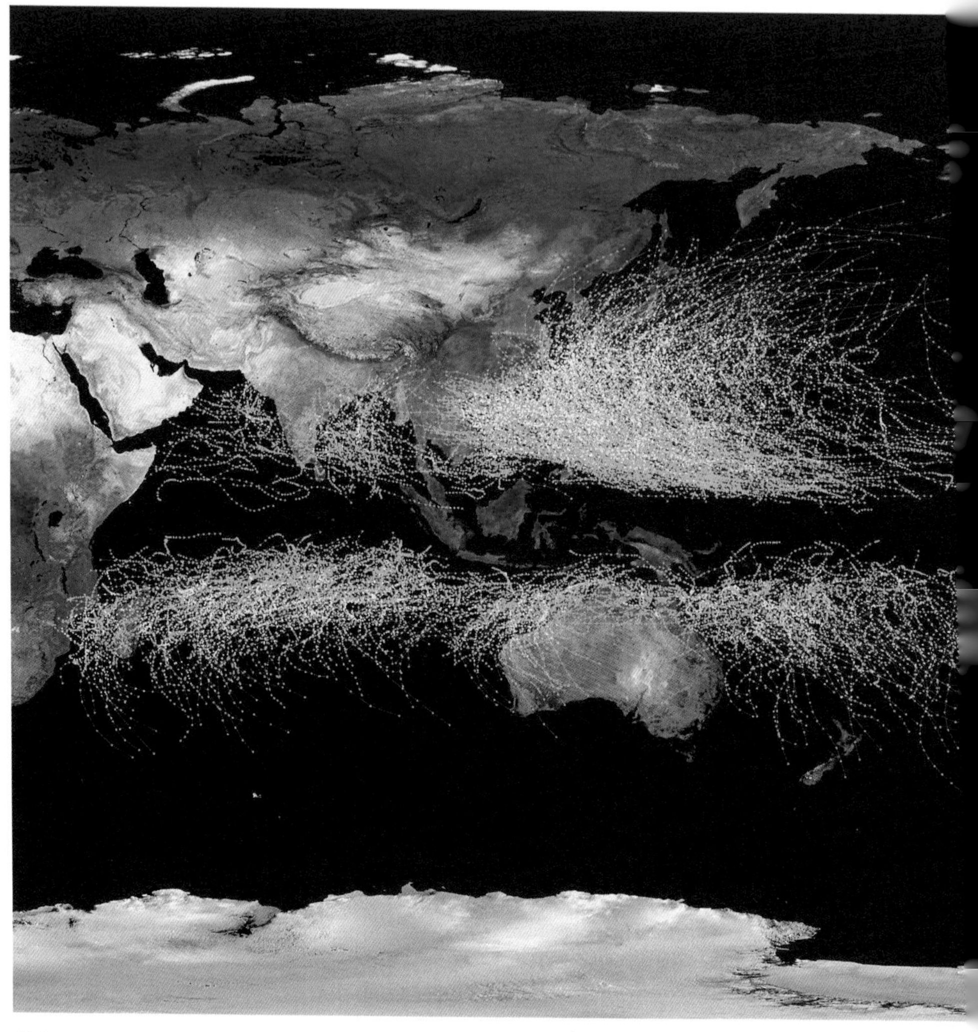

Plate 13

The tracks of all tropical cyclones that formed worldwide from 1985 to 2005.
Source: Wikipedia Commons.

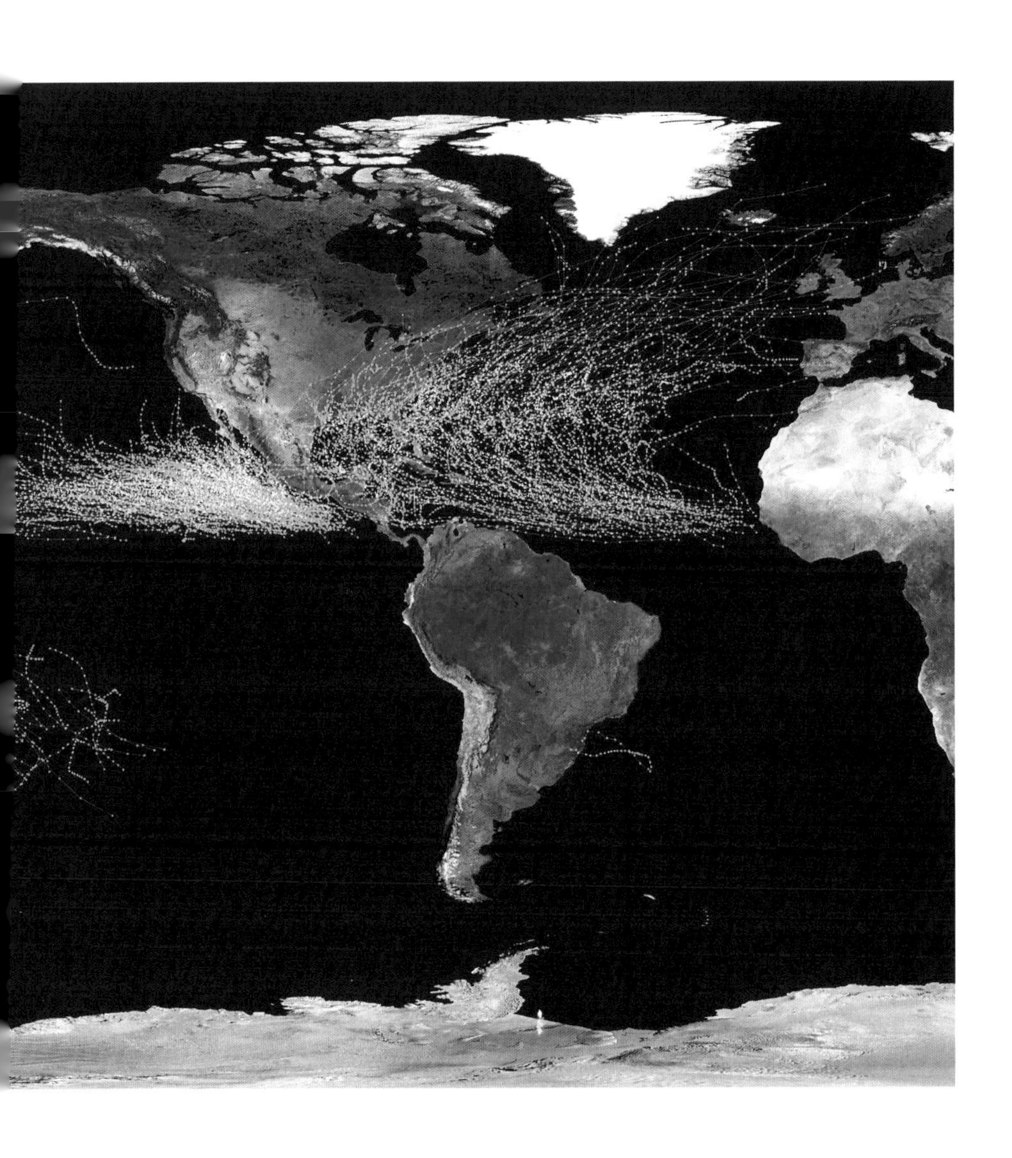

Plate 14
Regeneration of burned forest in East Kalimantan.
Source: Ferry Slik, reprinted with permission.

A

B

Plate 15

A, the experimental cut-and-burned (but not cultivated) Río Negro plot in 1982.
B, the same plot three years after the burn, with cultivated area in foreground.
Source: Christopher Uhl, reprinted with permission.

A

B

C

Plate 16
Photos of barren eroded land (*A*), a *Eucalyptus* plantation after 35 years (*B*), and mixed-species plantings (*C*) in 2004, 45 years after the initiation of the Xiaoliang reforestation experiment. Source: Hai Ren, reprinted with permission.

Plate 17

One-year-old experimental planting of balsa (*Ochroma pyramidale*) in a field infested with invasive bracken fern (*Pteridium caudatum*) for over 30 years in Chiapas, Mexico. Balsa trees have reached a height of 6 meters and a basal area of 4.1 square meters per hectare and are shading out the invasive fern understory.
Source: David Douterlungne, reprinted with permission. Douterlungne et al. (2010).

A

B

C

D

Plate 18

Assisted natural regeneration practices in *Imperata* grasslands of the Philippines: *A*, locating and liberating native tree seedlings; *B*, trainee learning pressing lodging technique; *C*, view of site prior to restoration; *D*, view of site six years after assisted natural regeneration. Source: Photos courtesy of Bagong Pagasa Foundation and FAO, reprinted with permission.

reported a considerably lower estimate of 0.35–0.6 petagrams of carbon per year. The differences in these estimates apparently reflect differences in rates of deforestation used and estimates of carbon lost from soils (Houghton 2010).

Pan et al. (2011) separated global forest sinks into intact and regrowth forest sinks based on an assessment of carbon stocks covering 95% of the world's forests. Further, they assessed three major fluxes in the tropics: carbon uptake by intact forests, carbon uptake of forest regrowth following human disturbances, and carbon loss from deforestation. The global carbon sink formed by tropical regrowth forests was estimated to be 1.7 ± 0.5 petagrams of carbon per year from 2000 to 2007 based on Food and Agriculture Organization (FAO) data. Emissions due to land-use change in the tropics declined between the 1990s and 2000–2007 as a result of decreasing rates of deforestation and increasing forest regrowth. More than half of carbon losses due to tropical deforestation were offset by carbon storage in regrowth forests following abandoned agriculture, logging, or forest clearance. At a global scale, the carbon sink from regrowing tropical forests was estimated to account for 43% of the total forest carbon sink, but only 14% of the total forest area from 2000 to 2007.

The estimated carbon sink in regrowth forests varies considerably across tropical regions. Since 1990, the greatest estimated increase in carbon stocks has been in the Americas (Pan et al. 2011). Caution is needed here, as estimates of gross carbon emissions from deforestation in South America from 2000 to 2007 by Pan et al. (2011) are roughly twice those of Eva et al. (2012), who used satellite-based assessments instead of FAO national statistics. Through reforestation and protection of existing regrowth forests, the global carbon sink in tropical forest regrowth can be expanded (Brown et al. 1993; Chazdon 2008a). This is the goal of the United Nations Collaborative Program on Reducing Emissions from Deforestation and Forest Degradation in Developing Countries, or REDD+, which will be discussed further in chapters 14 and 15 (Gibbs et al. 2007).

Despite the high degree of uncertainly in estimating carbon sinks and rates of deforestation (Ramankutty et al. 2007), there is no disagreement that regenerating forests in the tropics are major sinks for carbon (Yang et al. 2010). But this trend may not continue to hold. Regrowth vegetation in the tropics is being recleared at high rates (Chazdon, Peres, et al. 2009), and multiple cycles of forest clearing and burning can compromise forest regeneration (see box 7.1 and box 9.2). Climate change may also negatively affect biomass accumulation in regenerating vegetation. These factors could reduce future carbon storage by forest regeneration in the tropics. Major incentives are needed to promote forest regeneration and to strengthen carbon sinks in both intact and regrowth forests (see box 14.2).

after abandonment in central Amazonia (Gomes and Luizão 2012). The loss of nutrients and carbon from soils following deforestation can be exacerbated by years of extensive land use, particularly on sandy soils. These losses can be mitigated through land management practices that protect topsoil, prevent erosion, and reduce soil tilling.

11.3 Accumulation of Carbon and Nutrients during Forest Regeneration

Forest regrowth reverses losses of soil carbon and nutrients over time, as aboveground biomass and nutrient stocks accumulate and return depleted stocks to soil. Early deployment of leaf area and rapid turnover of leaf tissue in regenerating stands is a key factor in the replenishment of soil quality. Long-term accumulation of biomass and nutrients in forest vegetation ultimately leads to the recovery and retention of carbon and nutrients in the soil and in the entire forest ecosystem.

11.3.1 Leaf Biomass, Leaf Area Index, and Leaf Litter Production

Accumulations of carbon and nutrients are strongly linked during forest regeneration; nutrients are required for production of plant tissues, and plant production is required for nutrient cycling and organic matter deposition in soils. The concentration of nitrogen in leaves is five to eight times greater than in wood. During fallow periods, the largest source of nutrients in surface soils is through the decomposition of leaf litter. In seasonally dry deciduous forests in the lowlands of eastern Guatemala, litter fall in shifting cultivation fallows increased with age of vegetation to a maximum of 10 megagrams per hectare per year in a 14-year-old stand (Ewel 1976). This level was similar to litter fall in old-growth forests in the region. In contrast, total leaf litter production did not vary with forest age in Atlantic coastal forests on Ilha Grande, Brazil. Second-growth forests at 5 and 25 years old produced similar amounts of litter, 9.9 and 8.7 megagrams per hectare per year, respectively, to old-growth forests (10 Mg ha^{-1} yr^{-1}), despite a strong successional gradient in tree basal area (Oliveira 2008).

Production of nutrient-rich leaves and their subsequent decomposition as litter increase aboveground biomass stocks and replenish lost nutrients in topsoil. As young fallows or pastures regenerate, the fraction of biomass allocated to leaves and fine roots is initially high and then declines as woody structures develop (Ewel 1971; Brown and Lugo 1990). In fallows in Sarawak, Malaysia, leaves composed 3.4%–4.5% of the total aboveground biomass during the first decade; this percentage declined to 2.6% in 17-year-old second-growth forest (Kenzo et al. 2010).

Leaf area index (LAI)—the total leaf area per ground area—increases rapidly during early stages of succession, significantly reducing light availability at ground level (Kalacska et al. 2004). In the fallows studied by Kenzo et al. (2010) in Malaysia, an LAI of 5.0 was reached by 17 years. In contrast, former pastures in central Amazonia attained an LAI of only 3.2 after 12–14 years of regeneration. The recovery of LAI is slower following intensive pasture use than following shifting cultivation (Feldpausch et al. 2005). Vertical profiles of LAI using airborne scanning waveform LiDAR (light detection and ranging) from the Laser Vegetation Imaging Sensor at La Selva Biological Station in northeast Costa Rica showed that mean LAI increased from 2.3 in younger second-growth forests (6–17 yr.) to 5.2 in older stands (18–34 yr.), and reached highest levels at 5.6 in old-growth forest (Tang et al. 2012).

11.3.2 Aboveground Biomass Accumulation and Carbon Storage

During the stand initiation stage, total aboveground biomass accumulates linearly with time since abandonment (Szott et al. 1999). In the upper Río Negro of the Amazon basin, biomass increased linearly for 40 years during fallow regeneration (see box 11.1; Saldarriaga et al. 1988). At this point, aboveground biomass reached 50% of old-growth forest levels. After 40 years, canopy dominants began to senesce and die and new canopy trees were recruited, causing a plateau in forest biomass. Based on these data, Saldarriaga et al. (1988) estimated that old-growth aboveground biomass values are attained after about 190 years of succession in this shifting cultivation landscape.

As woody stems grow in height and diameter, basal area increases and woody tissue becomes the dominant component of stand-level biomass (Brown and Lugo 1990). Because a higher fraction of the total biomass is now invested in nonproductive tissues, the rate of biomass gain declines. Variation in the allocation of biomass to foliage and LAI were key factors affecting the rate of accumulation of biomass in fallows in Malaysia (Kenzo et al. 2010). The linear phase of biomass accumulation appears to be longer during forest succession on nutrient-poor soils, as nutrients can limit the production of plant tissues. LAI increased more slowly in former pastures than in shifting cultivation fallows in central Amazonia, slowing down the rate of aboveground biomass accumulation (Feldpaush et al. 2005).

After the stand initiation phase, biomass growth slows down and may decline as the initial cohort of short-lived pioneer trees dies (Chazdon et al. 2007). The rate of biomass growth eventually reaches a plateau, where mortality of pioneer trees is balanced by recruitment of new trees. These processes extend the time period required to reach the biomass and carbon stocks of old-growth stands, as biomass accumulation reflects changes in the size distribution and basal area of trees. Total stand biomass and carbon stocks are highly sensitive to the presence of very large trees (Brown et al. 1995). In old-growth forest in northeastern Costa Rica, trees greater than or equal to 10 centimeters in diameter at breast height (DBH) contributed less than 2% of the individual stems sampled but more than 25% of the estimated aboveground biomass (D. B. Clark and D. A. Clark 1996).

A substantial component of the total biomass during early stages of forest regeneration following shifting cultivation exists as residual biomass from incomplete burning of the former forest and sparing of selected (remnant) trees (Fearnside et al. 2007). The quantity of residual wood following forest conversion to other land uses strongly affects carbon stocks in pastures and cultivated fields. In the Los Tuxtlas region of Mexico, the total aboveground biomass ranged from 7 to 48 megagrams per hectare in pastures and from 5 to 42 megagrams per hectare in cornfields. In Pará and Rondônia, Brazil, from 48% to 64% of the aboveground biomass in second-growth forests was residual biomass from the original forest (Hughes, Kauffman, Jaramillo 2000). In Amazonian pastures less than 20 years old, residual wood comprised 47%–87% of total aboveground biomass (Kauffman et al. 1998). In 9- and 11-year-old fallows in the Río Negro chronosequence, residual biomass from the forest prior to clearing contributed 42% and 30% percent of total aboveground biomass, respectively (Saldarriaga et al. 1988).

Carbon stocks in vegetation are computed directly from stand-level live aboveground biomass, which in turn is estimated based on allometric equations relating harvested tree biomass to tree diameter, tree height, and specific gravity of wood (Alves et al. 1997; Chave et al. 2005; Nogueira et al. 2008). Most estimates of carbon stocks assume that carbon is 50% of biomass, but recent studies indicate that carbon content of wood, the largest component of biomass, varies significantly among species with a range from 42% to 52% (Elias and Potvin 2003; Martin and Thomas 2011). Carbon concentrations in different components of aboveground biomass in second-growth forests of the Los Tuxtlas region of Mexico varied from 41% to 48% (table 11.1; Hughes, Kauffman, Jaramillo 1999).

Biomass estimation is an imprecise science, as robust data on allometric relationships are available for few vegetation zones and assemblages (Chave et al. 2005; van Breugel et al. 2011). Biomass estimation of second-growth forests in Rondônia, Brazil, (Alves et al. 1997) and northeastern Costa Rica (Letcher and Chazdon 2009b) varied substantially when different allometric equations were used. Another source of error in biomass estimation is that

Table 11.1. *Concentrations of nitrogen (N), carbon (C), sulfur (S), and phosphorus (P) in components of aboveground biomass in second-growth forests of the Los Tuxtlas region, Mexico*

Vegetation/soil component	C (%)	N (%)	S (%)	P (%)
Litter	45 ± 0.6	1.5 ± 0.06	0.18 ± 0.02	0.09 ± 0.01
Seedlings	43 ± 0.4	1.5 ± 0.09	0.20 ± 0.01	0.14 ± 0.01
Grass	41 ± 1.8	1.0 ± 0.11	0.26 ± 0.06	n/a
Palm	47 ± 0.3	0.7 ± 0.13	0.11 ± 0.02	0.10 ± 0.02
Leaves of trees < 10 cm DBH	45 ± 0.5	2.6 ± 0.11	0.31 ± 0.01	0.19 ± 0.01
Leaves of trees > 10 cm DBH	47 ± 0.4	2.5 ± 0.14	0.27 ± 0.03	0.20 ± 0.03
Wood of trees > 10 cm DBH	48 ± 0.2	0.3 ± 0.02	0.04 ± 0.01	0.04 ± 0.01
Wood of trees < 10 cm DBH	47 ± 0.2	0.5 ± 0.05	0.06 ± 0.01	0.10 ± 0.01

Source: Hughes, Kauffman, Jaramillo (1999, table 5).

wood specific gravity data often rely on global databases or information from congeneric species collected in other regions, without accounting for variation within species across regions or successional stages (Williamson and Wiemann 2010). Estimates of rates of biomass and carbon accumulation during forest regrowth should be scrutinized carefully, as many sources of error can lead to inaccurate estimates (see box 11.2; Kauffman et al. 2009).

Climate, soil texture and mineralogy, and the duration and type of land use all influence rates of biomass and carbon accumulation during regrowth of tropical forests (Uhl, Buschbacker, and Serrão 1988; Moran et al. 2000; Zarin et al. 2001). During the first 10 years of fallow growth, aboveground biomass accumulates at faster rates in the humid tropics (4–15 Mg ha^{-1} yr^{-1}) than in the drier seasonal tropics (1–8 Mg ha^{-1} yr^{-1}; Szott et al. 1999). Within 12–14 years, biomass of regenerating forests on former pastures in central Amazonia reached 25%–50% of old-growth forests levels (Feldpausch et al. 2004), whereas 18-year-old forests regrowing following moderate land use in Rondônia, western Amazonia, attained 40%–60% of the old-growth forest levels (Alves et al. 1997). Regenerating forests accumulate carbon more rapidly on nonsandy soils than on sandy soils and in areas with a shorter dry season (Zarin et al. 2005).

Aboveground biomass increases rapidly during succession in seasonally dry tropical forests and can approach values of old-growth forests in 50 to 80 years (Becknell et al. 2012). Rates of biomass accumulation during succession are strongly influenced by mean annual precipitation. In a comparison

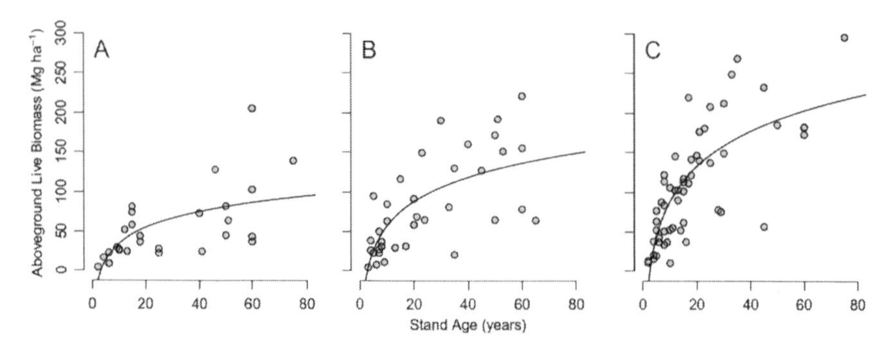

Figure 11.4
The rate of aboveground biomass accumulation increases with annual rainfall in tropical dry forests. Aboveground biomass as a function of log-transformed stand age for sites with mean annual precipitation of 500–1,000 millimeters per year (*A*), 1,000–1,500 millimeters per year (*B*), and 1,500–2,000 millimeters per year (*C*). Source: Becknell et al. (2012, fig. 4).

of biomass accumulation across 178 regenerating tropical dry forests, Becknell et al. (2012) found that biomass accumulation after 40 years of regeneration reached 50–100, 100–150, and over 200 megagrams per hectare in dry, intermediate, and moist forests, respectively (fig. 11.4).

During the first 20 years of succession, rates of aboveground carbon accumulation and aboveground biomass are generally higher in moist and humid forests than in dry forests, and on former agricultural land than on former pastures. Cleared lands with no cultivation showed higher rates of carbon accumulation than sites used for agriculture (Fearnside and Guimarães 1996; Silver et al. 2000; Marín-Spiotta et al. 2008). Zarin et al. (2005) found no systematic differences in carbon accumulation between former pastures and shifting cultivation in the Amazon basin but documented a strong legacy of fire history. Rates of biomass accumulation during forest regeneration across the Amazon basin were strongly affected by the number of previous burns, which often serves as a proxy for the number of shifting cultivation cycles. In the humid region of Los Tuxtlas, Mexico, rates of mean annual aboveground biomass accumulation in regenerating forests were inversely related to the duration of land use in each site (Hughes, Kauffman, Jaramillo 1999). Low rates of biomass accumulation in this region were linked to the reduced availability of soil nutrients following intensive land use.

Estimates of the time required to reach aboveground biomass and carbon storage of old-growth forests vary widely across geographic areas and forest types (table 11.2). These estimates rely primarily on data from relatively young second-growth forests. Based on airborne LiDAR data, estimated above-

Table 11.2. *Estimated time required to reach old-growth forest levels of aboveground biomass during forest regeneration in different regions and forest types*

Region	Time (yr.)	Reference
Rio Negro, Venezuela	144–189	Saldarriaga et al. (1988)
Brazilian Amazonia	100	Fearnside and Guimarães (1996)
Los Tuxtlas, México	73	Hughes, Kauffman, Jaramillo (1999)
Yucatán, México	55–95	Read and Lawrence (2003)
Central Amazonia	175	Gehring, Denich, Vleck (2005)
Yucatan, México	80	Vargas et al. (2008)

Note: Estimates are based on extrapolations modeled from observed chronosequence patterns.

ground carbon storage of 80- to 130-year-old forests on Barro Colorado Island, Panama, was still only 85% of old-growth values (Mascaro et al. 2011). Several chronosequence studies in humid forests of Panama, Costa Rica, and Puerto Rico have found higher levels of aboveground biomass in older second-growth forests than in old-growth forests (Denslow and Guzman 2000; Marín-Spiotta et al. 2007; Letcher and Chazdon 2009b). At least four factors can explain this "overshoot" of biomass during succession. Second-growth forests have fewer canopy gaps than old-growth forests (Montgomery and Chazdon 2001), and have a higher density of biomass-rich dicot trees and a lower density of biomass-poor subcanopy and canopy palms (Marín-Spiotta et al. 2007). Apparent overshoot of tree biomass during succession may also be an artifact of small plot sizes, which generally overestimate aboveground biomass (D. B. Clark and D. A. Clark 2000). Finally, second-growth forest biomass is often over-estimated when using equations based on old-growth forest biomass allometry (van Breugel et al. 2011).

11.3.3 Aboveground Nutrient Pools

In the humid tropics, woody plants are critical for restoring fertility in regenerating vegetation because of their deep and long-lived root systems (Ewel 1986). Root uptake pumps deeply leached nutrients up to surface soils and into plant tissues. Rapid uptake of available nutrients in vegetation can substantially reduce nutrient stocks in the surface soils (0–15 cm). In young regenerating forests on former pastures in eastern Amazonia, soil nutrient concentrations were poorly related to regeneration of vegetation but were related to intensity of prior land use. After 8 years of regeneration, total ecosystem nutrient stocks were higher on sites where pastures were abandoned shortly after formation and lower in bulldozed areas (Buschbacher et al. 1998). Aboveground pools of nutrients were rapidly restored during the first 14 years of regeneration on former pastures in central Amazonia, with the bulk of nutrients stored in wood tissue. Overall, total ecosystem nutrient stocks increased during succession, with the exception of phosphorus (Feldpausch et al. 2004).

In regenerating pastures in Los Tuxtlas, Mexico, aboveground pools of nutrients also increased with forest age (Hughes, Kauffman, Jaramillo 1999). Soil nitrogen pools increased with age since abandonment, whereas extractable phosphorus in soils up to 45 centimeters in depth declined dramatically after 6 years, as also observed in central Amazonia (Feldpausch et al. 2004). Soil nitrogen pools to up to one meter in depth remained relatively stable across the Mexican chronosequence and did not differ significantly between former pasture or cultivated fields or with duration of land use (Hughes, Kauffman, Jaramillo 1999). These results may reflect the capacity of young volcanic soils of this region (andosols) to sequester large quantities of soil organic matter. Concentrations of phosphorus in mineral soils were approximately 10 times higher in second-growth forest soils in the Los Tuxtlas region than in forest soils of southwestern Amazonia (Kauffman et al. 1995).

11.3.4 Soil Carbon Stocks and Nutrient Pools

Soil provides the nutrient capital for forest regrowth after clearance. Soil organic matter stabilizes soil aggregates, increases the water-holding capacity of the soil, provides an energy source for microbial decomposers, and increases soil fertility (Brown and Lugo 1990; Guariguata and Ostertag 2001). During forest regeneration, belowground carbon stocks in topsoils recover more rapidly than do aboveground stocks because belowground biomass is usually not completely eliminated (Powers and Peréz-Aviles 2012). Virtually nothing is known about the dynamics of carbon pools at depths below one meter in regenerating tropical forests.

Prior land use strongly influences rates of accumulation of soil carbon during forest regeneration. During the first 20 years of succession, soils accumulate carbon in the top 50 centimeters almost twice as rapidly after abandonment of pasture than cultivated fields (Silver et al. 2000; Marín-Spiotta et al. 2008). After burning of seasonally dry tropical forests in Yucatán Peninsula, soil carbon stocks returned to 90% of old-growth forest levels within 18 years, but aboveground stocks required 50 years. Carbon pools in the soil remained high following fires, as these forests were not used for agriculture.

In recovering stands 5–29 years old, belowground carbon stocks were significantly higher than aboveground stocks (Vargas et al. 2008).

Contradictory trends in soil carbon have been observed during forest regeneration in different contexts, as soil depth, soil texture, soil type, soil mineralogy, and land use can influence soil organic carbon storage and the dynamics of different fractions of organic carbon (Paul et al. 2008; Neumann-Cosel et al. 2011). Soil carbon shows no significant variation with second-growth forest age in several studies (Marín-Spiotta et al. 2008; Schedlbauer and Kavanagh 2008; Kauffman et al. 2009; Neumann-Cosel et al. 2011; Marín-Spiotta and Sharma 2012). Studies by Brown and Lugo (1990), Erickson et al. (2001), Lopez-Ulloa et al. (2005), and Rhoades et al. (2000) all showed increasing levels of soil organic carbon during forest regeneration. The age of *Acacia mangium* and *Eucalyptus urophylla* plantations and naturally regenerating forest stands in South and North Vietnam also significantly affected soil properties; older sites had lower soil bulk density, lower pH, higher soil carbon concentration, and higher extractable phosphorus levels (Sang et al. 2012). Soil type, prior land use, and the nature of regenerating vegetation all influence changes in soil carbon during forest regeneration.

Biomass accumulation in vegetation during forest regrowth provides inputs of organic matter into soil, leading to coordinated increases in soil carbon and nutrient supply during regeneration. Factors that limit aboveground biomass accumulation will therefore also reduce leaf production and nutrient supply in soils. The effects of climate and soils on ecosystem processes during forest regeneration are understood far better than the effects of species composition and functional traits of dominant tree species.

11.4 Nutrient Cycling and Nutrient Limitation

Growth of young forests may be limited by nutrient availability on highly weathered, nutrient-poor soils and in areas where land-use practices (particularly frequent burning) have depleted soil nutrient stocks (Tanner et al. 1990; Gehring et al. 1999). In eastern Amazonia, Davidson et al. (2004) experimentally fertilized plots in a young regenerating forest on moderately to heavily used pasture. After two years, the rates of tree biomass accumulation nearly doubled in the nitrogen-only and nitrogen-plus-phosphorus treatments. The fertilization effect was short-lived, however, and within five years of application, mean rates of woody biomass accumulation in unfertilized plots matched rates in the fertilized plots. Phosphorus and nitrogen addition favored a few responsive species over nonresponsive species and reduced species diversity compared to untreated control plots (Siddique et al.

2010). Experimental addition of leaf litter in second-growth forests in northeastern Costa Rica also led to a short-term increase in litter production and nutrient inputs but did not significantly influence basal area increments of trees (Wood et al. 2009).

As in primary succession, where forest growth is initially limited by nitrogen availability and later becomes limited by phosphorus availability, nutrient limitations may change during secondary succession (Vitousek et al. 1993; Davidson et al. 2007). Nitrogen fixation by legumes during early stages of succession allows rapid recovery of nitrogen levels in shifting cultivation fallows in Amazonia (Gehring, Vlek, et al. 2005; Gehring et al. 2008). A study of three chronosequences in eastern Amazonia showed that young successional forests have a conservative nitrogen cycle, but nitrogen availability in soils increases during succession, leading to increasing foliar nitrogen concentrations and higher nitrogen-to-phosphorus ratios in leaf litter. Decreasing foliar concentrations of phosphorus with increasing forest age, in contrast, suggest that phosphorus becomes more limiting later in succession (Feldpausch et al. 2004; Davidson et al. 2007).

During forest succession in southern Yucatán Peninsula in Mexico, the concentration of nitrogen in litter tends to increase, whereas the concentration of phosphorus decreases with forest age (Read and Lawrence 2003). Phosphorus cycling changes dramatically during forest regrowth, leading to increased efficiency of phosphorus utilization. Nitrogen inputs to the ecosystem increase during forest regrowth; aboveground pools of phosphorus remain constant while total leaf area and rates of litter production increase, therefore reducing the phosphorus concentration in litter (Lawrence and Foster 2002). Experimental fertilization using nitrogen and phosphorus treatments in second-growth forests in Yucatán increased seedling species richness and survival (Ceccon et al. 2003). Further, addition of nitrogen and phosphorus led to increased phosphorus concentration in leaves of dominant tree species and to higher nutrient inputs in leaf litter in 10- and 60-year-old second-growth forests in northeastern Yucatán (Campo and Dirzo 2003; Campo et al. 2007).

Experimental studies in a range of tropical forests suggest that multiple nutrients may influence carbon cycling during forest regeneration as well as in old-growth forests (Townsend et al. 2011). Molybdenum has been shown to limit nitrogen fixation in lowland forests in Panama (Barron et al. 2009), and nitrogen-fixing plants have enhanced abilities to extract soil phosphorus (Houlton et al. 2008). Nutrient limitations may shift seasonally, spatially, and temporally during forest regeneration (Townsend et al. 2011). Tree species grown in plantations have been shown to differ in their rates of nitrogen uptake and ability to mobilize nitrogen from soil organic matter (Russell and Raich 2012). The complex set of interactions that influence nutrient and carbon cycling in regenerating and old-growth tropical forests is far from being completely understood.

11.4.1 *Root Traits and Mycorrhizal Colonization during Forest Regeneration*

At least 80% of tropical trees have symbiotic relationships with mycorrhizal fungi (Bâ et al. 2012), which allow higher rates of absorption of phosphorus and nitrogen in soils of low fertility. The most common and oldest mycorrhizal association in tropical plant species is between plant roots and endomycorrhizal fungi, whose hyphae penetrate the root cortex and form arbuscules or vesicles inside the root tissue (Janos 1980). Hyphae of ectomycorrhizal fungi, in contrast, form a sheath surrounding fine roots but do not penetrate the root tissue. Ectomycorrhizal associations are widespread in cold-temperate and boreal latitudes but also occur within tropical trees in a number of economically and ecologically important plant families and subfamilies, including the Caesalpinioideae (Fabaceae) and the Dipterocarpaceae. Proximity to conspecific adult trees has been shown to enhance seedling survival in some, but not all, ectomycorrhizal species examined (Newbery et al. 2000; Onguiene and Kuyper 2002; Bâ et al. 2012).

Few studies have examined variation in the abundance and composition of mycorrhizal propagules in soils and the consequences of mycorrhizal associations for tree establishment and growth during natural regeneration. Conversion of forests to pasture or cultivation may alter the composition of microbial and fungal communities in the soil, potentially affecting mycorrhizal colonization rates early in succession (Janos 1980; Allen et al. 2003; Haug et al. 2010). In a 27-year dry forest chronosequence in Oaxaca, however, hyphal length, arbuscular mycorrhizal colonization, and spore density showed no variation with stand age (Guadarrama et al. 2008). In Chamela dry forest in Jalisco, Mexico, forest clearing and burning followed by shifting cultivation had no significant effect on mycorrhizal spore abundance, species richness, or root colonization (Aguilar-Fernández et al. 2009). Using inoculated soil from early or late successional stages, several experimental studies have examined the effects of inoculation on seedling growth. Huante et al. (2012) found that enhancement of seedling growth was similar using inoculum from either pasture or old-growth forest in Chamela dry forest region of Mexico. In another seasonally dry forest in Yucatán, Mexico, Allen et al. (2003) found that inoculum from early successional stages was more beneficial to plant growth than inoculum from late successional stages.

During early stages of forest regeneration, colonization of roots by arbuscular mycorrhizae can be important for supporting rapid growth rates of pioneer species. Early successional tree species showed higher rates of arbuscular mycorrhizal colonization than late successional species in both greenhouse and field studies in southern Brazil (Zangaro et al. 2003). Response to mycorrhizal inoculation was inversely related to seed weight. In three different regions of Brazil, soils showed continuous increases in fertility during succession, and mycorrhizal colonization decreased as succession advanced (Zangaro et al. 2012). Fine root mass and root diameter increased during succession, whereas specific root length, root hair length, and spore density decreased. The higher carbon costs of maintaining mycorrhizal relationships may be one reason for the decline in colonization during succession, as light becomes more limiting and slow-growing species predominate.

Root mass, root length, and arbuscular mycorrhizal colonization were higher in an old-growth forest compared to a 40-year forest in Xishuangbanna, southwest China, whereas spore density and specific root length were higher in the second-growth forest. The percentage of root length colonized by mycorrhizae was negatively correlated with the diameter of coarse and fine roots (Muthukumar et al. 2003). In a dry forest chronosequence in northeast Costa Rica, fine root biomass varied significantly with soil fertility measures but not with stand age (Powers and Peréz-Aviles 2012).

11.5 Hydrology and Water Balance

Tropical forest vegetation functions as a conduit for water movement from soils to the atmosphere. The transpiration from forests has important implications for regional climate. As much as 25%–56% of the Amazon basin rainfall is derived from "recycling" water evaporated within the basin (Eltahir and Bras 1996). The cycling of water from forest soils back to the atmosphere is disrupted following deforestation. When forests are cleared, rates of water-flow into streams increase significantly and frequently cause flooding and erosion. The removal of more than 33% of forest cover causes significant increases in annual rates of stream-flow over the next three years (Bruinjzeel 2004).

Regrowth of forests can rapidly restore many hydrological functions. Evapotranspiration rates in a 2.5-year-old regrowth forest in Pará, eastern Amazonia, were similar to old-growth forest rates in the same region (Hölscher et al. 1997). The water budget of a 3.5-year-old regrowth forest in Bragantina, eastern Amazonia, did not differ from that of old-growth forest growing under similar climate and edaphic conditions (Sommer et al. 2003). Evapotranspiration rates of 15- to 18-year-old forests in Pará, Brazil, were virtually identical to rates in neighboring old-growth forests during a four-year study period (Nepstad et al. 2001). Deep root systems of trees in both second-growth and old-growth forests draw water from depths of at least eight meters. Drought-resistant trees are more abundant in old-growth forests, leading to higher rates of leaf shedding in young second-growth forests during the dry season and increased flammability.

Rain falling on tropical forests directly penetrates to the forest floor as throughfall, flows down along branches and stems as stemflow, or is intercepted by foliage, canopy soil, or epiphytic mats. In a typical old-growth tropical forest, 80%–95% of incident rainfall infiltrates into the soil. Of this amount, about 1,000 millimeters per year are released to the atmosphere through transpiration, and the remainder flows into streams (Bruijnzeel 2004). Even after 40 years, second-growth tropical montane forests in Costa Rica have lower rates of rainfall interception compared to old-growth forests (Köhler et al. 2006). Stemflow accounted for as much as 17% of incident rainfall compared to only 2% in old-growth forests (Hölscher et al. 2004). These differences are mostly determined by differences in stem density and canopy structure. Because of the slow establishment of vascular and nonvascular canopy epiphytes, more than 100 years may be required for epiphyte mat weight and hydrological functions to reach old-growth forest values (Köhler et al. 2007).

The soil saturated hydraulic conductivity (K_s) determines the rate of water movement through the soil matrix under saturated conditions. When K_s is low, vertical percolation of water is impeded during high-intensity rain events, leading to runoff. Several studies have found lower K_s in pasture soils compared to forest soils in tropical regions (Hassler et al. 2011). Recovery of K_s after pasture abandoment took 15 years in a lowland Amazonian region (12.5 cm depth; Zimmermann et al. 2006) and 12 years in central Panama (0–6 cm depth; Hassler et al. 2011).

11.6 Conclusion

Lugo and Brown (1992) described regenerating tropical forests as exemplifing a "relentless tendency" to accumulate carbon. As they regenerate, forests accumulate much more than carbon. When trees grow, they accumulate nitrogen, phosphorus, calcium, and other nutrients in foliage, stems, branches, and roots and recycle them in litter and dead plant matter. Forests are the ultimate recyclers. They enrich soils with organic matter that increases fertility and sequesters carbon. They pump water from soils into the atmosphere. They increase in height and structural complexity, creating spatial arrays of microenvironments for diversification of plant and animal species. These ecosystem processes also occur in actively restored tropical forests and are the basis for the provision of valuable supporting and regulating ecosystem services in regenerating and restored tropical forests (see box 14.2), such as carbon sequestration (see box 11.2), nutrient cycling, and hydraulic regulation.

Based on studies conducted throughout the tropics, rates of accumulation of carbon and nutrients during forest regeneration appear to be more strongly influenced by climate and soil type than by species composition (Marín-Spiotta and Sharma 2012). The functional and structural changes that occur during succession are broadly similar across different regions and are strongly mediated by climate and soils. The effect of spatial variability in species composition and forest structure on carbon and nutrient cycling during forest regeneration is far less studied. The application of high-resolution remote-sensing methods to detect fine-scale variation in vegetation structure and composition (LiDAR, hyperspectral imagery, radar) provides an unprecedented opportunity to explore the role of species composition and forest structure in mediating successional changes in tropical forest ecosystems (Asner et al. 2011; Gallardo-Cruz et al. 2012; Tang et al. 2012).

CHAPTER 12

ANIMAL DIVERSITY AND PLANT-ANIMAL INTERACTIONS IN REGENERATING FORESTS

To plan a conservation policy for the humid tropics much more knowledge of secondary successions and secondary forests is required, but our present inadequate knowledge suggests that a considerable part of the existing forest flora and fauna may survive in the secondary forests and other seral communities of the future, especially if some relics of near-climax communities can be preserved in their neighbourhood as nuclei for re-colonization.—Paul W. Richards (1971, p. 178)

12.1 Overview

Complex interactions among species are the hallmark of tropical forest diversity. The manduvi tree (*Sterculia apetala*) of the Pantanal wetlands of Brazil presents a stunning example of a nexus of two types of specialized interactions between trees and charismatic birds (Pizo et al. 2008). These tall canopy trees in patches of semideciduous and gallery forest provide 95% of the nesting sites for the endangered hyacinth macaw (*Anodorhynchus hyacinthinus*). The fruits of the manduvi tree are eaten by 14 species of birds, but only 1 of these frugivores—the toco toucan (*Ramphastos toco*)—effectively disperses seeds far away from fruiting trees. Undispersed seeds that fall below the crown are consumed by peccaries and agoutis. Toco toucans account for 83% of all manduvi tree seed dispersal, increasing the likelihood that seedlings will be distributed over space and encounter favorable sites for survival and growth into adult trees. The foraging activities of toco toucans directly promote the growth of the manduvi trees required by the hyacinth macaw. But the story doesn't end here. The toco toucans also prey upon the eggs of the hyacinth macaw, accounting for about 53% of egg predation each year. The reproductive fates of all three species are tightly intertwined.

The interactions between plant species, dispersal vectors, pollinators, herbivores, and pathogens steer the course of successional trajectories. From the perspective of animal populations, the changes in vegetation structure and species composition during tropical forest regeneration determine the overall habitat quality of regrowth forests. Visitation, colonization, feeding, and breeding of animal species depend heavily on resources provided by plant species, such as food, shelter, roosting, nesting, and mating sites. As is always the case with species interactions, the dependency goes both ways.

The reproduction, colonization, establishment, and recruitment of plant species also depend heavily on animals that pollinate flowers, disperse seeds, consume seeds and leaves, bury seeds, and protect plants from ever-present herbivores.

Key mutualisms propel linked successional changes in plant and animal assemblages, joining the fates of interacting animal and plant populations. One set of mutualistic interactions facilitates the arrival of new actors that bring new mutualistic partners into the assemblage. These mutual dependencies create positive feedbacks on both plant and animal diversity in regenerating forests, leading to increasing species richness and functional diversification. For example, tropical forest regeneration often follows a nucleation model of succession, where patches of regenerating vegetation develop surrounding remnant or planted trees that attract seed-dispersing frugivores (see box 7.2). Pioneer tree species that produce small fleshy fruits consumed by numerous generalist frugivores serve as interspecific hubs, creating patches of diverse seed rain in old fields that generate spatial structure in regenerating forests (box 12.1; Franklin and Rey 2007; Cole et al. 2010). Dispersal at pioneer tree hubs in old fields can be considered a form of directed dispersal, as fleshy-fruited pioneer trees essentially function as "nurse" species that enhance seed germination and seedling establishment of seeds deposited there (Carlo et al. 2007). Even if remnant trees are cut later, they can leave a legacy of enhanced seed rain in the seed bank in pastures (Howe et al. 2010).

Dependencies between partners also create vulnerabilities for both plant and animal populations if key mutualists become rare, are extirpated locally, or are unable to colonize early successional habitats or later stages of forest regeneration (Babweteera and Brown 2009). Fruit production of two *Ficus* species that colonized the recently formed Anak Krakatau volcano became limited by a shortage of species-specific pollinating wasps (see box 6.1; Compton et al. 1994).

Mutualisms vary along a spectrum from obligate to facultative, and the species involved can be generalists or specialists in the interaction. Some birds, for example, are obligate frugivores, but they feed on a large number of species. Further, these relationships are usually asymmetric, with one partner more dependent on the interaction than the other. The extent to which mutualisms involve specialists and generalists strongly affects the vulnerability and resilience of ecosystems. Large-seeded woody species—such as the manduvi tree—generally rely on only one or a few species of specialist frugivores for effective seed dispersal, whereas small-seeded species are dispersed by many species of generalist frugivores (Loiselle and Blake 2002; Wheelwright et al. 1984). Larger-seeded species are therefore more vulnerable to lack of dispersal and reduced ability to colonize regenerating areas or remnant forest fragments if large frugivores are extirpated (Cordeiro et al. 2009; Moran et al. 2009; Sethi and Howe 2009).

Large-seeded, old-growth specialist tree species are declining from isolated forest fragments surrounded by cleared land and early successional habitats and are being replaced by successional tree species (Martínez-Garza and Howe 2003; Tabarelli et al. 2008). This scenario has played out in Samoa, Tonga, and other Pacific islands, where large birds and bats (flying foxes) are key dispersers of native trees (Cox and Elmqvist 2000). Only one frugivorous pigeon species (*Ducula pacifica*) remains in Tonga, and seeds larger than three centimeters in diameter have only a single vertebrate disperser left—the flying fox (*Pteropus tonganus*; McConkey and Drake 2002). Small-seeded species, in contrast, can be effectively dispersed by a number of different frugivore species; this functional redundancy favors their resilience following disturbances.

Old fields and young regrowth forests provide different types and amounts of resources (food and shelter) for animals than do older second-growth, logged, or old-growth forests (DeWalt et al. 2003; Bowen et al. 2007). Therefore, assemblages of different animal taxa change dramatically during forest regeneration and are closely linked to changes in forest structure and plant species composition. Species that require tree cavities for nesting, such as the hyacinth macaw, do not nest in young forests, although they might forage there. Manduvi trees do not develop nest cavities until they are at least 60 years old (Santos et al. 2006). Animal taxa vary widely in their abundance in early stages of forest regeneration due to differences in mobility, life history, and specialization on particular resources (Harvey et al. 2006; Barlow, Gardner, et al. 2007; Chazdon, Peres, et al. 2009). Phyllostominae bats were more abundant in mature forests and in *Cecropia*-dominated regenerating forests than in old fields and *Vismia*-dominated regeneration (Bobrowiec and Gribel 2009).

This chapter focuses on the current state of knowledge about how animal diversity and composition change during forest regeneration and interactions between plants and animals that drive successional processes. How do mutualistic interactions and partners change as plant and animal communities assemble during forest succession? The importance of plant-animal interactions during forest regeneration extends beyond the pollination and seed dispersal phases. Many species of animals serve as obligate pollinators for most tropical forest plants. The majority of tree fruits are removed and consumed by frugivores and are then dispersed through animal digestion, seed

Box 12.1. Interaction Networks and Mobile Links

Mutualistic interactions among multiple species can be visualized as a network in which individual species are the "nodes" and interactions between species are "links," or "edges" (Jordano 1987; Bascompte and Jordano 2007). The structure of these networks essentially describes the architecture of species interactions. Understanding the structure of interaction networks can provide insight into their stability or fragility in response to the loss or gain of mutualist partners in dynamic tropical landscapes (Montoya et al. 2006; Mello et al. 2011b).

A nested network structure emerges when specialists interact with generalists rather than with other specialists (Bascompte and Jordano 2007). Plant-animal mutualistic networks are highly nested, with a core set of interacting generalist species. Interacting groups of generalists provide functional redundancy, promoting stability if a species is temporarily or permanently lost from the assemblage. In addition, the asymmetrical structure of dependencies promotes the persistence of specialists. Pollination networks are likely to be more modular (with distinct subnetworks) than seed dispersal networks, because flowers can restrict floral visitors through morphological barriers. Fruit traits are less restrictive than floral traits (Donatti et al. 2011).

Mello et al. (2011b) compared bat-fruit and bird-fruit networks based on 17 published data sets. Overall, frugivorous birds interacted with more plant species than frugivorous bats. Bat-fruit networks were more nested, and observed links constituted a higher fraction of the total possible number of links between species. Frugivorous bats have a less diversified diet than frugivorous birds, leading to more generalized interactions among bats within subnetworks compared to birds. Analyses of nine bat frugivore networks in Neotropical forests indicated an average of four modules, with bat species grouped together with species in their preferred plant genera. Redundancy within modules and complementarity among modules leads to high robustness to removal of either plant or bat nodes in bat-plant networks (Mello et al. 2011a). But the smaller overall size of bat-plant networks was associated with a higher vulnerability to extinction than bird-plant networks when species were randomly removed in simulations (Mello et al. 2011b).

The largest plant-frugivore network to date was assembled for Parque Estadual Intervales (PEI) in southeastern Brazil. Within 42,000 hectares of old-growth and second-growth forest, 207 species of fruiting plants were dispersed by 110 animal species. On average, each plant has 5.4 animal partners and each frugivore has 10.2 plant partners. The network is highly asymmetrical, with specialists often dependent upon generalists, leading to a high degree of nestedness (Silva et al. 2007). Simulated changes in the abundance of a single species will rarely propagate throughout the entire PEI forest network, but a random disturbance that eliminates a "super-generalist" will likely affect many species in the network.

The plant-frugivore network of Kakamega Forest in Kenya consists of 88 frugivores and 33 plant species (Schleuning, Blüthgen, et al. 2011). Old-growth and second-growth plants showed similar degrees of specialization on particular frugivore species. Canopy plants were found to depend mostly on obligate frugivores that foraged on many species, midstory plants depended most on partial frugivores, and understory plants depended most on opportunistic frugivores whose choice of diet was more restricted. These differences led to more generalized plant-frugivore interactions in the canopy and more specialized interactions in lower strata.

Plant species that are visited by many frugivore species (hubs) are promising candidates for reforestation programs, as they will likely attract more seed-dispersing species and thereby accelerate forest regeneration (Silva et al. 2007). But frugivores present in the forest interior do not necessarily disperse seeds into former pastures or clearings. None of the top three generalists in an avian frugivore-plant network for second-growth forests in Puerto Rico showed significant activity in former pastures. The gray kingbird (*Tyrannus dominicensis*), which had a low importance in the forest network, was responsible for the highest deposition of seed in pasture sites and for the most crossings between the pasture, where it feeds primarily on insects, and forest patches, where it eats fruit (Carlo and Yang 2011).

Birds such as the gray kingbird serve as mobile links, connecting different landscape elements and connecting multiple interaction networks (Gilbert 1980; Lundberg and Moberg 2003). The frugivorous Samoan flying fox (*Pteropus samoensis*), native to Samoa and other Pacific islands, is the only remaining species capable of pollinating and dispersing seed of a large proportion of the island's canopy tree species. This near threatened species is also the primary pollinator of a liana (*Freycinetia reineckei*). Loss of mobile links impacts multiple interaction networks and can have cascading effects in several different habitats. Mobile links, while being maintained by larger forest fragments in the region, can provide critically important services such as long-distance seed dispersal in disturbed and regenerating forests.

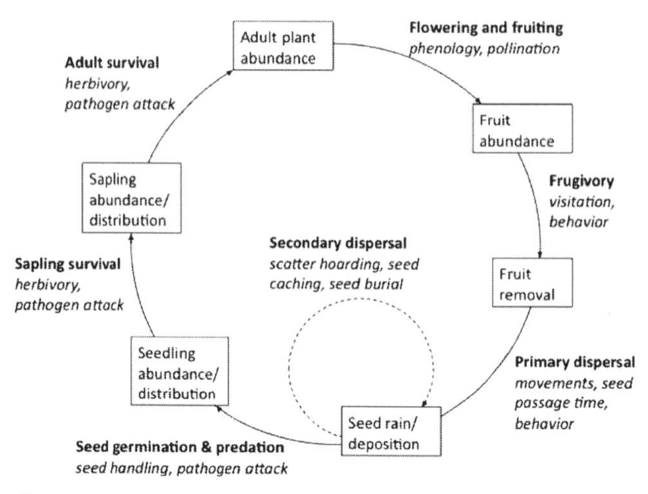

Figure 12.1

Species interactions occur at all phases of a plant's life cycle. General processes (in bold) and specific processes (in italics) determine the abundance of fruits, dispersed seeds, seedlings, saplings, and adult plants. Source: Redrawn based on Wang and Smith (2002, fig. 1).

handling, and movement. Seed dispersal by a diversity of primary and secondary dispersal agents, however, does not ensure seedling germination and survival (Wang and Smith 2002; Carlo and Yang 2011). Animals and pathogens attack seeds, seedlings, and leaves and wood of mature plants, affecting their growth and survival (fig. 12.1).

Seed dispersal and pollination distances have important consequences for the genetic composition of populations during forest regeneration. Long-distance dispersal is particularly important for colonization of former agricultural land or following natural disturbances (Nathan and Muller-Landau 2000) and for reducing effects of density-dependent seedling mortality (Comita, Muller-Landau, et al. 2010; Metz et al. 2011). An increasing number of studies provide molecular evidence for long-distance dispersal of pollen, seeds, and established seedlings in wind- and animal-dispersed tree species (Jones et al. 2005; Sezen et al. 2005; Hardesty et al. 2006). Colonization events during initial stages of succession can create genetic bottlenecks if founders are dominated by a small number of genotypes (Sezen et al. 2005, 2007; Davies et al. 2010). Ensuring genetic diversity in regenerating populations of tree species requires both the maintenance of high genetic diversity in source populations and conservation of animals that pollinate flowers and disperse seeds across long distances.

12.2 Animal Diversity in Regenerating Forests

Dunn (2004b) conducted the first meta-analysis of species richness and species composition of different animal taxa in different ages of second-growth forests. Based on 39 data sets covering a wide range of taxa, Dunn observed that standardized metrics of species richness increased asymptotically to mature forest levels during the first 30 years of forest regrowth. These same studies further revealed that the composition of species of birds and ants did not recover as rapidly as species richness. A growing number of studies have indicated that animal diversity increases during succession and that species composition recovers slowly due to the gradual influx of old-growth specialists (Bowen et al. 2007; Chazdon, Peres, et al. 2009; Karthik et al. 2010).

Studies of 15 taxonomic groups in the Jari region of Brazil showed highly idiosyncratic patterns of occurrence in regenerating forests compared to old-growth forests; 95% of old-growth orchid bee species occurred in second-growth, but less than 60% of old-growth lizards, dung beetles, leaf-litter amphibians, and birds were found (Barlow, Gardner et al. 2007). Dent and Wright (2009) examined 65 studies that compared faunal diversity (birds, herpetofauna, invertebrates, and mammals) in tropical old-growth forests and forests regenerating following complete clearance. On average, 58% of the species present in second-growth forests also occurred in old-growth forests. Moreover, the compositional similarity between old-growth and young regenerating forests (as measured by the Sørensen index) was similar to levels of similarity evaluated only among old-growth sites. In a subset of studies, similarity between regenerating and old-growth forest increased significantly with increasing time since abandonment. This increase in similarity arises from an increasing percent of old-growth species that are present in regenerating forests over time, reaching 80% in forests older than 50 years. Species composition of forests regenerating following intensive agriculture or pasture was less similar to old-growth forests than those recovering from shifting cultivation or clearance without cultivation. Not surprisingly, regenerating forests adjacent to old-growth forests had higher levels of similarity to old-growth forests than forests that were not adjacent. These findings mirror successional patterns of composition observed with tree species.

12.2.1 Vertebrates

Body size and flying ability are important factors affecting vertebrate diversity in regenerating forests. Large-bodied canopy frugivores and understory insectivores are adversely affected by habitat disturbance and are often restricted to old-growth forests (Karthik et al. 2009). Second-growth forests in the Jari region of Brazil supported a higher abundance of ungulate browsers and small-bodied primates than old-growth forests but a lower abundance of large-bodied birds and large primates (Parry et al. 2007). Nonflying vertebrates generally show slower rates of species influx into regenerating forests than flying taxa (Chazdon, Peres, et al. 2009). Of the 30 species of nonflying mammals detected in old-growth forests in the humid forest region of Los Tuxtlas, Mexico, only 11 were detected in 25- to 35-year-old forests and 8 in 5- to 15-year-old forests (Estrada 1994). Amphibian species richness and composition, however, did not vary with stand age among regenerating forest sites and old-growth sites in northeast Costa Rica (Hilje and Aide 2012).

A high proportion of old-growth forest birds are found in regenerating forests. Based on a review of 10 studies, Karthik et al (2009) reported that 70% of old-growth forest birds had recovered within 25 years of abandonment of shifting cultivation fields. In central Sulawesi, Indonesia, young second-growth and old-growth forests had similar species richness of birds, but the number of endemic bird species was significantly lower in second-growth forest (Waltert et al. 2004). In tropical lowland forests of Costa Rica, species richness of birds was higher in four-year-old regrowth compared to older second-growth (25–35 yr.) and old-growth forest. The high species richness of birds in young second growth was attributed to several factors including close proximity to old-growth forest source areas, high production of fruits and flowers in second-growth for frugivorous and nectarivorous birds, and the importance of young second growth for seasonal migrants. Migrants accounted for 15%–30% of the species and 6%–44% of the captures in these forests (Blake and Loiselle 2001). Migratory birds wintering in Panama commonly have small gape widths and are concentrated in second-growth forests where small fruits abound (Martin 1985).

Studies in Amazonia present contrasting patterns of bird species richness during forest regeneration. Species richness was similar in 7- to 35-year-old forests and old-growth forests on the western bank of the Río Negro, although species composition differed (Borges 2007). In contrast, 14- to 19-year-old second-growth forests in Jarí, Brazil, supported lower species richness of birds than old-growth forests in the same area (Barlow, Mestre, et al. 2007). These geographic differences in bird diversity in regenerating forests may, at least in part, reflect differences in the concentration of migratory bird species. Amazonia is south of the main wintering areas for most of the North American passerine bird species (Stotz et al. 1992), and migrants account for only a small proportion of birds in central Amazonian habitats. Nocturnal birds showed similar species richness and composition between old-growth and second-growth in terra firme forests in central Amazonia (Sberze et al. 2010).

Bat species richness recovers rapidly during forest regeneration. Species richness, diversity, and abundance of bats did not vary significantly across four stages of succession in tropical evergreen forests of Agua Blanca State Park in Tabasco, Mexico, but most of the rare species were captured only in the old-growth forest sites (Castro-Luna et al. 2007). Species richness of mammals was similar between six-year-old fallows and intact areas of mature forest in the Selva Lancandona region of Chiapas, Mexico (Medellín and Equihua 1998). Of the 34 species recorded, only 9 species were restricted to mature forest. In a dry forest region of Jalisco, Mexico, diversity and abundance of Phyllostomid bats did not differ significantly among different successional stages, but nectar-feeding species were more abundant in early successional forests, and rare species (species represented by only 1 individual) were sampled only in mature forest (Avila-Cabadilla et al. 2009). In former clearings near Iquitos, Peru, the species richness of Phyllostomid bats was less affected by habitat than species abundance. Among the 37 species found in old-growth forest, all occurred in fallows and young second-growth areas, and 32 species occurred in both. Frugivores and nectarivores were particularly abundant in former agricultural areas (Willig et al. 2007).

12.2.2 *Arthropods*

During forest regeneration in Kinabalu National Park in Sabah, Malaysia, the abundance and diversity of arthropod assemblages became increasingly similar to mature forest assemblages as succession progressed (Floren and Linsenmair 2001). Young regrowth forests showed simpler ant assemblages, but the rank-abundance curve of 40-year-old forests approached that of mature forest (fig. 12.2). As succession progressed, these curves became flatter as a result of increasing species richness and increasing evenness in abundance among species. Ant diversity recovered rapidly during forest regeneration in Puerto Rico, with similar species composition between forests 25–35 years old and more than 60 years old (Osorio-Pérez et al. 2007). Species richness of arboreal ants in regenerating seasonally dry forests in southeast Brazil did not vary significantly with forest age, but species composition showed a clear successional gradient toward recovery within 25 years (Neves et al. 2010). Species diversity and functional diversity of ground-foraging ants increased rapidly during succession in Atlantic Forest sites in Rio Cachoeira Nature Reserve, Paraná, Brazil. But species composition in second-growth forests remained distinct from old-growth forest, with composition of soil

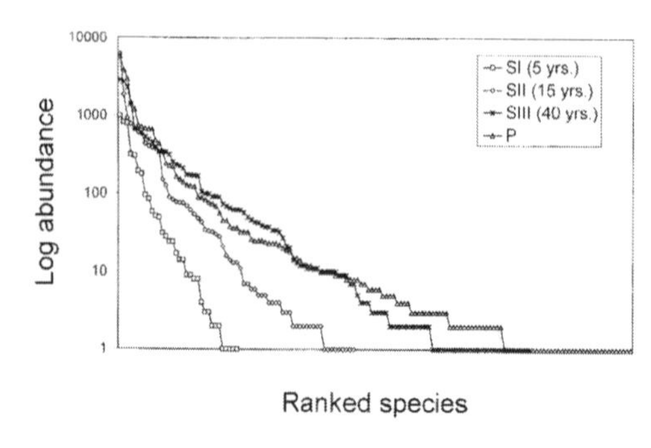

Figure 12.2
Rank-abundance plots for all ants in three ages of regenerating forest (5, 15, and 40 years old) and mature forest (P). Source: Floren and Linsenmair (2001, fig. 4).

assemblages recovering more slowly than litter assemblages (Bihn et al. 2008; Bihn et al. 2010).

Species richness of arboreal ants was substantially higher in an old-growth forest compared to a 10-year-old second growth forest in Papua New Guinea. Based on a complete census of ant nests on all trees in one-hectare plots, 80 species of nesting ants were found in old growth compared to only 42 in the second-growth forest (Klimes et al. 2012). Mean ant species richness per tree (alpha diversity) did not differ significantly between forest types. Differences in vegetation structure were more important than differences in tree diversity in influencing successional changes in species richness of arboreal ants.

The species richness and composition of litter beetles in coastal submontane second-growth forests in southern Brazil were similar to old-growth forests within 35–50 years (Hopp et al. 2010). The rapid recovery of leaf litter beetle assemblages during succession in this region was attributed to high rates of dispersal, short generation time, suitable microhabitats, and existence of large old-growth forest patches within the study area. Decreases in mean monthly temperature and other microclimatic changes associated with the development of forest structure were causally related to increased diversity of leaf litter beetles during succession in these sites (Ottermanns et al. 2011). Microclimatic changes also promote rapid colonization of dung beetle populations in second-growth forests (box 12.2).

12.2.3 *Conservation of Animal Biodiversity in Regenerating Forests*

Regenerating forests increase in animal diversity over time and can provide suitable habitats for most species that are also found in old-growth forests. Moreover, regenerating forests surrounding old-growth areas (such as in buffer zones of conservation areas) show more rapid recovery of conservation value if they are given protected status (Chazdon, Peres, et al. 2009). The dependence of biodiversity recovery in regenerating forest on surrounding old-growth areas highlights the critical need to protect remaining old-growth forest areas in intact landscapes as well as old-growth fragments in human-modified landscapes.

Although metrics of faunal diversity during forest succession indicate slow and steady recovery over time, young regenerating tropical forests are missing many species that lived in the same forest area before clearance (Dent and Wright 2009; Gibson et al. 2011). Nonflying taxa, rare species, endemic species, and taxa with specialized dietary or nesting requirements are less likely to colonize isolated and small areas of regenerating forests during early stages (Chazdon, Peres, et al. 2009). Consequently, animal assemblages in regenerating forests are often a nested subset of old-growth assemblages. Regenerating forests, particularly in the stand initiation stages, favor habitat generalists or disturbance-tolerant species, whereas taxa with highly specialized dietary or habitat requirements are often missing (Liebsch et al. 2008; Renner et al. 2006; Petit and Petit 2003). Several forest-specialist lizard species were absent from a 28-year old forest fragment in the Atlantic Forest in northeastern Brazil, despite the finding that lizard species richness, total abundance, and biomass were similar to old-growth sites (Guerrero and Rocha 2010). Furthermore, patches of regenerating forests are often small, isolated, and close to roads, further reducing their occupancy by rare species and species that cannot cross cleared habitat (Laurance and Gomez 2005; Lees and Peres 2009). Hunting pressures are also very high in regenerating forests that are adjacent to agricultural fields and human communities (Karthik et al. 2009). Hunting of large mammals can seriously disrupt seed dispersal of large-seeded mature forest species (Peres and Palacios 2007; Stoner, Riba, et al. 2007; Wang et al. 2007; Wright, Stoner, et al. 2007; Nuñez-Iturri et al. 2008).

In many studies that compare species diversity and composition of animal taxa between old-growth and second-growth forests, the second-growth forests sampled are very young, surrounded by intensive land uses, and have a smaller spatial extent than the interior of protected old-growth forest reserves. Missing old-growth species in young regenerating forests could partially reflect the effects of small forest area, edge effects, and forest fragmentation rather than forest successional stage. Even when sampling protocols are standardized across forest sites, these landscape factors can contribute to the lower levels of biodiversity observed in regenerating forests. Moreover, most studies rarely account for effects of prior land use, seasonality, or long-term temporal change (Gardner et al. 2009). Yet these results are consistently interpreted to mean that second-growth forests have intrinsically low conservation value (Gibson et al. 2011).

Box 12.2. Dung Beetles, Secondary Seed Dispersal, and Forest Regeneration

When mammalian frugivores pass viable seeds in their feces, seeds are highly vulnerable to predation by rodents or to attack by pathogens. But if seeds are quickly buried beneath the soil, they can be spared this fate. Dung beetles (Coleoptera) are attracted to mammal dung by their strong sense of smell. Adults and larvae of true dung beetles (subfamily Scarabaeinae) feed and raise their brood on mammal dung, with a preference for dung produced by herbivorous and omnivorous mammals. Shortly after they arrive at a dung pile, some types of beetles (rollers) process dung by rolling it into balls and burying these at a distance, others (tunnelers) bury dung in place, and still other kinds (dwellers) lay their eggs in the dung pile without burial (Hanski and Cambefort 1991). If dung is buried, seeds present in the dung are also inadvertently buried, thereby escaping predation by rodents (Estrada and Coates-Estrada 1991; Vulinec 2000; Andresen 2002). The interaction between dung beetles and the seeds they bury is a form of secondary seed dispersal, with important implications for seedling regeneration in tropical forests.

Burial of seeds by dung beetles significantly reduces the ability of rodents to locate and destroy seeds. In tropical forests of Los Tuxtlas, Mexico, seeds of 28 tree species were found in feces of howler monkeys (*Alouatta palliata*), but more than 90% of these seeds were destroyed by rodents that forage in dung piles. Twenty species of dung beetles are attracted to howler dung and remove fecal clumps within an average of 2.5 hours (Estrada and Coates-Estrada 1991). Burial by dung beetles doubled the probability of seedling establishment for 15 species in central Amazonia, Brazil (Andresen and Levey 2004). Secondary dispersal by rollers removes seeds from clumps and may reduce density-dependent effects on seedling establishment). Tunnelers, on the other hand, bury more species of seeds, including larger seeds, than rollers. Studies conducted in Mexico, Peru, Brazil, French Guiana, and Uganda showed that the percentage of seeds buried by dung beetles decreased as a function of seed size (Andresen and Feer 2005). Seeds more than 30 millimeters long are rarely buried by dung beetles.

Dung beetle assemblages are particularly sensitive to habitat perturbation and hunting due to their dependence on herbivorous mammal dung. Deforested areas contain few species of dung beetles compared to intact tropical forest. A meta-analysis of global patterns of dung beetle diversity in tropical forests revealed that land uses with a high degree of forest cover, including logged forests, second-growth forests, and agroforests, support dung beetle assemblages similar to those found in intact tropical forests (Nichols et al. 2007). Dung beetles are abundant and diverse in second-growth forests in five terra firme sites in the Amazon basin (Vulinec et al. 2006). In a dry forest region of Mexico, the species richness and composition of dung beetle assemblages were similar among old-growth and second-growth forests (Andresen 2008). In an exception to this generalization, surveys of dung beetles in second-growth forests in northeastern Brazilian Amazon showed an impoverished assemblage, depleted in larger-bodied old-growth species (Gardner et al. 2008). Second-growth forest fragments in Singapore also showed reduced species richness and lack of large species compared to old-growth fragments, particularly near urban areas with a low abundance of large mammals (Lee et al. 2009).

Following experimental isolation of forest fragments of different sizes in central Amazonia, population size and species richness of dung beetles declined in small forest fragments compared to continuous forest, reducing rates of dung decomposition in small forest fragments (Klein 1989; Andresen 2003). Fifteen years later, secondary regrowth of vegetation progressed substantially in areas surrounding the forest fragments, leading to the complete recovery of dung beetle assemblages (Quintero and Roslin 2005; Quintero and Halffter 2009).

Arboreal cover is the main factor regulating the distribution of Scarabaeinae in wet forests of Chiapas, Mexico (Navarette and Halffter 2008). Landscape transformation in cloud-forest landscapes in central Veracruz, Mexico, led to an increase in dung beetle diversity (Rös et al. 2011). Dung beetle assemblages in second-growth forest, pastures, and other modified habitats were enriched with species from the tropical Paleoamerican biogeographic region and depleted in species from the Mesoamerican montane region. Consequently, second-growth forests in this region have higher species richness of dung beetles than the original cloud forests that they replaced.

Small primates are important primary seed dispersers in young regrowth vegetation dominated by pioneer species in northeastern Peru. Dung beetles buried 24% of seeds present in the defecations of two tamarin species, significantly reducing predation pressure on dispersed seeds (Culot et al. 2010; Culot et al. 2009). The mustached tamarin (*Saguinus mystax*) and saddleback tamarin (*Saguinus fuscicollis*) disperse relatively large seeds of *Parkia panurensis*

Box 12.2 (*continued*)

from adult trees in old-growth forest into young second-growth vegetation (Knogge and Heymann 2003). Regeneration of *Parkia* in second-growth forests was enhanced by burial and secondary dispersal by dung beetles (Culot et al. 2011). Although dung beetle assemblages in this young forest had fewer and smaller species than old-growth assemblages, the ecological value of secondary dispersal during forest regeneration is very high. The interactions between arboreal cover, mammals, and dung beetles and the seeds they bury significantly affect seedling regeneration in tropical forests.

Information on patterns of animal diversity, function, and species composition in later stages of succession or in regenerating forests over time is sparse. The recording of species in surveys is a first step in understanding but does not reveal the extent to which populations are being sustained in these forests (Bowen et al. 2007; Chazdon, Harvey, et al. 2009; Gardner, Barlow, et al. 2007). Some animals may use regenerating forests to supplement their diet but not for nesting or breeding. Animals may also use second-growth forests as corridors that connect areas of old-growth forest.

12.3 Plant-Herbivore Interactions during Forest Regeneration

Studies of plant-animal interactions involving plant reproduction (frugivory, seed dispersal, and pollination) during tropical forest regeneration are far more numerous than studies of plant-herbivore interactions (Quesada et al. 2009). In tropical forests, herbivores remove between 10% and 30% of a plant's leaf area per year (Coley and Barone 1996). Plant-herbivore interactions can potentially affect patterns of tree abundance and community assembly during succession, as tree species vary in their tolerance to herbivory, specialist herbivores, and investments in plant defense (Bazzaz et al. 1987). Moreover, rates of herbivory and herbivore abundance and species richness can be strongly affected by the abundance and composition of surrounding vegetation, as well as abiotic conditions (Brown and Ewel 1987; Silva et al. 2012). Seedlings are particularly vulnerable to effects of herbivory as they have low leaf areas and limited ability to recover from damage compared to larger plants. Large seeds with hypogeal (belowground) storage organs are more capable of resprouting after defoliation than small seeds with epigeal (aboveground) storage organs (Harms and Dalling 1997).

Early successional plant species support higher densities of herbivores and sustain greater amounts of herbivory than late successional species (Lewinsohn et al. 2005). The resource availability hypothesis predicts that plants in early successional environments exhibit higher rates of photosynthesis and growth and lower investments in structural and chemical herbivore defense compared to plants in later successional, resource-limited environments (Coley et al. 1985). This hypothesis predicts that leaf herbivory damage will be highest in pioneer tree species during early successional stages and will decline later in succession due to replacement of fast-growing pioneer species by slow-growing shade-tolerant species with increased herbivore defenses. Early successional species have lower accumulated leaf damage because of shorter leaf life spans, whereas late successional species avoid herbivory due to greater mechanical and chemical defenses.

Despite community-level increases in species richness of herbivorous insects during forest regeneration, herbivore species richness per host plant does not appear to vary with successional status (Novotny 1994; Basset 1996; Leps et al. 2001). What factors, then, can explain the overall increase in herbivore species richness during forest regeneration? This increase could be caused by concerted changes in both plant and insect diversity during succession, or by increasing levels of host specialization (Lewinsohn et al. 2005). Although data are limited, host specificity does not appear to change markedly during forest succession (Basset 1996; Leps et al. 2001; Marquis 1991).

Several studies have compared successional variation in herbivore damage and leaf defense traits in tropical forests. Poorter et al. (2004) found that standing levels of herbivory declined in saplings across a successional gradient in Bolivia. Vasconcelos (1999) found no significant difference in standing levels of herbivore damage in canopy trees between old-growth and 7- to 10-year-old *Vismia*-dominated forests in central Amazonia, although the leaf-cutting ant *Atta laevigata* showed a clear preference for leaves from second-growth forests. An extensive study of nonflying herbivores in Papua New Guinea involved complete harvesting of all trees greater than or equal to 5 centimeters in diameter at breast height (DBH) in one hectare of a young (10-year-old) second-growth forest and one hectare of an old-growth forest (Whitfeld et al. 2012). Caterpillars (Lepidoptera) were twice as abundant in the young forest, whereas leaf miner abundance did not differ between forest types. The second-growth forest supported a greater number of insect herbivores per hectare, per tree, and per unit of leaf biomass (fig. 12.3). Second-growth trees had higher specific leaf area and leaf nitrogen content, and these leaf traits accounted for 30% of the variation in caterpillar abundance.

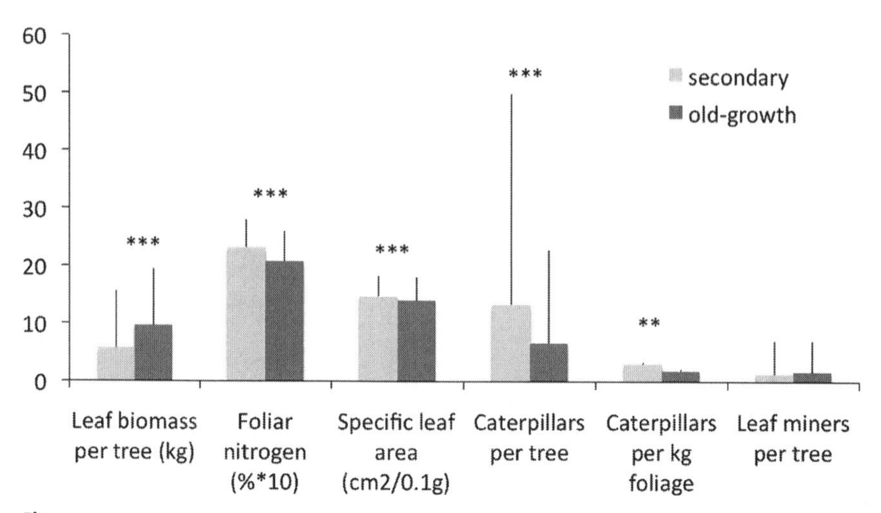

Figure 12.3
Plant traits and herbivore abundance in one-hectare plots in second-growth and old-growth forests in Papua New Guinea. *** $P < 0.0001$; ** $P < 0.01$. Source: Whitfeld et al. (2012, table 1).

12.4 Seed Dispersal and Predation during Forest Regeneration

Tropical forests are noted for the high diversity of frugivores that rely on year-round production of fruits and seeds. Globally, 50%–90% of the tree species in tropical forests produce fleshy fruits that are consumed by birds and mammals (Howe and Smallwood 1982). Dispersal of large fruits with large seeds requires specialist frugivores, whereas small fruits with small seeds can be dispersed by numerous generalist frugivores. Seed size also influences rates of seed predation and secondary dispersal, in addition to total seed production, seed dormancy, and seedling survival. These factors create complex interactions between fruits, frugivores, and seed fate throughout forest succession. Furthermore, these interactions link local forest dynamics with the structure and composition of the surrounding landscape.

12.4.1 *Frugivory and Seed Dispersal*

Unlike leaves, fleshy fruits are made to be eaten. Frugivores have diets consisting of at least 50% fruit (Terborgh 1986; Fleming et al. 1987). These animals can promote regeneration of the forests that they need for their own survival by moving seeds away from the parent (Janzen 1970; Connell 1971), by transporting seeds to suitable germination sites (Wenny and Levey 1998), and by enhancing rates of seed germination of ingested or handled seeds (Traveset et al. 2007). Seeds of fleshy fruits can be dispersed through ingestion and defecation (endozoochory), transport in an animal's mouth and deposition at feeding roosts (stomatochory), or transport on an animal's body (epizoochory). Animal-mediated seed dispersal can occur in two phases (diplochory) involving secondary dispersal by rodents, ants, or dung beetles after seeds have been dispersed by primary dispersal agents (see box 12.2).

Rates of fruit production are elevated in early colonizing species of former pastures and old fields because of high levels of light availability (Blake and Loiselle 2001), and fruits are produced over longer periods in early successional tree species compared to late successional species (Kang and Bawa 2003; Bentos et al. 2008). Overall, frugivore diversity is lower in Southeast Asia compared to Africa and the Neotropics (Fleming et al. 1987). Early successional forests in the Neotropics are havens for generalist frugivores (Blake and Loiselle 2001; DeWalt et al. 2003). Abundance of generalist bird species was the best predictor of the species richness of dispersed seeds in a human-modified landscape in southern Costa Rica (Pejchar et al. 2008). As the structural and taxonomic diversity of vegetation increases during regeneration, the availability of roosting and feeding resources increases, enticing more species of small- and large-bodied frugivores to become regular visitors and residents. Visiting birds and bats arrive bearing the fruits and seeds of old-growth species from surrounding areas (Wunderle 1997). The progression of vegetation during regeneration facilitates the colonization of new species of fruiting trees in a positive feedback.

The frequency of animal dispersal among tree species increased from 86% in second-growth forests to 90% in old-growth forests, but relative abundance was similar across forest types; 73%–77% of all trees were animal dispersed (Chazdon et al. 2003). Early successional forests are dominated by small-seeded species, whereas large-seeded species increase in importance during later successional stages. The predominance of animal dispersal increases during forest regeneration (del Castillo and Pérez Ríos 2008; Liebsch et al. 2008). In the Atlantic Forest of southern Bahia, Brazil, the proportion of animal-dispersed tree species increased from 76% to 89% across a successional chronosequence (Piotto et al. 2009). Increases in animal-dispersed species during succession are associated with increases in seed size and fruit size and in the proportion of shade-tolerant species (Tabarelli and Peres 2002; del Castillo and Pérez Ríos 2008; Piotto et al. 2009). In six chronosequences following shifting cultivation in southeastern Brazil, the percentage of vertebrate-dispersed plant species was positively correlated with the age of the forest plot. The percentage of small-seeded plant species (< 0.6 cm) declined significantly as forest age increased, whereas the percentage of medium-seeded species (0.6 to 1.5 cm) was positively correlated with forest age (fig. 12.4*A*).

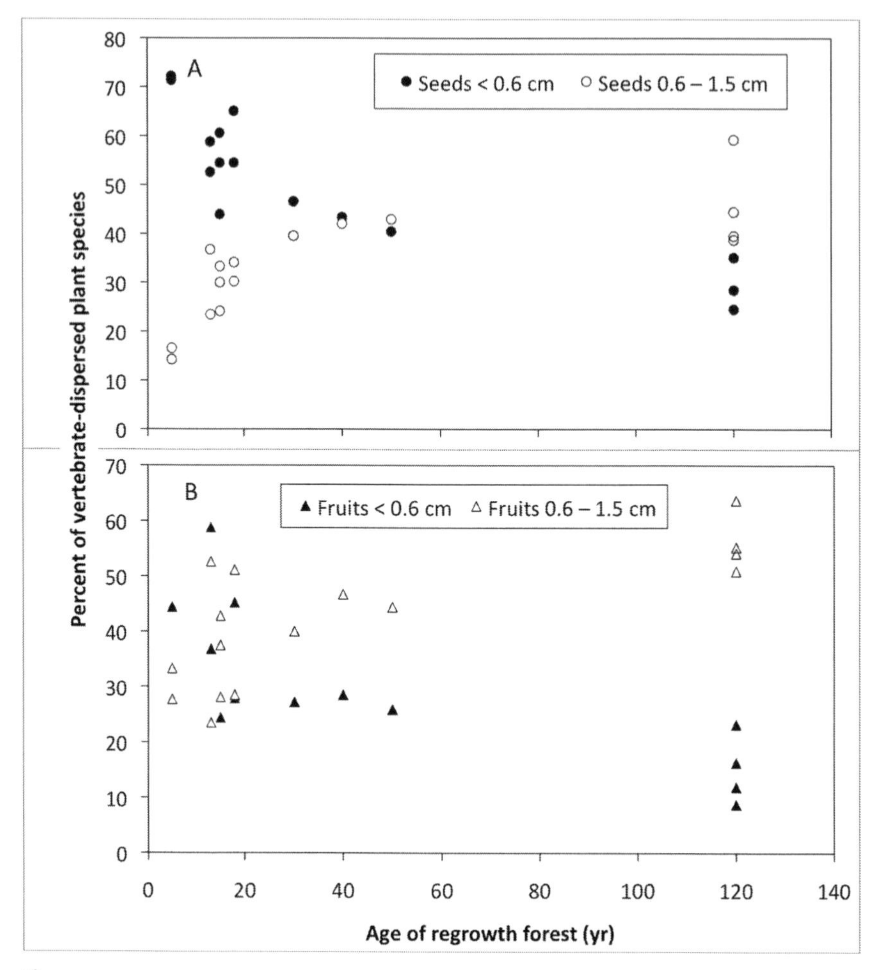

Figure 12.4

The percentage of vertebrate-dispersed species in different ages of forest regrowth following shifting cultivation in southeastern Atlantic Forest of Brazil. Old-growth forests are assumed to be 120 years old. Species are classified in two different seed size (*A*) and fruit size (*B*) categories. Species with small seeds and fruits (< 0.6 cm) decline with forest age, whereas species with medium-sized seeds and fruits (0.6–1.5 cm) increase with forest age. Source: Drawn from data in Tabarelli and Peres (2002, table 3).

Similar results were observed for fruit size in relation to forest age (fig. 12.4*B*; Tabarelli and Peres 2002).

Lack of seed dispersal to old fields can be a major obstacle to forest regeneration (Wijdeven and Kuzee 2000; Martínez-Garza et al. 2009; Corlett 2011). Dispersal limitation during succession is most pronounced for large-seeded species with fleshy fruits or wingless dry fruits (Corlett 2011). During mid and late successional stages, the major barriers to seed dispersal are distance from seed sources of rare or infrequently fruiting trees, availability of dispersal agents, and time. Even after 100 years of regeneration, forests in the central catchment reserve of Singapore were depleted of many tree species found in contiguous old-growth forests due to dispersal failure (Turner et al. 1997). A similar trend is observed in 19- to 62-year-old forests of the Atlantic Forest region, where large-seeded species compose only 5%–8% of the species recorded in the seed rain compared to 31% in adjacent old-growth forests (Costa et al. 2012). Populations of frugivores that eat large-seeded fruits are declining or have been extirpated locally.

Most frugivorous animals will not venture into large clearings unless they encounter suitable resources for feeding or perching (Silva et al. 1996). Open conditions in clearings expose frugivores to increased predation risk and have low availability of fleshy fruits. These factors lead to a steep decline in animal-mediated seed dispersal with increasing distance from the forest edge (Gorchov et al. 1993; Holl 1999; Cubiña and Aide 2001; Ganade 2007). Isolated remnant trees, shrubs, and pioneer trees in former pastures and cultivated fields attract frugivores, leading to higher densities of seed deposition and recruited seedlings (see box 7.2). In Kibale National Park, Uganda, seed rain below isolated trees in a deforested area was 90-fold greater than in adjacent grassland (Duncan and Chapman 1999).

Frugivorous bats play a particularly important role in seed dispersal during both primary and secondary succession (see box 6.1; Muscarella and Fleming 2007; Whittaker and Jones 1994; Foster et al. 1986). Species that often dominate early stages of succession in the Neotropics are concentrated in four of the five top plant families consumed by frugivorous bats: Solanaceae, Moraceae, Piperaceae, and Clusiaceae (Muscarella and Fleming 2007). At least 549 plant species in 191 genera and 63 families are dispersed by bats in the Neotropics (Lobova et al. 2009). In humid forest areas of Chiapas, Mexico, bats disperse more seeds than birds in early successional habitats (Medellín and Gaona 1999). Birds tend to deposit seeds beneath fruiting trees or carry fruit to perches or nests where they drop or defecate seeds (Corlett 1998). Bats deposit seeds below fruiting trees, but unlike birds, they also defecate small seeds during flight (Charles-Dominique 1986). Bat-dispersed seeds are therefore more likely than bird-dispersed seeds to reach former clearings or pastures that lack perching or nesting sites (Silva et al. 1996; Muscarella and Fleming 2007).

Neotropical and Paleotropical frugivorous bats overlap more in their diet with nonflying arboreal mammals than with birds (Hodgkison et al. 2003;

Lobova et al. 2009). Although some fruits of *Piper* (Piperaceae), *Vismia* (Hypericaceae), *Lycianthes* (Solanaceae), *Solanum* (Solanaceae), and *Cecropia* (Urticaceae) are eaten by both birds and bats in the Neotropics, frugivorous birds and bats generally consume and disperse different species of fruits in early second-growth habitats (Palmeirim et al. 1989; Gorchov et al. 1995). In young regenerating stands dominated by *Cecropia* or *Vismia* species in central Amazonia, 89% of the seeds dispersed by birds during the day were *Miconia* seeds (Melastomataceae), whereas 72% of the seeds dispersed by bats during the night were *Vismia* seeds (see plate 6; Wieland et al. 2011).

In the Paleotropics, bat-dispersed plants rarely dominate early stages of succession, with the exception of *Ficus* and *Musanga* species. Early successional fleshy fruits in the Paleotropics are dispersed primarily by birds (Ingle 2003; Gonzales et al. 2009). Bulbils (*Pycnonotis*) are ubiquitous birds in nonforest habitats and are the most important seed dispersal agents in deforested areas of tropical and subtropical East Asia (Corlett 1998). Dietary overlap between birds and bats in early second-growth vegetation is lower among Neotropical frugivores than among Paleotropical frugivores (Fleming et al. 1987). In subtropical Australian rain forests, frugivorous bats can serve as dispersal agents for many of the plant species that are also dispersed by birds (Moran et al. 2009).

Frugivore body size is the major determinant of the ability to feed upon fruit of a given size. Old-growth tree species with large seeds (and therefore large fruits) generally require large-bodied frugivores to disperse their seeds (Wheelwright 1985), but these frugivores, including several hornbill species, are now extinct in Singapore (Corlett 1992; Corlett and Turner 1997). Consequently, old second-growth forests in Singapore are composed almost entirely of small-fruited species (Corlett 1991; Turner et al. 1997).

In highly disturbed tropical landscapes in the Neotropics, the seed-dispersing fauna often comprises a subset of the original fauna, dominated by small- to medium-sized rodents, small birds, and bats that generally disperse small-seeded pioneer species (Melo et al. 2006). This altered species composition creates a major filter for seedling recruitment in successional forest landscapes (Tabarelli and Peres 2002; Corlett and Turner 1997; Duncan and Chapman 1999). Some frugivorous bats are effective dispersers of large-fruited species, such as the tent-roosting species *Artibeus watsonii* in humid Neotropical forests. Stomatochorous seed dispersal beneath leaf tent roosts of this species significantly increased the species richness and abundance of large-seeded species in regenerating forests of northeastern Costa Rica (Lopez and Vaughan 2004; Melo et al. 2009).

Small-bodied primates can be important dispersers of large seeds in regenerating forests (see box 12.2). Mixed-species groups of tamarins (*Saguinus mystax* and *S. fuscicollis*) dispersed 63 tree species with large seeds (> 1 cm long) in a regenerating forest adjacent to old-growth forest in northeastern Peru (Culot et al. 2010). In the Una Biological Reserve in the Atlantic Forest of southern Bahia, Brazil, the endemic golden-headed lion tamarin (*Leontopithecus chrysomelas*) is an important seed dispersal agent for at least 24 species of trees and epiphytes in 13 families, with seeds up to 2.3 centimeters in length (Catenacci et al. 2009). Tamarins disperse seeds into three successional stages and cacao agroforestry (cabruca) habitat.

Given the predominance of animal-dispersed species in tropical forests, it is not surprising that forest regeneration depends heavily on seed dispersal by generalist and specialist frugivores. These frugivores, in turn, depend on the forests to provide food and shelter.

12.4.2 *Secondary Seed Dispersal and Postdispersal Seed Fate*

Although primary seed dispersal by animals may successfully move seeds far away from parent trees, this mode of seed delivery may not provide suitable conditions for germination or seedling establishment (Wang and Smith 2002). Seeds deposited in vertebrate feces fall directly on soil surfaces or leaf litter, often in dense clusters, and are highly vulnerable to mortality from seed predators and pathogens, intense competition, or desiccation (Vander Wall and Longland 2004). Three groups of animals process, move, and consume seeds present in feces: scatter-hoarding rodents, ants, and dung beetles (see box 12.2). In some defaunated tropical forests, scatter-hoarding rodents can disperse large seeds in the absence of larger vertebrate dispersal agents (Cao et al. 2011). Potential benefits of secondary dispersal are enhancement of germination success and escape from predation. Carnivorous ground-dwelling ants are strongly attracted to lipid-rich diaspores (seeds and fruit together as a dispersal unit) and find them within eight minutes of deposition. Secondary dispersal by ants reduces the likelihood of fungal infection and can increase germination success by 19% to 63% in Atlantic Rain Forests of southeastern Brazil (Pizo et al. 2005).

Few studies have been conducted on secondary seed dispersal by ants or scatter-hoarding rodents in regenerating forests. In secondary montane forests of southern Costa Rica, secondary seed dispersal by scatter-hoarding rodents was rare (Cole 2009). Ants serve as both seed predators and secondary dispersal agents of small-seeded early successional species dispersed by birds or bats. Removal of seeds from frugivore feces by rodents and ants is usually assumed to represent seed predation, but a small fraction of seeds of *Miconia* spp. removed by *Pheidole* ants at La Selva Biological Station in Costa Rica were deposited on nutrient-rich refuse piles and successfully germinated (Levey and Byrne 1993). Litter ants from two subfamilies (Myrmicinae and Ponerinae) removed, on average, 45% of seeds from four animal-dispersed pioneer species in an experimental study in Panama (Fornara and Dalling 2005).

Following dispersal, seeds are subject to attack by vertebrates, invertebrates, and fungi. Rates of seed predation in pastures often exceed 50% (Scariot et al. 2008; Nepstad et al. 1996; Myster 2008b). High rates of seed predation can strongly influence early stages of succession in pastures (Nepstad et al. 1991; Holl and Lulow 1997). Small seeds dispersed into pastures are consumed by ants and other insects, whereas medium-sized seeds are often consumed by rodents (Myster 2008b). Rates of seed predation were higher in shifting agriculture fallows than in mature forest gaps (Uhl 1987) or closed-canopy sites (Hammond 1995), and the removal rate of seeds declined with age since abandonment (Peña-Claros and de Boo 2002). In regenerating forests, mammals appear to be more important seed predators than insects for large-seeded species such as palms (Peña-Claros and de Boo 2002; Notman and Villegas 2005). For tree and liana species in Peruvian lowland forests, rates of seed and seedling predation did not differ significantly between young fallows and old-growth forest areas (Notman and Gorchov 2001).

Declines in vertebrate abundance due to hunting can alter levels of both seed dispersal and seed predation, leading to changes in species composition of regenerating vegetation. Hunting can reduce the abundance of medium- and large-bodied mammals, while increasing the abundance of small rodents (Donatti et al. 2009). In defaunated forests in Mexico, small-seeded species were heavily attacked by abundant small rodents, whereas large-seeded species escaped predation and formed dense seedling carpets around fruiting trees (Dirzo et al. 2007). Declines in populations of mammalian seed predators in hunted areas may lead to increased recruitment of large-seeded species or species that are less susceptible to insect predation (Guariguata et al. 2002; Stoner, Vulinec, et al. 2007). In central Panama, hunting increased the relative abundance of seeds of lianas, species with large seeds, species dispersed by bats, and species with mechanical dispersal (Wright, Hernandéz, Condit 2007).

Reduced populations of seed predators in unprotected areas have led to increased abundance of recruited seedlings and saplings, particularly of large-seeded species. After more than 40 years of protection from hunting, populations of terrestrial game mammals, such as the agouti (*Agouti paca*) and collared peccary (*Pecari tajacu*), are noticeably higher at La Selva Biological Station in northeastern Costa Rica than in surrounding regions (Guariguata et al. 2000; Hanson et al. 2006). Seeds of palm species that are not mechanically or chemically defended are particularly favored by peccaries (Beck 2005; Kuprewicz 2013). Effects of hunting extend beyond changes in abundance, as large mammals and birds also play a critical role in seed dispersal and the genetic structure of tree populations (Sezen et al. 2005; Wright, Stoner, et al. 2007).

12.5 Pollination in Regenerating Forests

Over 90% of the flowering plant species in tropical forests rely on animals for pollination and sexual reproduction (Ollerton et al. 2011). Insects perform most of the pollination, but wind and vertebrates are also important vectors of pollen movement. Few studies have compared pollination modes of tropical forests at different successional stages. In evergreen wet forests of northeastern Costa Rica, 70% of the species and 85% of the individual trees in second-growth forests are insect pollinated (Chazdon et al. 2003). Bird pollination accounts for 7% of the trees in second-growth forests, but many of these are *Inga* species that are also pollinated by hawkmoths (Sphingidae). Mammal pollination, on the other hand, is less frequent in second-growth than old-growth forests (fig. 12.5). Similar trends have also been observed in second-growth forests in the Atlantic Forest region of Brazil, where vertebrate pollination is completely lacking (Lopes et al. 2009; Kimmel et al. 2010).

Early successional habitats in wet Neotropical regions support many non-woody species that are pollinated by hummingbirds, such as *Heliconia* species and vascular epiphytes (Piacentini and Varassin 2007). The staggered flowering peaks of sympatric *Heliconia* species maintain continuous visitation by hummingbirds that compete for nectar and forage opportunistically on any available flowers. Corollas of *Heliconias* in early successional areas are short enough to permit nectar access to most hummingbirds while being long enough to exclude most insects. *Heliconia* species found in older forests have morphologically specialized flowers to attract larger, specialized hummingbirds (Stiles 1975; Feinsinger 1978).

The theme of more generalist pollination systems in early successional vegetation is also observed in insect-pollinated species. Bees are the most important group of pollinators in tropical forests. Social and nonsocial bees pollinate 50% of the 270 plant species studied in a lowland dipterocarp forest

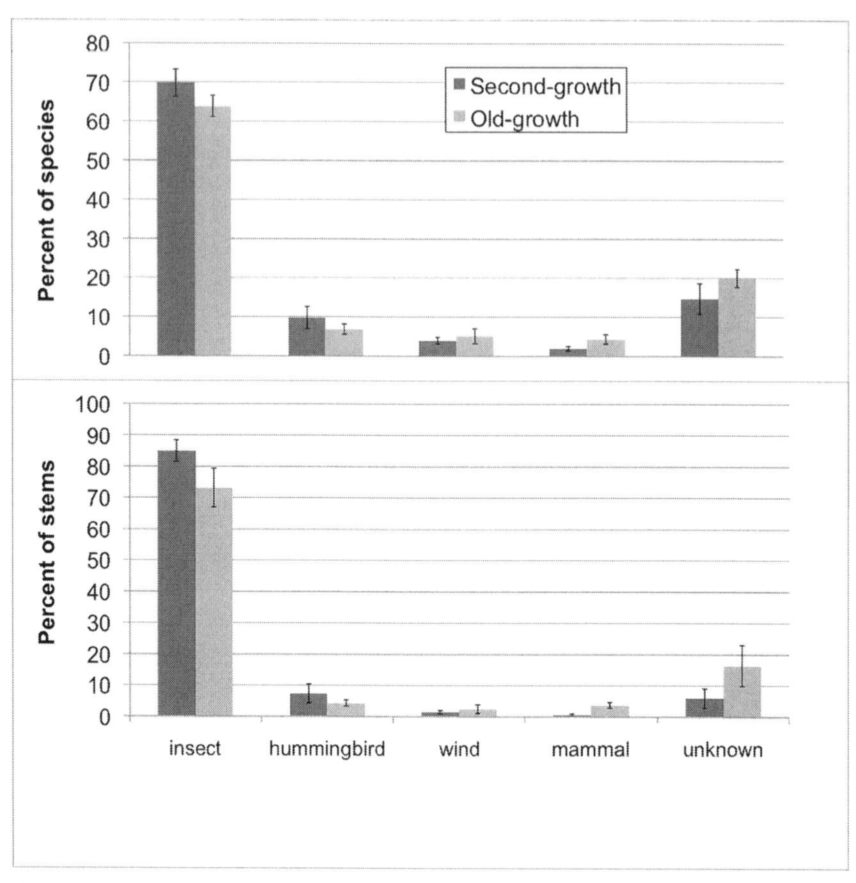

Figure 12.5
The percent of species and stems of trees greater than or equal to five centimeters in DBH with different modes of pollination in five second-growth and three old-growth forests in northeastern Costa Rica. Second-growth forests ranged in age from 15 to 25 years old and were all regrowing on former pasture. Bars are means and error bars are one standard deviation. Source: Redrawn from data in Chazdon et al. (2003).

in Sarawak, Malaysia (Momose et al. 1998) and 42% of the tree species studied at La Selva Biological Station in Costa Rica (Kress and Beach 1994). Although pollination networks have not been studied across a series of successional ages, it is likely that a core group of generalist pollinators persists throughout succession (see box 12.1), with specialists increasing in number during later successional stages. The structure of a bromeliad-hummingbird pollination network was studied in second-growth areas of the Atlantic Forest in Paraná, Brazil (Piacentini and Varassin 2007). The network of 12 species of bromeliads and 10 species of hummingbirds was highly asymmetric and nested, with specialist hummingbirds relying on abundant generalist partners and specialist bromeliads relying on generalist hummingbirds for pollination. Most plant species in a tropical dry evergreen forest in India exhibited a generalized pollination system involving diverse insects including social bees, solitary bees, wasps, moths, and flies (Nayak and Davidar 2010).

Deforestation and forest fragmentation in tropical regions have been clearly associated with the loss of reproductive trait diversity and the disruption of mutualisms between plants and animals on all continents (Cordeiro and Howe 2003; Vamosi et al. 2006; Girão et al. 2007; Moran et al. 2009). Little research has addressed how these mutualisms can be restored through natural regeneration (Kaiser-Bunbury et al. 2010; Menz et al. 2011). Remnant forest fragments and landscape composition likely play a large role in supporting populations of mutualist partners during regeneration of second-growth forests. The interactions between plants and animals are key determinants of successional pathways and are mobile links that connect regrowing forests with forest fragments and areas of intact forest within tropical landscape mosaics.

12.6 Conclusion

Much of our understanding of animal diversity and plant-animal interactions during forest regeneration is based on detailed studies of a small number of species, taxonomic groups, and forest sites. Available studies show a wide spectrum of variation in colonization and diversity patterns during forest regeneration. Although the potential for recovery of animal populations and plant-animal interactions during forest regeneration is high, this potential is often not realized (Chazdon, Peres, et al. 2009; Gibson et al. 2011). Animal diversity may fail to recover in areas that are isolated from old-growth forests or forest remnants, have a history of intensive land use, or have experienced local extirpation of vertebrate fauna. These barriers can be reduced through efforts to promote forest regeneration adjacent to existing forest fragments and to protect remaining old-growth forest areas. The structure and composition of the landscape matrix is critically important for animals that disperse and/or consume seeds and pollinate plants in second-growth forests (Turner and Corlett 1996).

The potential of second-growth forests to support many animal species that occur in old-growth forests increases over time, a trend that has also been observed for tree species (Chazdon. Peres, et al. 2009). More time is needed for the arrival of endemic species and specialists, as their ecological requirements may not be met during early stages of succession. The process of recovery of animal communities during forest regeneration is closely linked to the recovery of plant diversity. Phylogenetic studies provide a promising tool for investigating relationships between plant and animal diversity during forest regeneration, particularly when combined with long-term data on changes in species relative abundance and diversity.

CHAPTER 13

TROPICAL REFORESTATION PATHWAYS

Attempts to plant trees to accelerate regeneration or influence its direction should be based on a thorough understanding of the likely pathways of regeneration without interventions.—Sayer et al. (2004, p. 8)

13.1 Overview

Over 350 million hectares of deforested land in the tropics are plagued by infertile soils, erosion, weed infestation, or recurrent fire as a consequence of unsustainable land-use practices. The land degradation scenario does not uniformly apply to all land where forests are cleared or selectively harvested. As the previous chapters of this book affirm, most deforested areas have the potential to regenerate naturally (Lamb et al. 2005). But the quality and rate of regeneration of deforested lands are highly variable. In severe cases of soil degradation and loss of agricultural productivity, forest regeneration can be accelerated and enhanced through appropriate human interventions. Land degradation defines one extreme along a spectrum of possible states following deforestation and human land use. At the other end of the spectrum are degraded forests, standing forests whose structure, biomass, or species composition have been temporarily or permanently reduced due to logging, hunting, forest farming (agroforestry), or burning (Lamb 2011). At intermediate positions along this gradient are lands cleared for shifting agriculture, permanent cultivation, or pastures that have the potential to regenerate naturally following abandonment. In the context of reforestation, the distinction between degraded and nondegraded land is neither simple nor superficial.

Restoration is the process of returning a forest to its "original" condition. Defining precisely what the "original" condition is (or was) can be challenging, if not impossible, for several reasons. First, forests are in a constant state of flux due to phases of disturbance and recovery. Second, old-growth forests no longer exist in many regions of the tropics. Third, due to irreversible effects of climate change, some ecosystems and species assemblages can no longer be re-created (Jackson and Hobbs 2009). Fourth, many present-day tropical forests show legacies of prior human activities, calling into question precisely what their "original" condition was or should be. Even though forests cannot be restored completely to their original state, deforested lands can become reforested though spontaneous (passive) successional processes or through the planting of trees and assisted regeneration. If *forest* is defined as a type of natural land cover, then *reforestation* is the process of regenerating that natural land cover.

In previous chapters, this book has focused on natural regeneration (passive reforestation) on deforested land across the tropics through spontaneous successional processes. Extensive areas of former pastures and croplands in Puerto Rico and Costa Rica have naturally regenerated to second-growth forests since the 1950s (Aide et al. 1995; Arroyo-Mora, Sánchez-Azofeifa, Kalacska, and Rivard 2005). This chapter focuses on approaches to *active reforestation* or *ecological restoration* in the tropics. (Forest landscape restoration [FLR] is discussed in chapter 14.) Ecological restoration projects use tree planting, soil amelioration, direct seeding, or enrichment planting to accelerate natural regeneration and restore natural forest cover, with a primary focus on encouraging regeneration of native species (Chazdon 2008a; Lamb 2011). Active reforestation involves deliberate human intervention aimed at overcoming specific barriers to natural forest regeneration (Rey Benayas et al. 2008). Active reforestation is explicitly directed by human goals, but in practice, reforestation methods involve a combination of both passive and active processes.

The short- and long-lived pioneer species that are the main players in establishment of early successional forests are key resources for active reforestation efforts (see box 5.1). These species are adapted to colonize, establish, and grow in conditions where vegetation has been damaged, cleared, burned, or washed away. As discussed in chapter 10 (sec. 10.4), their functional traits evolved to maximize the capture of carbon and nutrients to favor rapid assimilation and growth. Successional pioneers are readily dispersed by wind and frugivorous animals. They catalyze the colonization and establishment of slower-growing species that build new forests over decades and centuries. Pioneer species are essential partners in the work of reforestation.

13.2 Reforestation Goals and Decisions

The ultimate goal of tropical forest restoration is to create a self-sustaining assemblage that includes natural species turnover and that enables resilient responses to local disturbance regimes (Walker, Walker, and del Moral 2007). The planning and practice of ecological restoration is therefore closely linked to an understanding of successional processes (Kageyama and Castro 1989). "Practicing restoration outside the framework of succession," Walker, Walker, and del Moral (2007, p. 4) stated, "may be likened to building bridges without attention to the laws of physics." Making appropriate decisions regarding the type of species and methods used in reforestation requires an understanding of local environmental conditions, successional processes, growth requirements of tree species, species interactions, and economic value of tree species.

Section 2 of the Society for Ecological Restoration's *International Primer on Ecological Restoration* defined ecological restoration as "the process of assisting the recovery of an ecosystem that has been degraded, damaged, or destroyed" (SER 2004, p. 3). Ecological restoration is designed to steer a degraded ecosystem toward a successional trajectory that will bring it closer to its original ecosystem properties and species composition. These efforts represent tiers 2 and 3 of a restoration "staircase," focusing on areas with a low to moderate state of degradation (fig. 13.1; Chazdon 2008a). In highly degraded areas, restoration is a multistep process, as rehabilitation of soils is required before ecologically based restoration can begin. Tree planting is often used to "jump-start" natural regeneration in anticipation that passive reforestation will take care of the job in later stages (Lugo 1997). In other cases, restoration projects can bypass the initial phase of pioneer-dominated succession

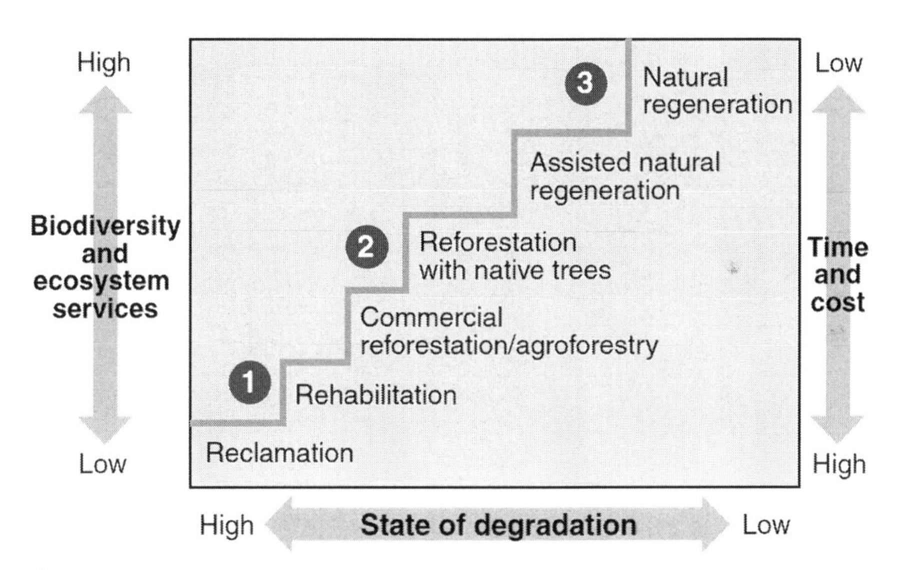

Figure 13.1
The restoration staircase. Depending on the state of land degradation of an initially forested ecosystem, a range of management approaches can at least partially restore levels of biodiversity and ecosystem services given adequate time (years) and financial investment (capital, infrastructure, and labor). Outcomes of particular restoration approaches are (1) restoration of soil fertility for agricultural or forestry use, (2) production of timber and nontimber forest products, or (3) recovery of biodiversity and ecosystem services. Source: Chazdon (2008a, fig. 1).

by planting dispersal-limited old-growth tree species directly into pastures to increase their abundance in the landscape matrix (Martínez-Garza and Howe 2003). All of these activities can be considered reforestation pathways.

The goals of reforestation initiatives are heterogeneous. Many different groups of people have a stake in tropical reforestation, including foresters, shifting cultivators, smallholder farmers, ecologists, and conservation biologists (fig. 13.2; Clewell and Aronson 2006). If these goals are carefully aligned with the natural regeneration potential in the landscape, reforestation is more likely to be successful (Holl and Aide 2011). Reforestation goals can be grouped into three general categories: (1) production of commercial or domestic timber and nontimber products, (2) natural regeneration of forests and recovery of ecosystem services, and (3) biodiversity conservation.

Different stakeholder groups prioritize some goals over others, leading to potential conflicts. But multiple stakeholder groups share each of the goals, suggesting that partnerships among groups can be formed to plan reforesta-

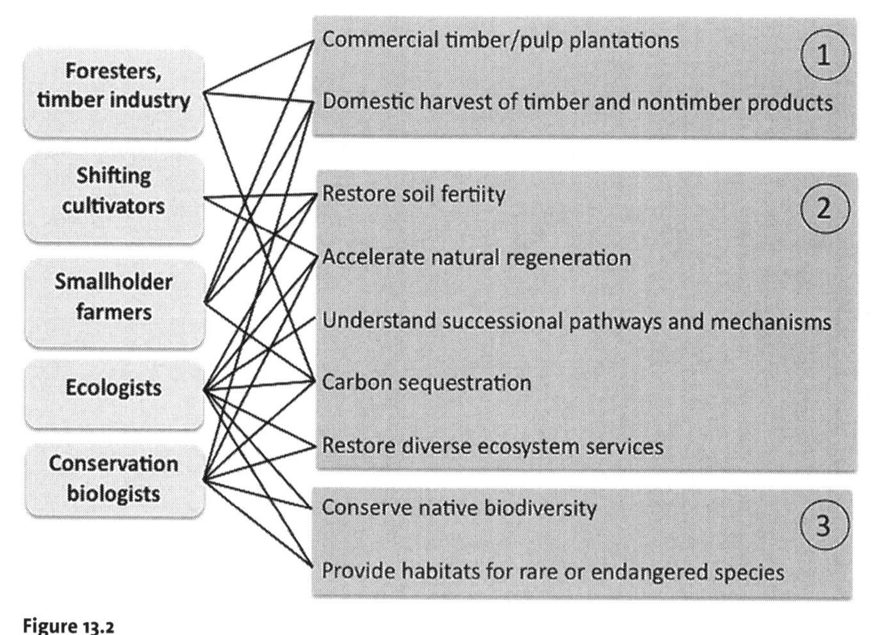

Figure 13.2

Diverse stakeholder groups (*left*) and their goals in tropical reforestation (*right*). Reforestation goals are grouped in three major arenas: (1) commercial and domestic use of forest products, (2) natural regeneration and ecosystem services, and (3) biodiversity conservation. Although each stakeholder group has a unique set of goals and priorities, different types of stakeholders overlap considerably in reforestation goals, suggesting that strong alliances can be formed across these diverse groups.

tion activities that fulfill common agendas. Foresters, smallholders, and conservation biologists all support tree planting for domestic use of timber and nontimber products. Ecologists and shifting cultivators share an interest in reforestation as a way to restore soil fertility and accelerate natural regeneration. Foresters, ecologists, and smallholders value the carbon sequestration potential of regenerating and restored forests and producing timber. Ecologists and conservation biologists view reforestation as a way to provide habitats for rare and endangered species. By acknowledging these shared goals, stakeholders can unite to achieve their collective vision.

13.2.1 Afforestation versus Reforestation

By 2007, over 57% of the reforestation projects financed by the Indonesian government's restoration fund were allocated to planting over one million hectares of industrial timber and pulp wood plantations in Indonesian Borneo (Kettle 2010). These plantations are often established in areas of "degraded forest," which include naturally regenerating fallows, selectively logged forests, and essentially any forest that has been burned. This issue goes way beyond semantics. Whether large-scale commercial tree plantations should be considered a reforestation pathway has been debated for over a decade (Sayer et al. 2004; Brockerhoff et al. 2008; Lindenmayer et al. 2012). Commercial monoculture plantations are needed to supply timber, pulp, and biofuels, just as farms are needed to supply food for people and domesticated animals. But monoculture tree plantations are not forests any more than cornfields or pastures are grasslands. Here, I distinguish between reestablishing tree cover via *afforestation* through managed commercial plantations and *reforestation* through active or passive modes of forest regeneration. The United Nations Food and Agriculture Organization (FAO) Forest Resource Assessment (FRA) defined afforestation as "the act of establishing forests through planting and/or deliberate seeding on land that is not classified as forest," whereas it defined reforestation as the "re-establishment of forest through planting and/or deliberate seeding on land classified as forest, for instance after a fire, storm or following clearfelling" (FAO 2010, p. 95). It is important to note that rates of reforestation published in the 2010 FRA do not include natural or assisted regeneration and are based on data contributed by only 68% of the countries reporting.

The ultimate goal of reforestation, as discussed here, is to restore a deforested area to a forested area by promoting the establishment of native biodiversity, fostering the development of multiple ecosystem services, and permitting sustainable forest uses, including selective harvesting and management of timber and nontimber products. Commercial monoculture plantations without silvicultural management can promote understory regeneration, with modest benefits for biodiversity and ecosystem services (Brockerhoff et al. 2008; Bremer and Farley 2010; Tomimura et al. 2012). But intensively managed commercial tree plantations—particularly those managed for the production of pulp, biofuels, or carbon sequestration—are simply not compatible with facilitating natural forest regeneration or restoring forest biodiversity and multiple ecosystem functions (Lindenmayer et al. 2012; Hall et al. 2012).

In the case of deforested lands, the decision must be made whether to embark on a reforestation pathway at all, or whether to use the land to expand agriculture or commercial plantations, which may avoid further deforestation. Wilcove and Koh (2010) proposed using financial incentives to encourage the use of degraded land for expanding oil palm plantations in Southeast Asia as a way to spare forests from clearing. To be effective for conservation, schemes like this must be based on clear definitions of what constitutes degraded land and careful analysis of alternative land uses. If financial incentives are improperly conceived and executed, far more land will be condemned as degraded land than restored, benefiting only the oil palm industry.

Reforestation decisions and approaches should require active participation of local residents, whose environment and sustenance will be greatly affected (Chazdon, Harvey, et al. 2009; Lamb 2011; Newton et al. 2012). Land that may be well suited for passive reforestation or ecological restoration is often condemned as "degraded" to promote development of single-species (usually nonnative) tree plantations, with negative social and ecological consequences. In rural areas of Ha Tinh Province in Vietnam, deforested areas that supplied diverse timber and nontimber products heavily used by local populations were designated as "bare hills" by government authorities to promote the establishment of *Eucalyptus* and *Acacia* plantations that supply only low-quality timber and firewood for local consumption (McElwee 2009).

13.2.2 *Active versus Passive Reforestation*

A central issue in planning tropical reforestation efforts is whether active intervention is needed, and if so, what form it should take. This decision should ideally be based on an understanding of the natural regeneration potential within the area, soil conditions, the specific goals of the reforestation program, the spatial scale of the reforestation project, and available funding to support a long-term program with periodic assessments (Holl and Aide 2011; Cantarello et al. 2011). Natural regeneration may be the most successful approach within a larger landscape or regional scale, with more active reforestation approaches applied selectively within localized areas that have lost their potential for natural regeneration due to soil degradation or dispersal limitation (Holl and Aide 2011). A two-phase approach involving initial establishment of single-species tree plantations followed by introduction of native species (or mixtures of native and exotic species) after soil amelioration can lead to successful reforestation on severely degraded land (Lugo 1997). The fast-growing exotic *Acacia mangium* (Fabaceae) is widely planted as a nurse tree in Borneo to improve soil fertility on sandy soils (Norisada et al. 2005). Topsoil or litter amendments are required for restoration of areas where soil has been removed, such as mines or bulldozed areas (Lamb and Gilmour 2003; Parrota and Knowles 1999).

Because ecological restoration approaches aim to remove or reduce existing barriers to natural regeneration pathways, effective local restoration prescriptions require a diagnosis specific to each local area. Ecological barriers to natural regeneration fall into four general categories: (1) poor soil conditions due to erosion and loss of topsoil, (2) inadequate colonization of species due to dispersal limitation, (3) dominance by weedy or invasive species of grasses or ferns, or (4) altered microclimatic conditions. Reforestation approaches need to address specific ecological barriers to natural regeneration within the local site (Griscom and Ashton 2011). These constraints vary across geographic and floristic regions. For example, constraints on seed dispersal and seedling establishment among species in the Dipterocarpaceae suggest that restoration of dipterocarp forests in Southeast Asia after shifting agriculture will provide more benefits following enrichment planting, careful tending of planted trees, and establishment of a nurse canopy, rather than from natural regeneration (Kettle 2010). In other forest areas, natural regeneration can be impeded by recurrent fire, grazing, harvesting of litter, or other disturbances that prevent establishment of woody vegetation. Ecological restoration efforts are doomed to fail if the major local barriers to natural regeneration are poorly understood. Many costly restoration projects have been lost to fire, low seedling survival, and inadequate monitoring and management.

As a starting point, Lamb (2011) suggested a series of nine draft principles to guide ecological restoration of tropical forests:

1. Forests can be restored by protecting sites from future disturbances, eliminating impediments to colonization, and ameliorating environmental conditions.
2. Species used to initiate a succession will have a strong influence upon later pathways. Early establishing species should grow quickly to capture resources, suppress growth of weeds, attract seed dispersers, and foster successional development.
3. Succession will proceed more slowly in heavily degraded sites and in

more seasonal environments. In more degraded sites, fewer original species will be able to occupy the site and tolerate site conditions. In these cases, exotic species may be needed initially to facilitate establishment of native species.

4. It is not necessary to mimic natural succession precisely; species from different successional stages can be planted at the same time to ensure rapid canopy closure and control of weeds.
5. Planting species with a range of longevities will ensure continued development of canopy gaps and opportunities for recruitment.
6. It is better to use a number of species in initial planting than only a few species. More species will be more likely to produce a more structurally and functionally complex forest that is self-sustaining and resilient.
7. Some late successional species will not be able to establish early and will need to be sown or planted at later stages.
8. Landscape context is important. Nonplanted species will only colonize if there is relatively intact forest nearby and visitation by seed dispersing fauna.
9. New plantings are likely to be attractive to seed-dispersing wildlife when they offer perch trees, are structurally complex, and provide food resources. Carefully designed mixtures with these attributes are likely to be more effective than random mixtures.

These principles emerge from numerous case studies in different countries and ecological settings. If active reforestation is planned, another major decision is the use of single versus multiple species, or the choice of native versus exotic species in plantings (Lamb 2011). In addition to these considerations, the use of species exotic to the region is a major issue in planning restoration projects (Lugo 1997; Ewel and Putz 2004; Brockerhoff et al. 2008). In cases of severe soil degradation, exotic species can be used effectively to initiate natural regeneration, restore soil fertility, and improve nutrient cycling, as in the Xiaoliang region of Guangdong Province, China (box 13.1). Once soil conditions improve and natural regeneration has taken hold, these exotic species can be harvested, although some species, such as *Gmelina arborea* and *Tectona grandis* (teak), resprout following harvesting.

Ecosystem functioning of naturally regenerating forests can compare favorably to similarly aged actively reforested areas in regions with similar land-use history, vegetation zone, and landscape surroundings, suggesting that active interventions may not always be needed (this situation is discussed further in sec. 13.6). Naturally regenerating forests offer the added bonus of protecting native biodiversity at a substantially reduced cost. There is an urgent need to develop "rapid assessment criteria" for evaluating the natural regeneration potential of an area or region and for assessing the relative success of specific methods in meeting specific reforestation goals. Rodrigues et al. (2011) described five minimal diagnostic criteria that can be assessed to determine the natural regeneration potential of a site:

1. Are conditions within the local site favorable for establishment and growth of native plant species?
2. Are sufficient propagules available in the seed bank or in cohorts of sprouts, seedlings, and saplings of native tree and shrub species?
3. Are different life-forms and successional groups present among naturally regenerating plants?
4. Are forest fragments located sufficiently close to the restoration area to contribute to spontaneous seed rain of diverse species?
5. Are seed-dispersing animals present within the area?

If the answer to all of these questions is yes, the restoration site is likely to exhibit self-propagating natural regeneration with little or no human intervention. These criteria can be used as general ecological guidelines, but each case should be evaluated individually based on historic, landscape, social, economic, and political considerations. Overall, the natural regeneration potential of tropical dry forests may be greater than wet forests due to the higher percentage of wind-dispersal trees and higher abundance of species with resprouting ability (Holl and Aide 2011; Cantarello et al. 2011).

The spatial scale of focal areas and the composition of the surrounding landscape also strongly influence the potential for passive reforestation, as seed dispersal may limit regeneration in areas distant from seed sources. Mixed plantings of wind- and animal-dispersed native trees in fenced enclosures in pastures accelerated seedling establishment of animal-dispersed species of trees and shrubs that characterize later stages of succession (de la Peña-Domene et al. 2013). Late successional recruits dispersed by animals accumulated 10 times faster in planted plots compared to unplanted plots.

Active reforestation pathways, like passive pathways, are constrained by current and past land uses and isolation from seed sources or elimination of seed dispersing fauna. Forest restoration approaches need to be tailored to the ecological, social, and economic limitations and opportunities that are specific to each location. In some cases, soils must first be rehabilitated to support growth of native vegetation (Parrotta et al. 1997). In other cases, forest restoration simply requires prevention of fire (Durno et al. 2007; Omeja et al. 2011), construction of a fence to keep cattle out (Griscom and Ashton 2011),

Box 13.1. Reforestation of Degraded Land in Guangdong, China

Over 100 years of logging and harvesting for firewood in the Xiaoliang region of Guangdong Province, China, transformed extensive areas of tropical monsoon rain forest into a wasteland (Yu et al. 1994; Ren et al. 2007). By 1959, lateritic soils had become barren and heavily eroded, with low organic matter, low water-holding capacity, and high bulk density (see plate 16*A*). More than 100 centimeters of topsoil had been lost in some areas. Some fragments of second-growth forest persisted in the area. Investigators from the South China Institute of Botany, the Xiaoliang Water Conservancy Unit, and Academia Sinica initiated a pioneering long-term experiment in 1959 to determine whether forest could return to this barren landscape and whether soil fertility could be restored. Initially, they established plantations of *Pinus massoniana* and *Eucalyptus exserta*, with the primary goal of controlling soil erosion and restoring soil fertility (see plate 16*B*). Trees grew well initially, but growth rates declined after 8–10 years (Yu et al. 1994). Harvesting of litter and fuel by local residents prevented the accumulation of soil organic matter in the plantations. The untreated area showed no sign of any natural regeneration. The reforestation plan was not working very well.

After several failed attempts to establish broadleaf tree seedlings beneath the plantations, the research team felled 20 hectares of the *Eucalyptus* plantation in 1974 and planted mixtures of 10 species of native trees. This time, seedlings were fertilized and carefully tended, and collection of litter or fuel was prohibited. Over time, more species were planted, including several nonnative nitrogen-fixing species. Species were planted in mixtures using eight different combinations. By 1980, 320 species, including 75 legumes had been planted in the experimental area (Yu et al. 1994; Ren et al. 2007).

The mixed-species plantings grew well and attracted birds that brought in more native species of trees from remnant forest patches. In one 1.4-hectare block where 41 tree species were planted by 1980, 119 species were recorded at

removal of exotic species (Durigan and Melo 2011), or no intervention at all (Holl and Aide 2011). Planted agroforests can be part of a rotational shifting cultivation system involving long fallow periods and tree production (Michon et al. 2007; Cairns 2007; Vieira et al. 2009). Reforested areas and agroforests can provide critical habitats for plant and animal species and can be integral components of landscape-level restoration and conservation (Bhagwat et al.

this site in 1982. Thirteen of these species were bird-dispersed (Yu et al. 1994). Native species achieved the size of nonnative legume species within 15 years. From 1959 to 2004, soil organic matter in the mixed plantation increased from 0.64% to 2.95% and total soil nitrogen doubled. By 1986, the mixed forest effectively reduced surface runoff and soil erosion to zero, whereas soil erosion continued to occur in the *Eucalyptus* plantation (Ren et al. 2007; Liu and Li 2010). Soil organic matter still remained lower than in residual forests of the region, however, and soil experts estimated that up to 150 years would be required to fully restore soil fertility (Yu et al. 1984). The increase in soil fertility in plantations was sufficient to enable cultivation of commercial fruit trees in some areas.

After 45 years, the untreated area still remained barren and eroded, with no sign of recovery. The pine plantation died completely in 1964, a victim of insect attacks and heat stress from high soil surface temperatures (Ren et al. 2007). The *Eucalyptus* plantations persisted, but the understory was poorly developed. Only the mixed-species plantings had a well-developed forest structure and diverse composition of trees, shrubs, and herbs. Dominant tree and shrub species were shared with naturally regenerating forest ecosystems in the region. In 2004, 100 species of birds were recorded. Diversity of soil animals increased rapidly, as soil organic matter increased, and increasing microbial diversity in the soil promoted nutrient cycling. Thirty years after the mixed forest was initiated, this "artificial" forest began to take on a natural life of its own, with a complex trophic structure including ants, termites, spiders, locusts, beetles, butterflies, insectivorous birds, tree frogs, and rodents. Top predators and large mammals have not yet arrived (plate 16*C*; Ren et al. 2007).

Since 2007, the *Eucalyptus* plantations established in 1959 have been cut down to plant new, fast-growing, short-rotation (3–5 years) *Eucalyptus* cultivars for the production of paper (lignum) throughout the region (Liu and Li 2010). The 20-hectare mixed-species forest of Xiaoliang has now become a tiny island in a sea of *Eucalyptus* trees.

2008; Parrotta 2010; Reid et al. 2012). Abandoned commercial tree plantations can also give rise to diverse natural regeneration if there are nearby forest fragments. Fifteen years after abandonment of a third-rotation plantation of *Eucalyptus saligna* in Parque das Neblinas in Sao Paolo State, Brazil, 44% of the tree species in neighboring forest fragments were found as regenerating saplings and small trees (Onofre et al. 2010).

13.3 Reforestation through Management of Forest Fallows

Long before the word *reforestation* even existed, traditional shifting cultivators across the tropics practiced reforestation and fallow management as an integral component of long-fallow agricultural systems (see box 2.1; Gómez-Pompa 1987; Denevan et al. 1984; Michon et al. 2007; Wangpakapattanawong et al. 2010). Today, these practices are still evident in some regions, although shifting cultivation systems have changed dramatically due to shortages of land; pressures to grow commercial crops, intensify production, and reduce fallow periods; and government policies that restrict forest use and prevent forest clearing (Padoch et al. 2007; Rerkasem et al. 2009; Ziegler et al. 2011). Smallholder shifting cultivators in the Peruvian Amazon adopted practices to improve fallows to restore agricultural productivity to land that is declining in soil fertility (Marquardt et al. 2013). Repeated, short-fallow cycles can result in severe degradation of soils and formation of fire-prone grasslands or other forms of arrested succession (see box 7.1; Durno et al. 2007; Ramakrishnan 2007).

Well-managed fallows are an effective form of reforestation (Kammesheidt 2002; Michon et al. 2007). Formerly cultivated fields are not abandoned but continue to be tended, managed, and planted, often using silvicultural techniques that have proven effective for many decades or even centuries. This management ensures ecological and cultural continuity between the original forest, swidden field, and fallow vegetation. What begins as an agricultural system is transformed into an agroforestry or *taungya* system in which trees interplanted with crops grow up alongside naturally established successional vegetation. When clearing and burning a patch of forest to plant crops, farmers in many tropical regions actively promote forest regeneration by leaving selected remnant trees to encourage resprouting of fast-growing trees during the cultivation phase (table 13.1). These "enriched fallows," "domestic" forests, "man-made" forests, "forest gardens," and "agroforests" all represent pathways of forest restoration that simultaneously provide functioning forest ecosystems and timber and nontimber products that are selectively harvested by local farmers and their families.

Enriched fallows provide essential foods, medicines, construction materials, and hunting grounds for shifting cultivators. Fruiting trees attract numerous frugivores, including commonly hunted animals such as small deer (Nations and Nigh 1980; Denevan et al. 1984; Posey 1985; Barrera-Bassols and Toledo 2005). In the past, shifting cultivators obtained all of their material and spiritual needs from their heterogeneous forest habitat, supplementing the food supply provided by ephemeral cultivated fields with numerous resources provided by different stages of fallow vegetation enriched by planting

Table 13.1. *Principal tree species planted or selectively managed in enriched fallow systems in tropical regions of the world*

Region	Principal tree species planted and harvested	Reference
Indonesia	Rubber (*Hevea brasiliensis*)	Dove (1993); Penot (2007)
	Damar resin (*Shorea javanica*)	Michon (2005)
	Benzoin resin (*Styrax* spp.)	García-Fernández and Casado (2005)
	Mixed fruit trees	De Jong (2002); Michon (2005)
	Rattan (*Calamus caesius*)	Weinstock (1983); García-Fernández and Casado (2005)
Thailand	Tea (*Camellia sinensis* var. *assamica*)	Sasaki et al. (2007)
	Firewood, timber, fruit, medicinal plants	Durno et al. (2007)
Vietnam	Fruit, medicinal plants	Dao et al. (2001)
	Bamboo (*Neohouzeaua dulloa*, other species)	Ty (2007)
Xishuangbanna, southwestern China	Tea, rattan, fruit, spice trees	Xu (2007)
Laos PDR	Benzoin resin, cardamon (*Elettaria* spp.)	Michon (2005)
	Paper mulberry (*Broussonetia papyrifera*)	Fahrney et al. (2007)
	Teak (*Tectona grandis*)	Hansen et al. (2007)
Australia	Fruit trees	Hynes and Chase (1982)
Pacific-Melanesia	Fruit, nut trees	Clarke and Thaman (1993); Hviding and Bayliss-Smith (2000)
	Casuarina oligodon	Clarke (1966); Bourke (2007)
	Gnetum gnemon, *Areca catechu*, *Metroxylon sagu*, *Pandanus* spp.	Kennedy and Clarke (2004)
Pacific-Polynesia and Micronesia	Fruit trees	Clarke and Thaman (1993)
Southern Yucatán Peninsula, Mexico	Fruit, timber, spice trees	Chowdhury (2007)
Lowland Chiapas, Mexico	Fruit, timber, spice trees	Nations and Nigh (1980); Levy Tacher et al. (2002); Diemont et al. (2011)

(*continued*)

Table 13.1. (*continued*)

Region	Principal tree species planted and harvested	Reference
Petén, Guatemala	Fruit, timber, other useful trees	Atran et al. 1993
Eastern Peru	Fruit, timber, other useful species	Denevan et al. (1984); Unruh (1990)
Napo-Amazon floodplain, lowland Peru	*Maquira coriaceae*, other timber species	Padoch and Pinedo-Vásquez (2006)
Brazilian Amazonia	Babaçu palm (*Attalea phalerata*)	Hecht et al. (1988)
	Brazil nut (*Bertholettia excelsa*)	Posey (1985)

Source: Adapted from Lamb (2011, table 5.6).

key tree resources and tending numerous naturally colonizing or resprouting species. These efforts led to high concentrations of useful species in local forests, which are still evident today (Posey 1985; Hecht et al. 1988). Shifting cultivation fallows still function as a safety net for the rural poor in many tropical countries (Rerkasem et al. 2009). Smallholder farmers often employ a mixed strategy of permanent and shifting cultivation.

Management of fallows begins even before forest areas are cleared for burning and cultivation. During forest cutting, trees are managed to promote rapid fallow regeneration (Rerkasem et al. 2009). Resprouting is considered to be the most effective and more common method of tree regeneration in shifting cultivation systems (Kammesheidt 1999; Wangpakapattanawong et al. 2010). Often large trees are protected from cutting to remain as seed trees in swidden fields. The Lawa of northern Thailand leave all trees with a diameter above 12 centimeters and selectively prune large trees to initially reduce shading of planted crops (Schmid-Vogt 1997). Resprouting is encouraged by leaving stumps at heights of 0.5 to 1 meters that are undamaged by burning of slash. The Hani of Mensong in Xishuangbanna, Yunan, China, practice long-fallow shifting cultivation with enriched fallows. Particular tree species attractive to birds are planted in fallows, ensuring diverse seed rain and regeneration. Their traditional practices foster complex vegetation structure, higher diversity of vegetation, and higher diversity of bird species than in the Jinuo area, where traditional swidden practices have broken down (Wang and Young 2003).

Lacandon Mayans are well known for their highly developed enriched fallow system (Nations and Nigh 1980; Toledo et al. 2003; Diemont and Martin 2009). Following four to six years of polyculture cropping, they sow balsa tree (*Ochroma pyramidale*) seeds in their fields. The leaf litter of young *Ochroma* trees suppresses weeds, inhibits nematodes, and enriches soil organic matter (Levy Tacher and Golicher 2004). *Ochroma* plantings prevent invasion of the aggressive fern *Pteridium caudatum*, which can arrest succession (see box 7.1 and plate 17; Douterlungne et al. 2010). Traditional ecological knowledge of the Lacandon people provides useful guidance for the restoration of tropical forests in this region. Over 30 tree species are planted in cropped fields or protected and nurtured following spontaneous colonization to enrich the soil and suppress weeds (Diemont et al. 2006; Diemont et al. 2011). Common to all of these Mayan systems is the use of distinct successional stages in fallow management, the use of species with properties that enrich soil or facilitate succession, and direct human consumption of forest products at all successional stages.

Strong parallels between agroforestry systems, managed long-fallow systems, and reforestation approaches have been noted (Michon et al. 2007; Vieira et al. 2009). Enriched fallow systems mimic and promote vegetation changes that occur during natural regeneration). Traditional ecological knowledge serves to identify species with proven potential for survival and growth that facilitate forest regeneration, support local wildlife, and supply local populations with a wide variety of timber and nontimber products, including medicinal species (Vocks 1996; Chazdon and Coe 1999).

Indigenous knowledge of forest succession can be directly applied to reforestation programs. Based on their knowledge of forest ecology, the Akha people in the mountains of northern Thailand developed a way to facilitate regeneration of forest around their village (Durno et al. 2007). They transformed *Imperata*-dominated fallows into forest by preventing fire and promoting regeneration from resprouting rootstocks and seeds (see plate 18 and box 13.2). In fact, villagers found natural forest regeneration to be more effective than reforestation (planting trees) in restoring forest cover. Durno and her colleagues noted that "the process of natural succession, and the hill farmers' intimate understanding of it, can today be harnessed to regenerate and conserve watershed forests, which are critical to the environmental health of the entire country" (2007, p. 135).

13.4 Ecological Forest Restoration in the Tropics

Ecological restoration approaches vary widely in cost, biodiversity levels supported, time for forest development, and specialized research re-

Box 13.2. Assisted Natural Regeneration of Grasslands in the Philippines

By 1997, the Philippines had lost 80% of its forest cover (Lasco et al. 2001). Over two million hectares of former forest areas in the Philippines have been converted to grasslands, dominated by aggressive, fire-resistant grasses such as *Imperata cylindrica* and *Saccharum spontaneum* (see box 7.1). If protected from fire, these grasslands will slowly, naturally grow back to forest (Durno et al. 2007). But this slow successional process can be accelerated. For over 30 years, assisted natural regeneration approaches have successfully restored forests on grasslands in the Philippines (Ganz and Durst 2003; Shono et al. 2007). This approach can effectively and rapidly reestablish forest cover in ecologically vulnerable areas, such as steep slopes in critical watersheds, and in areas where the restoration of biological diversity is urgently needed, such as within conservation areas and biological reserves (Sajise 2003). Furthermore, the cost of assisted natural regeneration is half that of conventional forest restoration methods involving tree planting (Durst et al. 2011).

The goal of assisted natural regeneration is to accelerate the establishment, growth, and survival of native tree species that naturally colonize grassland. This approach requires the proximity of remnant forest areas and a minimum density of naturally occurring tree seedlings that establish beneath the dense grass canopy. Typically, 200 to 800 seedlings per hectare are required (Shono et al. 2007). A density of 800 seedlings is sufficient to produce a closed canopy within three years, shading out the grasses.

Assisted regeneration as practiced in the Philippines has four key elements: controlling fire through development of fire breaks, restricting grazing, suppressing growth of *Imperata* and other aggressive grasses, and nurturing the growth of naturally regenerating tree seedlings. The approach requires active

quirements (table 13.2; Kettle 2010). Early afforestation projects focused heavily on plantations of commercial species, such as *Eucalyptus* and *Pinus* species, to accelerate natural regeneration, as these were the only species available for planting and little research was required. Assisted natural regeneration costs far less than planting seedlings, and requires little research, but leads to slower rates of forest structural development. Natural regeneration poses greater risks than tree planting to smallholders, as land in early participation of local people. Agroforestry plantings and enrichment planting of large-seeded old-growth tree species can also be implemented to enhance livelihoods for local people and to promote native biodiversity.

Shono et al. (2007) provided a detailed description of the methodology for assisted natural regeneration. First the area is surveyed, and naturally woody seedlings are located (see plate 17*A*). Seedlings are marked and monitored for survival. The growth of each seedling is promoted by weeding competing vegetation within a 0.5-meter radius and fertilizing if needed. Grass cover is suppressed throughout the area by pressing or trampling grasses using a board or by rolling a barrel on slopes (see plate 17*B*). In this process, the base of the grass stem is bent, but not broken, so as to prevent tillering. This process should be repeated two to three times a year until the grasses die. Most importantly, the site needs to be protected from fire (using firebreaks), and fencing should be used to keep grazing animals away (see plate 17*C*).

The Philippine forestry agency has been slow to adopt this approach, instead promoting conventional plantation forestry approaches using fast-growing exotic species, such as *Gmelina arborea*, *Acacia mangium*, *Pterocarpus indicus*, and *Swietenia macrophylla* (Lasco and Pulhin 2006). Not only are these conventional methods more costly, but they also show a poor rate of success (often < 30%) due to fire, failed maintenance, and political corruption (Lasco et al. 2001).

In 2006, a three-year project on assisted natural regeneration was launched through a partnership among the United Nation's Food and Agriculture Organization (FAO), the Philippines Department of Environment and Natural Resources, and the Bagong Pagasa Foundation. The Philippine government, impressed with the low-cost and high success of this approach, has allocated special funding to support assisted natural regeneration on over 9,000 hectares of forest land. Local communities will benefit as well, as these projects have allowed them to become partners in helping their forests regrow.

stages of succession may appear abandoned and is more susceptible to fire. Planting seedlings of native species is now a popular reforestation method, as knowledge of silviculture and nursery practices rapidly expands. The planting of five native tree species in degraded lands of Kibale National Park, Uganda, significantly enhanced forest recovery, tree biomass, and dispersal and regeneration of nonplanted native species within 10 years (Omeja et al. 2010). Practiced widely in the Atlantic Forest region of Brazil, high-diversity plantings of

Table 13.2. *Merits of different ecological forest restoration approaches*

Reforestation approach	Costs (labor and capital)	Biodiversity	Time for forest development	Research input required
Monoculture of commercial nurse trees	High	Low to medium	Fast	Low
Assisted natural regeneration (without enrichment planting)	Low	Low to medium	Slow to medium	Low
Assisted natural regeneration (with enrichment planting)	Low to medium	Medium	Medium	Low to medium
Framework species method	Medium to high	Medium	Medium	High
High-diversity planting of native tree seedlings	High	High	Fast	High
Direct sowing of forest tree seeds	Low to medium	High	Fast	High
Nucleation plantings	Medium	Medium to high	Medium to fast	High

Source: Shono et al. (2007).

native species promote rapid development of forests with relatively high levels of biodiversity but with a high price tag (box 13.3; Rodrigues et al. 2009).

Competition with aggressive exotic species and reduced soil fertility can be major barriers to natural regeneration in the tropics. In these cases, nurse trees are frequently planted to ameliorate local site conditions and to promote establishment of naturally regenerating native species (Parrotta et al. 1997). To be effective, nurse species must live long enough for grasses or other weeds to be excluded and new tree colonists establish (Lamb 2011). Nurse trees can be native or exotic species that can tolerate extreme conditions, exclude weeds, and improve soil fertility—for example, the planting of balsa trees (*Ochroma pyramidale*) to suppress bracken fern (*Pteridium aquilinum*) in Chiapas, Mexico (see plate 17; Douterlungne et al. 2010). Plantations of nitrogen-fixing legumes *Acacia mangium* and *Paraserianthes falcataria* facilitated natural regeneration of 63 species in 24 families in *Imperata cylindrica* grasslands in South Kalimantan, Indonesia (Otsamo 2000). The nurse tree approach is relatively inexpensive but relies on seed dispersal from surrounding areas of natural vegetation. In some cases, nurse trees can become an impediment to tree recruitment and need to be harvested or thinned (Lamb 2011).

The framework species approach uses a larger number of species to jump-start succession. From 20 to 30 species are planted to initiate the succession and provide the framework for future successional development. The mixture of species includes short-lived and longer-lived pioneers (see box 5.1) and species that produce fruit attractive to animals. This method is well suited to sites that are close to existing natural forests to ensure seed dispersal. In seasonally dry tropical regions, some of the planted species should be tolerant of fire and capable of resprouting (Lamb 2011).

The maximum diversity approach to forest restoration uses an even larger number of species in initial plantings. Seedlings of up to 80–100 species are planted, including rare species and a diversity of life-forms and functional types (Goosem and Tucker 1995; Rodrigues et al 2009). Many of the species are adapted to later successional stages, possessing large seeds that are poorly dispersed, and otherwise would be unlikely to reach restored areas. This method ensures relatively high species diversity and functional diversity in restored areas and provides a rapid increase in diversity compared to other planting methods. Both seeds and seedlings can be used in plantings.

Direct seeding and nucleation are promising approaches to high biodiversity restorations with lower cost than establishing tree plantations; both methods are receiving increasing research attention. Direct sowing of seeds reduces costs of transport and labor considerably and does not require development of nursery propagation techniques for poorly studied species (see table 13.2). Parrotta and Knowles (1999) used a seed mix of 27 native species collected from nearby rain forest to restore a 17-hectare area following bauxite mining in Brazilian Amazonia. Topsoil was first spread across the site. Several studies have shown relatively high germination and survival rates using direct seeding in former pastures, particularly for large-seeded species (Camargo et al. 2002; Cole et al. 2011). Tree species with intermediate-sized or large seeds, round or oval seeds, low or medium moisture content, and relatively thick seed coats had higher rates of seedling survival following direct seeding in *Imperata* grasslands in southern Thailand (Tunjai and Elliott 2012). Rates of

Box 13.3 Reforestation in the Atlantic Forest Region of Brazil

The Atlantic Forest was once an expanse of evergreen and seasonally dry forests covering 1.5 million square kilometers between 3° and 30° S along the Brazilian Atlantic coast. Thousands of years ago, shifting cultivators were the first people to clear Atlantic Forest to plant crops (Dean 1995; Oliveira 2007). The arrival of Portuguese colonists in 1500 led to a far more devastating pattern of deforestation, habitat fragmentation, and forest degradation (Dean 1995; Rodrigues et al. 2009; Tabarelli et al. 2010). Today, slightly less than 12% of the original forest cover remains (Ribeiro et al. 2009). This region is home to 70% of Brazil's population, creating enormous demands for land, water, and forest products. It is recognized as a global biodiversity hot spot on the basis of its high levels of biodiversity, high rates of local endemism, and high levels of deforestation (Tabarelli et al. 2010).

Reforestation is a major enterprise in the Atlantic Forest region, supported by legal and political instruments and joint ventures between government agencies, research agencies, and private enterprises (Wuethrich 2007). In 1980, new federal environmental legislation required reforestation in areas affected by hydroelectric dams. In 2000, for areas deforested after 1965, all landowners became legally obligated to restore forest on 20% of their land (80% in the Amazon region). This obligation created a huge demand for reforestation projects and stimulated diverse approaches (Durigan and Melo 2011). In 2006, the federal government of Brazil established a restoration fund to encourage projects and research in the Atlantic Forest region (Rodrigues et al. 2009).

Approaches to reforestation in Brazil's Atlantic Forest region have changed dramatically from the 1970s and early 1980s when *Pinus* and *Eucalyptus* were planted for timber (Rodrigues et al. 2009; Brancalion et al. 2010; Durigan and Melo 2011). Following the lead of Kageyama and Castro (1989), researchers have developed approaches to reforestation based on successional theory that attempt to accelerate or enhance natural regeneration. The planting of native species has become widespread, with a focus on reestablishing self-sustaining forests that are on a path to develop similar levels of biodiversity to the remnant forest fragments in the surrounding landscape.

Goals for reforesting particular sites generally include removing or reducing human impact, creating forest structure to provide permanent shade, maintaining or increasing the number of woody species, promoting colonization of nonarboreal native species, providing shelter and food to support local fauna, and managing invasive grasses (Rodrigues et al. 2009). Reforested areas are being designed to create forest corridors between existing biological reserves, to protect water resources, and to provide new habitats for endangered primates endemic to the region, such as the golden lion tamarin, (*Leontopithecus rosalia*; Sansavero et al. 2011).

Some researchers have concluded that reforestation of highly diverse Atlantic Forest is most successful when a high diversity of species is initially planted (Brancalion et al. 2010; Aronson et al. 2011). Under Resolution SMA 21, passed in 2001 by the state government of São Paulo, planting a minimum of 30 species is required for all reforestation projects in "high-diversity" forest types. The resolution has since been revised, with the most recent version in 2007 requiring that at least 80 species are present at the end of the restoration process, including naturally regenerating species. High species diversity is a form of "insurance" for reforestation projects, as some species fail to colonize, establish, or persist in the new forest (Wuethrich 2007). In some cases, however, high diversity of regeneration in restored sites can be achieved when few species are initially planted, suggesting that site factors can limit regeneration more than colonization (Durigan et al. 2010).

Throughout the Atlantic Forest region, restoration projects use a mix of pioneer species and a high diversity of nonpioneer species, sometimes planted in alternate lines to promote long-term development of forest structure and increase functional diversity (Nave and Rodrigues 2007). During the first 10–20 years, the abundance of successional pioneers declines, while shade-tolerant species increase in size and abundance.

The Atlantic Forest Restoration Pact (AFRP) was formed in 2009 to launch one of the most ambitious ecological reforestation programs in the world. The goal of the AFRP is to restore a total of 15 million hectares of degraded lands in the Brazilian Atlantic Forest by 2050. Members of the AFRP include 96 nongovernmental organizations, 34 government institutions, 25 private companies, and 7 research institutions. Reforestation activities will potentially generate over three million jobs and will require investment of US$77 billion (Calmon et al. 2011). Incentives are provided to encourage small landowners to comply with reforestation programs. If the program is successful, at least 30% of the forest cover present before the European conquest will be restored (Aronson et al. 2007). These efforts are expected to reduce the risk of extinction for many endemic species whose home has all but disappeared.

seed predation can be high and seed quality is highly variable. Based on their study in Laos, Sovu et al. (2010) recommended burying seeds to prevent seed predation and desiccation. Direct seeding without burial led to poor seedling establishment and high seed wastage in three sites in the Wet Tropics in northeast Queensland. Larger seeded species showed higher rates of establishment, but only in treatments when seeds were buried 5–20 millimeters deep. Direct sowing may be more appropriate as a supplement to planted seedlings or as a means of enriching sites after the initial stages of regeneration (Doust et al. 2006; Cole et al. 2011). It is not an appropriate method in the absence of weed control. Furthermore, large amounts of seed are needed, potentially impacting natural populations of seed-producing trees and restricting the number of species that can be used (Lamb 2011).

Applied nucleation approaches are based on principles of succession of nonforested landscapes (Reis et al. 2010; Corbin and Holl 2012). Seedlings of nurse species or foundation species are planted in clusters or "islands" rather than in a uniform pattern, mimicking nucleation processes around shrubs or remnant trees during early stages of succession (see box 7.2; Yarranton and Morrison 1974). Because fewer trees are planted, costs are reduced to a quarter or a third of typical restoration plantation costs (Rey Benayas et al. 2008; Holl et al. 2011). Applied nucleation plantings (also called "islets") create a more heterogeneous environment than uniform plantations and may therefore favor a higher diversity of regenerating species (Rey Benayas et al. 2008; Corbin and Holl 2012). Restoration nuclei appear to have a minimum size threshold to attract seed dispersers. Cole et al (2011) found that nuclei greater than 64 square meters were more effective for facilitating diverse seed rain in southern Costa Rica than smaller nuclei (fig. 13.3). Fenced plantings of trees in active pastures are an effective means to establish small forest patches to facilitate connectivity among forest fragments in the surrounding landscape (de la Peña-Domene et al. 2013).

The framework species and nurse tree approaches have been the most widely adopted ecological restoration approaches in the tropics. The stand initiation phase develops through planted tree seedlings rather than through natural colonization of seedlings from surrounding areas or from the seed bank. During reforestation, the pioneer tree species that colonize naturally or that are planted strongly influence the dynamics, structure, and composition of the developing forest cover. Planting tree seedlings on unused or degraded land is expected to result in more rapid and predictable tree establishment, promoting seed dispersal and establishment of diverse species of native trees that otherwise would not be able to colonize the site.

Few studies have examined long-term reforestation trajectories following

Figure 13.3
View of experimental reforestation treatments on degraded pasture land near Las Cruces Biological Station in southern Costa Rica (Cole et al. 2010). Three treatments in one study site can be seen: plantation (P), control (C), and tree islands (I). Photo was taken four years after planting. Source: Rebecca Cole, reprinted with permission.

the establishment of different pioneer species, nurse tree species, or framework species (Powers et al. 1997; Erskine et al. 2007). These assessments generally ignore the socioeconomic impacts of restoration projects (Dinh Le et al. 2012). The success of ecological restoration projects requires continuous maintenance of experimental treatments and careful monitoring of both treated and untreated areas to determine whether treatments actually lead to the desired ecological outcomes, which can include seed rain, regeneration of natural vegetation, restoration of ecosystem functions, or return of native biodiversity. One difficulty in using monitoring results to assess restoration objectives is that the time scale for natural regeneration is from decades to centuries, far longer than the typical time scale for restoration projects (usually 3–5 years). Prior to 2010, none of the major reforestation programs in the Wet Tropics region of Queensland, Australia, incorporated a formal monitoring program (Kanowski et al. 2008). Many restoration projects lack a formal monitoring program because they are funded on a short-term basis, rely on community groups for volunteer labor, and lack expertise among practitioners and volunteers. Two-thirds of the restoration plantings established in North Queensland between 1997 and 2002 were in suboptimal condition presumably due to lack of ongoing maintenance. Within 5–10 years of establishment, the initial investments in these projects were at serious risk of being forfeited (Kanowski et al. 2010). A comprehensive toolkit for monitoring restoration projects has been developed by Kanowski et al. (2010), which includes detailed criteria for monitoring site conditions, forest structure, plant species composition, carbon sequestration, and bird species composition.

13.5 Recovery of Biodiversity during Reforestation

A small number of projects using mixed plantings of native species have now reached a sufficient age to permit assessments of biodiversity and ecosystem services in comparison with other land uses, including old-growth forests and unplanted areas undergoing natural regeneration in the same region. Extensive studies in tropical regions of northern Queensland, Australia, have shown that diverse restoration plantings of native tree species can develop closed canopies and complex vegetation structure within 10 years of establishment and can support a moderate diversity of rain forest fauna (Kanowski et al. 2003, 2006; Catterall et al. 2012). Dense ecological restoration plantings in the Wet Tropics region can achieve canopy closure within 3–5 years of establishment on favorable sites (Goosem and Tucker 1995). Animal diversity is expected to increase during ecological restoration in much the same way as occurs during natural regeneration. Colonization of restoration sites by forest specialists is strongly mediated by changes in forest structure and understory microclimate and by spatial patterns of vegetation in the surrounding landscape (Erskine et al. 2007). In seasonally dry tropical regions, where fire naturally takes place in forests, burning of restored and rehabilitated areas may be needed to enhance colonization of native species (Brady and Noske 2010).

Increasing canopy height, canopy cover, and litter cover with age of restoration plantings correlated strongly with beetle species richness and compositional similarity with rain forest areas in the Wet Tropics region. Older sites (6–17 yr. since planting) and sites adjacent to forest had beetle species composition most similar to local rain forest areas (Grimbacher and Catterall 2007). Presence of rain forest reptiles in these reforestation plantings required a threshold of 60% canopy cover; reforested sites that were completely isolated from forested areas are unlikely to be recolonized by rain forest reptiles if habitat corridors are not established (Kanowski et al. 2006).

Ten years after planting, ecological restoration sites averaged about half the number of rain forest-dependent birds typically found in local rain forest areas (Catterall et al. 2012). As plantings increased in age from 1 to 24 years, open-habitat species declined in richness and abundance, while rain forest-dependent species increased. Regional rain forest endemic species were the slowest to recolonize restoration sites and were half as likely to occupy older reforested sites as nonendemic forest species.

Studies in the Kakamega Forest of Kenya also illustrate the recolonization of trees and forest-dependent bird species in 60- to 70-year-old mixed indigenous plantings. Species composition of birds in mixed plantings was similar to natural forests of the region. Richness and abundance of seedlings of late successional tree species showed high recovery in mixed-species plantings and monoculture tree plantations (Farwig et al. 2008, 2009). Despite their value as supplemental habitat, however, planted forests are not adequate replacements for lost old-growth forests. Mixed native tree plantings are used as habitat by three diurnal primate species (*Colobus guereza, Cercopithecus mitis, C. ascanius*) but support lower group densities (Fashing et al. 2012).

Few studies have compared species richness of plants or animals in restoration plantings with naturally regenerating areas matched for age and environmental conditions. In Luquillo Experimental Forest in Puerto Rico, small unmanaged pine (*Pinus caribaea*) and mahogany (*Swietenia macrophylla*) plantations showed lower numbers of woody species in the understory compared to paired second-growth forests of similar age. But after 50 years, species richness in the understory of mahogany plantations approached that of paired second-growth forests (Lugo 1992). Thirty-year-old plantations of native *Alnus acuminata* in the Colombian Andes had similar vegetation structure and basal area as naturally regenerated forests of the same age, but plantations had one-third fewer species per plot, low species turnover among plots, and distinct species composition from the regional species pool (Murcia 1997). Bird abundance and species composition did not differ significantly between 40-year-old restoration plantings of exotic Chinese ash (*Fraxinus chinensis*) and naturally regenerating forests (Durán and Kattan 2005). These examples illustrate that biodiversity following natural regeneration may recover at least as rapidly as in restoration plantations.

Higher diversity of tree species planted in restoration plantations does not necessarily lead to higher diversity of regenerating trees in the understory. In the Caribbean lowlands of Costa Rica, 15- to 16-year-old mixed-species restoration plantations of native tree species showed similar abundance, species richness, and species composition of woody understory regeneration compared to single-species plantations (Butler et al. 2008). Species richness of woody understory regeneration was inversely correlated with percent canopy openness across all plantations and was least developed in unplanted areas undergoing natural regeneration.

13.6 Recovery of Ecosystem Properties during Forest Restoration

A major impetus for ecological restoration of tropical forests is to restore soil fertility to enhance provision of ecosystem goods and services lost through deforestation and forest degradation (Montagnini and Finney 2011). Ecosystem services are defined broadly as the benefits that humans receive from ecosystems at local, regional, or global scales. These benefits include timber and nontimber forest products, restoration of soil fertility, above- and belowground carbon storage, regulation of hydrological flows, biodiversity conservation, recreation, and ecotourism (Rey Benayas et al. 2009). Although active reforestation using native species plantations can be designed to enhance a wide array of ecosystem services, important trade-offs in ecosystem services must be recognized (Hall et al. 2012). Silvicultural management in restoration plantations may increase timber productivity, but it is likely to reduce total carbon storage and alter the composition of plant and animal biodiversity.

Ecological reforestation of degraded pastures using tree islands in Costa Rica increased litter inputs, tree regeneration, and carbon storage faster than in unplanted sites (Celetano, Zahawi, Finegan, Ostertag, et al. 2011; Celetano, Zahawi, Finegan, Casanoves, et al. 2011). Plantations of the same four species recovered these properties even faster. Three types of restoration treatments in Kibale National Park, Uganda, led to rapid rates of aboveground biomass accumulation after 12–32 years but differed in costs of establishment and maintenance (Omeja et al. 2012). The least expensive approach, fire protection through cutting and maintaining fire lanes, cost approximately US$500 per square kilometer per year and was highly effective in promoting natural regeneration and biomass accumulation. Restoration treatments also varied in their social ramifications; conflicts with the local community arose from the hiring of outside laborers to harvest pine plantations and from degradation of local roads from the hauling of the timber.

Several studies have compared carbon storage and nutrient cycling in naturally regenerating forests and monoculture tree plantations, controlling for forest age, prior land use, and environmental conditions. Lugo (1992) found that unmanaged plantations in Luquillo Experimental Forest in Puerto Rico had higher aboveground biomass, higher root biomass, greater depth of root penetration, and faster rates of nutrient cycling in soils than paired naturally regenerating forests. But the contribution of understory biomass to total stand biomass was lower in plantations than in naturally regenerating forests of similar age. A 20-year-old pine plantation and second-growth forest that regrew on the same farmland in Puerto Rico showed similar levels of soil organic carbon (0–25 cm depth), but the forest soil accumulated more of the heavy fraction of organic carbon, increasing the potential for long-term carbon storage (Li et al. 2005). Rates of leaf litter decomposition and total root biomass were also higher in the naturally regenerating forest. A meta-analysis of 48 studies from across the tropics showed that aboveground biomass accumulation was significantly higher in monoculture plantations than in natural regeneration, but these differences were small and diminished in sites more than 18 years old (Bonner et al. 2013).

The ecosystem properties of reforestation plantations are also influenced by the number and characteristics of species planted. In mixed-species plantations, carbon storage in vegetation is higher than single-species plantations, matching for site age and environmental conditions. In the Wet Tropics region of Queensland, Australia, carbon stocks were higher in diverse restoration plantings than in commercial tree conifer monocultures and mixed-species plantations of similar age (Kanowski and Catterall 2010). The higher levels of aboveground carbon in restoration plantings were due to higher density of plantings, larger diameter of trees, and higher wood density than conifers in monoculture plantations. In Costa Rica, aboveground biomass in 15- to 16-year-old mixed plantations was higher than in single-species plantations (Piotto et al. 2010). The number of tree species in young reforestation plantations in Sardinilla, Panama, did not significantly affect aboveground carbon storage, but soil respiration and decomposition rates of coarse woody debris and leaf litter were significantly higher in monocultures (Potvin et al. 2011). Annual rates of stand-level transpiration were 14%–56% higher in mixed-species plantations compared to monocultures, which were partially attributed to the larger diameters and higher basal area of trees grown in mixtures (Kunert et al. 2012).

In eastern subtropical Australia, both active and passive reforestation led to recovery of a wide range of soil properties. Nitrification, pH, bulk density, plant-available nitrate, and plant-available phosphate were similar in replanted restoration sites and old-growth rain forest sites. Among the nine soil properties measured, only fine root biomass remained similar between replanted sites and pasture sites. Restoration pathways had differential effects on soil properties and early seedling growth of pioneer species (Paul et al. 2010a, 2010b). Regrowth dominated by the exotic camphor laurel (*Cinnamomum camphora*) tree covers over 25% of this region. Camphor-dominated regrowth (20–40 yr. old) showed slower recovery of soil properties compared to ecological restoration plantings of diverse tree species after 12–20 years. Removing camphor laurel trees by girdling restored soil properties to values more similar to those observed in adjacent rain forest areas and promoted regeneration of diverse rain forest tree seedlings (Kanowski and Catterall

2010). Managing extensive areas of camphor laurel regrowth is a promising option for reforestation that is far more cost-effective than restoration plantings (Paul et al. 2010b).

Changes in soil properties accompanied reforestation in controls (natural regeneration) and two types of restoration plantings in the Atlantic Forest region of São Paulo, Brazil (Nogueira et al. 2011). Nutrient cycling, soil organic matter, and soil porosity all increased over 10 years in high-diversity plantings of seedlings of 40 species and low-diversity plots established through direct seeding of five pioneer trees species. Restoration treatments did not differ significantly in soil carbon stocks, but the low-diversity treatment promoted faster changes in soil properties as a consequence of the faster tree growth and more rapid stand development. Experimental reforestation studies in northeast Costa Rica have indicated that six native tree species varied significantly in their effects on surface soil properties (0–15 cm depth), even 15 years after planting. Soil organic carbon in surface soils was significantly correlated with fine root ingrowth, but not with rates of nonwoody litterfall, suggesting that microbial dynamics associated with root traits are strong determinants of soil carbon storage during reforestation (Russell et al. 2007). Short-term recovery of soil carbon and nitrogen cycling in fenced areas of pasture in Veracruz, Mexico, was enhanced in plantings, including leguminous trees (Roa-Fuentes et al. 2013).

13.7 Conclusion

The Xiaoliang experience (see box 13.1) provides several valuable lessons for the reforestation of severely degraded land. The most valuable lesson is the need to prevent continuing degradation of previously forested land through sound management of tropical forest ecosystems. Successful reforestation requires overcoming two major limitations: (1) the harsh physical environment and (2) limited biodiversity and dispersal into the site. Overcoming these limitations requires human assistance, often in multiple stages. The choice of species initially planted influences the successional trajectory and the spontaneous colonization of native species. Improper management of reforestation plantings and lack of understory development can lead to failure in the establishment of native species and continued soil erosion.

The development of effective and affordable reforestation approaches and metrics to quantify their short-term and long-term effects on tropical forest structure, biodiversity, and ecosystem function is in the experimental phases. Menz et al. (2013, p. 527) succinctly described the current state of restoration ecology research: "Restoration ecology is not a magic bullet that provides instant ecosystems of the desired type, but an emerging science less than four decades old." Few studies have described long-term trajectories of forest change during reforestation or compared different active approaches with natural regeneration over more than a few years. The role of species interactions (including rhizosphere interactions) during reforestation is also poorly understood and critically important. Incorporating local knowledge and traditional practices with proven success in particular regions will enable development of more effective methods to restore biodiversity and ecosystem functions through tree planting and management of existing second-growth forests. These approaches are also more likely to be compatible with sustainable local livelihoods. Although technical issues still impede reforestation efforts in many areas, social issues (economic, political, behavioral) pose the major limitations to reforestation worldwide today (Kettle et al. 2011).

It is tempting to think of reforestation as the reverse of deforestation, but this is not the case. Reforestation pathways, whether passive or active, lead to the development of new forest ecosystems and species assemblages that did not previously exist and that represent "works in progress." To some, these "new" forests generate hope for the future; to others, they are mere shadows of the forests that previously occupied the land: "The 'achievable' is an ever-shifting and ever-negotiating n-dimensional hyperspace produced by the intrinsic traits of a specific wildland interwoven with the mosaic of social energies and agendas brought to bear on it" (Janzen 1997, p. 275).

CHAPTER 14
REGENERATING FORESTS IN TROPICAL LANDSCAPES

Globalization is often viewed as a driver of deforestation, but there are contexts where it promotes forest recovery.—Hecht et al. (2006, p. 208)

14.1 Overview

In 1978, shortly before the civil war began in El Salvador, only 18% of forest cover remained in the Cutumayo basin. Aerial bombing campaigns by the El Salvadoran federal army during the 1980s forced the surviving villagers of Cinquera to abandon their homes and fields. In 1992, the residents returned to their destroyed village to reconstruct their lives following the devastating civil war. During their exile, formerly cultivated fields were transformed into 5,300 hectares of secondary tropical dry forest. The village was colonized by 70 species of trees and 224 species of vertebrates. After their return, the villagers formed the Association for the Reconstruction and Municipal Development of Cinquera and developed their new forest as a protected natural area that is now the foundation for economic development of the entire area (Herrador Valencia et al. 2011). In 2004, forest cover in the basin reached 61%. The regrowth of their forest has given their lives a new beginning and a new purpose.

During the 1980s, one-sixth of El Salvador's population migrated out of the country as the civil war escalated. By the end of the war, farming was no longer a dominant sector of the economy. The fraction of the country's revenues in traditional agricultural exports fell from 81% in 1970 to 11% in 2000. Forests began regrowing spontaneously in many areas of El Salvador. During the 1990s, the rate of forest regrowth in El Salvador (5.8%) was higher than the rate of deforestation (2.88%). Loss of farming income was replaced with remittances sent by family members who had migrated abroad, mostly to the United States. The rate of forest regeneration was highest in those regions where residents received the largest remittances (Hecht et al. 2006; Hecht et al. 2007; Hecht and Saatchi 2007).

Despite their contrasting climates, biotas, and environmental conditions, and their distinct social, economic, and political histories, El Salvador, India, Vietnam, Bhutan, Costa Rica, and Puerto Rico all share a common trend. These countries, among others in Europe and North America, have more forest cover now than they did 15 or 20 years ago—they have undergone forest transitions (Mather 1992; Walker 1993; Rudel et al. 2005). The land-use changes that mark a forest transition come about through reductions in deforestation and increases in reforestation and forest regeneration. But just as they showed diverse paths of development and deforestation, these countries illustrate diverse paths toward reforestation (Grainger 2010).

Some forest transition pathways are more likely to lead to higher quality reforestation than others (Lambin and Meyfroidt 2010; Lamb 2011). Land-use change statistics used to define forest transitions generally do not distinguish between afforestation and reforestation. In Vietnam, for example, approximately half of the increase in forest area from 1987 to 2006 was attributed to monoculture plantations of fast-growing exotic tree species (Meyfroidt and Lambin 2009). In contrast, virtually all of the forest cover increase in Puerto Rico from 1948 to 1990 was due to spontaneous natural regeneration (box 14.1; Rudel et al. 2000).

Spontaneous forest regrowth fundamentally requires a reduction of anthropogenic pressure on the land. This reduction can follow a wide variety of socioeconomic changes, including rural-to-urban migration (Aide and Grau 2004; López et al. 2006), emigration to other countries (Schmook and Radel 2008), abandonment of farming or ranching on marginal lands (Arroyo-Mora, Sánchez-Azofeifa, Rivard, et al. 2005), adoption of agroforestry, or development of forest-friendly land uses, which include ecotourism, privately owned conservation areas, or ecological reforestation projects (Kull et al. 2007; Sloan 2008). Although decisions to abandon farms or grazing lands are primarily based on socioeconomic factors, biophysical conditions such as topography and soil fertility can play an important role, particularly when forest conversion to agriculture occurred on lands poorly suited for agricultural use. Reforestation requires a confluence of socioeconomic, political, and ecological factors.

Earlier chapters of this book focus on the ecological processes of forest regeneration within landscape patches following natural disturbances or agricultural abandonment. Here, I widen the view to consider how tropical forests regenerate within a larger spatial context of human-modified landscapes, where natural or seminatural habitats coexist within a matrix of different types and extents of agricultural land uses (Perfecto et al. 2009). This wide-angle view brings the links between human activities and forest regeneration into sharper focus, emphasizing the coupled nature of ecological and social systems. Deforestation and reforestation pathways emerge from the coupled dynamics of ecological and social systems. Although I explicitly focus here on the *causes* of forest regeneration in the tropics, the social and economic *consequences* deserve a broad and in-depth treatment of their own.

Box 14.1. Socioecological Drivers of Forest Regrowth in Puerto Rico

When deforestation reached peak levels in the late 1930s, forests covered less than 10% of Puerto Rico, a Caribbean island with a total land area of 9,104 square kilometers. At this time, 90% of the island was in agricultural land use and 72% of the population lived in rural areas. Agriculture accounted for 45% of the gross national product. Life in Puerto Rico changed dramatically in 1948, when the United States invested millions of dollars in the Puerto Rican economy in an ambitious development program called Operation Bootstrap. Factories were established, using imported raw materials from the United States and exporting finished products back to the United States at lower labor costs. Manufacturing and tourism jobs attracted workers who left their former agricultural jobs in rural areas. Spontaneous forest regrowth followed abandonment of sugarcane fields, coffee plantations, and pastures. By 1980, agriculture had declined to below 5% of the gross national product (Grau et al. 2003). Forests were growing on over 42% of the island in 1991, despite continued population growth (Helmer et al. 2002).

This widespread transformation from an agricultural economy to a manufacturing economy in Puerto Rico provided an unprecedented opportunity to investigate the social and ecological factors associated with forest regrowth at different spatial scales across rough topography, different land uses, and varied ecological life zones. Spontaneously reforested areas were concentrated in the coffee-growing areas of the western central highlands where smallholders predominated, agriculture was labor intensive, lands were marginal for production, and migration rates of workers to urban areas were high. Small farms showed a 2.7 times higher probability of reforestation (Rudel et al. 2000; Yackulic et al. 2011). The small size of the farms (< 30 ha) and infrequent use of fire apparently facilitated rapid recovery of forest structure on former agricultural lands (Grau et al. 2003). Industrialization played a major role in attracting workers away from the agricultural labor force, promoting agricultural decline and spontaneous forest regeneration (Rudel et al. 2000).

Factors affecting reforestation in Puerto Rico shifted over time. From 1977 to 1991, ecological factors such as high slopes and life zone were more important predictors of forest regrowth in Puerto Rico than were socioeconomic factors (Yackulic et al. 2011). From 1991 to 2000, forest cover continued to increase but at a slower rate, as development pressures on available land increased (Parés-Ramos et al. 2008; Crk et al. 2009; Yackulic et al. 2011). In contrast to previous decades, population density and the proximity of protected areas became more important predictors of forest regrowth within individual municipalities and barrios (Yackulic et al. 2011). The percentage of forest cover within a 100-meter radius was the strongest predictor of reforestation from 1991 to 2000. Steep slopes, northern aspects, and distance to primary roads were associated with forest regeneration. Soils of intermediate agricultural potential were associated with the highest rates of regeneration, suggesting that these areas were less profitable for agriculture but had high potential for forest growth (Crk et al. 2009).

Both social and ecological factors influenced forest regeneration in Puerto Rico. Within the municipality of Luquillo in the northeast, proximity to Luquillo Experimental Forest (LEF) significantly promoted forest regeneration between 1936 and 1988. In 1936, sugarcane and pastures each occupied about one-third of this landscape, whereas dense forest occupied only 15% of the area. By 1988, sugarcane completely disappeared from the landscape, pasture occupied only 25% of the area, urban areas occupied 10%, and dense forests covered 54%. Areas within one kilometer of the LEF boundary showed greater regeneration of forest cover compared to areas more than two kilometers away. Dense forest patches and riparian corridors within the landscape served as nuclei for forest regeneration (Thomlinson et al. 1996).

By 1980, within 30–40 years of regrowth, regenerating forests of Puerto Rico were similar to preexisting forests in their physiognomy and canopy structure but differed substantially in their species composition, abundance of large trees, and soil structure. In regenerating forests, endemic tree species were less abundant, and nonnative tree species persisted as dominant canopy species. In 1980, nonnative species accounted for 11% of the basal area and 2% of the tree species in Puerto Rico's regenerating forests. Nonnative tree species played an important role in the reforestation in Puerto Rico, and they remain a distinguishing feature of these emerging new forests (Lugo and Helmer 2004).

14.2 Land-Use Transitions and Forest Transitions

Land-cover change is not a simple, linear, irreversible process. The tiles composing a landscape mosaic shift in color over time. Consider an idealized landscape mosaic composed of three colors of tiles: green for old-growth forests, old second-growth forests, or mature plantations; yellow for forests in early stages of succession (young fallows) or young plantations; and white for agricultural land use (crops or pasture). Summing up the number of tiles that change from green to white over a given time interval yields the rate of deforestation, whereas summing up the number of tiles that change from white to yellow yields the rate of reforestation. Three other types of land-use transitions can occur within the same time interval. Tiles can change from yellow to green, reflecting forest growth or succession; tiles can change from yellow to white, indicating clearance of fallows for repeated agricultural use; and tiles can change from green to yellow, reflecting logging or other forms of forest degradation that do not pass through an agricultural stage.

At a given spatial and temporal scale (number of tiles per unit time) *land-use transitions* within landscape mosaics become *forest transitions* when the rate of reforestation exceeds the rate of deforestation. But these metrics are deceptively simple—in reality, mosaic landscapes have more than three colors of tiles. As previously discussed in chapter 5, section 5.4, definitions of forest, secondary forest, and forest degradation vary widely across regions and among agencies and researchers (e.g., see FAO 2011).

The rate of reforestation within a landscape does not reveal information about changes in species composition or species characteristics, which are particularly important for understanding effects on biodiversity, ecosystem services, and rural livelihoods (Rudel et al. 2005; Chazdon 2008a). Reforestation takes three major forms: natural (spontaneous) regeneration, agroforestry (or enriched fallows), or tree plantations. These reforestation modalities vary in their importance in different regions of the tropics. Spontaneous regeneration has been the most common modality in Latin America, whereas reforestation has been mostly achieved through smallholder agroforestry in Africa and through a combination of forest plantations and spontaneous regeneration in Asia (Rudel 2012).

The dynamics of reforestation and deforestation can be assessed at multiple spatial and temporal scales. A landscape is an array of ecosystems and land-use types at spatial scales extending from tens to hundreds of kilometers. Just as forest patches within the landscape undergo periods of deforestation, clearing, and regeneration, entire landscapes can also be viewed as dynamic entities, undergoing phases of forest conversion, agricultural land use, and forest regrowth (Lambin et al. 2003). And just as successional pathways are contingent on prior events, trajectories of land use and land cover are contingent on the biophysical, political, social, economic, and historical context (Rudel et al. 2005; Perz 2007; Lambin and Meyfroidt 2010).

Reforestation and deforestation are distinct processes with unique sets of causal factors (Rudel et al. 2005; Lambin and Geist 2006; Meyfroidt and Lambin, 2009). At the local patch level, rates of forest regrowth following natural and anthropogenic disturbance are always slower than rates of forest clearance. At landscape, regional, or national scales, however, both reforestation and deforestation can simultaneously occur at different rates and under different geographic, social, and economic circumstances (Tucker et al. 2005; Elmqvist et al. 2007; Morse et al. 2009; Sloan 2008; Bray 2010). At regional or national scales, measures of the net outcome of reforestation and deforestation provide limited information about the particular factors that drive landscape change.

Drivers of reforestation and forest transitions are best investigated at smaller spatial and temporal scales where dynamics can be quantified within particular landscapes or geographic units. Inconsistencies in the reporting of global forest area by the United Nations Food and Agriculture Organization (FAO) create challenges for charting trajectories of forest regrowth at the national level (Grainger 2010). The complexity of forest cover dynamics often extends beyond regional or national borders, with increasing forest cover in one region or country accompanied by importation of wood and agricultural products from other regions (Meyfroidt and Lambin 2009; Mansfield et al. 2010; R. Walker 2012). From 1990 to 2005, the 40% increase in forest cover in the Atlantic Forest biome of Brazil was enabled, at least in part, by expanding agricultural and timber production in the Amazon biome of Brazil (Pfaff and Walker 2010). Among seven tropical countries that have undergone a forest transition recently (reforestation rate exceeds deforestation rate), increases in forest cover were associated with net displacement of deforestation in other countries (Meyfroidt et al. 2010).

14.3 The Landscape Context of Forest Regeneration

The spatial distribution of regrowth within tropical forest landscapes is not random, in large part because the distribution of prior deforestation, land use, and land abandonment are not random (Helmer 2000). Three major landscape features are associated with forest regeneration: dissected terrain in upland areas, proximity to existing forest patches, and distance from roads (table 14.1). Access and suitability for agriculture strongly influence the

Table 14.1. *Biophysical and landscape factors that are expected to favor spontaneous forest regeneration in tropical regions*

Factor	Explanation	Reference
High annual precipitation	Promotes tree regeneration and reduces fire frequency	Daly et al. 2003; Brandeis et al. 2007
Steep slopes, high elevation	Marginal land for agriculture	Helmer et al. 2008; Crk et al. 2009
Low or intermediate soil fertility	Marginal land for agriculture	Chinea 2002; Arroyo-Mora, Sánchez-Azofeifa, Rivard, et al. 2005
Less surrounding pasture	Marginal land for grazing	Helmer et al. 2008
Serpentine soils	Marginal land for agriculture	Helmer et al. 2008
Total forest cover	Facilitates seed dispersal, colonization, conservation of wildlife populations	Thomlinson et al. 1996; Helmer et al. 2008; Crk. et al. 2009
Poor access, far from roads	Agricultural land more likely to be abandoned	Thomlinson et al. 1996
Proximity to old-growth forest fragments or protected areas	Facilitates seed dispersal, colonization, conservation of wildlife populations	Thomlinson et al. 1996; Helmer et al. 2008; Crk et al. 2009

Source: Modified from Yackulic et al. (2011, table 1).

spatial patterns of land abandonment that lead to forest regeneration as well as the persistence of old-growth forest remnants in the landscape (Helmer et al. 2008). Within a largely deforested landscape, remnant patches of old-growth forest tend to be restricted to areas marginal for agricultural development, on steep slopes and higher elevations (Helmer 2000; Rudel 2012). Based on a literature review of remote-sensing and ground-based studies, Asner et al. (2009) examined documented cases of forest regrowth in tropical regions for periods of at least 10 years. In about 70% of the cases studied, forest regrowth occurred in upland areas with rugged topography. Steep slopes and ravines in hilly or mountainous terrain are generally farmed by smallholders and are often of marginal quality for agriculture or grazing. Abandonment of farms in uplands occurs when intensified agriculture expands into relatively flat lowland areas, attracting labor, infrastructure, and capital investment (Rudel et al. 2009).

Spontaneous natural regeneration requires suitable soil conditions for pioneer tree establishment and continuous dispersal of seeds from vegetation elements within the surrounding landscape. As a consequence, the spatial location, extent, and quality of remnant forest fragments within landscapes are good predictors of the locations of regenerating forests following abandonment of agriculture. These structural and compositional features of landscapes, which can collectively be termed *land architecture*, reflect the historical and current patterns of deforestation, forest fragmentation, land use, and abandonment within an area (Turner 2010a).

When the international price of beef fell dramatically in the early 1980s, smallholders in the dry forest region of Guanacaste in northwestern Costa Rica abandoned their cattle pastures in marginal areas with poor accessibility and low productivity on steep slopes with shallow soils (fig. 14.1). From 1979 to 2005, forest cover in this region increased from 24% to 48% (Calvo-Alvarado et al. 2009). Most of the current forest in this region is regrowth forest located in areas that are marginally productive for cattle ranching. Between 1986 and 2000, forest regrowth was most pronounced on soils in marginal land classes characterized by slopes above 50%, shallow soils, low fertility, excessive external drainage, high susceptibility to erosion, and long dry periods (Arroyo-Mora, Sánchez-Azofeifa, Rivard, et al. 2005).

As often occurs in land-use transitions, forest regrowth was concentrated on the flanks of existing fragments of old-growth forest. Areas where pasture regenerated to forest in Guanacaste were farther from population centers and roads but closer to large patches of existing forest (Daniels 2010). In a montane region of Costa Rica, the probability of occurrence of second-growth forest increased at high elevation, on steeper slopes, further from roads, in areas of low human population density, and in forest reserves (Helmer 2000). Existing forest patches serve as nuclei for regeneration following land abandonment in adjacent areas.

As a consequence of socioeconomic changes in the Republic of Palau following the takeover by the United States in 1944, forests returned to abandoned agricultural fields, grasslands, and plantations (Endress and Chinea 2001). Over 92% of the areas where forest regeneration occurred were located within 100 meters of an existing forested area. A similar pattern was observed in areas undergoing forest recovery in Puerto Rico (see box 14.1; Thomlinson et al. 1996; Crk et al. 2009). These landscape patterns can be largely explained by precipitous declines in seed dispersal with increasing distance from the forest edge.

Landscape factors affect species composition in shifting cultivation fallows. Landscape structure variables, such as patch density and shape index,

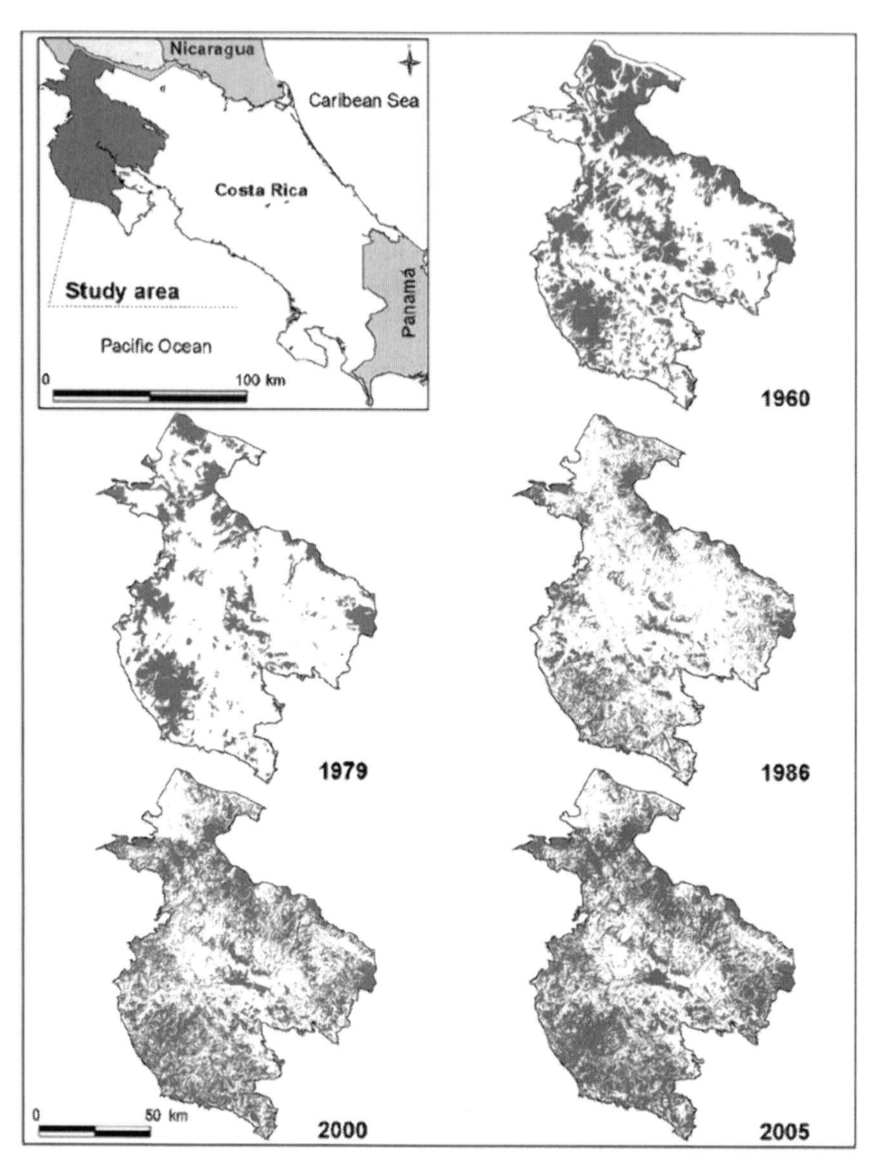

Figure 14.1
Changes in forest cover in the Chorotega region of Costa Rica from 1960 to 2005. In 1979, forest cover began increasing in the region following abandonment of cattle pastures. Source: Calvo-Alvarado et al. (2009, fig. 3).

were significant predictors of tree species density in patches of regenerating dry forest following shifting cultivation in the Yucatán Peninsula, Mexico (Hernández-Stefanoni et al. 2011). In shifting cultivation areas of southern Cameroon, patterns of fragmentation, local forest cover, and distance from forest can have a stronger effect on species abundance patterns in fallows than local plot characteristics (Robiglio and Sinclair 2011). Species richness of early successional species in fallows declined with increasing levels of landscape-level deforestation and fragmentation in regions of Borneo (Lawrence 2004) and Mexico (Dalle and de Blois 2006). Agricultural intensification can also lead to weed infestations that reduce tree establishment (see box 7.1; Robiglio and Sinclair 2011).

Changes in overall landscape structure due to spontaneous forest regeneration can mitigate some of the negative effects of deforestation and forest fragmentation on biodiversity. Increases in forest cover are associated with the reduction of cleared or open areas and increases in forest patch size, connectivity between forest patches, and the number of forest patches (Sitzia et al. 2010). In 1980, 11 forest fragments were experimentally isolated from blocks of continuous old-growth forest in a study of the effects of forest fragmentation near Manaus, Brazil. Over time, areas of active pasture surrounding the forest fragments declined, while areas of regenerating forest increased. By 2007, all of the fragments became reconnected to old-growth forest through second-growth forests at least six years old (Stouffer et al. 2011). By enhancing landscape connectivity, the development of a matrix of regenerating forests surrounding the forest fragments encouraged recolonization of some bird species that had earlier gone extinct from individual fragments. The enhanced recolonization appears to have reduced extinction rates of species that are particularly vulnerable to fragmentation. The development of second-growth forests in the landscape matrix surrounding forest fragments provides a "lifeline" for these vulnerable species and can eventually lead to increasing areas of suitable habitat for forest-requiring birds.

Widespread edge effects in a hyperfragmented landscape can lead to the replacement of shade-tolerant trees with light-demanding generalist species that proliferate in human-modified landscapes (Tabarelli et al. 2010). This scenario appears to be occurring in the northern portion of the Atlantic Forest region in Brazil, where 48% of the forest fragments are less than 10 hectares (Lôbo et al. 2011; Melo et al. 2013). The proliferation of native generalist tree species may lead to biotic homogenization and population declines of forest-requiring species.

Despite recent advances in forest detection within particular regions, we lack robust information regarding the extent and dynamics of successional forests worldwide (Asner et al. 2011; Castillo et al. 2012; Gallardo-Cruz et al. 2012). Multispectral satellite imagery poorly detects regrowth forests beyond 15–20 years of age, as their spectral features are similar to those of old-growth or old second-growth forests (Moran et al. 1994; Nelson et al. 2000). Most studies of forest regeneration based on single-time analyses can only reliably detect young stages of regrowth. Multitemporal studies are required to identify older stages of forest regeneration and to examine landscape dynamics in detail (Helmer et al. 2000; da Conceição Prates-Clark et al. 2009). Because natural regeneration tends to occur in small, isolated areas in a spatially diffuse process, it is very difficult to detect using most remote-sensing methods (Kolb and Galicia 2012).

The age structure of regenerating forests is an important determinant of landscape-level biodiversity, as local species richness of plant and animal taxa increases over time during forest regrowth (Chazdon, Peres, et al. 2009; Dent and Wright 2009). Young second-growth forests are far more likely to be recleared for land development or for cultivation than older forests (Etter et al. 2005; Helmer et al. 2008). Deforestation rates in the lower Mekong basin in Southeast Asia from 1993 to 1997 were three times higher in second-growth forests compared to old-growth forests (Heinimann et al. 2007). Repeated clearing of young forests leads to degradation of soils and arrested succession, with negative consequences for biodiversity and ecosystem services. A more balanced age structure of regenerating forests maximizes biodiversity conservation at the landscape level and minimizes land degradation (Chazdon, Peres, et al. 2009).

14.4 Socioecological Drivers of Tropical Reforestation

Multiple factors and processes act to favor or constrain forest regrowth at different spatial and organizational scales (fig. 14.2; Perz and Almeyda 2010). Each level of factors is influenced by factors at higher hierarchical levels. In areas where smallholders control land use, the quality and quantity of forest cover are determined by household-level decisions that are influenced by land tenure status, life cycles, migration history, access to labor and land, and soil quality. These household-level decisions are influenced by municipal, regional, and national policies as well as by global markets, trade policies, and environmental conventions. Within household farms in the northern Ecuadorean Amazon, changes in land use and land cover from 1990 to 1999 could be explained by a combination of household demographic and socioeconomic variables that are related to resource endowments on farms, geographic accessibility to other farms and market towns, commodity prices, labor availability, and characteristics of local markets and towns (Walsh

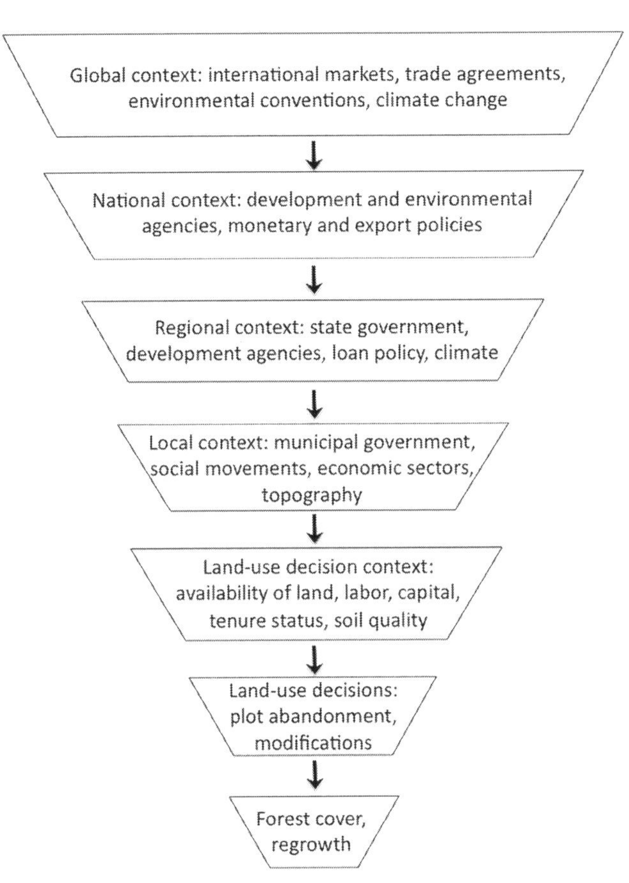

Figure 14.2
Causal factors affecting forest regeneration and land-use change can be viewed within a hierarchical framework. This framework illustrates top-down causation patterns, but causation can also occur in the reverse direction. Source: Redrawn from Perz and Almeyda (2010, fig. 4.1).

et al. 2008). In a peasant village in northeastern Peru, the extent and age of second-growth forests, orchards, and crops varied with the size of initial land holdings during earlier land claims (Coomes et al. 2011). In our globalized economy, migration, employment outsourcing, and trade relationships link the fate of forests, agricultural production, and human populations across national borders (Meyfroidt et al. 2010; R. Walker 2012).

Reforestation is a long-term process that requires long-term financial support and human capital. Dinh Le et al. (2012) presented a framework for assessing reforestation success in developing countries of the tropics that integrates

Table 14.2. *Socioeconomic drivers of reforestation success*

Drivers	Explanation
Livelihood planning	Project must address livelihood needs of people in the area and ensure their participation and interests in the project.
Local participation and involvement	Key actors participate, and local ecological knowledge and practice are incorporated.
Socioeconomic incentives	Direct or indirect economic benefits and services should be provided to local communities.
Financial and economic viability	Required for long-term sustainability of reforestation projects.
Payments for environmental services	Ensures that reforestation is an attractive alternative for landowners who might otherwise choose not to participate .
Social equity	Market and nonmarket costs and benefits must be shared by all stakeholders.
Corruption	Can cause failure of reforestation projects.
Degree of dependency on traditional forest products	Reforestation is more likely to be successful if forest goods are valued and supplies from natural forests are declining.
Marketing prospects	Marketing plans lead to good production outcomes for reforestation projects.
Knowledge of markets for timber and other forest products and services	Reforestation is more successful if there is a known market for forest goods and services.
Addressing underlying causes of forest loss and degradation	Reforestation will not be successful if underlying causes of forest loss are not addressed.

Source: Based on Dinh Le et al (2012, table 8).

socioeconomic, institutional, technical, and biophysical drivers. Socioeconomic drivers of reforestation success are based on participation and livelihood enhancement of local people (table 14.2). Institutional drivers of reforestation success are also critically important, such as effective governance, tenure security, mechanisms for conflict resolution, long-term maintenance of reforested sites, forestry support programs, and community leadership. Technical drivers include tree species selection, quality of seeds or seedlings, appropriate planting time, postestablishment silviculture, and site quality. It is easy to see why so many reforestation projects fail, as many conditions in different arenas need to be satisfied and properly aligned to meet both short-term and long-term reforestation objectives. Successful reforestation is not accomplished by simply planting trees. Sayer et al. (2004) emphasized that involving stakeholders in defining reforestation objectives and investing in communities and local institutions will enhance the environmental and social benefits of reforestation programs. Monitoring suitable indicators and applying principles of ecosystem management and common property management further enhance the likelihood of successful outcomes.

Many factors can promote forest expansion, leading to a range of forest transition pathways across regions (table 14.3; Nagendra and Southworth 2010). In many cases, more than one pathway or mechanism is required to explain forest expansion (Perz 2007; Nagendra 2009; Rudel 2010; Lambin and Meyfroidt 2010). Reforestation can proceed as part of a deliberate plan at the local, regional, or national level or can happen as an unintended consequence of economic and political change (Rudel 2012). In the former case, forest cover expands through an endogenous socioecological feedback mechanism, whereby land-use decisions are made in response to land degradation, scarcity of forest products within the region, or conservation initiatives (Lambin and Meyfroidt 2010). This "forest scarcity pathway" is also driven by increases in prices of forest products that induce landowners to plant trees (Rudel et al. 2005).

14.4.1 *Forest Transitions in Asia*

Forest transitions in China, India, and Vietnam are largely the outcome of deliberate planning involving all levels of government, with strong involvement of their respective national forestry agencies (Rudel 2012). Adoption and enforcement of policies to expand forest cover require effective governance (Mather and Needle 1999). A forestry-based approach to reforestation is most effective when management practices are tailored to local conditions, thus requiring community-wide participation and local forest management capacity, resources, and institutions (Boissiére et al. 2009; Nagendra 2009). China adopted national reforestation policies designed to increase the production of forest products and to reduce flooding and soil erosion caused by excessive deforestation (Zhang and Song 2006). About 80% of China's increase in forest cover is attributed to plantation forests (Song and Zhang 2010). Even with this investment in production forestry, imports of wood products in China led to significant outsourcing of deforestation in other countries (Meyfroidt et al. 2010). The forest transition in India is also largely due to an increase in community-managed tree plantations, which accounted for 51% of the forest area in 2000 (Rudel 2005; Grainger 2010). Unlike China, however,

Table 14.3. *Socioeconomic drivers associated with forest regeneration as observed in particular regions, with examples of case studies*

Driver	Effect on forest regeneration	Region or country	Reference
Depopulation of rural areas due to demand for nonfarm labor and out-migration	+	Puerto Rico	Rudel et al. (2000); Grau et al. (2003)
Availability of off-farm labor	+	Puerto Rico; Michoacan, Mexico	Rudel et al. (2000); Klooster (2003)
Agricultural intensification and abandonment of marginal lands	+/–	Vietnam, Mexico	Tachibana et al. (2001); García-Barrios et al. (2009)
Rural depopulation due to armed conflict	+	El Salvador	Hecht et al. (2006); Hecht and Saatchi (2007)
Advanced history of rural settlements	+	Amazonia	Perz and Skole (2003)
Retention of traditional agricultural practices	+	Amazonia	Perz and Skole (2003)
Scarcity of forest products (and increase in price)	+	Vietnam, India	Rudel et al. (2005); Meyfroidt and Lambin (2008)
Changes in forest policy at regional or national level (logging bans, incentives for reforestation)	+	Honduras, Vietnam	Southworth and Tucker (2001); Meyfroidt and Lambin (2009)
Changes in conservation policy (creation of protected areas, biosphere reserves, biological corridors; payments for ecosystem services)	+	Costa Rica	Kull et al. (2007); Timm et al. (2009); Morse et al. (2009); Montagnini and Finney (2011)
Development of ecotourism	+	Costa Rica, Belize, South Africa	Kull et al. (2007); Blangy and Mehta (2006)
Increased international migration and remittances from family members	+	El Salvador, Honduras, Mexico	Hecht and Saatchi (2007)
Well-defined property rights	+	Vietnam, Madagascar	Tachibana et al. (2001); Elmqvist et al. (2007)

Note: The symbol + represents a positive effect, and the symbol – represents a negative effect.

forest expansion in India has not displaced deforestation abroad (Meyfroidt et al. 2010).

In 1988, Vietnam embarked on an ambitious program to reforest five million hectares of degraded land when forest cover reached a low of 25%–31%. By 2005, forest cover increased to 39%, with 8.1% contributed by plantations, which were predominantly of exotic species (fig. 14.3; Meyfroidt and Lambin 2008). Natural regeneration was concentrated in mountainous regions, whereas plantations were established in midelevation and lowland areas (Meyfroidt and Lambin 2010). A national land allocation policy was implemented across the country, giving households the responsibility to preserve existing forests and to manage forest regrowth on deforested lands. But uneven land allocation has left many poor households deprived of land tenure rights and deprived of the scrubby forests that they relied upon for firewood and other products. (McElwee 2009). Local communities have not been permitted to participate in management of local conservation areas. Villagers have been relocated and forced to abandon traditional shifting agriculture practices (Boissiére et al. 2009). Moreover, a logging ban in all natural forests in northern provinces led to increased legal and illegal imports of timber from neighboring countries, such as Laos and Cambodia (Meyfroidt and Lambin 2009, 2011).

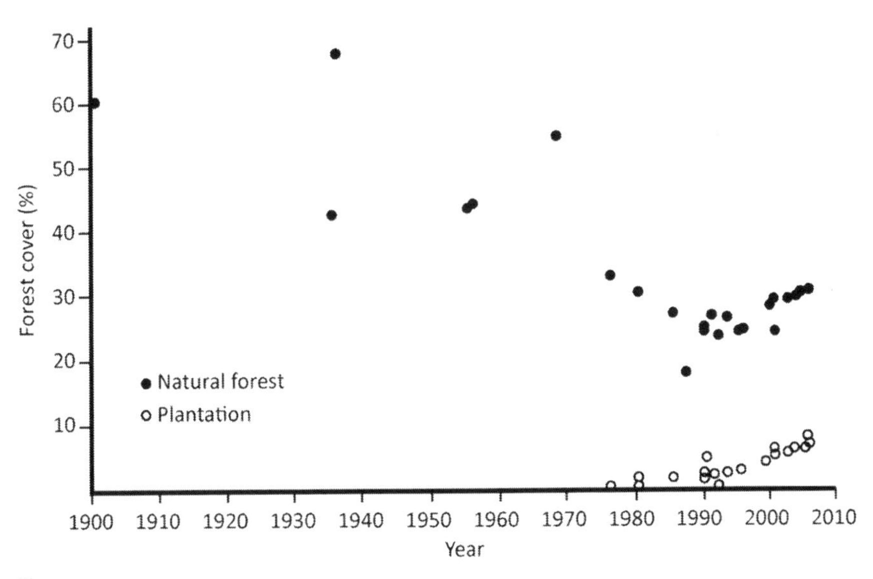

Figure 14.3
Changes in forest cover in Vietnam from 1900 to 2005. Source: Redrawn from Meyfroidt and Lambin (2008, fig. 3).

14.4.2 *Forest Transitions in Latin America and the Caribbean*

In the cases discussed above, forest cover expanded despite increases in rural populations. But forest expansion can also occur as an unintended consequence of economic policies at the national or global level that reduce population size in rural areas, leading to abandonment of agriculture. This "economic development" pathway applies well to the forest transition in Puerto Rico (see box 14.1).

Rural out-migration in the Sierra Norte of northern Oaxaca, Mexico, has led to the abandonment of up to 60% of the community's agricultural lands since 1980 (Robson 2009; Robson and Berkes 2011). The introduction of coffee plantations at lower altitudes drew farm labor, leading to abandonment of traditional maize cultivation at higher elevations. Entire towns were displaced (del Castillo and Blanco-Macías 2007). Naturally regenerating forests and swidden cultivation fallows expanded as the population's dependence on local food supplies diminished. Abandonment of maize shifting cultivation is also common in the dry forest region of Nizanda in southeastern Oaxaca (Lebrija-Trejos et al. 2008). Consequently, these areas provide excellent opportunities to study replicated chronosequences following abandonment of shifting cultivation fallows in montane and dry forest regions of Oaxaca (del Castillo and Blanco-Macías 2007; Lebrija-Trejos, Meave, et al. 2010; del Castillo and Pérez Ríos 2008).

Abandonment of agricultural land is closely linked to the regional marginalization of the agricultural economy, diversification of incomes of rural landowners, and low prices for agricultural commodities (Aide and Grau 2004; Walsh et al. 2008). In Panama, forest cover increased from 1992 to 2000, while the proportion of workers employed in agriculture, fishing, and hunting declined (Wright and Samaniego 2008). The agricultural sector contributed only 7% of all economic activity in 2007. In northeastern Costa Rica, declines in beef prices drove pasture abandonment in marginal upland areas, which led to a shift to a more urbanized and service-oriented economy with diversified sources of income (see fig. 14.1; Daniels 2010; McClennan and Garvin 2012). Payments to farmers for environmental services, fire prevention, and establishment of protected areas also promoted forest regrowth in this region.

Increasing availability of nonfarm jobs through tourism is an important component of the changing socioecological picture in some tropical regions undergoing forest transitions. Jobs provided by tourism development in the Cancun-Tulum corridor may have contributed to farm abandonment in Quintana Roo, Mexico (Bray and Klepeis 2005). In the Guabo Valley of Pacific central Costa Rica, natural regeneration of former pastures since 1992 has been driven by the development of ecotourism and by land purchases by expatriates and nongovernmental organizations to promote conservation (Kull et al. 2007). Similar trends are underway in the Osa Peninsula of Costa Rica (Sierra and Russman 2006; Zambrano et al. 2010).

Another factor contributing to off-farm income during forest transitions is the growing importance of remittances (Hecht and Saatchi 2007). Remittances from family members that have migrated to the United States are an important source of income for rural residents in many regions of Mexico, El Salvador, and Nicaragua and have contributed to forest expansion (Klooster 2003; Hecht 2010; Turner 2010a). Remittances can also effectively stabilize rural populations when accompanied by agricultural intensification (García-Barrios et al. 2009).

Forest expansion can also occur in regions with stable or even growing populations. In these cases, former annual crop fields or fallows are modified into agroforests or managed forests that support rural livelihoods (Rudel 2010). This forest transition pathway is most commonly observed in smallholder farms or communally held lands that support forest-dependent livelihoods within multifunctional and diversified landscapes. In Michoacán, Mexico, forest regeneration provides firewood for Purepecha villagers who have diversified their livelihoods (Klooster 2003). Coffee agroforestry provides an important pathway for expanding forest cover in Mexico and Central America (Bray 2010). In the highlands of western Honduras, forest cover increased 23% from 1987 to 2000 due to expansion of shade coffee production (Redo et al. 2009). In Candelaira Loxicha in Oaxaca, Mexico, sugarcane cultivation was replaced by shade coffee plantations from 1930 to 1970, dramatically increasing forest cover, which now composes 66% of the land area (Aguilar-Støen et al. 2011).

An analysis of forest transitions across all of Central America from 2001 to 2010 clearly demonstrates that socioecological drivers of forest transitions vary across geographic scales and across climatic zones (Redo et al. 2012). Based on analysis of MODIS (moderate resolution imaging spectroradiometer) satellite imagery, the Central American region showed a net decrease in forest cover during this period, despite a net increase in forest cover in Costa Rica and Panama. Deforestation was concentrated in moist forest zones along the Caribbean slopes of Nicaragua and the Petén in northern Guatemala, whereas reforestation was concentrated in coniferous and dry forest areas in Honduras, Nicaragua, Costa Rica, and El Salvador. The Human Development Index was significantly positively associated with net increases in moist forest and with the stable distribution of forest cover at the national scale. This association suggests that factors affecting per capita income, education, and health underlie forest transition trends among Central American countries. Countries with the highest rates of deforestation also showed the highest rates of reforestation, but in distinct biomes. These trends reflect a confluence of social and economic factors operating at multiple geographic scales, including rural-to-urban migration, increases in shade coffee production in highlands, international trade in agricultural and forest commodities, and national conservation and forestry policies.

Aide et al. (2013) extended this analysis to 45 countries (16,050 municipalities) of Latin America and the Caribbean. Dynamics of deforestation and reforestation varied widely across forest zones, with deforestation concentrated in moist forest regions, and reforestation concentrated in xeric shrublands in northeast Brazil and north-central Mexico (fig. 14.4). Overall, population size was a poor predictor of forest transitions at the municipality level. Considering only those municipalities in moist forest zones where a significant change

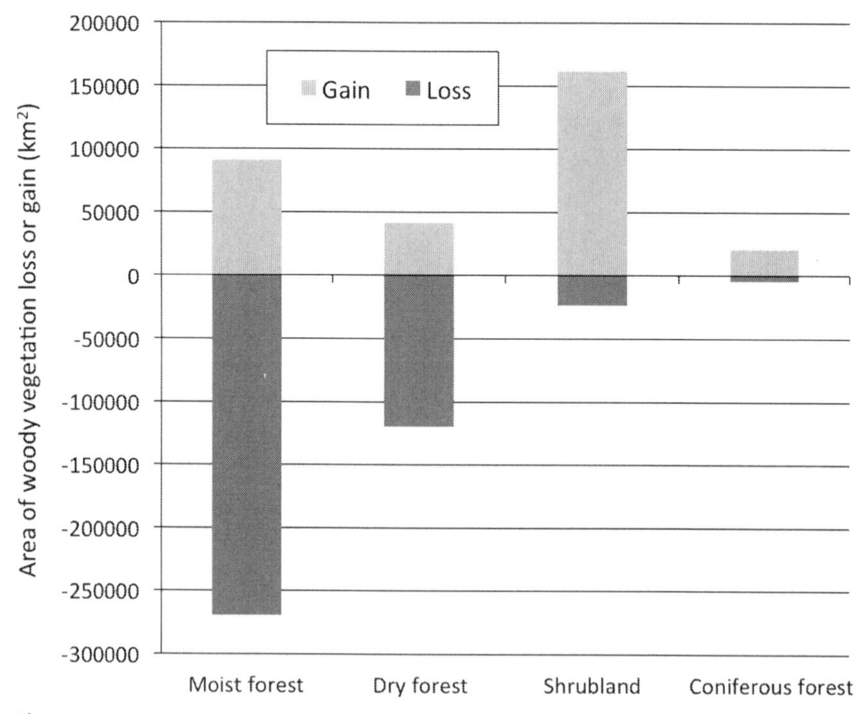

Figure 14.4
Gain (reforestation) and loss (deforestation) of woody vegetation in four vegetation zones between 2001 and 2010 across all countries of Latin America and the Caribbean. Deforestation was highest in moist forest zones, whereas reforestation was highest in dry shrublands. Source: Aide et al. (2013, table S4).

in woody vegetation was observed, areas undergoing reforestation were at higher elevations than those undergoing deforestation. Deforestation was concentrated in zones with low population density (median of 17 people km^{-2}) compared to areas undergoing reforestation (median of 42 people km^{-2}). At least for moist forests, deforestation is occurring in frontier zones with low population density and in lowland areas suitable for mechanized agriculture. Reforestation, on the other hand, is occurring in areas of higher population density where forests have already been cleared for agriculture.

14.4.3 *Forest Transitions in Protected Areas and Community-Managed Forests*

Although forest scarcity and changing economic policies are the major drivers of reforestation in most tropical regions, conservation activities at the local, national, and international level can also promote forest regrowth (Harvey et al. 2008; Nagendra 2010). Forest regeneration within national parks can offset continuing deforestation or promote net reforestation within park boundaries. One success story is the Cabo Blanco Nature Reserve at the southern tip of the Nicoya Peninsula in northwestern Costa Rica. In 1963, when the park was established, 15% of the park area was in old-growth forest and 85% was in cultivated fields and pastures. Most of the Nicoya Peninsula had been deforested. Transitional dry-moist tropical forest quickly regenerated within the park boundaries. The 45-year-old second-growth forest now reaches heights of 20–35 meters, with similar structure to remnant old-growth forest, and supports a diverse native fauna consisting of mammalian frugivores, seed dispersers, and several species of mammalian predators (Timm et al. 2009).

Establishment of protected areas does not always prevent forest clearing. Deforestation is often displaced to surrounding areas (Dewi et al. 2013). In the Blue and John Crow Mountains National Park in Jamaica, 14% was deforested from 1991 to 2002, whereas 11% of the park area experienced forest regrowth. The gross rate of deforestation was similar to pre-park levels, but reforestation reduced the net rate of deforestation by 68% (Chai et al. 2009). When Kibale Forest Reserve in western Uganda became a national park in 1993, pine plantations within the park were harvested and subsequently regenerated to forest. Forest cover within the park increased from 86% to 91% in 2003. Forest fragments increased in size and decreased in number (Hartter et al. 2010). Despite continued deforestation in the densely populated areas surrounding the park, reduced encroachment in Kibale led to net reforestation within the park's boundaries. Reduced deforestation rates and increased reforestation rates have also been observed in national parks in Costa Rica (Sánchez-

Azofeifa et al. 2003), Honduras (Southworth et al. 2004), Peru (Oliveira et al. 2007), and Mexico (Mas 2005). But three studies of protected areas of Sumatra, China, and Guatemala found similar or higher rates of deforestation after establishment (Liu et al. 2001; Hayes 2006; Gaveau et al. 2007).

Rates of net deforestation in community-managed forests of Nepal, Butan, and Mexico can be as low or lower than in protected areas (Nagendra 2007; Bray and Velázquez 2009; Meyfroidt and Lambin 2010). A meta-analysis of studies of 40 protected areas and 33 community-managed forests throughout the tropics revealed that average deforestation rates were lower in the community-managed forests (Porter-Bolland et al. 2012). Forest protection and forest expansion can occur in tropical regions with communal land use, well-developed conservation policies and institutions, government ownership of land, and active natural resource management.

14.5 Enhancing Forest Regeneration and Human Livelihoods in the Landscape Matrix

Reversing the tide of deforestation and land degradation in tropical regions is not simply a matter of creating the right ecological conditions. The solution to this "wicked problem" must stem from a comprehensive analysis of environmental, social, economic, political, and cultural factors. Garrity (2004, pp. 14–15) has described the appropriate approach well: "It is futile to attempt to conserve tropical forests without addressing the needs of local poor people, nor is it desirable." Protection of remaining old-growth forests must remain a high conservation priority, even if remnant forest areas are small and isolated (Schleuning, Farwig, et al. 2011). Even if all remaining old-growth forest areas could be protected forever from further deforestation or degradation, these areas would be insufficient to retain existing levels of forest biodiversity (Chazdon, Peres, et al. 2009). We must rely on these forests and the surrounding matrix habitats to retain the ecological memory needed to restore forest cover and biodiversity in the landscape (Bengtsson et al. 2003).

We must adopt a broader view of biodiversity conservation and ecosystem service provision that encompasses entire landscapes and that ensures sustainable livelihoods and food security of local communities (Sayer et al. 2004; DeFries and Rosenzweig 2010; Schroth and McNeely 2011). This challenge requires that we focus on the landscape matrix in which old-growth forests are now embedded and in which people live and farm (Perfecto and Vandermeer 2010; Newton et al. 2012). As many as 90% of tropical species live in human-modified, working, or mosaic landscapes (Garrity 2004; Perfecto et al. 2009). Forest regeneration and expanding tree cover within the landscape matrix must be an integral part of the solution to alleviate rural poverty, restore environmental services, and provide habitats for forest-requiring species. How can this matrix be transformed to enhance conservation of forest-requiring species, to provide products and environmental services, and to support populations that have jobs and food security? Can economic, social, or political incentives promote increasing forest cover and sustainable rural livelihoods in the tropics (box 14.2)?

14.5.1 Restoring Tree Cover in Agroforests and Silvopastoral Systems

Reforestation is often viewed as being in direct conflict with agricultural production. Most smallholders cannot afford to allow forest regrowth on their land without receiving some material benefits in the form of food products, firewood, or other goods. One promising approach toward restoring forest cover in landscape mosaics is through the production of tree crops for local use or export that are grown under forest canopies, such as coffee and cacao. Compared to annual crops or treeless pastures, agroforestry systems are intensified ecologically, providing a higher diversity of habitats for biodiversity, and offering greater environmental services (Ranganathan et al. 2008). In many ways, agroforestry systems resemble enriched fallow systems. They can serve as a transitional stage between agricultural crop production and forest restoration while providing economic benefits and food security to farmers (Vieira et al. 2009). Agroforestry can also serve as an important route to land ownership (Kalame et al. 2011).

Ecological intensification through agroforestry or taungya systems permits a transition from traditional enriched long-fallow shifting cultivation systems to more permanent cultivation that can potentially be integrated into the regional, national, or global economy. How this integration takes place is a critical aspect of long-term economic and ecological sustainability for smallholders in the tropical countryside. Agroforestry systems can mimic successional stages, with gradual addition of late successional species and increasing structural complexity over time (Ewel 1986). An agro-successional system in Cameroon successfully replaces persistent *Imperata cylindrica* grasslands with cacao plantations within 10–20 years while increasing soil organic matter and the diversity of forest trees, fruit trees, and oil palms. On average, 25 tree species were planted along with cacao (Jagoret et al. 2012; Jagoret et al. 2011). The forest gardens, orchards, and enriched successional systems that were more widely used by indigenous cultures throughout the world's tropics (see box 2.1; Schulz et al. 1994) provide relevant models for the development of modern agroforestry systems and transitional forms of agro-successional restoration in human-managed landscapes (Vieira et al. 2009).

Box 14.2. Incentives for Reforestation and Forest Regeneration in Tropical Countries

Reforestation and forest regeneration are high priorities for many tropical countries. But most farmers cannot afford to change land use without compensation for the loss of income due to abandonment of agricultural activities or to cover costs of planting trees and monitoring changes in on-farm tree cover. Voluntary programs can provide financial incentives to farmers or farming communities for planting trees, promoting natural regeneration on farms, or increasing tree cover in pastures or crop fields. Carefully designed regional and national programs for payments for ecosystem services (PES) have achieved considerable success in reducing deforestation, protecting water resources, promoting reforestation, protecting forest reserves, and developing biological corridors in tropical regions (Pagiola et al. 2010; Daniels et al. 2010). Continued funding to support these programs presents an enormous financial challenge for tropical countries.

The Regional Integrated Silvopastoral Ecosystem Management Project is pioneering the use of PES to promote adoption of silvopastoral practices in Nicaragua, Colombia, and Costa Rica (Pagiola et al. 2007; Montagnini and Finney 2011). The project was financed by the Global Environment Facility (GEF) of the World Bank. In the Matiguás-Rio Blanco region of Nicaragua, participating landowners are contracted to receive annual payments over a four-year period, based on a point system in which forms of tree cover, forest regeneration, and pasture improvement rank higher than degraded pasture or annual crop cultivation. Payments are received after land-use changes have been carefully monitored. Farmers of different income levels participate and benefit from the PES program. As a result of PES, farmers increased the use of silvopastoral systems in over 24% of the total project area. Forest cover increased by 31% over the landscape, and landscape connectivity within farms increased significantly. In the Colombian project site, the shift toward silvopastoral systems led to dramatic improvements of water quality where riverbanks were reforested and protected from livestock entry. In all regions, land-use changes resulted in significant carbon sequestration and reduced soil erosion and weed infestation in pastures.

The Scolel Té agroforestry project sells voluntary carbon credits to institutions within Mexico and in other countries that wish to offset their own emissions (see fig 14.5; De Jong et al. 2007; Soto-Pinto et al. 2010; Paladino 2011).

A local nongovernmental organization administers funds and field activities. Farmers and their *ejidos* (communal farming associations) receive financial incentives and technical assistance for increasing tree cover in their fields and pastures. Initiated in 1997, the project supports 51 farming communities and 2,400 families in central and northern Chiapas and in northeast Oaxaca. Scolel Té was designed from the beginning to benefit small, poor, peasant families.

Government-based PES programs in Costa Rica and Ecuador provide multiple benefits for biodiversity, carbon storage, and water resource management. The pioneering PES program in Costa Rica, established through the fourth Forestry Law in 1996, bundles several environmental services together—biodiversity, watershed function, scenic beauty, and greenhouse-gas mitigation through carbon storage and sequestration—and pays landowners for 5–10 years to protect old-growth and second-growth forests or to establish plantations of native tree species. Payments are disbursed from a national fund based on a fossil fuel tax. A logging ban mandated in the same forestry law has worked in concert with the PES program to reduce deforestation rates within Costa Rica and to help bring about a forest transition at the national level. Payments for reforestation are twice the value per hectare than for conservation and have incentivized the establishment of approximately 40,000 hectares of native tree plantations. In the La Selva–San Juan Biological Corridor in northeast Costa Rica, two-thirds of the landowners that receive PES would not otherwise have developed reforestation plantations on their property (Morse et al. 2009). The highly successful Socio Bosque program in Ecuador is a government-based bundled PES program with nationwide participation by indigenous communities and smallholders (de Koning et al. 2011). Reforestation programs are not yet a component of the program but will be incorporated in future plans.

Several factors contribute to the success of these projects. First, the regional-scale projects did not displace deforestation outside of the project area, achieving the goal of net increases in carbon sequestration and other environmental services. Second, the participants (or their communities) all have secure land tenure and user rights. Third, the economic benefits reach the families and communities that protect existing forests or modify land-use practices to increase forest cover. Fourth, landowners fully participate in project decisions and are empowered by participation. Finally, payments to farmers are linked to a monitoring program with specific performance targets.

Successful PES projects promote community participation and planning,

Box 14.2 (*continued*)

local capacity building, ownership of carbon benefits, livelihood security, and sustainable development co-benefits. They are targeted to achieve specific goals and are often small-scale projects developed with active participation of local communities. Mi Bosque (My Forest) initiated in 1990 in 11 municipalities in western Guatemala became the first project in Central America to mitigate climate change through community-based reforestation, agroforestry, and forest conservation (Hall et al. 2012; CARE 2007).

Coffee and cacao forests are key elements of forest cover in many tropical regions (Garcia et al. 2010; Bray 2010; Schroth et al. 2011). In the Sierra Norte of Pueblo, Mexico, shade coffee forests are key to a recent forest transition. Coffee forests support the livelihoods of more than 20 million people, and tend to be located in regions of high biodiversity. Coffee agroforests promote local self-sufficiency, food security, and local self-governance (Toledo and Moguel 2012). They also provide critically important habitats for biodiversity conservation in mosaic landscapes (Moguel and Toledo 1999; Philpott et al. 2008). In the Atlantic Forest of southern Bahia, Brazil, traditional cacao agroforests and second-growth forests cover 45% of the landscape and provide critical habitats for several endangered endemic mammal species such as the maned sloth (*Bradypus torquatus*) and the golden-headed lion tamarin (*Leontopithecus chrysomelas*; Catenacci et al. 2009; Oliveira et al. 2010; Cassano et al. 2011). In agricultural landscapes of coastal Ecuador where mature forests have nearly disappeared, managed coffee forests provide a refuge for tree and bird species and serve as seed sources for regenerating forests in abandoned coffee plantations (Lozada et al. 2007). Shade coffee plantations also serve as refuges for biodiversity in many regions of Latin America (Perfecto et al. 1996; Philpott et al. 2008).

Agroforestry systems are critical reservoirs for biodiversity in many tropical regions where natural forests remain only as small fragments (Bhagwat et al. 2008). In Bangladesh, where less than 5% of land remains in natural forests, homegarden agroforestry systems account for about 80% of the total annual harvest of forest products (Bardhan et al. 2012). Trees are used for timber, fruit, firewood, or medicinal products, providing supplies and income to smallholder farmers.

In addition to providing food security and forest products, and enhancing biodiversity and watershed protection, agroforestry and silvopastoral systems can be important mechanisms for increasing carbon sequestration in the tropics (De Jong et al. 1995; Soto-Pinto et al. 2010). Approximately 27% of the land area of Latin America and the Caribbean is devoted to cattle production (Murgueitio et al. 2011). Intensive silvopastoral systems are a form of agroforestry that integrate high-density plantings of trees and shrubs into pastures. Silvopastoral systems are now being encouraged by payment for environmental services (PES) projects in several regions of Latin America (see box 14.2; Ibrahim et al 2010; Montagnini and Finney 2011). These systems are successfully being adopted in several regions of Colombia, Mexico, and Panama (Murgueitio et al. 2011). The Scolel Té project in Chiapas and Oaxaca, Mexico, has been actively participating in the voluntary carbon market since 1997 through carbon sales under a PES scheme (fig. 14.5; Soto-Pinto et al. 2010). Carbon storage in pastures can be significantly enhanced by planting trees or living fences, by intercropping trees with cultivated maize, or by planting timber trees in agricultural fallows (Soto-Pinto et al. 2010).

CERTIFICATE OF PLAN VIVO CARBON OFFSET

One Tonne of CO_2

Certificate Number: A02 0375

From: Scolel Té Project, Chiapas, Mexico

To: New York City Mayor Michael R. Bloomberg

To offset travel to Washington

Presented by the World Bank on Urban Sector Day, Sustainable Development Week, 21 February 2008

Signed for and on behalf of BioClimate Research and Development Limited

Plan Vivo c/o BR&D (BioClimate Research and Development Limited). BR&D is a not-for-profit company, registered in Scotland
BR&D is shortly to become the Plan Vivo Foundation. Registered office: 17 Croft Street Dalkeith, Midlothian, Edinburgh, EH22 3BA
www.planvivo.org

Figure 14.5
Carbon certificate issued to New York mayor Michael Bloomberg by Plan Vivo in February 2008. Source: Paladino (2011), reprinted with permission.

14.5.2 *Forest Regeneration in Biological Corridors*

Proximity to existing forest fragments or large forest tracts favors rapid recovery of forest structure and promotes diverse colonization of plant and animal species (see table 14.1). Forest regeneration in mosaic landscapes is most effective for conservation and environmental services when it leads to enlargement of existing forest fragments or creates biological corridors that link formerly isolated fragments. In an Atlantic Forest region of southeastern São Paolo, Brazil, second-growth forest corridors increased the species richness of small mammals in forest fragments (Pardini et al. 2005). Cocoa forests and second-growth forests form corridors that connect small forest patches to a large forest area in Una Biological Preserve, creating a forested matrix that supports diverse assemblages of ferns, butterflies, leaf litter reptiles, amphibians, bats, birds, and small mammals.

Ecotourism can provide a strong impetus for reforestation in biological corridors or in former agricultural lands. Reforestation in biological corridors is being carried out in many small-scale projects associated with private reserves, ecotourist lodges, and government-managed conservation areas (Blangy and Mehta 2006). Many of these projects are focused on reestablishing forest cover to promote the conservation and genetic diversity of endangered wildlife species. Restored forests will bring tourists as well as wildlife. The Monte Pascoal-Pau Brasil Ecological Corridor Reforestation Project in the Atlantic Forest of Brazil is also a certified carbon sequestration project (Alexander et al. 2011).

Forest regeneration can increase the conservation value in the landscape matrix, creating corridors and habitat patches, and expanding buffer zones surrounding existing reserves (Chazdon, Harvey, et al. 2009; Perfecto et al. 2009; Schroth and McNeely 2011). But large and intact forest areas are critically important for the conservation of many species. Many forest-interior species do not find suitable habitats in mosaic landscapes (Chazdon, Peres, et al. 2009; Gardner et al. 2009; Mahood et al. 2012), suggesting that there may be a threshold of forest cover, patch size, or patch density that will support these sensitive species in mosaic landscapes (Pardini et al. 2010; Melo et al. 2013).

14.5.3 *Forest Landscape Restoration*

The concept of forest landscape restoration (FLR) offers a new approach to reestablish key ecosystem functions across an entire landscape (Dudley et al. 2005; Mansourian et al. 2005). The goal of FLR is to promote a wider range of options for the delivery of forest-related goods and services at the landscape scale, rather than to return forest landscapes to their "original" state (Maginnis et al. 2007). Of particular importance is the restoration of landscape heterogeneity and structural complexity that serve to maintain biodiversity and ecosystem services (Newton et al. 2012). Site-level reforestation activities are planned to accommodate landscape-level objectives. Stakeholder participation is an essential component of decision making, to reach goals of long-term sustainability through consensus and compromise (Newton and Tejedor 2011). Forest landscape restoration initiatives are underway in regions of Brazil, India, Malaysia, Mali, New Caledonia, and Tanzania, as well as some temperate forest regions. Natural forest regeneration is a major component of FLR, in addition to targeted tree planting and agroforestry. Through the implementation of FLR, the restoration of functioning forest landscapes is directly linked to the well-being of the resident caretakers.

14.5.4 Forest Regeneration in Relation to Global Carbon Mitigation Programs

Tropical forest regeneration is a major global carbon sink (see box 11.2). At the 16th meeting of the United Nations Framework Convention on Climate Change (UNFCC) in Cancun in December 2010, the parties agreed to a series of rules to structure the carbon mitigation program known as REDD+ (Reducing Emissions from Deforestation and Forest Degradation "plus" conservation, sustainable management of forests, and enhancement of forest carbon stocks; Agrawal et al. 2011). These rules are designed to safeguard concerns about conservation of natural forests and biological diversity, forest governance, participation of rural stakeholders, respect for the knowledge and rights of indigenous peoples and local communities, and reduction of displaced deforestation (Grabowski and Chazdon 2012; Scriven 2012). These principles were affirmed at the 17th UNFCC Conference of Parties in Durban, South Africa, in 2011. Billions of dollars from industrialized nations could be transferred to developing tropical countries each year through REDD+, essentially creating a massive, international-scale PES program. The foundational principle of REDD+ is that market mechanisms can be applied to the purchase of carbon to maintain standing forests and to enhance forest regrowth (Hall 2012). Many tropical countries are actively developing national strategies or action plans for REDD+ readiness.

Angelsen and Rudel (2013) argued that the design and application of REDD+ policies should be tailored to the stage of forest transition and types of landscapes within a particular region or country. For example, in countries with few or no remaining old-growth forests, such as El Salvador, REDD+ programs should incentivize active and passive forms of reforestation. On the other hand, in countries with large stocks of old-growth forest, such as

Surinam, REDD+ policies should aim to protect existing of old-growth forest and promote sustainable forest use. In the case of countries or regions that are experiencing rapid deforestation, such as Indonesia, REDD+ programs should focus on slowing down or halting rates of deforestation and promoting reforestation around existing old-growth forest areas.

The bundling of co-benefits and the safeguards that are being put into place present a positive picture for REDD+ policies. But serious concerns remain regarding potential negative effects of REDD+ on several fronts (Huettner 2012; Phelps et al. 2012). First, there may be clear trade-offs between conservation and carbon sequestration goals (Alexander et al. 2011; Phelps et al. 2012). Monoculture plantations of nonnative tree species can enhance forest carbon stocks but provide little biodiversity value; these plantations are rapidly replacing natural forest regeneration in many regions. Second, lack of land tenure and user rights excludes landless rural people and indigenous groups from participation and revenue sharing. In the absence of land reform, REDD+ policies are likely to favor land-rich farmers, providing few or no benefits to poor smallholders or peasants. Local livelihoods are undermined through loss of access to forest areas (Mustalahti et al. 2012). Local stakeholder participation is essential for success of REDD+ policies. Third, REDD+ policies may lead to land grabbing by public and private investors, driving up prices of land and crops and increasing food insecurity for rural peoples. Fourth is the concern that the decentralization of forest governance that has been achieved in many developing countries will be reversed by REDD+ policies (Agrawal et al. 2008). Finally, corruption and poor governance can also lead to unfair distribution of REDD+ benefits within participating countries (Huettner 2012). These issues will need to be resolved if REDD+ is to deliver benefits to tropical forests and tropical peoples worldwide.

14.6 Conclusion

Changing economic and environmental conditions have created an upswing in reforestation in many regions of the world. Rates of deforestation are indeed declining in many tropical countries, creating the potential for forest transitions. These scenarios provide hope of reversing trends in declining forest cover, declining forest quality, declining biodiversity, and declining environmental services that have dominated the past few centuries. But despite these changes, the accumulated area of forest loss still far exceeds the areas of forest regeneration throughout the tropics. Recent reforestation may have caught up with recent deforestation in some cases, but these dynamics do not erase decades or centuries of forest transformations. Historical "deforestation debt" may never be repaid.

This chapter highlights several issues that have sparked new research. One of these is the role of globalization in forest transitions. Forces of globalization have driven much of the conversion of old-growth forests to pasture and industrial agriculture. At the same time, forces of globalization have led to the expansion of regenerating forests (Meyfroidt et al. 2010; Rudel 2012). Reforestation in one region is often associated with deforestation in another, in what resembles an international-scale shifting cultivation system. New remote-sensing tools and applications may soon make it possible to assess different modalities of forest regeneration at regional scales. The inability to detect older stages of natural regeneration or to distinguish tree plantations and agroforestry practices from young stages of natural regeneration continues to limit our understanding of the nature of forest transitions and the social and economic drivers of forest regeneration.

CHAPTER 15

SYNTHESIS: THE PROMISE OF TROPICAL FOREST REGENERATION IN AN AGE OF DEFORESTATION

To say that without conservation measures all primary tropical forests will be 'destroyed' by the end of the century is not to assert that in the year 2000 there will be no forests of any kind, or that the whole existing forest flora and fauna will disappear, but it is likely that, except on steep slopes and other inaccessible or uncultivable sites, all forest of a climax or near-climax type will disappear. There will be large areas of cultivation under tree and other crops and also doubtless more or less artificial forests with a very reduced number of tree species managed in the interests of timber or cellulose production. There will also probably be large areas of secondary forests and other seral communities. What will these be like and how much of the present forest flora and fauna will survive in them? The answer to this question is important for any conservation policy and it can be answered only by studying the secondary communities which can be seen so abundantly in all parts of the tropics today. Unfortunately far too little attention has been given to secondary successions and the biota of seral communities.—Paul W. Richards (1971, p. 176)

15.1 The Power of Forest Regeneration

Paul Richards (1971) had remarkable insight into what the future held for tropical forests. Over 40 years ago, he foresaw that regenerating forests would become the predominant form of tropical forest cover. He clearly recognized the importance of studying second-growth forests and understanding the dynamics of plant and animal species during forest regeneration. The knowledge that tropical forests are dynamic and resilient systems is empowering. It is time to use this power to help tropical forests regrow wherever and whenever possible. Forests can regrow in many ways. Forest regeneration and ecological restoration can have positive consequences for the billions of people who depend on forests for their livelihoods and well-being. I wrote this book primarily to convey this urgent message. We now have an opportunity to use existing old-growth forest areas as leverage in an unprecedented effort to expand the boundaries of remnant forests, to build biological corridors, to enhance biodiversity and ecosystem services on unproductive or unused lands, and to create diverse and multifunctional landscapes. But if we do not act soon, this task will become even more challenging, expensive, and risky.

The power of forest regeneration is strong medicine to counter the maladies of global environmental degradation. I recently revisited three areas of regenerating forest in northeastern Costa Rica, where my collaborators, students, and I have been following vegetation dynamics for over 15 years. Despite similar overall structures and shared common species, each forest is marching along on its own successional trajectory. In a 50-year-old plot, giant *Vochysia ferruginea* trees are dying, creating recruitment opportunities for tree saplings that have been waiting in the understory for this very moment. In a 36-year-old plot, stilt palms *Iriartea deltoidea* and *Socratea exorrhiza* are pushing their crowns up and into the canopy, filling formerly empty subcanopy spaces. The leaves of understory *Geonoma cuneata* palms are now large enough to serve as roosting sites for fruit bats that disperse large-seeded tree seeds from neighboring old-growth forest areas. Entering a hidden patch of 18-year-old forest, I was greeted by howler monkeys and lekking manakins. This young forest is packed with a high diversity of tree seedlings and saplings planted by birds and bats who came to visit remnant trees left when the original forest was logged and converted to pasture. These young trees foretell the future of this forest, just as these young forests foretell the future of the landscape mosaic that surrounds them.

Between 1997 and 2002, four international workshops were held in Peru, Malaysia, Kenya, and Cameroon with the goal of highlighting the management potential of tropical secondary forests. The proceedings of these workshops uniformly emphasized the low priority and lack of recognition accorded to secondary forests in national budgets, research agendas, and policy agendas of government agencies and nongovernmental organizations. The sustainable management of secondary forests is often ignored in policy making and resource allocation largely because of a lack of awareness of the value of goods and services that secondary forests produce (FAO 2003).

A similar view was emphasized in a report by the International Tropical Timber Organization (ITTO), which presented 49 principles and 160 recommendations to guide the restoration, management, and rehabilitation of degraded and secondary forests in the tropics (ITTO 2002). This report highlighted a large gap in policies to restore, manage, and rehabilitate degraded and secondary forests (fig. 15.1). Guidelines exist to establish and manage planted tropical forests, on the one hand, and to sustainably manage old-growth forests, on the other. But the forests that lie between these extremes—the regenerating forests that are discussed in this book—are in serious need of attention from policy and management sectors. This neglect may stem from the lack of integration between the forestry sector and the agricultural sector and between commercial development and conservation

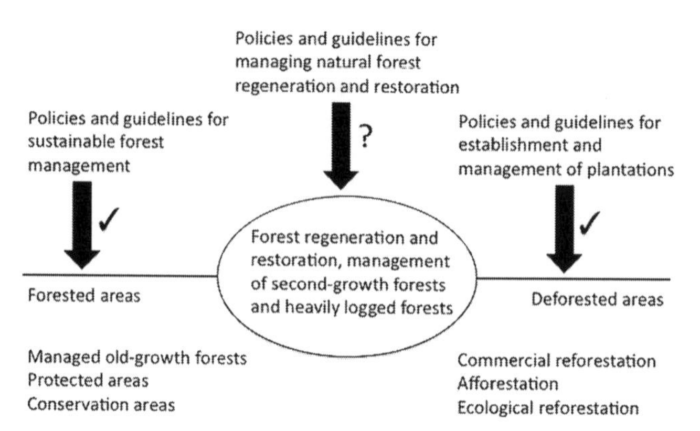

Figure 15.1
A graphical depiction of the gap in policies and guidelines for managing natural forest regeneration and reforestation. Existing policies and guidelines pertain to management of forested and deforested areas in the tropics but do not address the management of second-growth forests, heavily logged forests, natural regeneration, or reforested areas. Source: Redrawn from ITTO (2002, fig. 1).

interests. Regenerating forests are indeed a nexus between these distinct constituencies. They demand their own policy and management framework that is linked with forestry, agricultural, and national land-use policies, with the active participation of local stakeholders. Indigenous practices provide an excellent starting point, as fallow management has been an integral component of indigenous cultures for millennia.

15.2 Tropical Forest Change and Resilience

We look toward a future when rates of forest regeneration will overcome rates of forest loss (Meyfroidt and Lambin 2011). The literature abounds in books and articles about deforestation, degradation, and destruction in the tropics. This book has focused on turning these trends around through reforestation, regeneration, restoration, and recovery. Understanding the process of forest regeneration reveals the potential for forest change and resilience following human disturbances, natural disturbances, and climate change (Whitmore 1998c). Below I review the main points and messages of the previous chapters.

15.2.1 *Lessons from Forest History*

We can gain important insights into the potential for future forest regeneration by looking into the past. Forests are not static systems. Forest assemblages are constantly changing, as species adapt and populations respond to changing biotic and abiotic conditions and to a wide range of natural and anthropogenic disturbances. Tropical forests are complex adaptive systems, where disturbance and heterogeneity is the norm (Chazdon and Arroyo 2013). They grow, get cleared or burned or blown down, and grow back again. When forests regrow, they do not completely resemble the forests they previously were, even if they gradually acquire the structural and compositional features of old forests. The process of regrowth is slow and can take centuries—as long as the age of the oldest trees in the forest. Old-growth tree species grow slowly, and some long-lived pioneer trees can persist for centuries. Many of the tropical forests that we think of as "primary forests" are actually undergoing late stages of secondary succession and still have cohorts of long-lived pioneer species.

15.2.2 *Tropical Forests Are Resilient to Many Types of Disturbance*

Forest disturbances are a natural component of the forest growth cycle. Forests have regrown following complete obliteration of islands, river valleys, and mountainsides, without any human intervention. Through the process of succession, forests constantly renew themselves. Species of plants and animals are specialized to colonize and exploit small and large forest disturbances. Some species have adaptations that enable persistence following forest cutting, burning, and severe wind damage. Nutrients and carbon lost during deforestation are gradually recovered during forest regrowth. By studying the ecology of these species and their assemblages, we gain insight into the attributes of species and species interactions that promote forest resilience, such as functional redundancy, generalist species interactions, and nested interaction networks. Further, we can identify appropriate native species for reforestation in areas that fail to colonize naturally following intensive land use. Practices of harvesting, gardening, and management of useful species and soil enhancement have been an integral part of indigenous management systems for thousands of years, built upon an understanding and appreciation of forest regeneration processes. These practices offer models for sustainable forest use today. From an ecological perspective, we understand the fundamental pathways of vegetation succession, and we can identify how local and landscape conditions cause variation in rates of change in species composition, functional diversity, and ecosystem processes during forest regeneration.

15.2.3 *Humans Are an Integral Part of Tropical Forest Dynamics*

During the Holocene, tropical forests reassembled their composition in the presence of humans. Hunting and the domestication of plants impacted tropical forests and were associated with forest burning. Early feats of human engineering altered seasonal wetlands, hillsides, and savanna boundaries. Many forest areas that were heavily impacted in the past have since regrown, with few superficial signs of past disturbance today.

Human impacts are far more pervasive today than in the past and pose tremendous challenges for existing and future tropical forests. Climate change and invasive species are impacting tropical forest growth and species interactions, even in areas that are not being subjected to logging, fragmentation, or clearance for cultivation (Feeley, Wright, et al. 2007; Butt et al. 2013). Increasing intensity of land use, reduction of fallow periods, and proliferation of invasive species restrict the potential for forest regeneration throughout many regions.

The process of forest regeneration is intimately linked with human activities. Forest regeneration requires lasting partnerships between humans and their landscapes. Heavily impacted forest areas rely on people to restore forest structure and function. To adapt a phrase from Hillary Clinton, "It takes a village to raise a forest." But reforestation can cost much more than a village can afford.

15.2.4 *Forest Regeneration Reflects the Surrounding Landscape*

Forest regeneration relies upon seed dispersal from patches of forest vegetation, remnant trees, or intact forest areas. Extensively cleared and fragmented areas of the landscape will therefore have reduced potential for natural regeneration. Tropical regions with a long history of clearing, agricultural use, and forest fragmentation may have lost many forest-dependent species of both plants and animals (Tabarelli et al. 2012; Melo et al. 2013). In contrast, recently modified landscapes that still retain a significant amount of old-growth forest and have intact populations of seed-dispersing animals show a high potential for forest regeneration.

Interactions between plant and animal species are essential for functioning tropical forest ecosystems and play an important role in community assembly during forest regeneration. These interactions include herbivory, seed predation, seed dispersal, and pollination. Forest regeneration is enhanced in regions with intact faunas and where old-growth forests are close by. Animal diversity increases during forest regeneration, along with increases in plant diversity and complexity of forest structure, creating a positive feedback loop that further enhances forest succession.

15.3 The Current and Future Value of Regenerating Tropical Forests

If forest regeneration and restoration are to achieve their potential to transform tropical landscapes from a degraded to a revitalized state, recognizing their current and future value is crucial. Today, fallowed fields and unused agricultural lands are frequently viewed as wastelands or degraded habitats when compared to older forests with high levels of biodiversity and high carbon stocks. The future value of these abandoned lands is rarely considered in conservation planning. Many young or intervened forests are unrecognized and underappreciated as valuable ecosystems. The very nature of the term *abandoned land* implies that nobody cares about the land anymore.

This picture is slowly changing, as international and national programs focus efforts on reforestation and restoration. Reforestation provides many co-benefits, such as income opportunities for rural households, enhancement of biodiversity conservation, sustainable production of raw materials for forest industries, and a wide array of supporting and regulating ecosystem services (Barr and Sayer 2012; Chazdon 2008a). Target 15 of the Convention of Biological Diversity's Strategic Plan 2011–2010 stated that "by 2020, ecosystem resilience and the contribution of biodiversity to carbon stocks have been enhanced, through conservation and restoration, including restoration of at least 15 per cent of degraded ecosystems, thereby contributing to climate change mitigation and adaptation and to combating desertification" (Secretariat of the Convention on Biological Diversity 2010, p. 2). An analysis by the International Union for the Conservation of Nature estimated that restored forests on 150 million hectares of deforested land will have an economic value of US$85 billion per year, based solely on the market value of stored carbon, timber, and nontimber products. The Bonn Challenge is a global movement to restore 150 million hectares of degraded and deforested land by 2020. The governments of Costa Rica and El Salvador have each committed to restoring up to 1 million hectares of tropical forest. The National Mission for a Green India aims to increase forest and tree cover on 5 million hectares of forest and nonforest land and improve the quality of forest cover on another 5 million hectares. In 1998, Vietnam launched a major effort to reforest 5 million hectares (McElwee 2009).

Active forms of reforestation are more strongly promoted by government agencies and nongovernmental organizations than passive reforestation (Kammesheidt 2002). This is particularly true in the Asia-Pacific region, where "reforestation" programs are controlled by state forestry bureaucracies with strong commercial interests and close ties to government and military officials (Barr and Sayer 2012). State reforestation policies often lead to perverse incentives to convert second-growth and selectively logged forests to short-rotation industrial tree plantations (Lindenmayer et al. 2012). These practices are highly subject to corruption, are counter to the objectives of ecologically based reforestation efforts, and fail to protect the interests of rural people who depend on forests for their subsistence.

In cases where conditions are inimical for natural regeneration, planting trees appropriate to the area (including carefully selected nonnative species) can initiate the process of reforestation. But in many cases, natural regeneration can do the job of restoring biodiversity and ecosystem services at least as effectively as planted forests. Several studies have shown that carbon and nutrient stocks in soils and vegetation do not substantially differ between plantations and second-growth forests of similar age. And, of course, natural regeneration is far less costly than planting and maintaining trees. Investments in the form of social capital are more urgently needed for natural regeneration than are financial investments. Natural regeneration promotes recovery of biodiversity and species interactions, as native species are adapted for growth in local areas and can be supported by networks of mutualists and mobile links that facilitate the arrival, survival, and reproduction of forest generalists and specialists.

If natural regeneration can achieve these goals as well or better than planted forests, why is this mode of reforestation less widely adopted and incentivized? Although natural regeneration is by far the "cheapest" form of reforestation, it does not fit a centralized, industrial model of production and does not provide corporate profits. Policies for management of secondary forests are lacking in most countries (ITTO 2002). Successional pathways are intrinsically variable, and outcomes are far less predictable and uniform than with planted forests. Consequently, natural regeneration is viewed as a more risky approach to reforestation. Young, naturally regenerating forests are simply not valued as highly as planted forests. Early successional forests—commonly known as "jungles"—are viewed as nasty, impenetrable tangles of vegetation where snakes, thorny vegetation, and stinging insects lurk, whereas plantations are neat, orderly, civilized, and safe. Natural regeneration is also difficult for landowners to justify, as unused land presents an open invitation to occupation by landless people and may be perceived as "improper" land-use practice. Regrowth forests are often referred to as "degraded" forests, as they have reduced diversity and function relative to old-growth forests. Regenerating tropical forests continue to be misunderstood, understudied, and unappreciated for what they really are—young, self-organizing forest ecosystems that are undergoing construction.

15.3.1 *Assessing the Conservation Value of Regenerating Forests*

Recent debate has focused on the extent to which successional forests provide suitable habitats for the regeneration of plants and animals that specialize in old-growth forests and for locally endemic species of high conservation value (Wright and Muller-Landau 2006; Laurance 2007; Gardner, Barlow, et al. 2007; Chazdon, Peres, et al 2009; Gibson et al. 2011). Resolving this debate presents several challenges (Gardner et al. 2009):

1. Limited age classes of regenerating forest are sampled or available for study.
2. Old-growth forests used as baseline comparisons are assumed to be stable in species composition and environmental conditions during the study period.
3. Classification of species as old-growth specialists is often based on poor data or limited qualitative assessment.
4. Regenerating forests and old-growth forests used as baselines for comparison tend to occupy different ecological zones, elevations, soil types, slope conditions, and levels of human access.
5. True replication of forest types being compared is often lacking.
6. Old-growth forests may no longer exist within the region, restricting baseline comparisons to older regenerating forests.
7. Edge effects confound effects of stand age on biodiversity patterns.
8. Recovery of biodiversity in regenerating forests is highly dependent on the quality of the landscape matrix and is not a simple function of stand age.

Almost all available information on changes in species composition during succession is based on short-term studies involving sites less than 50 years old, and few studies track changes in regenerating forests over time. Although the young age of recovering forests is a stark reality in many areas, this limitation biases our understanding of the potential recovery of biodiversity during later stages of succession. Secondary forests are often recleared before there is an opportunity to assess long-term successional trajectories and species gain in these forests (Chazdon, Peres, et al. 2009; Williams-Linera et al. 2011; Melo et al. 2013). Comparisons of species composition in successional and old-growth forests should take into account observed changes in species composition or population abundance during the same time interval in old-growth areas that serve as reference points. This temporal information is often not available. Moreover, the classification of old-growth specialists in a particular region or landscape is often highly subjective and is rarely based on robust statistical procedures (Chazdon et al. 2011).

An additional problem arises because, in many regions, the old-growth forests used as reference points and for classifying old-growth forest specialists do not occur within the same elevational zone or soil type as second-growth forests due to landscape variation in deforestation and land abandonment (Helmer et al. 2008; Crk et al. 2009). The absence of certain old-growth species from second-growth forests could be attributed to differences in forest elevation or soil type rather than to forest successional stage. Regenerating forests often occur in small forest fragments, surrounded by nonforest land uses. For this reason, the status of biodiversity in forests is often strongly affected by effects of forest fragmentation or small fragment size, apart from effects of forest age or successional status. Finally, the recovery of biodiversity in regenerating forests depends on the composition and structure of the surrounding landscape matrix and is not a simple function of stand age. For all of these reasons, assessments of conservation value in regenerating forests are fraught with methodological pitfalls.

The landscape context of regenerating forests also affects their local and regional role in conserving biodiversity. In a landscape with high levels of cover by intact, old-growth forests, secondary forests contribute relatively little to overall biodiversity conservation. But in matrix landscapes with a small number of remnant patches of forest, regenerating forest patches play a critical role in providing forest habitats and stepping stones or corridors between other forest areas (Harvey et al. 2008; Chazdon, Harvey, et al. 2009; Tabarelli et al. 2012).

15.3.2 Ecosystem Services of Regenerating Forests

Aboveground biomass, soil carbon, nutrient cycling, and hydrological functions recover rapidly during forest regeneration. These processes are the basis for the provision of many goods and services that directly benefit people. Regenerating and restored forests offer a wide range of ecosystem goods and services (table 15.1; Rey Benayas et al. 2009; Parotta 2010). The carbon sequestration potential of tropical regenerating forests is a major focus of global programs to mitigate carbon emissions, such as REDD+ (see boxes 11.2 and 14.2). Naturally regenerating forests contain a high abundance of useful tree species, many with commercial value for timber and nontimber products (Finegan 1992; Chazdon and Coe 1999; Apel and Sturm 2003). Commercial timber species composed between 35% and 40% of all species and more than 70% of the basal area in second-growth forests in northeastern Costa Rica (Vílchez Alvarado et al. 2008).

Table 15.1. *Ecosystem services provided by natural regeneration and reforestation in tropical forest watersheds*

Type of ecosystem services	Description
	A. Provisioning services
Water	Safe and clean drinking water; stable water supply for domestic use, irrigation, industrial purposes, or generation of electricity
Food and medicine	Gathering of fish, game, fruits, medicinal plants
Raw materials	Materials used by local people for construction, fuel, and fiber (wood, firewood, charcoal, thatch, cordage, crafts)
	B. Regulating and supporting services
Carbon storage	Sequestration of carbon in vegetation and soils to offset carbon emissions
Mitigation of extreme weather events	Flood control and storm protection
Air quality regulation	Interception of dust, chemical, and other elements by land cover
Erosion prevention	Prevention of damage from erosion and siltation
Pest regulation	Maintain natural heterogeneity and populations of natural enemies to regulate populations of insect pests
Pollination and seed dispersal	Maintain populations of insects, birds, and mammals that pollinate flowers and disperse seeds
Nutrient cycling	Prevent nutrient losses and maintain stocks of nutrients in soils for plant uptake
	C. Cultural services
Aesthetic beauty and spiritual experience	Provide environments to appreciate nature, enjoy surroundings, and inspire spiritual or religious experiences
Cultural heritage	Share cultural heritage and local ecological knowledge with family and friends; folklore; special events
Educational value	Outings and formal field classes in botany, ecology, and forestry
Recreation	Enjoy sports, recreational fishing, exercise, swimming, hiking, and social events in natural settings

Source: Based on Cremaschi et al. (2013) and Brancalion et al. (2013).

15.4 New Approaches to Promoting Forest Regeneration

Bringing forests back to deforested landscapes is a challenging prospect, but one we cannot ignore. The drivers of reforestation are complex and emerge from biophysical, socioeconomic, and cultural factors. This book has mostly addressed the ecological factors that affect forest regeneration following natural and human disturbances. Resource availability and species availability are the two major ecological factors that determine successional

Box 15.1. Raising Forests and Hope through Reforestation and Restoration Concessions

A new approach to promoting forest regeneration in tropical landscapes is emerging that follows the principles of forest landscape restoration. The approach draws from the recently developed concept of conservation concessions, which have been initiated in several countries with the assistance of conservation organizations (Milne and Niesten 2009). Unlike conservation concessions, which are designed to restrict harvesting and forest conversion in community-owned forest land through rental agreements, *reforestation concessions* provide financial support, infrastructure, and training to local communities with the goal of bringing back forests through ecological reforestation projects. Local stakeholders are actively involved in planning, executing, and monitoring local reforestation programs. The costs of reforestation and rental of previously deforested land owned by communities or governments are raised through global participation of individuals and organizations who provide financial support for these reforestation and conservation efforts while receiving offsets for their own carbon emissions.

In some ways, the concept of reforestation concessions is similar to the model of community-supported agriculture (Sharp et al. 2002), but rather than producing local vegetables, farms produce young tropical forests, a global public good. Shareholders are not limited to members of the local community but are drawn from across the world as partners, investors, and scientific researchers in a long-term reforestation project. Institutional and government support would be needed to provide training on establishment, management, and monitoring of ecological reforestation and natural regeneration concessions. Priority for natural regeneration concessions should be assigned to unused former agricultural lands close to remaining forest fragments that show a high pathways. Geographic variation in climate and soils also influences natural regeneration and vegetation traits, such as resprouting ability and rates of nutrient uptake. Although climate change is not new to the tropics, we have little insight into how forest regeneration in tropical regions will respond to rapid climate change (Whitmore 1998c). It is likely that increasing temperatures and changes in rainfall and seasonality will affect plant and animal species in all types of tropical forests (Wright et al. 2009).

potential for natural regeneration. Priority for ecological reforestation concessions, in contrast, should be granted in unproductive agricultural land in areas that present significant obstacles to natural regeneration.

New approaches are also being developed to restore degraded, formerly logged forests. The world's first ecosystem restoration concession was recently established in Harapan Forest in lowland Sumatra on a former logging concession, part of the government forest estate. A private company, composed of a consortium of organizations (Burung Indonesia, the Royal Society for the Protection of Birds, and BirdLife International), manages the concession under a license fee to the Indonesian government (Normile 2010; Rands et al. 2010). The license is for 35 years, after which time the forest may be harvested or protected for conservation. The project has received additional funding from the governments of Denmark and Germany.

More than 40,000 people living around the forest will be involved in reforestation and conservation efforts. The Harapan Rainforest Project aims to restore approximately 100,000 hectares of logged and disturbed lowland rain forest. Plans are being developed for experimental reforestation treatments in open areas and in areas with low and high second-growth forest cover. Nurseries are being established to grow seedlings of native tree species to plant in open areas and in logged forests. Tree planting and nurseries provide technical training and livelihoods for indigenous Bathin Sembilan families. The project also includes long-term monitoring of vegetation and fauna in logged, restored, and unlogged forest areas.

Appropriately, *Harapan* is the Indonesian word for "hope." As the world's first reforestation concession, Harapan Rainforest Project provides hope that successful approaches for landscape-scale forest restoration in the lowlands of Sumatra will provide working models for restoring, managing, and conserving tropical forests throughout the world.

Ultimately, natural forest regeneration requires a reduction of human pressure on land, whether this occurs by deliberate planning or as a consequence of other actions and policies. This reduction should not compromise the means for sustenance and livelihoods of human populations, particularly those living in rural areas. Reducing human pressure on remaining old-growth forests is also an essential part of this challenge, as forest regeneration will not occur in the absence of remaining old-growth forest areas. Regrowth forests will not replace old-growth forests. These two types of forest function together in landscapes to protect much of the forest-requiring biodiversity in the tropics and to provide essential ecosystem services.

Resource availability and species availability are directly affected by the intensity of land use and by the location and forest types present in the overall landscape. Through these interactions, social systems directly impact the rates and quality of active and passive forms of reforestation. Land-use and land-cover change are, in turn, products of social and economic policies at different governmental levels. Governmental policies regarding land use are often poorly informed by ecological considerations, particularly those regarding forest regeneration (Menz et al. 2013). Policies regarding forestry, land use, and conservation need to be based on solid scientific information about the biophysical potential for forest regeneration as well as on the values of biodiversity conservation and ecosystem goods and services that are provided by naturally regenerating forests and native species plantations.

The objective of increasing forest cover throughout the tropics, either through passive spontaneous regeneration or through planned reforestation projects, is based on the recognition that forests need people, just as people need forests. Successful reforestation is a partnership that benefits people and forests (box 15.1). Facilitated, participatory, multistakeholder processes are fundamental components of reforestation policies (Boedhihartono and Sayer 2012). The social, economic, and institutional drivers of reforestation success are also the foundations of human well-being, social harmony, and effective governance (Dinh Le et al. 2012). Our forests reflect our values and our dreams. The forests of the tropics deserve a second chance, and so do we.

REFERENCES

Achard, F., H. J. Stibig, H. D. Eva, E. J. Lindquist, A. Bouvet, O. Arino, and P. Mayaux. 2010. Estimating tropical deforestation from Earth observation data. Carbon 1:271–87.

Ackerly, D. D. 1996. Canopy structure and dynamics: Integration of growth processes in tropical pioneer trees. Pages 619–58 *in* S. S. Mulkey, R. L. Chazdon, and A. P. Smith, eds. Tropical forest plant ecophysiology. Chapman & Hall, New York.

Adams, R. E. W., W. E. Brown, and T. P. Culbert. 1981. Radar mapping, archeology, and ancient Maya land use. Science 213:1457–63.

Adum, G. B., M. P. Eichhorn, W. Oduro, C. Ofori-Boateng, and M. O. Rödel. 2013. Two-stage recovery of amphibian assemblages following selective logging of tropical forests. Conservation Biology 27: 354–63.

Agrawal, A., A. Chhatre, and R. Hardin. 2008. Changing governance of the world's forests. Science 320:1460–62.

Agrawal, A., D. Nepstad, and A. Chhatre. 2011. Reducing emissions from deforestation and forest degradation. Annual Review of Environment and Resources 36:373–96.

Aguiar, A. V., and M. Tabarelli. 2010. Edge effects and seedling bank depletion: The role played by the early successional palm *Attalea oleifera* (Arecaceae) in the Atlantic Forest. Biotropica 42:158–66.

Aguilar-Fernández, M., V. Jaramillo, L. Varela-Fregoso, and M. Gavito. 2009. Short-term consequences of slash-and-burn practices on the arbuscular mycorrhizal fungi of a tropical dry forest. Mycorrhiza 19:179–86.

Aguilar-Støen, M., A. Angelsen, and S. R. Moe. 2011. Back to the forest: Exploring forest transitions in Candelaria Loxicha, Mexico. Latin American Research Review 46:194–216.

Aide, T., and J. Cavelier. 1994. Barriers to lowland tropical forest restoration in the Sierra Nevada de Santa Marta, Colombia. Restoration Ecology 2:219–29.

Aide, T. M., M. L. Clark, H. R. Grau, D. López-Carr, M. A. Levy, D. Redo, M. Bonilla-Moheno, G. Riner, Andrade-Núñez, and M. Muñiz. 2013. Deforestation and reforestation of Latin America and the Caribbean (2001–2010). Biotropica 45:262–71.

Aide, T. M., and H. R. Grau. 2004. Globalization, migration, and Latin American ecosystems. Science 305:1915–16.

Aide, T. M., J. K. Zimmermann, L. Herrera, M. Rosario, and M. Serrano. 1995. Forest recovery in abandoned tropical pastures in Puerto Rico. Forest Ecology and Management 77:77–86.

Ainsworth, A., and J. B. Kauffman. 2009. Response of native Hawaiian woody species to lava-ignited wildfires in tropical forests and shrublands. Plant Ecology 201:197–209.

Aldrich, P. R., and J. L. Hamrick. 1998. Reproductive dominance of pasture trees in a fragmented tropical forest mosaic. Science 281:103–5.

Alexander, S., C. R. Nelson, J. Aronson, D. Lamb, A. Cliquet, K. L. Erwin, C. M. Finlayson, et al. 2011. Opportunities and challenges for ecological restoration within REDD+. Restoration Ecology 19:683–89.

Allen, E. B., M. E. Allen, L. Egerton-Warburton, L. Corkidi, and A. Gómez-Pompa. 2003. Impacts of early- and late-seral mycorrhizae during restoration in seasonal tropical forest, Mexico. Ecological Applications 13:1701–17.

Allen, J. A. 1997. The impact of Pleistocene hunters and gatherers on the ecosystems of Australia and Melanesia: In time with nature? Pages 22–38 *in* P. V. Kirch and T. L. Hunt, eds. Historical ecology in the Pacific Islands: Prehistorical environmental and landscape change. Yale University Press, New Haven, CT.

Alvarez-Añorve, M. Y., M. Quesada, G. A. Sánchez-Azofeifa, L. D. Avila-Cabadilla, and J. A. Gamon. 2012. Functional regeneration and spectral reflectance of trees during succession in a highly diverse tropical dry forest ecosystem. American Journal of Botany 99:816–26.

Alvarez-Buylla, E., and M. Martínez-Ramos. 1992. Demography and allometry of *Cecropia obtusifolia*, a Neotropical pioneer tree—an evaluation of the climax-pioneer paradigm for tropical rain forests. Journal of Ecology 80:275–90.

Alvarez-Clare, S., and K. Kitajima. 2007. Physical defence traits enhance seedling survival of Neotropical tree species. Functional Ecology 21:1044–54.

Alves, D. S., J. V. Soares, S. Amaral, E. M. K. Mello, S. A. S. Almeida, O. F. DaSilva, and A. M. Silveira. 1997. Biomass of primary and secondary vegetation in Rondônia, Western Brazilian Amazon. Global Change Biology 3:451–61.

Andresen, E. 2002. Dung beetles in a Central Amazonian rainforest and their ecological role as secondary seed dispersers. Ecological Entomology 27:257–70.

———. 2003. Effect of forest fragmentation on dung beetle communities and functional consequences for plant regeneration. Ecography 26:87–97.

———. 2008. Dung beetle assemblages in primary forest and disturbed habitats in a tropical dry forest landscape in Western Mexico. Journal of Insect Conservation 12:639–50.

Andresen, E., and F. Feer. 2005. The role of dung beetles as secondary seed dispersers and their effect on plant regeneration in tropical rainforests. Pages 331–49 *in* P.-M. Forget, J. E. Lambert, P. E. Hulme, and S. B. Vander Wall, eds. Seed fate: Predation, dispersal, and seedling establishment. CABI Publishing, Cambridge, MA.

Andresen, E., and D. J. Levey. 2004. Effects of dung and seed size on secondary dispersal, seed predation, and seedling establishment of rain forest trees. Oecologia 139:45–54.

Angelsen, A., and T. K. Rudel. 2013. Designing and implementing effective REDD+ policies: A forest transition approach. Review of Environmental Economics and Policy 7:91–113.

Ansell, F. A., D. P. Edwards, and K. C. Hamer. 2011. Rehabilitation of logged rain forests: Avifaunal composition, habitat structure, and implications for biodiversity friendly REDD+. Biotropica 43:504–11.

Anselmetti, F. S., D. A. Hodell, D. Aziztegui, M. Brenner, and M. F. Rosenmeier. 2007. Quantification of soil erosion rates related to ancient Maya deforestation. Geology 35:915–18.

Apel, U., and K. Sturm. 2003. The use of forest succession for establishment of production forest in northeastern Viet Nam. Pages 39–49 *in* H. C. Sim, S. Appanah, and P. B. Durst, eds. Bringing back the forests: Policies and practices for degraded lands and forests. Food and Agriculture Organization of the United Nations Regional Office for Asia and the Pacific, Bangkok, Thailand.

Aronson, J., P. H. S. Brancalion, G. Durigan, R. R. Rodrigues, V. L. Engel, M. Tabarelli, J. Torezan, S. Gandolfi, A. C. G. de Melo, and P. Y. Kageyama. 2011. What role should government regulation play in ecological restoration? Ongoing debate in São Paulo State, Brazil. Restoration Ecology 19:690–95.

Aronson, J., S. J. Milton, and J. N. Blignaut. 2007. Restoring natural capital: Science, business and practice. Island Press, Washington, D.C.

Arroyo-Kalin, M. 2012. Slash-burn-and-churn: Landscape history and crop cultivation in pre-Columbian Amazonia. Quaternary International 249:4–18.

Arroyo-Mora, J. P., G. A. Sánchez-Azofeifa, M. E. R. Kalacska, and B. Rivard. 2005. Secondary forest detection in a Neotropical dry forest landscape using Landsat 7 ETM+ and IKONOS imagery. Biotropica 37:498–507.

Arroyo-Mora, J. P., G. A. Sánchez-Azofeifa, B. Rivard, J. C. Calvo, and D. H. Janzen. 2005. Dynamics in landscape structure and composition for the Chorotega region, Costa Rica, from 1960 to 2000. Agriculture, Ecosystems & Environment 106:27–39.

Arroyo-Mora, J. P., M. Kalacska, R. Chazdon, D. L. Civco, G. Obando-Vargas, and A. A. Sanchun Hernández. 2008. Assessing recovery following selective logging of lowland tropical forests based on hyperspectral imagery. Pages 193–212 *in* M. Kalacska and G. A. Sánchez-Azofeifa, eds. Hyperspecral remote sensing of tropical and sub-tropical forests. CRC Press, New York.

Ashton, P., S. Gamage, I. Gunatilleke, and C. Gunatilleke. 1997. Restoration of a Sri Lankan rainforest: Using Caribbean pine *Pinus caribaea* as a nurse for establishing late-successional tree species. Journal of Applied Ecology 34:915–25.

Asner, G. P., and G. Goldstein. 1997. Correlating stem biomechanical properties of Hawaiian canopy trees with hurricane wind damage. Biotropica 29:145–50.

Asner, G. P., D. E. Knapp, E. N. Broadbent, P. J. C. Oliveira, M. Keller, and J. N. Silva. 2005. Selective logging in the Brazilian Amazon. Science 310:480–82.

Asner, G. P., J. Mascaro, H. C. Muller-Landau, G. Vieilledent, R. Vaudry, M. Rasamoelina, J. S. Hall, and M. Van Breugel. 2011. A universal airborne LiDAR approach for tropical forest carbon mapping. Oecologia 168:1147–60.

Asner, G. P., G. V. N. Powell, J. Mascaro, D. E. Knapp, J. K. Clark, J. Jacobson, T. Kennedy-Bowdoin, A. Balaji, G. Paez-Acosta, and E. Victoria. 2010. High-resolution forest carbon stocks and emissions in the Amazon. Proceedings of the National Academy of Sciences 107:16738–42.

Asner, G. P., T. K. Rudel, T. M. Aide, R. Defries, and R. Emerson. 2009. A contemporary assessment of change in humid tropical forests. Conservation Biology 23:1386–95.

Athens, J. S., H. D. Tuggle, J. V. Ward, and D. J. Welch. 2002. Avifaunal extinctions, vegetation change, and Polynesian impacts in prehistoric Hawai'i. Archaeology in Oceania 37:57–78.

Atran, S. 1993. Itza Maya tropical agro-forestry. Current Anthropology 34:633–700.

Aubréville, A. 1938. La forêt coloniale: Les forêts de l'Afrique occidentale française. Annales Académie Sciences Coloniale 9:1–245.

Avila-Cabadilla, L. D., K. E. Stoner, M. Henry, and M. Y. A. Anorve. 2009. Composition, structure and diversity of phyllostomid bat assemblages in different successional stages of a tropical dry forest. Forest Ecology and Management 258:986–96.

Aweto, A. O. 2013. Shifting cultivation and secondary succession in the tropics. CAB International, Boston, MA.

Azevedo-Ramos, C., O. de Carvalho Jr., B. D. do Amaral. 2006. Short-term effects of reduced-impact logging on eastern Amazon fauna. Forest Ecology and Management 232:26–35.

Bâ, A. M., R. Duponnois, B. Moyersoen, and A. G. Diedhiou. 2012. Ectomycorrhizal symbiosis of tropical African trees. Mycorrhiza 22:1–29.

Babweteera, F., and N. Brown. 2009. Can remnant frugivore species effectively disperse tree seeds in secondary tropical rain forests? Biodiversity and Conservation 18:1611–27.

Bailey, R. C., G. Head, M. Jenike, B. Owen, R. Rechtman, and E. Zechenter. 1989. Hunting and gathering in tropical rain forest: Is it possible? American Anthropologist 91:59–82.

Baker, P. J., and S. Bunyavejchewin. 2009. Fire behavior and fire effects across the forest landscape of continental Southeast Asia. Pages 311–34 *in* M. A. Cochrane, ed. Tropical fire ecology: Climate change, land use and ecosystem dynamics. Springer, New York.

Baker, P. J., S. Bunyavejchewin, C. D. Oliver, and P. S. Ashton. 2005. Disturbance history and historical stand dynamics of a seasonal tropical forest in western Thailand. Ecological Monographs 75:317–43.

Baker, P. J., S. Bunyavejchewin, and A. P. Robinson. 2008. The impacts of large-scale, low-intensity fires on the forests of continental South-east Asia. International Journal of Wildland Fire 17:782–92.

Baker, T., O. Phillips, Y. Malhi, S. Almeida, L. Arroyo, A. Di Fiore, T. Erwin, N. Higuchi, T. Killeen, and S. Laurance. 2004. Increasing biomass in Amazonian forest plots. Philosophical Transactions of the Royal Society B: Biological Sciences 359:353–65.

Bakker, M. A., G. Carreno-Rocabado, and L. Poorter. 2011. Leaf economics traits

predict litter decomposition of tropical plants and differ among land use types. Functional Ecology 25:473–83.
Balch, J. K., D. C. Nepstad, and L. M. Curran. 2009. Pattern and process: Fire-initiated grass invasion at Amazon transitional forest edges. Pages 481–502 *in* M. Cochrane, ed. Tropical fire ecology: Climate change, land use and ecosystem dynamics. Springer-Praxis, Heidelberg, Ger.
Balch, J. K., D. C. Nepstad, L. M. Curran, P. M. Brando, O. Portela, P. Guilherme, and J. D. Reuning-Scherer. 2011. Size, species, and fire behavior predict tree and liana mortality from experimental burns in the Brazilian Amazon. Forest Ecology and Management 261:68–77.
Balée, W. 1988. Indigenous adaptation to Amazonian palm forests. Principes 32:47–54.
———. 1989. The culture of Amazonian Forests. Advances in Economic Botany 7:1–21.
———. 1993. Indigenous transformation of Amazonian forests: An example from Maranhão, Brazil. L'Homme 33:231–54.
Balée, W., and D. G. Campbell. 1990. Evidence for the successional status of liana forest (Xingu River basin, Amazonian Brazil). Biotropica 22:36–47.
Baraloto, C., and P.-M. Forget. 2007. Seed size, seedling morphology, and response to deep shade and damage in Neotropical rain forest trees. American Journal of Botany 94:901–11.
Baraloto, C., P. M. Forget, and D. E. Goldberg. 2005. Seed mass, seedling size and Neotropical tree seedling establishment. Journal of Ecology 93:1156–66.
Baraloto, C., C. E. T. Paine, L. Poorter, J. Beauchene, D. Bonal, A. M. Domenach, B. Hérault, S. Patiño, J. C. Roggy, and J. Chave. 2010. Decoupled leaf and stem economics in rain forest trees. Ecology Letters 13:1338–47.
Bardhan, S., S. Jose, S. Biswas, K. Kabir, and W. Rogers. 2012. Homegarden agroforestry systems: An intermediary for biodiversity conservation in Bangladesh. Agroforestry Systems 85:29–34.
Barker, G., H. Barton, M. Bird, P. Daly, I. Datan, A. Dykes, L. Farr, et al. 2007. The 'human revolution' in lowland tropical Southeast Asia: The antiquity and behavior of anatomically modern humans at Niah Cave (Sarawak, Borneo). Journal of Human Evolution 52:243–61.
Barlow, J., T. A. Gardner, I. S. Araujo, T. C. Avila-Pires, A. B. Bonaldo, J. E. Costa, M. C. Esposito, et al. 2007. Quantifying the biodiversity value of tropical primary, secondary, and plantation forests. Proceedings of the Natational Academy of Sciences 104:18555–60.
Barlow, J., T. A. Gardner, A. C. Lees, L. Parry, and C. A. Peres. 2012. How pristine are tropical forests? An ecological perspective on the pre-Columbian human footprint in Amazonia and implications for contemporary conservation. Biological Conservation 151:45–49.
Barlow, J., B. O. Lagan, and C. A. Peres. 2003. Morphological correlates of fire-induced tree mortality in a Central Amazonian forest. Journal of Tropical Ecology 19:291–99.

Barlow, J., L. A. M. Mestre, T. A. Gardner, and C. A. Peres. 2007. The value of primary, secondary and plantation forests for Amazonian birds. Biological Conservation 136:212–31.

Barlow, J., and C. A. Peres. 2004. Ecological responses to El Niño induced surface fires in central Brazilian Amazonia: Management implications for flammable tropical forests. Philosophical Transactions of the Royal Society B: Biological Sciences 359:367–80.

———. 2006a. Consequences of cryptic and recurring fire disturbances for ecosystem structure and biodiversity in Amazonian forests. Pages 225–40 *in* W. Laurance and C. A. Peres, eds. Emerging threats to tropical forests. University of Chicago Press, Chicago.

———. 2006b. Effects of single and recurrent wildfires on fruit production and large vertebrate abundance in a central Amazonian forest. Biodiversity and Conservation 15:985–1012.

———. 2008. Fire-mediated dieback and compositional cascade in an Amazonian forest. Philosophical Transactions of the Royal Society B: Biological Sciences 363:1787–94.

Barlow, J., J. M. Silveira, and M. A. Cochrane. 2010. Fire scars on Amazonian trees: Exploring the cryptic fire history of the Ilha de Maracá. Biotropica 42:405–9.

Barr, C. M., and J. A. Sayer. 2012. The political economy of reforestation and forest restoration in Asia-Pacific: Critical issues for REDD+. Biological Conservation 154:9–19.

Barrera-Bassols, N., and V. M. Toledo. 2005. Ethnoecology of the Yucatec Maya: Symbolism, knowledge and management of natural resources. Journal of Latin American Geography 4:9–41.

Barreto, M., C. J. Souza, R. Nogueron, A. Anderson, and R. Salomao. 2006. Human pressure on Brazilian Amazon forests. World Resources Institute, Washington, DC.

Barron, A. R., D. W. Purves, and L. O. Hedin. 2011. Facultative nitrogen fixation by canopy legumes in a lowland tropical forest. Oecologia 165:511–20.

Barthlott, W., V. Schmit-Neuerburg, J. Nieder, and S. Engwald. 2001. Diversity and abundance of vascular epiphytes: A comparison of secondary vegetation and primary montane rain forest in the Venezuelan Andes. Plant Ecology 152:145–56.

Barton, H., and V. Paz. 2007. Subterranean diets in the tropical rain forests of Sarawak, Malaysia. Pages 50–77 *in* T. P. Denham, J. Iriarte, and L. Vrydaghs, eds. Rethinking agriculture: Archaeological and ethnoarchaeological perspectives. Left Coast Press, Walnut Creek, CA.

Bascompte, J., and P. Jordano. 2007. Plant-animal mutualistic networks: The architecture of biodiversity. Annual Review of Ecology, Evolution, and Systematics 38:567–93.

Basset, Y. 1996. Local communities of arboreal herbivores in Papua New Guinea: Predictors of insect variables. Ecology 77:1906–19.

Bayliss-Smith, T. 2007. The meaning of ditches: Interpreting the archaeological record from New Guinea using insights from ethnography. Pages 126–48 *in* T. P. Denham, J. Iriarte, and L. Vrydaghs, eds. Rethinking agriculture: Archaeological and ethnoarchaeological perspectives. Left Coast Press, Walnut Creek, CA.

Bayliss-Smith, T. P., E. Hviding, and T. C. Whitmore. 2003. Rainforest composition and histories of human disturbance in the Solomon Islands. AMBIO 32:346–52.

Bazzaz, F. A. 1979. The physiological ecology of plant succession. Annual Review of Ecology and Systematics 10:351–71.

———. 1991. Regeneration of tropical forests: Physiological responses of pioneer and secondary species. Pages 91–118 *in* A. Gómez-Pompa, T. C. Whitmore, and M. Hadley, eds. Rain forest regeneration and management. UNESCO/Parthenon, Paris.

Bazzaz, F. A., N. R. Chiariello, P. D. Coley, and L. F. Pitelka. 1987. Allocating resources to reproduction and defense. BioScience 37:58–67.

Bazzaz, F. A., and S. T. A. Pickett. 1980. Physiological ecology of tropical succession: A comparative review. Annual Review of Ecology and Systematics 11:287–310.

Beach, T., N. Dunning, S. Luzzadder-Beach, D. E. Cook, and J. Lohse. 2006. Impacts of the ancient Maya on soils and soil erosion in the central Maya Lowlands. Catena 65:166–78.

Beach, T., S. Luzzadder-Beach, N. Dunning, J. Jones, J. Lohse, T. Guderjan, S. Bozarth, S. Millspaugh, and T. Bhattacharya. 2009. A review of human and natural changes in Maya Lowland wetlands. Quaternary Science Reviews 28:1710–24.

Beard, J. S. 1955. The classification of tropical American vegetation types. Ecology 36:89–100.

Beck, H. 2005. Seed predation and dispersal by peccaries throughout the Neotropics and its consequences: A review and synthesis. Pages 77–100 *in* P. M. Forget, J. E. Lambert, P. E. Hulme, and S. B. Vander Wall, eds. Seed fate. CABI Publishing, Cambridge, MA.

Becknell, J. M., L. Kissing Kucek, and J. S. Powers. 2012. Aboveground biomass in mature and secondary seasonally dry tropical forests: A literature review and global synthesis. Forest Ecology and Management 276:88–95.

Behling, H. 1997. Late Quaternary vegetation, climate and fire history of the *Araucaria* forest and campos region from Serra Campos Gerais, Paraná State (South Brazil). Review of Palaeobotany and Palynology 97:109–21.

Behling, H., and H. Hooghiemstra. 2000. Holocene Amazon rainforest-savanna dynamics and climatic implications: High-resolution pollen record from Laguna Loma Linda in eastern Colombia. Journal of Quaternary Science 15:687–95.

Bellingham, P. J. 1991. Landforms influence patterns of hurricane damage: Evidence from Jamaican montane forests. Biotropica 23:427–33.

———. 2008. Cyclone effects on Australian rain forests: An overview. Austral Ecology 33:580–84.

Bellingham, P., V. Kapos, N. Varty, J. Healey, E. Tanner, D. Kelly, J. Dalling, L. Burns,

D. Lee, and G. Sidrak. 1992. Hurricanes need not cause high mortality: The effects of Hurricane Gilbert on forests in Jamaica. Journal of Tropical Ecology 8:217–23.

Bellingham, P. J., E. V. J. Tanner, and J. R. Healey. 1995. Damage and responsiveness of Jamaican montane tree species after disturbance by a hurricane. Ecology 76:2562–80.

———. 2005. Hurricane disturbance accelerates invasion by the alien tree *Pittosporum undulatum* in Jamaican montane rain forests. Journal of Vegetation Science 16:675–84.

Bellwood, P. 1995. Austronesian prehistory in Southeast Asia: Homeland, expansion and transformation. Pages 103–18 *in* P. Bellwood, J. J. Fox, and D. Tryon, eds. The Austronesians: Historical and comparative perspectives. Australian National University Press, Canberra, ACT, Australia.

———. 1997. Prehistory of the Indo-Malaysian Archipelago. Rev. ed. University of Hawai'i Press, Honolulu.

———. 2006. Asian farming diasporas? Agriculture, languages, and genes in China and Southeast Asia. Pages 96–118 *in* M. T. Stark, ed. Archaeology of Asia. Blackwell Publishing, Malden, MA.

Benavides, A. M., J. H. D. Wolf, and J. F. Duivenvoorden. 2006. Recovery and succession of epiphytes in upper Amazonian fallows. Journal of Tropical Ecology 22:705–17.

Bengtsson, J., P. Angelstam, T. Elmqvist, U. Emanuelsson, C. Folke, M. Ihse, F. Moberg, and M. Nystrôm. 2003. Reserves, resilience and dynamic landscapes. AMBIO 32:389–96.

Benítez-Malvido, J., M. Martínez-Ramos, and E. Ceccon. 2001. Seed rain vs. seed bank, and the effect of vegetation cover on the recruitment of tree seedlings in tropical successional vegetation. Pages 1–18 *in* G. Gottsberger and S. Liede, eds. Life forms and dynamics in tropical forests. J. Cramer, Stuttgart.

Bennett, E. L., and M. T. Gumal. 2001. The interrelationships of commercial logging, hunting, and wildlife in Sarawak: Recommendations for forest management. Pages 359–74 *in* R. Fimbel, A. Grajal, and J. G. Robinson, eds. The cutting edge: Conserving wildlife in logged tropical forest. Columbia University Press, New York.

Bentos, T. V., R. C. G. Mesquita, and G. B. Williamson. 2008. Reproductive phenology of Central Amazon pioneer trees. Tropical Conservation Science 1:186–203.

Bentos, T. V., R. C. G. Mesquita, J. L. C. Camargo, and G. B. Williamson. 2013. Seed and fruit tradeoffs—the economics of seed packaging in Amazon pioneers. Plant Ecology & Diversity. doi:10.1080/17550874.2012.740081.

Bentos, T. V., H. E. M. Nascimento, and G. B. Williamson. 2013. Tree seedling recruitment in Amazon secondary forest: Importance of topography and gap micro-site conditions. Forest Ecology and Management 287:140–46.

Berkes, F., and C. Folke, eds. 1998. Linking social and ecological systems: Management practices and social mechanisms for building resilience. Cambridge University Press, Cambridge.

Berry, N., O. Phillips, S. Lewis, J. Hill, D. Edwards, N. Tawatao, N. Ahmad, D. Magintan, C. Khen, and M. Maryati. 2010. The high value of logged tropical forests: Lessons from northern Borneo. Biodiversity and Conservation 19:985–97.
Berry, N. J., O. L. Phillips, R. C. Ong, and K. C. Hamer. 2008. Impacts of selective logging on tree diversity across a rainforest landscape: The importance of spatial scale. Landscape Ecology 23:915–29.
Bhagwat, S. A., K. J. Willis, H. J. B. Birks, and R. J. Whittaker. 2008. Agroforestry: A refuge for tropical biodiversity? Trends in Ecology & Evolution 23:261–67.
Bihn, J. H., G. Gebauer, and R. Brandl. 2010. Loss of functional diversity of ant assemblages in secondary tropical forests. Ecology 91:782–92.
Bihn, J. H., M. Verhaagh, M. Brandle, and R. Brandl. 2008. Do secondary forests act as refuges for old growth forest animals? Recovery of ant diversity in the Atlantic forest of Brazil. Biological Conservation 141:733–43.
Binford, M. W., M. Brenner, T. J. Whitmore, A. Higuera-Gundy, E. S. Deevey, and B. Leyden. 1987. Ecosystems, paleoecology and human disturbance in subtropical and tropical America. Quaternary Science Reviews 6:115–28.
Bischoff, W., D. A. Newbery, M. Lingenfelder, R. Schnaeckel, G. H. Petol, L. Madani, and C. E. Ridsdale. 2005. Secondary succession and dipterocarp recruitment in Bornean rain forest after logging. Forest Ecology and Management 218:174–92.
Bishop, P., D. Penny, M. Stark, and M. Scott. 2003. A 3.5 ka record of paleoenvironments and human occupation at Angkor Borei, Mekong Delta, Southern Cambodia. Geoarchaeology 18:359–93.
Bitencourt, A. J. V., and P. M. Krauspenhar. 2006. Possible prehistoric anthropogenic effect on *Araucaria angustifolia* (Bert.) O. Kuntze expansion during the late Holocene. Revista Brasileira de Paleontologia 9:109–16.
Blake, J. G., and B. A. Loiselle. 2001. Bird assemblages in second-growth and old-growth forests, Costa Rica: Perspectives from mist nets and point counts. Auk 118:304–26.
Blanc, L., M. Echard, B. Herault, D. Bonal, E. Marcon, J. Chave, and C. Baraloto. 2009. Dynamics of aboveground carbon stocks in a selectively logged tropical forest. Ecological Applications 19:1397–1404.
Blangy, S., and H. Mehta. 2006. Ecotourism and ecological restoration. Journal for Nature Conservation 14:233–36.
Blaser, J., A. Sarre, D. Poore, and S. Johnson. 2011. Status of tropical forest management 2011. International Tropical Timber Organization, Yokohama, Japan.
Bloor, J. M. G., and P. J. Grubb. 2003. Growth and mortality in high and low light: Trends among 15 shade-tolerant tropical rain forest tree species. Journal of Ecology 91:77–85.
Bobrowiec, P., and R. Gribel. 2009. Effects of different secondary vegetation types on bat community composition in Central Amazonia, Brazil. Animal Conservation 13:204–16.
Bock, C. 1882. The head hunters of Borneo: A narrative of travel up the Mahakkam

and down the Barito; also Journeyings in Sumatra. 2nd ed. Sampson Low, Martson, Searle & Rivington, London.

Boedhihartono, A. K., and J. A. Sayer. 2012. Forest landscape restoration: Restoring what and for whom? Pages 309–23 *in* J. A. Stanturf, D. Lamb, and P. Madsen, eds. Forest landscape restoration: Integrating natural and social sciences. Springer, New York.

Boissiére, M., D. Sheil, I. Basuki, M. Wan, and H. Le. 2009. Can engaging local people's interests reduce forest degradation in Central Vietnam? Biodiversity and Conservation 18:2743–57.

Bonal, D., C. Born, C. Brechet, S. Coste, E. Marcon, J. C. Roggy, and J. M. Guehl. 2007. The successional status of tropical rainforest tree species is associated with differences in leaf carbon isotope discrimination and functional traits. Annals of Forest Science 64:169–76.

Bond, W. J., and B. W. van Wilgen. 1996. Fire and plants. Springer, New York.

Bongers, F., L. Poorter, W. D. Hawthorne, and D. Sheil. 2009. The intermediate disturbance hypothesis applies to tropical forests, but disturbance contributes little to tree diversity. Ecology Letters 12:798–805.

Bonnell, T. R., R. Reyna-Hurtado, and C. A. Chapman. 2011. Post-logging recovery time is longer than expected in an East African tropical forest. Forest Ecology and Management 261:855–64.

Bonner, M. T., S. Schmidt, and L. P. Shoo. 2013. A meta-analytical global comparison of aboveground biomass accumulation between tropical secondary forests and monoculture plantations. Forest Ecology and Management 291:73–86.

Borges, S. H. 2007. Bird assemblages in secondary forests developing after slash-and-burn agriculture in the Brazilian Amazon. Journal of Tropical Ecology 23:469–77.

Boucher, D. 1990. Growing back after hurricanes. BioScience 40:163–66.

Boucher, D. H., J. H. Vandermeer, I. Granzow de la Cerda, M. A. Mallona, I. Perfecto, and N. Zamora. 2001. Post-agriculture versus post-hurricane succession in southeastern Nicaraguan rain forest. Plant Ecology 156:131–37.

Boucher, D. H., J. H. Vandermeer, M. A. Mallona, N. Zamora, and I. Perfecto. 1994. Resistance and resilience in a directly regenerating rain forest—Nicaraguan trees of the Vochysiaceae after Hurricane Joan. Forest Ecology and Management 68:127–36.

Bourke, R. M. 2007. Managing the species composition of fallows in Papua New Guinea by planting trees. Pages 379–88 *in* M. Cairns, ed. Voices from the forest: Integrating indigenous knowledge into sustainable upland farming. Resources for the Future, Washington, DC.

Bowen, M. E., C. A. McAlpine, A. P. N. House, and G. C. Smith. 2007. Regrowth forests on abandoned agricultural land: A review of their habitat values for recovering forest fauna. Biological Conservation 140:273–96.

Bowler, J. M., H. Johnston, J. M. Olley, J. R. Prescott, R. G. Roberts, W. Shawcross, and N. A. Spooner. 2003. New ages for human occupation and climatic change at Lake Mungo, Australia. Nature 421:837–40.

Bowman, D. M. J. S. 1998. The impact of aboriginal landscape burning on the Australian biota. New Phytologist 140:385–410.

———. 2000. Australian rainforests: Islands of green in a land of fire. Cambridge University Press, Cambridge.

Brady, C. J., and R. A. Noske. 2010. Succession in bird and plant communities over a 24-year chronosequence of mine rehabilitation in the Australian monsoon tropics. Restoration Ecology 18:855–64.

Brancalion, P. H., I. V. Cardozo, A. Camatta, J. Aronson, and R. R. Rodrigues. 2013. Cultural ecosystem services and popular perceptions of the benefits of an ecological restoration project in the Brazilian Atlantic Forest. Restoration Ecology. doi:10.1111/rec.12025.

Brancalion, P. H. S., R. R. Rodrigues, S. Gandolfi, P. Y. Kageyama, A. G. Nave, F. B. Gandara, L. M. Barbosa, and M. Tabarelli. 2010. Instrumentos legais podem contribuir para a restauração de florestas tropicais biodiversas. Revista Arvore 34:455–70.

Brandeis, T. J., E. H. Helmer, and S. N. Oswalt. 2007. The status of Puerto Rico's forests, 2003. Resource Bulletin SRS-119. US Forest Service Southern Research Station, Asheville, NC.

Bray, D. 2010. Forest cover dynamics and forest transitions in Mexico and Central America: Towards a "Great Restoration"? Pages 85–120 *in* H. Nagendra and J. Southworth, eds. Reforesting landscapes: Linking pattern and process. Springer, New York.

Bray, D. B., and P. Klepeis. 2005. Deforestation, forest transitions, and institutions for sustainability in Southeastern Mexico, 1900–2000. Environment and History 11:195–223.

Bray, D., and A. Velázquez. 2009. From displacement-based conservation to place-based conservation. Conservation and Society 7:11–14.

Bremer, L. L., and K. A. Farley. 2010. Does plantation forestry restore biodiversity or create green deserts? A synthesis of the effects of land-use transitions on plant species richness. Biodiversity and Conservation 19:3893–3915.

Brienen, R. J. W., E. Lebrija-Trejos, M. van Breugel, E. A. Pérez-García, F. Bongers, J. A. Meave, and M. Martínez-Ramos. 2009. The potential of tree rings for the study of forest succesion in southern Mexico. Biotropica 41:186–95.

Brncic, T. M., K. J. Willis, D. J. Harris, M. W. Telfer, and R. M. Bailey. 2009. Fire and climate change impacts on lowland forest composition in northern Congo during the last 2580 years from palaeoecological analysis of a seasonally flooded swamp. Holocene 19:79–89.

Brncic, T. M., K. J. Willis, D. J. Harris, and R. Washington. 2007. Culture or climate? The relative influences of past processes on the composition of the lowland Congo rainforest. Philosophical Transactions of the Royal Society B: Biological Sciences 362:229–42.

Brockerhoff, E. G., H. Jactel, J. A. Parrotta, C. P. Quine, and J. Sayer. 2008.

Plantation forests and biodiversity: Oxymoron or opportunity? Biodiversity and Conservation 17:925–51.

Brokaw, N. V. L. 1982. The definition of treefall gap and its effect on measures of forest dynamics. Biotropica 14:158–60.

———. 1987. Gap-phase regeneration of three pioneer tree species in a tropical forest. Journal of Ecology 75:9–19.

Brook, E. 2009. Palaeoclimate: Atmospheric carbon footprints? Nature Geoscience 2:170–72.

Brown, B. J., and J. J. Ewel. 1987. Herbivory in complex and simple tropical successional ecosystems. Ecology 68:108–16.

Brown, I. F., L. A. Martinelli, W. W. Thomas, M. Z. Moreira, C. Cid Ferreira, and R. A. Victoria. 1995. Uncertainty in the biomass of Amazonian forests: An example from Rondônia, Brazil. Forest Ecology and Management 75:175–89.

Brown, K. A., and J. Gurevitch. 2004. Long-term impacts of logging on forest diversity in Madagascar. Proceedings of the National Academy of Sciences 101:6045–49.

Brown, S., C. A. S. Hall, W. Knabe, J. Raich, M. C. Trexler, and P. Woomer. 1993. Tropical forests—their past, present, and potential future role in the terrestrial carbon budget. Water Air and Soil Pollution 70:71–94.

Brown, S., and A. E. Lugo. 1990. Tropical secondary forests. Journal of Tropical Ecology 6:1–32.

Bruijnzeel, L. A. 2004. Hydrological functions of tropical forests: Not seeing the soil for the trees? Agriculture Ecosystems & Environment 104:185–228.

Buckley, B., K. Anchukaitis, D. Penny, R. Fletcher, E. Cook, M. Sano, L. Nam, A. Wichienkeeo, T. Minh, and T. Hong. 2010. Climate as a contributing factor in the demise of Angkor, Cambodia. Proceedings of the National Academy of Sciences 107:6748.

Budowski, G. 1965. Distribution of tropical American rain forest species in the light of successional processes. Turrialba 15:40–42.

———. 1970. The distinction between old secondary and climax species in tropical Central American lowland forests. Tropical Ecology 11:44–48.

Burney, D. A., and T. F. Flannery. 2005. Fifty millennia of catastrophic extinctions after human contact. Trends in Ecology & Evolution 20:395–401.

Burslem, D., and M. D. Swaine. 2002. Forest dynamics and regeneration. Pages 577–83 *in* R. L. Chazdon and T. C. Whitmore, eds. Foundations of tropical forest biology: Classic papers with commentaries. University of Chicago Press, Chicago.

Burslem, D., and T. C. Whitmore. 1999. Species diversity, susceptibility to disturbance and tree population dynamics in tropical rain forest. Journal of Vegetation Science 10:767–76.

Burslem, D., T. C. Whitmore, and G. C. Brown. 2000. Short-term effects of cyclone impact and long-term recovery of tropical rain forest on Kolombangara, Solomon Islands. Journal of Ecology 88:1063–78.

Buschbacher, R., C. Uhl, and E. A. S. Serrão. 1988. Abandoned pastures in eastern

Amazonia. II. Nutrient stocks in the soil and vegetation. Journal of Ecology 76:682–99.
Bush, M. B., and P. A. Colinvaux. 1988. A 7000-year pollen record from the Amazon lowlands, Ecuador. Vegetatio 76:141–54.
———. 1994. Tropical forest disturbance: Paleoecological records from Darién, Panama. Ecology 75:1761–68.
Bush, M. B., P. E. de Oliveira, M. C. Miller, E. Moreno, and P. A. Colinvaux. 2004. Amazonian paleoecological histories: One hill, three watersheds. Palaeogeography, Palaeoclimatology, Palaeoecology 214:359–93.
Bush, M. B., M. C. Miller, P. E. de Oliveira, and P. A. Colinvaux. 2000. Two histories of environmental change and human disturbance in eastern lowland Amazonia. Holocene 10:543–53.
Bush, M. B., D. R. Piperno, P. A. Colinvaux, P. E. de Oliveira, L. A. Krissek, M. C. Miller, and W. E. Rowe. 1992. A 14,300-yr paleoecological profile of a lowland tropical lake in Panama. Ecological Monographs 63:251–76.
Bush, M. B., and M. R. Silman. 2007. Amazonian exploitation revisited: Ecological asymmetry and the policy pendulum. Frontiers in Ecology and the Environment 5:457–65.
———. 2008. Amazonia: A cultural landscape? Pages 51–61 *in* G. Nelson and I. Hronsky, eds. An international forum on sustainability. Arisztotelész, Budapest.
Bush, M. B., M. R. Silman, M. B. de Toledo, C. Listopad, W. D. Gosling, C. Williams, P. E. de Oliveira, and C. Krisel. 2007. Holocene fire and occupation in Amazonia: Records from two lake districts. Philosophical Transactions of the Royal Society B: Biological Sciences 362:209–18.
Bush, M. B., M. R. Silman, C. McMichael, and S. Saatchi. 2008. Fire, climate change and biodiversity in Amazonia: A Late-Holocene perspective. Philosophical Transactions of the Royal Society B: Biological Sciences. 363:1795–1802
Butler, R., F. Montagnini, and P. Arroyo. 2008. Woody understory plant diversity in pure and mixed native tree plantations at La Selva Biological Station, Costa Rica. Forest Ecology and Management 255:2251–63.
Butt, N., Y. Malhi, M. New, M. J. Macía, S. L. Lewis, G. Lopez-Gonzalez, W. F. Laurance, S. Laurance, R. Luizão, and A. Andrade. 2013. Shifting dynamics of climate-functional groups in old-growth Amazonian forests. Plant Ecology & Diversity. doi:10.1080/17550874.2012.715210.
Cairns, M. A., ed. 2007. Voices from the forest: Integrating indigenous knowledge into sustainable upland farming. Resources for the Future, Washington, DC.
Calmon, M., P. H. S. Brancalion, A. Paese, J. Aronson, P. Castro, S. C. da Silva, and R. R. Rodrigues. 2011. Emerging threats and opportunities for large-scale ecological restoration in the Atlantic Forest of Brazil. Restoration Ecology 19:154–58.
Calvo-Alvarado, J., B. McLennan, A. Sánchez-Azofeifa, and T. Garvin. 2009. Deforestation and forest restoration in Guanacaste, Costa Rica: Putting conservation policies in context. Forest Ecology and Management 258:931–40.
Camargo, J. L. C., I. D. K. Ferraz, and A. M. Imakawa. 2002. Rehabilitation of

degraded areas of Central Amazonia using direct sowing of forest tree seeds. Restoration Ecology 10:636–44.
Campbell, B. M., T. Lynam, and J. C. Hatton. 1990. Small-scale patterning in the recruitment of forest species during succession in tropical dry forest, Mozambique. Vegetatio 87:51–57.
Campbell, D. G., A. Ford, K. S. Lowell, J. Walker, J. K. Lake, C. Ocampo-Raeder, A. Townesmith, and M. Balick. 2006. The feral forests of the Eastern Peten. Pages 21–55 *in* W. Balee and C. L. Erickson, eds. Time and complexity in historical ecology. Columbia University Press, New York.
Campo, J., and R. Dirzo. 2003. Leaf quality and herbivory resposnes to soil nutrient addition in secondary tropical dry forests of Yucatán, Mexico. Journal of Tropical Ecology 19:525–30.
Campo, J., E. Solís, and M. G. Valencia. 2007. Litter N and P dynamics in two secondary tropical dry forests after relaxation of nutrient availability constraints. Forest Ecology and Management 252:33–40.
Canham, C., J. Thompson, J. Zimmerman, and M. Uriarte. 2010. Variation in susceptibility to hurricane damage as a function of storm intensity in Puerto Rican tree species. Biotropica 42:87–94.
Cannon, C. H., D. R. Peart, and M. Leighton. 1998. Tree species diversity in commercially logged Bornean rainforest. Science 281:1366–68.
Cantarello, E., A. C. Newton, R. A. Hill, N. Tejedor-Garavito, G. Williams-Linera, F. Lopez-Barrera, R. H. Manson, and D. J. Golicher. 2011. Simulating the potential for ecological restoration of dryland forests in Mexico under different disturbance regimes. Ecological Modelling 222:1112–28.
Cao, L., Z. Xiao, Z. Wang, C. Guo, J. Chen, and Z. Zhang. 2011. High regeneration capacity helps tropical seeds to counter rodent predation. Oecologia 166:997–1007.
Capers, R. S., and R. L. Chazdon. 2004. Rapid assessment of understory light availability in a wet tropical forest. Agricultural and Forest Meteorology 123:177–85.
Capers, R. S., R. L. Chazdon, A. R. Brenes, and B. V. Alvarado. 2005. Successional dynamics of woody seedling communities in wet tropical secondary forests. Journal of Ecology 93:1071–84.
Carcaillet, C., H. Almquist, H. Asnong, R. Bradshaw, J. Carrion, M. Gaillard, K. Gajewski, J. Haas, S. Haberle, and P. Hadorn. 2002. Holocene biomass burning and global dynamics of the carbon cycle. Chemosphere 49:845–63.
CARE. 2007. CARE in Guatemala: Reversing land degradation and building carbon stocks. Accessed July 24, 2012. http://www.careclimatechange.org/files/carbon/CaseStudy_Guatemala.pdf.
Carlo, T., J. Aukema, and J. Morales. 2007. Plant-frugivore interactions as spatially explicit networks: Integrating frugivore foraging with fruiting plant spatial patterns. Pages 369–90 *in* A. J. Dennis, E. W. Schupp, R. J. Green, and D. A.

Westcott, eds. Seed dispersal: Theory and its application in a changing world. CAB International, Wallingford, UK.
Carlo, T. A., and S. Yang. 2011. Network models of frugivory and seed dispersal: Challenges and opportunities. Acta Oecologica 37:619–24.
Carpenter, F. L., S. P. Mayorga, E. G. Quintero, and M. Schroeder. 2001. Land-use and erosion of a Costa Rican Ultisol affect soil chemistry, mycorrhizal fungi and early regeneration. Forest Ecology and Management 144:1–17.
Carreño-Rocabado, G., M. Peña-Claros, F. Bongers, A. Alarcón, J. C. Licona, and L. Poorter. 2012. Effects of disturbance intensity on species and functional diversity in a tropical forest. Journal of Ecology 100:1453–63.
Carrière, S. M., P. Letourmy, and D. B. McKey. 2002. Effects of remnant trees in fallows on diversity and structure of forest regrowth in a slash-and-burn agricultural system in southern Cameroon. Journal of Tropical Ecology 18:375–96.
Carto, S., A. J. Weaver, R. Hetherington, Y. Lam, and E. C. Wiebe. 2009. Out of Africa and into the ice age: On the role of global climate change in the late Pleistocene migration of early modern humans out of Africa. Journal of Human Evolution 56:139–51.
Cascante-Marín, A., J. H. D. Wolf, J. G. B. Oostermeijer, J. C. M. den Nijs, O. Sanahuja, and A. Duran-Apuy. 2006. Epiphytic bromeliad communities in secondary and mature forest in a tropical premontane area. Basic and Applied Ecology 7:520–32.
Cassano, C. R., M. C. M. Kierulff, and A. G. Chiarello. 2011. The cacao agroforests of the Brazilian Atlantic forest as habitat for the endangered maned sloth *Bradypus torquatus*. Mammalian Biology 76:243–50.
Castillo, M., B. Rivard, A. Sánchez-Azofeifa, J. Calvo-Alvarado, and R. Dubayah. 2012. LIDAR remote sensing for secondary tropical dry forest identification. Remote Sensing of Environment 121:132–43.
Castillo-Campos, G., G. Halffter, and C. E. Moreno. 2008. Primary and secondary vegetation patches as contributors to floristic diversity in a tropical deciduous forest landscape. Biodiversity and Conservation 17:1701–14.
Castro-Arellano, I., S. J. Presley, L. N. Saldanha, M. R. Willig, and J. M. Wunderle Jr. 2007. Effects of reduced impact logging on bat biodiversity in *terra firme* forest of lowland Amazonia. Biological Conservation 138:269–85.
Castro-Luna, A. A., V. J. Sosa, and G. Castillo-Campos. 2007. Bat diversity and abundance associated with the degree of secondary succession in a tropical forest mosaic in south-eastern Mexico. Animal Conservation 10:219–28.
Catenacci, L. S., K. M. De Vleeschouwer, and S. L. G. Nogueira-Filho. 2009. Seed dispersal by golden-headed lion tamarins *Leontopithecus chrysomelas* in Southern Bahian Atlantic Forest, Brazil. Biotropica 41:744–50.
Catterall, C. P., A. N. D. Freeman, J. Kanowski, and K. Freebody. 2012. Can active restoration of tropical rainforest rescue biodiversity? A case with bird community indicators. Biological Conservation 146:53–61.

Ceccon, E., P. Huante, and J. Campo. 2003. Effects of nitrogen and phosphorus fertilization on the survival and recruitment of seedlings of dominant tree species in two abandoned tropical dry forests in Yucatan, Mexico. Forest Ecology and Management 182:387–402.

Celentano, D., R. A. Zahawi, B. Finegan, F. Casanoves, R. Ostertag, R. J. Cole, and K. D. Holl. 2011. Restauración ecológica de bosques tropicales en Costa Rica: Efecto de varios modelos en la producción, acumulación y descomposición de hojarasca. Revista de Biologia Tropical 59:1323–36.

Celentano, D., R. A. Zahawi, B. Finegan, R. Ostertag, R. J. Cole, and K. D. Holl. 2011. Litterfall dynamics under different tropical forest restoration strategies in Costa Rica. Biotropica 43:279–87.

Chai, S., and E. Tanner. 2011. 150 year legacy of land use on tree species composition in old secondary forests of Jamaica. Journal of Ecology 99:113–21.

Chai, S. L., E. Tanner, and K. McLaren. 2009. High rates of forest clearance and fragmentation pre-and post-National Park establishment: The case of a Jamaican montane rainforest. Biological Conservation 142:2484–92.

Chambers, J. Q., G. P. Asner, D. C. Morton, L. O. Anderson, S. S. Saatch, F. D. B. Espírito-Santo, M. Palace, and C. Souza. 2007. Regional ecosystem structure and function: Ecological insights from remote sensing of tropical forests. Trends in Ecology & Evolution 22:414–23.

Chambers, J. Q., R. I. Negron-Juarez, G. C. Hurtt, D. M. Marra, and N. Higuchi. 2009. Lack of intermediate-scale disturbance data prevents robust extrapolation of plot-level tree mortality rates for old-growth tropical forests. Ecology Letters 12:E22–E25.

Chambers, J. Q., R. I. Negron-Juarez, D. M. Marra, A. Di Vittorio, J. Tews, D. Roberts, G. H. Ribeiro, S. E. Trumbore, and N. Higuchi. 2013. The steady-state mosaic of disturbance and succession across an old-growth Central Amazon forest landscape. Proceedings of the National Academy of Sciences 110:3949–54.

Chambers, J. Q., A. L. Robertson, V. M. C. Carneiro, A. J. N. Lima, M.-L. Smith, L. C. Plourde, and N. Higuchi. 2009. Hyperspectral remote detection of niche partitioning among canopy trees driven by blowdown gap disturbances in the Central Amazon. Oecologia 160:107–17.

Chapman, C. A., and L. J. Chapman. 2004. Unfavorable successional pathways and the conservation value of logged tropical forest. Biodiversity and Conservation 13:2089–2105.

Chapman, C. A., L. J. Chapman, L. Kaufman, and A. E. Zanne. 1999. Potential causes of arrested succession in Kibale National Park, Uganda: Growth and mortality of seedlings. African Journal of Ecology 37:81–92.

Charles-Dominique, P. 1986. Inter-relations between frugivorous vertebrates and pioneer plants: *Cecropia*, birds and bats in French Guyana. Pages 119–35 *in* A. Estrada and T. H. Fleming, eds. Frugivores and seed dispersal. Dr. W. Junk, Dordrecht, Neth.

Charles-Dominique, P., P. Blanc, D. Larpin, M. P. Ledru, B. Riéra, C. Sarthou,

M. Servant, and C. Tardy. 1998. Forest perturbations and biodiversity during the last ten thousand years in French Guiana. Acta Oecologica 19:295–302.

Chave, J., C. Andalo, S. Brown, M. A. Cairns, J. Q. Chambers, D. Eamus, H. Folster, et al. 2005. Tree allometry and improved estimation of carbon stocks and balance in tropical forests. Oecologia 145:87–99.

Chave, J., R. Condit, S. Aguilar, A. Hernandez, S. Lao, and R. Perez. 2004. Error propagation and scaling for tropical forest biomass estimates. Philosophical Transactions of the Royal Society B: Biological Sciences 359:409–20.

Chave, J., R. Condit, H. C. Muller-Landau, S. C. Thomas, P. S. Ashton, S. Bunyavejchewin, L. L. Co, et al. 2008. Assessing evidence for a pervasive alteration in tropical tree communities. Plos Biology 6:455–62.

Chave, J., D. Coomes, S. Jansen, S. L. Lewis, N. G. Swenson, and A. E. Zanne. 2009. Towards a worldwide wood economics spectrum. Ecology Letters 12:351–66.

Chave, J., H. C. Muller-Landau, T. R. Baker, T. A. Easdale, H. ter Steege, and C. O. Webb. 2006. Regional and phylogenetic variation of wood density across 2456 Neotropical tree species. Ecological Applications 16:2356–67.

Chazdon, R. L. 1992. Photosynthetic plasticity of two rain forest shrubs across natural gap transects. Oecologia 92:586–95.

———. 2003. Tropical forest recovery: Legacies of human impact and natural disturbances. Perspectives in Plant Ecology Evolution and Systematics 6:51–71.

———. 2008a. Beyond deforestation: Restoring forests and ecosystem services on degraded lands. Science 320:1458–60.

———. 2008b. Chance and determinism in tropical forest succession. Pages 384–408 *in* W. Carson and S. A. Schnitzer, eds. Tropical forest community ecology. John Wiley & Sons, West Sussex, UK.

Chazdon, R. L., and J. P. Arroyo. 2013. Tropical forests as complex adaptive systems. Pages 35–59 *in* C. Messier, K. J. Puettmann, and K. D. Coates, eds. Managing world forests as complex adaptive systems in the face of global change. Routledge, New York.

Chazdon, R. L., S. Careaga, C. Webb, and O. Vargas. 2003. Community and phylogenetic structure of reproductive traits of woody species in wet tropical forests. Ecological Monographs 73:331–48.

Chazdon, R. L., A. Chao, R. K. Colwell, S.-Y. Lin, N. Norden, S. G. Letcher, D. B. Clark, B. Finegan, and J. P. Arroyo. 2011. A novel statistical method for classifying habitat generalists and specialists. Ecology 92:1332–43.

Chazdon, R. L., and F. G. Coe. 1999. Ethnobotany of woody species in second-growth, old-growth, and selectively logged forests of northeastern Costa Rica. Conservation Biology 13:1312–22.

Chazdon, R. L., R. K. Colwell, J. S. Denslow, and M. R. Guariguata. 1998. Statistical methods for estimating species richness of woody regeneration in primary and secondary rain forests of NE Costa Rica. Pages 285–309 *in* F. Dallmeier and J. Comiskey, eds. Forest biodiversity research, monitoring and modeling: Conceptual background and Old World case studies. Parthenon, Paris.

Chazdon, R. L., and N. Fetcher. 1984. Photosynthetic light environments in a lowland tropical rainforest in Costa Rica. Journal of Ecology 72:553–64.

Chazdon, R. L., and C. B. Field. 1987. Determinants of photosynthetic capacity in six rainforest *Piper* species. Oecologia 73:222–30.

Chazdon, R. L., B. Finegan, R. S. Capers, B. Salgado-Negret, F. Casanoves, V. Boukili, and N. Norden. 2010. Composition and dynamics of functional groups of trees during tropical forest succession in northeastern Costa Rica. Biotropica 42:31–40.

Chazdon, R. L., C. A. Harvey, O. Komar, D. M. Griffith, B. G. Ferguson, M. Martínez-Ramos, H. Morales, et al. 2009. Beyond reserves: A research agenda for conserving biodiversity in human-modified tropical landscapes. Biotropica 41:142–53.

Chazdon, R. L., S. G. Letcher, M. van Breugel, M. Martinez-Ramos, F. Bongers, and B. Finegan. 2007. Rates of change in tree communities of secondary Neotropical forests following major disturbances. Philosophical Transactions of the Royal Society B: Biological Sciences 362:273–89.

Chazdon, R. L., R. W. Pearcy, D. W. Lee, and N. Fetcher. 1996. Photosynthetic responses of tropical forest plants to contrasting light environments. Pages 5–55 *in* S. S. Mulkey, R. L. Chazdon, and A. P. Smith, eds. Tropical forest plant ecophysiology. Chapman & Hall, New York.

Chazdon, R. L., C. A. Peres, D. Dent, D. Sheil, A. E. Lugo, D. Lamb, N. E. Stork, and S. E. Miller. 2009. The potential for species conservation in tropical secondary forests. Conservation Biology 23:1406–17.

Chazdon, R. L., A. Redondo Brenes, and B. Vílchez Alvarado. 2005. Effects of climate and stand age on annual tree dynamics in tropical second-growth rain forests. Ecology 86:1808–15.

Chazdon, R. L., B. Vilchez Alvarado, S. G. Letcher, A. Wendt, and U. U. Sezen. 2014. Effects of human activities on successional pathways: Case studies from lowland wet forests of Northeastern Costa Rica. Pages 129–39 *in* S. Hecht, K. Morrison, and C. Padoch, eds. The social lives of forests. University of Chicago Press, Chicago.

Chinea, J. D. 2002. Tropical forest succession on abandoned farms in the Humacao Municipality of eastern Puerto Rico. Forest Ecology and Management 167:195–207.

Chokkalingam, U., and W. de Jong. 2001. Secondary forest: A working definition and typology. International Forestry Review 3:19–26.

Chowdhury, R. R. 2007. Household land management and biodiversity: Secondary succession in a forest-agriculture mosaic in southern Mexico. Ecology and Society 12:31. http://www.ecologyandsociety.org/vol12/iss2/art31/.

Clark, D. A. 2002. Are tropical forests an important carbon sink? Reanalysis of the long-term plot data. Ecological Applications 12:3–7.

———. 2004. Sources or sinks? The responses of tropical forests to current and future climate and atmospheric composition. Philosophical Transactions of the Royal Society B: Biological Sciences 359:477–91.

———. 2007. Detecting tropical forests' responses to global climatic and atmospheric change: Current challenges and a way forward. Biotropica 39:4–19.

Clark, D. A., S. Brown, D. W. Kicklighter, J. Q. Chambers, J. R. Thomlinson, J. Ni, and E. A. Holland. 2001. Net primary production in tropical forests: An evaluation and synthesis of existing field data. Ecological Applications 11:371–84.
Clark, D. A., and D. B. Clark. 1992. Life history diversity of canopy and emergent trees in a Neotropical rain forest. Ecological Monographs 62:315–44.
———. 2001. Getting to the canopy: Tree height growth in a Neotropical rain forest. Ecology 82:1460–72.
Clark, D. A., D. B. Clark, R. M. Sandoval, and M. V. Castro C. 1995. Landscape-scale variation in community structure of Neotropical rain forest palms: Edaphic and human effects. Ecology 76:2581–94.
Clark, D. A., S. C. Piper, C. D. Keeling, and D. B. Clark. 2003. Tropical rain forest tree growth and atmospheric carbon dynamics linked to interannual temperature variation during 1984–2000. Proceedings of the National Academy of Sciences 100:5852–57.
Clark, D. B. 1990. The role of disturbance in the regeneration of Neotropical moist forests. Pages 291–315 *in* K. S. Bawa and M. Hadley, eds. Reproductive ecology of tropical forest plants. UNESCO/Parthenon, Paris.
———. 1996. Abolishing virginity. Journal of Tropical Ecology 33:385–92.
Clark, D. B., and D. A. Clark. 1996. Abundance, growth and mortality of very large trees in Neotropical lowland rain forest. Forest Ecology and Management 80:235–44.
———. 2000. Landscape-scale variation in forest structure and biomass in a tropical rain forest. Forest Ecology and Management 137:185–98.
Clark, D. B., D. A. Clark, and S. F. Oberbauer. 2010. Annual wood production in a tropical rain forest in NE Costa Rica linked to climatic variation but not to increasing CO_2. Global Change Biology 16:747–59.
Clark, K. E., and C. Uhl. 1987. Fishing and fire in the history of the upper Río Negro region of Venezuela. Human Ecology 15:1–26.
Clarke, W. C. 1966. Extensive to intensive shifting cultivation: A succession from New Guinea. Ethnology 5:347–59.
Clarke, W. C., and R. R. Thaman, eds. 1993. Agroforestry in the Pacific Islands: Systems for sustainability. United Nations University Press, Tokyo.
Cleary, D. F. R., and A. Priadjati. 2005. Vegetation responses to burning in a rain forest in Borneo. Plant Ecology 177:145–63.
Clement, C. R. 1999. 1492 and the loss of Amazonian crop genetic resources. I. The relation between domestication and human population decline. Economic Botany 53:188–202.
Clement, C. R., and A. B. Junqueira. 2010. Between a pristine myth and an impoverished future. Biotropica 42:534–36.
Clement, C. R., J. M. McCann, and N. J. H. Smith. 2003. Agrobiodiversity in Amazonia and its relationship with dark earths. Pages 159–78 *in* J. Lehmann, D. Kern, B. Glaser, and W. Woods, eds. Amazonian Dark Earths: Origin, Properties, and Management. Kluwer Academic Publishers, Dordrecht, Neth.

Clement, R. M., and S. P. Horn. 2001. Pre-Columbian land-use history in Costa Rica: A 3000-year record of forest clearance, agriculture and fires from Laguna Zoncho. Holocene 11:419–26.
Clements, F. E. 1916. Plant succession: An analysis of the development of vegetation. Publication no. 242. Carnegie Institute, Washington, DC.
Clewell, A. F., and J. Aronson. 2006. Motivations for the restoration of ecosystems. Conservation Biology 20:420–28.
Cochrane, M. A. 2001. Synergistic interactions between habitat fragmentation and fire in evergreen tropical forests. Conservation Biology 15:1515–21.
———. 2003. Fire science for rainforests. Nature 421:913–19.
Cochrane, M. A., A. Alencar, M. D. Schulze, C. M. Souza, D. C. Nepstad, P. Lefebvre, and E. A. Davidson. 1999. Positive feedbacks in the fire dynamic of closed canopy tropical forests. Science 284:1832–35.
Cochrane, M. A., and W. F. Laurance. 2002. Fire as a large-scale edge effect in Amazonian forests. Journal of Tropical Ecology 18:311–25.
Cochrane, M. A., and M. D. Schulze. 1999. Fire as a recurrent event in tropical forests of the eastern Amazon: Effects on forest structure, biomass, and species composition. Biotropica 31:2–16.
Cockburn, P. F. 1974. The origin of Sook Plain, Sabah. Malaysian Forester 37:61–63.
Cohen, A. L., B. M. P. Singhakumara, and P. M. S. Ashton. 1995. Releasing rain forest succession: A case study in the *Dicranopteris linearis* fernlands of Sri Lanka. Restoration Ecology 3:261–70.
Cohen, A. S., J. Stone, K. Beuning, L. Park, P. Reinthal, D. Dettman, C. Scholz, T. Johnson, J. King, and M. Talbot. 2007. Ecological consequences of early Late Pleistocene megadroughts in tropical Africa. Proceedings of the National Academy of Sciences 104:16422–27.
Cole, R. J. 2009. Postdispersal seed fate of tropical montane trees in an agricultural landscape, Southern Costa Rica. Biotropica 41:319–27.
Cole, R., K. Holl, C. Keene, and R. Zahawi. 2011. Direct seeding of late-successional trees to restore tropical montane forest. Forest Ecology and Management 261:1590–97.
Cole, R. J., K. D. Holl, and R. A. Zahawi. 2010. Seed rain under tree islands planted to restore degraded lands in a tropical agricultural landscape. Ecological Applications 20:1255–69.
Coley, P. D., and J. A. Barone. 1996. Herbivory and plant defenses in tropical forests. Annual Review of Ecology and Systematics 27:305–35.
Coley, P. D., J. P. Bryant, and F. S. I. Chapin. 1985. Resource availability and antiherbivore defense. Science 230:895–99.
Colinvaux, P. A., and P. E. D. Oliveira. 2000. Palaeoecology and climate of the Amazon basin during the last glacial cycle. Journal of Quaternary Science 15:347–56.
Comita, L. S., H. Muller-Landau, S. Aguilar, and S. Hubbell. 2010. Asymmetric density

dependence shapes species abundances in a tropical tree community. Science 329:330–32.
Comita, L. S., J. Thompson, M. Uriarte, I. Jonckheere, C. D. Canham, and J. K. Zimmerman. 2010. Interactive effects of land-use history and natural disturbance on seedling dynamics in a subtropical forest. Ecological Applications 20:1270–84.
Compton, S., S. Ross, and I. Thornton. 1994. Pollinator limitation of fig tree reproduction on the island of Anak Krakatau (Indonesia). Biotropica 26:180–86.
Condit, R., R. Sukumar, S. P. Hubbell, and R. B. Foster. 1998. Predicting population trends from size distributions: A direct test in a tropical tree community. American Naturalist 152:495–509.
Connell, J. H. 1971. On the roles of natural enemies in preventing competitive exclusion in some marine animals and in rain forest. Pages 298–312 *in* P. J. Den Boer and G. R. Grandwell, eds. Proceedings of the Advanced Study Institute on dynamics of numbers in population, Oosterbeek, Wageningen, Holland.
———. 1978. Diversity in tropical rain forests and coral reefs. Science 199:1302–10.
Connell, J. H., M. D. Lowman, and I. R. Noble. 1997. Subcanopy gaps in temperate and tropical forests. Australian Journal of Ecology 22:163–68.
Coomes, D. A., and P. J. Grubb. 2003. Colonization, tolerance, competition and seed-size variation within functional groups. Trends in Ecology & Evolution 18:283–91.
Coomes, O. T., F. Grimard, and G. J. Burt. 2000. Tropical forests and shifting cultivation: Secondary forest fallow dynamics among traditional farmers of the Peruvian Amazon. Ecological Economics 32:109–24.
Coomes, O. T., Y. Takasaki, and J. M. Rhemtulla. 2011. Land-use poverty traps identified in shifting cultivation systems shape long-term tropical forest cover. Proceedings of the National Academy of Sciences 108:13925–30.
Corbin, J. D., and K. D. Holl. 2012. Applied nucleation as a forest restoration strategy. Forest Ecology and Management 265:37–46.
Cordeiro, N. J., and H. F. Howe. 2003. Forest fragmentation severs mutualism between seed dispersers and an endemic African tree. Proceedings of the National Academy of Sciences 100:14052–56.
Cordeiro, N. J., H. J. Ndangalasi, J. P. McEntee, and H. F. Howe. 2009. Dispersal limitation and recruitment of an endemic African tree in a fragmented landscape. Ecology 90:1030–41.
Corlett, R. T. 1991. Plant succession on degraded land in Singapore. Journal of Tropical Forest Science 4:151–61.
———. 1992. The ecological transformation of Singapore, 1819–1990. Journal of Biogeography 19:411–20.
———. 1994. What is secondary forest? Journal of Tropical Ecology 10:445–47.
———. 1998. Frugivory and seed dispersal by vertebrates in the Oriental (Indomalayan) Region. Biological Reviews 73:413–48.
———. 2007. The impact of hunting on the mammalian fauna of tropical Asian forests. Biotropica 39:292–303.

———. 2011. Seed dispersal in Hong Kong, China: Past, present and possible futures. Integrative Zoology 6:97–109.

Corlett, R., and I. Turner. 1997. Long-term survival in tropical forest remnants in Singapore and Hong Kong. Pages 333–45 *in* W. F. Laurance and R. O. Bierregaard Jr., eds. Tropical forest remnants: Ecology, management, and conservation of fragmented communities. University of Chicago Press, Chicago.

Cosgrove, R., J. Field, and A. Ferrier. 2007. The archaeology of Australia's tropical rainforests. Palaeogeography, Palaeoclimatology, Palaeoecology 251:150–73.

Costa, F., and W. Magnusson. 2002. Selective logging effects on abundance, diversity, and composition of tropical understory herbs. Ecological Applications 12:807–19.

Costa, J. B. P., F. P. L. Melo, B. A. Santos, and M. Tabarelli. 2012. Reduced availability of large seeds constrains Atlantic forest regeneration. Acta Oecologica-International Journal of Ecology 39:61–66.

Cox, P. A., and T. Elmqvist. 2000. Pollinator extinction in the Pacific Islands. Conservation Biology 14:1237–39.

Cremaschi, D. G., R. D. Lasco, and R. J. P. Delfino. 2013. Payments for watershed protection services: Emerging lessons from the Philippines. Journal of Sustainable Development 6:90–103.

Crews, T. E., K. Kitayama, J. H. Fownes, R. H. Riley, D. A. Herbert, D. Mueller-Dombois, and P. M. Vitousek. 1995. Changes in soil phosphorus fractions and ecosystem dynamics across a long chronosequence in Hawaii. Ecology 76:1407–24.

Crk, T., M. Uriarte, F. Corsi, and D. Flynn. 2009. Forest recovery in a tropical landscape: What is the relative importance of biophysical, socioeconomic, and landscape variables? Landscape Ecology 24:629–42.

Cubiña, A., and T. M. Aide. 2001. The effect of distance from forest edge on seed rain and soil seed bank in a tropical pasture. Biotropica 33:260–67.

Culot, L., M. C. Huynen, P. Gérard, and E. W. Heymann. 2009. Short-term post-dispersal fate of seeds defecated by two small primate species (*Saguinus mystax* and *Saguinus fuscicollis*) in the Amazonian forest of Peru. Journal of Tropical Ecology 25:229–38.

Culot, L., F. Lazo, M. C. Huynen, P. Poncin, and E. W. Heymann. 2010. Seasonal variation in seed dispersal by tamarins alters seed rain in a secondary rain forest. International Journal of Primatology 31:553–69.

Culot, L., D. J. Mann, F. J. J. Muñoz Lazo, M. C. Huynen, and E. W. Heymann. 2011. Tamarins and dung beetles: An efficient diplochorous dispersal system in the Peruvian Amazonia. Biotropica 43:84–92.

Curran, L. M., I. Caniago, G. D. Paoli, D. Astianti, M. Kusneti, M. Leighton, C. E. Nirarita, and H. Haeruman. 1999. Impact of El Niño and logging on canopy tree recruitment in Borneo. Science 286:2184–88.

Curran, T. J., R. L. Brown, E. Edwards, K. Hopkins, C. Kelley, E. McCarthy, E. Pounds, R. Solan, and J. Wolf. 2008. Plant functional traits explain interspecific differences

in immediate cyclone damage to trees of an endangered rainforest community in north Queensland. Austral Ecology 33:451–61.
da Conceição Prates-Clark, C., R. M. Lucas, and J. R. dos Santos. 2009. Implications of land-use history for forest regeneration in the Brazilian Amazon. Canadian Journal of Remote Sensing 35:534–53.
Dalle, S. P., and S. de Blois. 2006. Shorter fallow cycles affect the availability of noncrop plant resources in a shifting cultivation system. Ecology and Society 11:2. http://www.ecology andsociety.org/vol11/iss2/art2/.
Dalling, J. W. 1994. Vegetation colonization of landslides in the Blue Mountains, Jamaica. Biotropica 26:392–99.
Dalling, J. W., and T. Brown. 2009. Long term persistence of pioneer species in tropical rain forest soil seed banks. American Naturalist 173:531–35.
Dalling, J. W., and J. S. Denslow. 1998. Soil seed bank composition along a forest chronosequence in seasonally moist tropical forest, Panama. Journal of Vegetation Science 9:669–78.
Dalling, J. W., and S. P. Hubbell. 2002. Seed size, growth rate and gap microsite conditions as determinants of recruitment success for pioneer species. Journal of Ecology 90:557–68.
Dalling, J. W., M. D. Swaine, and N. C. Garwood. 1998. Dispersal patterns and seed bank dynamics of pioneer trees in moist tropical forest. Ecology 79:564–78.
Dalling, J. W., K. Winter, S. P. Hubbell, J. L. Hamrick, J. D. Nason, and D. A. Murawski. 2001. The unusual life history of *Alseis blackiana*: A shade-persistent pioneer tree? Ecology 82:933–45.
Daly, C., E. H. Helmer, and M. Quiñones. 2003. Mapping the climate of Puerto Rico, Vieques and Culebra. International Journal of Climatology 23:1359–81.
Daniels, A. E. 2010. Forest expansion in northwest Costa Rica: Conjuncture of the global market, land-use intensification, and forest protection. Pages 227–52 *in* H. Nagendra and J. Southworth, eds. Reforesting landscapes: Linking pattern and process. Springer, Dordrecht, Neth.
Daniels, A. E., K. Bagstad, V. Esposito, A. Moulaert, and C. M. Rodriguez. 2010. Understanding the impacts of Costa Rica's PES: Are we asking the right questions? Ecological Economics 69:2116–26.
D'Antonio, C., and P. Vitousek. 1992. Biological invasions by exotic grasses, the grass/fire cycle, and global change. Annual Review of Ecology and Systematics 23:63–87.
Dao, T. H., C. T. Tran, and T. C. Le. 2001. Agroecology. Pages 51–83 *in* T. C. Le and A. T. Rambo, eds. Bright peaks, dark valleys: A comparative analysis of environmental and social conditions and development trends in five communities in Vietnam's northern mountain region. National Political Publishing House, Hanoi.
Davidson, E. A., C. J. R. de Carvalho, A. M. Figueira, F. Y. Ishida, J. Ometto, G. B. Nardoto, R. T. Saba, et al. 2007. Recuperation of nitrogen cycling in Amazonian forests following agricultural abandonment. Nature 447:995–99.
Davidson, E. A., C. J. R. de Carvalho, I. C. G. Vieira, R. D. Figueiredo, P. Moutinho,

F. Y. Ishida, M. T. P. dos Santos, J. B. Guerrero, K. Kalif, and R. T. Saba. 2004. Nitrogen and phosphorus limitation of biomass growth in a tropical secondary forest. Ecological Applications 14:S150-S163.

Davies, S. J. 1998. Photosynthesis of nine pioneer *Macaranga* species from Borneo in relation to life history. Ecology 79:2292–308.

———. 2001. Tree mortality and growth in 11 sympatric *Macaranga* species in Borneo. Ecology 82:920–32.

Davies, S. J., S. Cavers, B. Finegan, C. Navarro, and A. J. Lowe. 2010. Genetic consequences of multigenerational and landscape colonisation bottlenecks for a Neotropical forest pioneer tree, *Vochysia ferruginea*. Tropical Plant Biology 3:14–27.

Davies, S. J., and H. Semui. 2006. Competitive dominance in a secondary successional rain-forest community in Borneo. Journal of Tropical Ecology 22:53–64.

Dean, W. 1995. With broadax and firebrand: The destruction of the Brazilian Atlantic Forest. University of California Press, Berkeley.

DeFries, R., and C. Rosenzweig. 2010. Toward a whole-landscape approach for sustainable land use in the tropics. Proceedings of the National Academy of Sciences 107:19627–32.

de Gouvenain, R., and J. Silander Jr. 2003. Do tropical storm regimes influence the structure of tropical lowland rain forests? Biotropica 35:166–80.

de Jong, B. H. J., E. Esquivel Bazán, and S. Quechulpa Montalvo. 2007. Application of the "Climafor" baseline to determine leakage: The case of Scolel Té. Mitigation and Adaptation Strategies for Global Change 12:1153–68.

de Jong, W. 2002. Forest products and local forest management in West Kalimantan, Indonesia: Implications for conservation and development. Tropenbos International, Wageningen, Neth.

de Koning, F., M. Aguinaga, M. Bravo, M. Chiu, M. Lascano, T. Lozada, and L. Suarez. 2011. Bridging the gap between forest conservation and poverty alleviation: The Ecuadorian Socio Bosque program. Environmental Science & Policy 14:531–42.

Delacourt, C., D. Raucoules, S. Le MouÈlic, C. Carnec, D. Feurer, P. Allemand, and M. Cruchet. 2009. Observation of a large landslide on La Reunion Island using differential Sar interferometry (JERS and Radarsat) and correlation of optical (Spot5 and Aerial) images. Sensors 9:616–30.

Delang, C. O., and W. M. Li. 2013. Ecological succession on fallowed shifting cultivation fields: A review of the literature. Springer, New York.

de la Peña-Domene, M., C. Martínez-Garza, and H. F. Howe. 2013. Early recruitment dynamics in tropical restoration. Ecological Applications 23:1124–34.

del Castillo, R. F., A. Blanco-Macías, and A. Newton. 2007. Secondary succession under a slash-and-burn regime in a tropical montane cloud forest: Soil and vegetation characteristics. Pages 158–80 *in* A. C. Newton, ed. Biodiversity loss and conservation in fragmented forest landscapes: The forests of montane Mexico and temperate South America. CAB International, Wallingford, UK.

del Castillo, R. F., and M. A. Pérez Ríos. 2008. Changes in seed rain during secondary succession in a tropical montane cloud forest region in Oaxaca, Mexico. Journal of Tropical Ecology 24:433–44.
Delegue, M.-A., M. Fuhr, D. Schwartz, A. Mariotti, and R. Nasi. 2001. Recent origin of a large part of the forest cover in the Gabon coastal area based on stable carbon isotope data. Oecologia 129:106–13.
del Moral, R., and S. Y. Grishin. 1999. Volcanic disturbances and ecosystem recovery. Pages 137–60 *in* L. R. Walker, ed. Ecosystems of disturbed ground. Elsevier, Amsterdam, Neth.
Denevan, W. M. 1988. Measurement of abandoned terracing from air photos: Colca Valley, Peru. Yearbook (Conference of Latin Americanist Geographers) 14:20–30.
———. 1992a. The native population of the Americas in 1492. 2nd ed. University of Wisconsin Press, Madison.
———. 1992b. The pristine myth: The landscape of the Americas in 1492. Annals of the Association of American Geographers 82:369–85.
———. 1996. A bluff model of riverine settlement in prehistoric Amazonia. Annals of the Association of American Geographers 86:654–81.
———. 2001. Cultivated landscapes of native Amazonia and the Andes. Oxford University Press, New York.
———. 2007. Pre-European human impacts on tropical lowland environments. Pages 265–78 *in* T. T. Veblen, K. R. Young, and A. R. Orme, eds. The physical geography of South America. Oxford University Press. Oxford.
Denevan, W. M., J. M. Treacy, J. B. Alcorn, C. Padoch, J. Denslow, and S. F. Paitan. 1984. Indigenous agroforestry in the Peruvian Amazon: Bora Indian management of swidden fallows. Interciencia 9:346–57.
Denham, T., R. Fullagar, and L. Head. 2009. Plant exploitation on Sahul: From colonisation to the emergence of regional specialisation during the Holocene. Quaternary International 202:29–40.
Denham, T. P., S. G. Haberle, and C. Lentfer. 2004. New evidence and interpretations for early agriculture in Highland New Guinea. Antiquity 78:839–57.
Denham, T. P., S. G. Haberle, C. Lentfer, R. Fullagar, J. Field, M. Therin, N. Porch, and B. Winsborough. 2003. Origins of agriculture at Kuk Swamp in the Highlands of New Guinea. Science 301:189–93.
Dennis, R., G. Applegate, K. Kartawinata, A. Hoffmann, and G. Von Gemmingen. 2001. Large-scale fire: Creator and destroyer of secondary forests in Western Indonesia. Journal of Tropical Forest Science 13:786–99.
Denslow, J. S. 1978. Secondary succession in a Colombian rainforest: Strategies of species response across a disturbance gradient. PhD diss., University of Wisconsin, Madison.
———. 1987. Tropical rainforest gaps and tree species diversity. Annual Review of Ecology and Systematics 18:431–52.
———. 1995. Disturbance and diversity in tropical rain forests: The density effect. Ecological Applications 5:962–68.

———. 1996. Functional group diversity and responses to disturbance. Pages 127–51 *in* G. H. Orians, R. Dirzo, and J. H. Cushman, eds. Biodiversity and ecosystem processes in tropical forests. Ecological Studies. Vol. 122. Springer, Berlin.
Denslow, J. S., and R. L. Chazdon. 2002. Ecosystem ecology in the tropics. Pages 639–45 *in* R. L. Chazdon and T. C. Whitmore, eds. Foundations of tropical forest biology: Classic papers with commentaries. University of Chicago Press, Chicago.
Denslow, J. S., and S. Guzman. 2000. Variation in stand structure, light and seedling abundance across a tropical moist forest chronosequence, Panama. Journal of Vegetation Science 11:201–12.
Dent, D. H., and S. J. Wright. 2009. The future of tropical species in secondary forests: A quantitative review. Biological Conservation 142:2833–43.
de Oliveira, R. R. 2002. Ação antrópica e resultantes sobre a estrutura e composição da Mata Atlânica na Ilha Grande RJ. Rodriguesia 53:33–57.
de Rouw, A. 1993. Regeneration by sprouting in slash and burn rice cultivation, Taï rain forest, Côte d'Ivoire. Journal of Tropical Ecology 9:387–408.
Devall, M. S., B. R. Parresol, and S. J. Wright. 1995. Dendroecological analysis of *Cordia alliodora, Pseudobombax septenatum* and *Annona spraguei* in central Panama. IAWA Journal 16:411–24.
DeWalt, S. J., S. K. Maliakal, and J. S. Denslow. 2003. Changes in vegetation structure and composition along a tropical forest chronosequence: Implications for wildlife. Forest Ecology and Management 182:139–51.
DeWalt, S. J., S. A. Schnitzer, and J. S. Denslow. 2000. Density and diversity of lianas along a chronosequence in a central Panamanian lowland forest. Journal of Tropical Ecology 16:1–19.
Dewi, S., M. van Noordwijk, A. Ekadinata, and J. L. Pfund. 2013. Protected areas within multifunctional landscapes: Squeezing out intermediate land use intensities in the tropics? Land Use Policy 30:38–56.
Diamond, J. 2007. Easter Island revisited. Science 317:1692–94.
———. 2009. Archaeology: Maya, Khmer and Inca. Nature 461:479–80.
Dickinson, M. B., D. F. Whigham, and S. M. Hermann. 2000. Tree regeneration in felling and natural treefall disturbances in a semideciduous tropical forest in Mexico. Forest Ecology and Management 134:137–51.
Diemont, S. A. W., J. L. Bohn, D. D. Rayome, S. J. Kelsen, and K. Cheng. 2011. Comparisons of Mayan forest management, restoration, and conservation. Forest Ecology and Management 261:1696–1705.
Diemont, S. A. W., and J. F. Martin. 2009. Lacandon Maya ecosystem management: Sustainable design for subsistence and environmental restoration. Ecological Applications 19:254–66.
Diemont, S. A. W., J. F. Martin, S. I. Levy-Tacher, R. B. Nigh, P. R. Lopez, and J. D. Golicher. 2006. Lacandon Maya forest management: Restoration of soil fertility using native tree species. Ecological Engineering 28:205–12.
Dillehay, T. D. 1997. Monte Verde: A Late Pleistocene settlement in Chile. Vol. 2:

The archaeological context and interpretation. Smithsonian Institution Press, Washington, DC.

Ding, Y., and R. G. Zang. 2009. Effects of logging on the diversity of lianas in a lowland tropical rain forest in Hainan Island, South China. Biotropica 41:618–24.

Ding, Y., R. Zang, S. G. Letcher, S. Liu, and F. He. 2012. Disturbance regime changes the trait distribution, phylogenetic structure and community assembly of tropical rain forests. Oikos 121:1263–70.

Ding, Y., R. Zang, S. Liu, F. He, and S. G. Letcher. 2012. Recovery of woody plant diversity in tropical rain forests in southern China after logging and shifting cultivation. Biological Conservation. 145: 225–33.

Dinh Le, H., C. Smith, J. Herbohn, and S. Harrison. 2012. More than just trees: Assessing reforestation success in tropical developing countries. Journal of Rural Studies 28:5–19.

Dirzo, R., E. Mendoza, and P. Ortiz. 2007. Size-related differential seed predation in a heavily defaunated Neotropical rain forest. Biotropica 39:355–62.

Don, A., J. Schumacher, and A. Freibauer. 2011. Impact of tropical land use change on soil organic carbon stocks—a meta analysis. Global Change Biology 17:1658–70.

Donatti, C. I., P. R. Guimarães, and M. Galetti. 2009. Seed dispersal and predation in the endemic Atlantic rainforest palm *Astrocaryum aculeatissimum* across a gradient of seed disperser abundance. Ecological Research 24:1187–95.

Donatti, C. I., P. R. Guimarães, M. Galetti, M. A. Pizo, F. Marquitti, and R. Dirzo. 2011. Analysis of a hyper-diverse seed dispersal network: Modularity and underlying mechanisms. Ecology Letters 14:773–81.

Doust, S. J., P. D. Erskine, and D. Lamb. 2006. Direct seeding to restore rainforest species: Microsite effects on the early establishment and growth of rainforest tree seedlings on degraded land in the wet tropics of Australia. Forest Ecology and Management 234:333–43.

Douterlungne, D., S. Levy-Tacher, D. Golicher, and F. Dañobeytia. 2010. Applying indigenous knowledge to the restoration of degraded tropical rain forest clearings dominated by bracken fern. Restoration Ecology 18:322–29.

Dove, M. R. 1993. Smallholder rubber and swidden agriculture in Borneo: A sustainable adaptation to the ecology and economy of the tropical forest. Economic Botany 47:136–47.

Drake, J. B., R. O. Dubayah, D. B. Clark, R. G. Knox, J. B. Blair, M. A. Hofton, R. L. Chazdon, J. F. Weishampel, and S. D. Prince. 2002. Estimation of tropical forest structural characteristics using large-footprint lidar. Remote Sensing of Environment 79:305–19.

Dranzoa, C. 1998. The avifauna 23 years after logging in Kibale National Park, Uganda. Biodiversity and Conservation 7:777–97.

Drew, A. P., J. D. Boley, Y. H. Zhao, M. H. Johnston, and F. H. Wadsworth. 2009. Sixty-two years of charge in subtropical wet forest structure and composition at El Verde, Puerto Rico. Interciencia 34:34–40.

Duan, W.-J., H. Ren, S.-L. Fu, Q.-F. Guo, and J. Wang. 2008. Pathways and determinants of early spontaneous vegetation succession in degraded lowland of South China. Journal of Integrative Plant Biology 50:147–56.

Dubayah, R. O., S. L. Sheldon, D. B. Clark, M. A. Hofton, J. B. Blair, G. C. Hurtt, and R. L. Chazdon. 2010. Estimation of tropical forest height and biomass dynamics using lidar remote sensing at La Selva, Costa Rica. Journal of Geophysical Research 115:1–17.

Dudley, N., S. Mansourian, and D. Vallauri. 2005. Forest landscape restoration in context. Pages 3–7 *in* S. Mansourian, D. Vallauri, and N. Dudley, eds. Forest restoration in landscapes: Beyond planting trees. Springer, New York.

Dull, R. A. 2008. Unpacking El Salvador's ecological predicament: Theoretical templates and "long-view" ecologies. Global Environmental Change 18:319–29.

Dull, R. A., R. J. Nevle, W. I. Woods, D. K. Bird, S. Avnery, and W. M. Denevan. 2010. The Columbian encounter and the Little Ice Age: Abrupt land use change, fire, and greenhouse forcing. Annals of the Association of American Geographers 100:1–17.

Duncan, R. S., and C. A. Chapman. 1999. Seed dispersal and potential forest succession in abandoned agriculture in tropical Africa. Ecological Applications 9:998–1008.

Dunn, R. R. 2004a. Managing the tropical landscape: A comparison of the effects of logging and forest conversion to agriculture on ants, birds, and lepidoptera. Forest Ecology and Management 191:215–24.

———. 2004b. Recovery of faunal communities during tropical forest regeneration. Conservation Biology 18:302–9.

Dunning, N., and T. Beach. 2000. Stability and instability in prehispanic Maya landscapes. Pages 179–202 *in* D. L. Lentz, ed. Imperfect balance: Landscape transformations in the Precolumbian Americas. Columbia University Press. New York.

Dunning, N., T. Beach, and D. Rue. 1997. The paleoecology and ancient settlement of the Petexbatun region, Guatemala. Ancient Mesoamerica 8:255–66.

Dupuy, J. M., and R. L. Chazdon. 2006. Effects of vegetation cover on seedling and sapling dynamics in secondary tropical wet forests in Costa Rica. Journal of Tropical Ecology 22:65–76.

———. 2008. Interacting effects of canopy gap, understory vegetation and leaf litter on tree seedling recruitment and composition in tropical secondary forests. Forest Ecology and Management 255:3716–25.

Durán, S. M., and G. H. Kattan. 2005. A test of the utility of exotic tree plantations for understory birds and food resources in the Colombian Andes. Biotropica 37:129–35.

Durigan, G., V. L. Engel, J. M. Torezan, A. C. G. Melo, M. C. M. Marques, S. V. Martins, A. Reis, and F. R. Scarano. 2010. Legal rules for ecological restoration: An additional barrier to hinder the success of initiatives? Revista Arvore 34:471–85.

Durigan, G., and A. C. G. Melo. 2011. An overview of public policies and research on ecological restoration in the state of São Paulo, Brazil. Pages 325–55 *in* E. Figueiroa, ed. Biodiversity conservation in the Americas: Lessons and policy. Editorial FEN, Universidad de Chile, Santiago.

Durno, J., T. Deetes, and J. Rajchaprasit. 2007. Natural forest regeneration from an *Imperata* fallow: The case of Pakhasukjai. Pages 122–36 *in* M. Cairns, ed. Voices from the forest: Integrating indigenous knowledge into sustainable upland farming. Resources for the Future, Washington, DC.

Durst, P. B., P. Sajise, and R. N. Leslie, eds. 2011. Forests beneath the grass. Proceedings of the regional workshop on advancing the application of assisted natural regeneration for effective low-cost restoration. Food and Agriculture Organization of the United Nations, Bangkok, Thailand.

Eaton, J. M., and D. Lawrence. 2009. Loss of carbon sequestration potential after several decades of shifting cultivation in the Southern Yucatan. Forest Ecology and Management 258:949–58.

Eden, M. J., W. Bray, L. Herrera, and C. McEwan. 1984. *Terra Preta* soils and their archaeological context in the Caquetá Basin of Southeast Colombia. American Antiquity 49:125–40.

Edwards, D. P., F. A. Ansell, A. H. Ahmad, R. Nilus, and K. C. Hamer. 2009. The value of rehabilitating logged rainforest for birds. Conservation Biology 23:1628–33.

Edwards, D. P., T. H. Larsen, T. D. S. Docherty, F. A. Ansell, W. W. Hsu, M. A. Derhé, K. C. Hamer, and D. S. Wilcove. 2011. Degraded lands worth protecting: The biological importance of Southeast Asia's repeatedly logged forests. Proceedings of the Royal Society B: Biological Sciences 278:82–90.

Edwards, D. P., P. Woodcock, F. A. Edwards, T. H. Larsen, W. W. Hsu, S. Benedick, and D. S. Wilcove. 2012. Reduced-impact logging and biodiversity conservation: A case study from Borneo. Ecological Applications 22:561–71.

Eggeling, W. J. 1947. Observations on the ecology of the Budongo Rain Forest, Uganda. Journal of Ecology 34:20–87.

Egler, F. 1954. Vegetation science concepts I. Initial floristic composition, a factor in old-field vegetation development with 2 figs. Plant Ecology 4:412–17.

Elias, M., and C. Potvin. 2003. Assessing inter- and intra-specific variation in trunk carbon concentration for 32 Neotropical tree species. Canadian Journal of Forest Research 33:1039–45.

Ellsworth, D., and P. Reich. 1996. Photosynthesis and leaf nitrogen in five Amazonian tree species during early secondary succession. Ecology 77:581–94.

Elmqvist, T., M. Pyykönen, M. Tengö, F. Rakotondrasoa, E. Rabakonandrianina, and C. Radimilahy. 2007. Patterns of loss and regeneration of tropical dry forest in Madagascar: The social institutional context. PLoS ONE 2:e402. doi:10.1371/journal.pone.0000402 2.

Eltahir, E. A. B., and R. L. Bras. 1996. Precipitation recycling. Reviews of Geophysics 34:367–78.

Emery, K. F. 2007. Assessing the impact of ancient Maya animal use. Journal for Nature Conservation 15:184–95.

Endress, B. A., and J. D. Chinea. 2001. Landscape patterns of tropical forest recovery in the Republic of Palau. Biotropica 33:555–65.
Erickson, C. L. 2000. An artificial landscape-scale fishery in the Bolivian Amazon. Nature 408:190–93.
———. 2003. Historical ecology and future explorations. Pages 455–500 *in* J. Lehmann, D. Kern, B. Glaser, and W. Woods, eds. Amazonian dark earths: Origin, properties, management. Kluwer Academic Publishers, Dordrecht, Neth.
———. 2006. The domesticated landscapes of the Bolivian Amazon. Pages 235–78 *in* W. Balee and C. L. Erickson, eds. Time and complexity in historical ecology. Columbia University Press. New York.
———. 2008. Amazonia: The historical ecology of a domesticated landscape. Pages 157–83 *in* H. Silverman and W. Isbell, eds. Handbook of South American archaeology. Springer, New York.
Erickson, C. L., and W. Balée. 2006. The historical ecology of a complex landscape in Bolivia. Pages 187–233 *in* W. Balee and C. L. Erickson, eds. Time and complexity in historical ecology. Columbia University Press, New York.
Erickson, H., M. Keller, and E. A. Davidson. 2001. Nitrogen oxide fluxes and nitrogen cycling during postagricultural succession and forest fertilization in the humid tropics. Ecosystems 4:67–84.
Erickson, C. L., and J. H. Walker. 2009. Precolumbian causeways and canals as landesque capital. Pages 233–52 *in* J. E. Snead, C. Erickson, and J. A. Darling, eds. Landscapes of movement: Trails, paths, and roads in anthropological perspective. University of Pennsylvania Museum of Archaeology and Anthropology, Philadelphia.
Erskine, P. D., C. P. Catterall, D. Lamb, and J. Kanowski. 2007. Patterns and processes of old field reforestation in Australian rainforest landscapes. Pages 119–43 *in* V. A. Cramer and R. J. Hobbs, eds. Old fields: Dynamics and restoration of abandoned farmland. Island Press, Washington, DC.
Espírito-Santo, F., M. Keller, B. Braswell, B. Nelson, S. Frolking, and G. Vicente. 2010. Storm intensity and old-growth forest disturbances in the Amazon region. Geophysical Research Letters 37:L11403.
Estrada, A., and R. Coates-Estrada. 1991. Howler monkeys (*Alouatta palliata*), dung beetles (Scarabaeidae) and seed dispersal: Ecological interactions in the tropical rain forest of Los Tuxtlas, Mexico. Journal of Tropical Ecology 7:459–74.
Estrada, A., R. Coates Estrada, and D. Meritt Jr. 1994. Non flying mammals and landscape changes in the tropical rain forest region of Los Tuxtlas, Mexico. Ecography 17:229–41.
Etter, A., C. A. McAlpine, D. Pullar, and H. Possingham. 2005. Modeling the age of tropical moist forest fragments in heavily-cleared lowland landscapes of Colombia. Forest Ecology and Management 208:249–60.
Eva, H. D., F. Achard, R. Beuchle, E. de Miranda, S. Carboni, R. Seliger, M. Vollmar, W. A. Holler, O. T. Oshiro, and V. Barrena Arroyo. 2012. Forest cover changes

in tropical South and Central America from 1990 to 2005 and related carbon emissions and removals. Remote Sensing 4:1369–91.
Evans, D., C. Pottier, R. Fletcher, S. Hensley, I. Tapley, A. Milne, and M. Barbetti. 2007. A comprehensive archaeological map of the world's largest preindustrial settlement complex at Angkor, Cambodia. Proceedings of the National Academy of Sciences 104:14277–82.
Ewel, J. J. 1971. Biomass changes in early tropical succession. Turrialba 21:110–12.
———. 1976. Litter fall and leaf decomposition in a tropical forest succession in eastern Guatemala. Journal of Ecology 64:293–308.
———. 1977. Differences between wet and dry successional tropical ecosystems. Geo-Eco-Trop 1:103–17.
———. 1980. Tropical succession: Manifold routes to maturity. Supplement, Biotropica 12:2–7.
———. 1983. Succession. Pages 217–23 *in* F. B. Golley, ed. Tropical rain forest ecosystems. Elsevier, Amsterdam, Neth.
———. 1986. Designing agricultural ecosystems for the humid tropics. Annual Review of Ecology and Systematics 71:245–71.
Ewel, J. J., and S. W. Bigelow. 1996. Plant life-forms and tropical ecosystem functioning. Pages 101–26 *in* G. H. Orians, R. Dirzo, and J. H. Cushman, eds. Biodiversity and ecosystem processes in tropical forests. Springer, New York.
Ewel, J. J., and F. E. Putz. 2004. A place for alien species in ecosystem restoration. Frontiers in Ecology and the Environment 2:354–360.
Fagan, M., and R. S. DeFries. 2009. Measurement and monitoring of the world's forests: A review and summary of remote sensing technical capability, 2009–2015. Resources for the Future, Washington, DC.
Fahrney, K., O. Boonaphol, B. Keoboualapa, and S. Maniphone. 2007. Indigenous management of paper mulberry in swidden rice fields and fallows in Northern Lao PDR. Pages 475–89 *in* M. Cairns, ed. Voices from the Forest: Integrating indigenous knowledge into sustainable upland farming. Resources for the Future, Washington, DC.
Fairbairn, A. S., G. S. Hope, and G. R. Summerhayes. 2006. Pleistocene occupation of New Guinea's highland and subalpine environments. World Archaeology 38:371–86.
Fairbridge, R. W. 1976. Shellfish-eating preceramic Indians in Coastal Brazil. Science 191:353–59.
Fairhead, J., and M. Leach. 2009. Amazonian dark earths in Africa. Pages 265–78 *in* W. I. Woods, ed. Amazonian dark earths: Wim Sombroek's vision. Springer, New York.
Fall, P. L., and T. D. Drezner. 2011. Plant dispersal, introduced species, and vegetation change in the South Pacific kingdom of Tonga. Pacific Science 65:143–56.
FAO (Food and Agriculture Organization of the United Nations). 2003. Workshop on tropical secondary forest management in Africa: Reality and perspectives—proceedings. FAO, Nairobi, Kenya.

———. 2006a. Choosing a forest definition for the Clean Development Mechanism. Forests and climate change working paper no. 4. FAO, Rome, Italy.
———. 2006b. Global Forest Resources Assessment 2005. Progress towards sustainable forest management. FAO forestry paper no. 147. Food and Agriculture Organization of the United Nations, Rome.
———. 2010. Global forest resources assessment 2010. FAO forestry paper no. 163. FAO, Rome, Italy.
———. 2011. Assessing forest degradation: Towards the development of globally applicable guidelines. Forest resources assessment working paper no. 177. FAO, Rome, Italy.
Farwig, N., N. Sajita, and K. Böhning-Gaese. 2008. Conservation value of forest plantations for bird communities in western Kenya. Forest Ecology and Management 255:3885–92.
———. 2009. High seedling recruitment of indigenous tree species in forest plantations in Kakamega Forest, western Kenya. Forest Ecology and Management 257:143–50.
Fashing, P. J., N. Nguyen, P. Luteshi, W. Opondo, J. F. Cash, and M. Cords. 2012. Evaluating the suitability of planted forests for African forest monkeys: A case study from Kakamega forest, Kenya. American Journal of Primatology 74:77–90.
Faust, F., G. Gnecco, H. Mannstein, and J. Stamm. 2006. Evidence for the postconquest demographic collapse of the Americans in historic CO_2 levels. Earth Interactions 10:1–14.
Fay, J. M. 1997. The ecology, social organization, populations, habitat and history of the western lowland gorilla (*Gorilla gorilla gorilla* Savage and Wyman 1847). Washington University, St. Louis, MO.
Fearnside, P. M., R. I. Barbosa, and P. M. L. de Alencastro Graça. 2007. Burning of secondary forest in Amazonia: Biomass, burning efficiency and charcoal formation during land preparation for agriculture in Apiaú, Roraima, Brazil. Forest Ecology and Management 242:678–87.
Fearnside, P. M., and W. M. Guimarães. 1996. Carbon uptake by secondary forests in Brazilian Amazonia. Forest Ecology and Management 80:35–46.
Fedick, S. L. 1995. Land evaluation and ancient Maya land use in the Upper Belize River area, Belize, Central America. Latin American Antiquity 6:16–34.
Feeley, K. J., S. J. Davies, P. S. Ashton, S. Bunyavejchewin, M. N. N. Supardi, A. R. Kassim, S. Tan, and J. Chave. 2007. The role of gap phase processes in the biomass dynamics of tropical forests. Proceedings of the Royal Society B: Biological Sciences 274:2857–64.
Feeley, K., S. J. Wright, M. Nur Supardi, A. Kassim, and S. Davies. 2007. Decelerating growth in tropical forest trees. Ecology Letters 10:461–69.
Feinsinger, P. 1978. Ecological interactions between plants and hummingbirds in a successional tropical community. Ecological Monographs 48:269–87.
Feldpausch, T. R., C. D. Prates-Clark, E. C. M. Fernandes, and S. J. Riha. 2007.

Secondary forest growth deviation from chronosequence predictions in central Amazonia. Global Change Biology 13:967–79.
Feldpausch, T. R., S. J. Riha, E. C. M. Fernandes, and E. V. Wandelli. 2005. Development of forest structure and leaf area in secondary forests regenerating on abandoned pastures in Central Amazônia. Earth Interactions 9:1.
Feldpausch, T. R., M. A. Rondon, E. C. M. Fernandes, S. J. Riha, and E. Wandelli. 2004. Carbon and nutrient accumulation in secondary forests regenerating on pastures in central Amazonia. Ecological Applications 14:S164-S176.
Felton, A., A. Felton, W. Foley, and D. Lindenmayer. 2010. The role of timber tree species in the nutritional ecology of spider monkeys in a certified logging concession, Bolivia. Forest Ecology and Management 259:1642–49.
Felton, A., J. Wood, A. M. Felton, B. Hennessey, and D. B. Lindenmayer. 2008. Bird community responses to reduced-impact logging in a certified forestry concession in lowland Bolivia. Biological Conservation 141:545–55.
Ferguson, B. G., J. Vandermeer, H. Morales, and D. M. Griffith. 2003. Post-agricultural succession in El Peten, Guatemala. Conservation Biology 17:818–28.
Filippelli, G., C. Souch, S. Horn, and D. Newkirk. 2010. The pre-Colombian footprint on terrestrial nutrient cycling in Costa Rica: Insights from phosphorus in a lake sediment record. Journal of Paleolimnology 43:843–56.
Fimbel, R. A., A. Grajal, and J. G. Robinson. 2001. Logging and wildlife in the tropics: Impacts and options for conservation. Pages 667–95 *in* R. Fimbel, A. Grajal, and J. G. Robinson, eds. The cutting edge: Conserving wildlife in logged tropical forest. Columbia University Press, New York.
Finegan, B. 1992. The management potential of Neotropical secondary lowland rain forest. Forest Ecology and Management 47:295–321.
———. 1996. Pattern and process in Neotropical secondary forests: The first 100 years of succession. Trends in Ecology and Evolution 11:119–24.
Finegan, B., and M. Camacho. 1999. Stand dynamics in a logged and silviculturally treated Costa Rican rain forest, 1988–1996. Forest Ecology and Management 121:177–89.
Fisher, J., G. Hurtt, R. Thomas, and J. Chambers. 2008. Clustered disturbances lead to bias in large-scale estimates based on forest sample plots. Ecology Letters 11:554–63.
Fleming, T. H., R. Breitwisch, and G. H. Whitesides. 1987. Patterns of tropical vertebrate frugivore diversity. Annual Review of Ecology and Systematics 18:91–109.
Floren, A., and K. E. Linsenmair. 2001. The influence of anthropogenic disturbances on the structure of arboreal arthropod communities. Plant Ecology 153:153–67.
Flores, B. M., M. T. F. Piedade, and B. W. Nelson. 2013. Fire disturbance in Amazonian blackwater floodplain forests. Plant Ecology & Diversity. doi:10.1080/17550874.2012.716086.
Flynn, D. F. B., M. Uriarte, T. Crk, J. B. Pascarella, J. K. Zimmerman, T. M. Aide, and

M. A. C. Ortiz. 2010. Hurricane disturbance alters secondary forest recovery in Puerto Rico. Biotropica 42:149–57.
Foody, G. M., and P. J. Curran. 1994. Estimation of tropical forest extent and regenerative stage using remotely sensed data. Journal of Biogeography 21:223–44.
Ford, A. 2008. Dominant plants of the Maya forest and gardens of El Pilar: Implications for paleoenvironmental reconstructions. Journal of Ethnobiology 28:179–99.
Ford, A., and R. Nigh. 2009. Origins of the Maya forest garden: Maya resource management. Journal of Ethnobiology 29:213–36.
Fornara, D., and J. W. Dalling. 2005. Post-dispersal removal of seeds of pioneer species from five Panamanian forests. Journal of Tropical Ecology 21:79–84.
Foster, B. L., and D. Tilman. 2000. Dynamic and static views of succession: Testing the descriptive power of the chronosequence approach. Plant Ecology 146:1–10.
Foster, D. R. 2000. Conservation lessons and challenges from ecological history. Forest History Today, Fall, 2–11.
Foster, D. R., D. H. Knight, and J. F. Franklin. 1998. Landscape patterns and legacies resulting from large, infrequent forest disturbances. Ecosystems 1:497–510.
Foster, D. R., G. Motzkin, and B. Slater. 1998. Land-use history as long-term broad-scale disturbance: Regional forest dynamics in central New England. Ecosystems 1:96–119.
Foster, D. R., D. A. Orwig, and J. S. McLachlan. 1996. Ecological and conservation insights from reconstructive studies of temperate old-growth forests. Trends in Ecology and Evolution 11:419–23.
Foster, R. B., J. Arce B., and T. S. Wachter. 1986. Dispersal and the sequential plant communities in Amazonian Peru floodplain. Pages 357–70 *in* A. Estrada and T. H. Fleming, eds. Frugivores and seed dispersal. W. Junk, Dordrecht, Neth.
Fox, J. E. D. 1976. Constraints on the natural regeneration of tropical moist forest. Forest Ecology and Management 1:37–65.
Franklin, J. 1989. Toward a new forestry. American Forests, November/December, 37–44.
Franklin, J., D. Drake, K. McConkey, F. Tonga, and L. Smith. 2004. The effects of Cyclone Waka on the structure of lowland tropical rain forest in Vava'u, Tonga. Journal of Tropical Ecology 20:409–20.
Franklin, J., and S. J. Rey. 2007. Spatial patterns of tropical forest trees in Western Polynesia suggest recruitment limitations during secondary succession. Journal of Tropical Ecology 23:1–12.
Fraser, J. A., A. Junqueira, N. C. Kawa, C. P. Moraes, and C. R. Clement. 2011. Crop diversity on anthropogenic dark earths in central Amazonia. Human Ecology 39:395–406.
Fredeen, A., and C. Field. 1996. Ecophysiological constraints on the distribution of Piper species. Pages 597–618 *in* S. Mulkey, R. Chazdon and A. Smith, eds. Tropical forest plant ecophysiology. Chapman & Hall, New York.

Frederick, C. 2007. Chinampas cultivation in the Basin of Mexico: Observations on the evolution of form and function. Pages 107–24 *in* T. Thurston and C. Fisher, eds. Seeking a richer harvest: The archaeology of subsistence, intensification, innovation, and change. Springer, New York.

Fredericksen, T. S., and W. Pariona. 2002. Effect of skidder disturbance on commercial tree regeneration in logging gaps in a Bolivian tropical forest. Forest Ecology and Management 171:223–30.

Fredriksson, G., L. Danielsen, and J. Swenson. 2007. Impacts of El Niño related drought and forest fires on sun bear fruit resources in lowland dipterocarp forest of East Borneo. Biodiversity and Conservation 16:1823–38.

Froyd, C. A., and K. J. Willis. 2008. Emerging issues in biodiversity and conservation management: The need for a palaeoecological perspective. Quaternary Science Reviews 27:1723–32.

Fukushima, M., M. Kanzaki, M. Hara, T. Ohkubo, P. Preechapanya, and C. Choo-charoen. 2008. Secondary forest succession after the cessation of swidden cultivation in the montane forest area in Northern Thailand. Forest Ecology and Management 255:1994–2006.

Gadgil, M., F. Berkes, and C. Folke. 1993. Indigenous knowledge for biodiversity conservation. AMBIO 22:151–56.

Gadgil, M., and M. D. S. Chandran. 1988. On the history of Uttara Kannada forests. Pages 47–58 *in* J. Dargavel, K. Dixon, and N. Semle, eds. Changing tropical forests: Historical perspectives on today's challenges in Asia, Australia, and Oceania. Centre for Resource and Environmental Studies. Canberra, ACT, Australia.

Galindo-González, J., S. Guevara, and V. J. Sosa. 2000. Bat- and bird-generated seed rains at isolated trees in pastures in a tropical rainforest. Conservation Biology 14:1693–1703.

Gallardo-Cruz, J. A., J. A. Meave, E. J. González, E. E. Lebrija-Trejos, M. A. Romero-Romero, E. A. Pérez-García, R. Gallardo-Cruz, J. L. Hernández-Stefanoni, and C. Martorell. 2012. Predicting tropical dry forest successional attributes from space: Is the key hidden in image texture? PLoS One 7:e30506.

Galvão, L. S., F. J. Ponzoni, V. Liesenberg, and J. R. dos Santos. 2009. Possibilities of discriminating tropical secondary succession in Amazonia using hyperspectral and multiangular CHRIS/PROBA data. International Journal of Applied Earth Observation and Geoinformation 11:8–14.

Ganade, G. 2007. Processes affecting succession in old fields of Brazilian Amazonia. Pages 75–92 *in* V. A. Cramer and R. J. Hobbs, eds. Old fields: Dynamics and restoration of abandoned farmland. Island Press, Washington, DC.

Ganade, G., and V. K. Brown. 2002. Succession in old pastures of central Amazonia: Role of soil fertility and plant litter. Ecology 83:743–54.

Gannon, M. R., and M. R. Willig. 1994. The effects of Hurricane Hugo on bats of the Luquillo Experimental Forest of Puerto Rico. Biotropica 26:320–31.

Ganz, D. J., and P. B. Durst. 2003. Assisted natural regeneration: An overview. Pages 1–4 *in* P. C. Dugan, P. B. Durst, D. J. Ganz, and P. J. McKenzie, eds. Advancing assisted natural regeneration (ANR) in Asia and the Pacific. Food and Agriculture Organization of the United Nations, Bangkok, Thailand.

Garcia, C. A., S. A. Bhagwat, J. Ghazoul, C. D. Nath, K. M. Nanaya, C. G. Kushalappa, Y. Raghuramulu, R. Nasi, and P. Vaast. 2010. Biodiversity conservation in agricultural landscapes: Challenges and opportunities of coffee agroforests in the Western Ghats, India. Conservation Biology 24:479–88.

García-Barrios, L., Y. M. Galván-Miyoshi, I. A. Valdivieso-Pérez, O. R. Masera, G. Bocco, and J. Vandermeer. 2009. Neotropical forest conservation, agricultural intensification, and rural out-migration: The Mexican experience. BioScience 59:863–73.

García-Fernández, C., and M. A. Casado. 2005. Forest recovery in managed agroforestry systems: The case of benzoin and rattan gardens in Indonesia. Forest Ecology and Management 214:158–69.

Gardner, T. A., J. Barlow, R. Chazdon, R. M. Ewers, C. A. Harvey, C. A. Peres, and N. S. Sodhi. 2009. Prospects for tropical forest biodiversity in a human-modified world. Ecology Letters 12:561–82.

Gardner, T. A., J. Barlow, L. W. Parry, and C. A. Peres. 2007. Predicting the uncertain future of tropical forest species in a data vacuum. Biotropica 39:25–30.

Gardner, T. A., M. I. M. Hernández, J. Barlow, and C. A. Peres. 2008. Understanding the biodiversity consequences of habitat change: The value of secondary and plantation forests for Neotropical dung beetles. Journal of Applied Ecology 45:883–93.

Gardner, T. A., M. A. Ribeiro-Junior, J. Barlow, T. Cristina, S. Avila-Pires, M. S. Hoogmoed, and C. A. Peres. 2007. The value of primary, secondary, and plantation forests for a Neotropical herpetofauna. Conservation Biology 21:775–87.

Garrity, D. P. 2004. Agroforestry and the achievement of the Millennium Development Goals. Agroforestry Systems 61:5–17.

Garrity, D. P., M. Soekardi, M. Van Noordwijk, R. De La Cruz, P. Pathak, H. Gunasena, N. Van So, G. Huijun, and N. Majid. 1997. The *Imperata* grasslands of tropical Asia: Area, distribution, and typology. Agroforestry Systems 36:3–29.

Garwood, N. 1996. Functional morphology of tropical tree seedlings. Pages 59–130 *in* M. D. Swaine, ed. The ecology of tropical forest tree seedlings. Parthenon, Paris.

Garwood, N., D. Janos, and N. Brokaw. 1979. Earthquake-caused landslides: A major disturbance to tropical forests. Science 205:997–99.

Gaveau, D. L. A., H. Wandono, and F. Setiabudi. 2007. Three decades of deforestation in southwest Sumatra: Have protected areas halted forest loss and logging, and promoted re-growth? Biological Conservation 134:495–504.

Gehring, C., M. Denich, M. Kanashiro, and P. L. G. Vlek. 1999. Response of secondary vegetation in Eastern Amazonia to relaxed nutrient availability constraints. Biogeochemistry 45:223–41.

Gehring, C., M. Denich, and P. L. G. Vlek. 2005. Resilience of secondary forest regrowth after slash-and-burn agriculture in central Amazonia. Journal of Tropical Ecology 21:519–27.

Gehring, C., F. H. Muniz, and L. A. Gomes de Souza. 2008. Leguminosae along 2–25 years of secondary forest succession after slash-and-burn agriculture and in mature rain forest of Central Amazonia 1. Journal of the Torrey Botanical Society 135:388–400.

Gehring, C., P. L. G. Vlek, L. A. G. de Souza, and M. Denich. 2005. Biological nitrogen fixation in secondary regrowth and mature rainforest of central Amazonia. Agriculture, Ecosystems & Environment 111:237–52.

Gerwing, J. J., and E. Vidal. 2002. Changes in liana abundance and species diversity eight years after liana cutting and logging in an eastern Amazonian forest. Conservation Biology 16:544–48.

Geurts, R., A. Lillo, and T. Bisseling. 2012. Exploiting an ancient signalling machinery to enjoy a nitrogen fixing symbiosis. Current Opinion in Plant Biology 15:438–43.

Ghazoul, J. 2004. Alien abduction: Disruption of native plant pollinator interactions by invasive species. Biotropica 36:156–64.

Ghazoul, J., and D. Sheil. 2010. Tropical rain forest: Ecology, diversity, and conservation. Oxford University Press, Oxford.

Giambelluca, T. W. 2002. Hydrology of altered tropical forest. Hydrological Processes 16:1665–69.

Gibbs, H. K., S. Brown, J. O. Niles, and J. A. Foley. 2007. Monitoring and estimating tropical forest carbon stocks: Making REDD a reality. Environmental Research Letters 2:045023 (13 pp.).

Gibson, L., T. M. Lee, L. P. Koh, B. W. Brook, T. A. Gardner, J. Barlow, C. A. Peres, C. J. A. Bradshaw, W. F. Laurance, and T. E. Lovejoy. 2011. Primary forests are irreplaceable for sustaining tropical biodiversity. Nature 478:378–81.

Gilbert, B., S. J. Wright, H. C. Muller-Landau, K. Kitajima, and A. Hernandéz. 2006. Life history trade-offs in tropical trees and lianas. Ecology 87:1281–88.

Gilbert, L. E. 1980. Food web organization and the conservation of Neotropical diversity. Pages 11–33 *in* M. Soulé and B. Wilcox, eds. Conservation biology. Cambridge University Press, Cambridge.

Gill, R. 2001. The great Maya droughts: Water, life, and death. University of New Mexico Press, Albuquerque.

Girão, L., A. Lopes, M. Tabarelli, and E. Bruna. 2007. Changes in tree reproductive traits reduce functional diversity in a fragmented Atlantic forest landscape. PLoS One 2:908.

Glaser, B. 2007. Prehistorically modified soils of central Amazonia: A model for sustainable agriculture in the twenty-first century. Philosophical Transactions of the Royal Society B: Biological Sciences 362:187–96.

Glaser, B., and J. J. Birk. 2012. State of the scientific knowledge on properties and genesis of Anthropogenic dark earths in Central Amazonia (*terra preta de índio*). Geochimica et Cosmochimica Acta 82:39–51.

Glaser, B., L. Haumaier, G. Guggenberger, and W. Zech. 2001. The *Terra Preta* phenomenon: A model for sustainable agriculture in the humid tropics. Naturwissenschaften 88:37–41.

Gloor, M., O. L. Phillips, J. J. Lloyd, S. L. Lewis, Y. Malhi, T. R. Baker, G. Lopez-Gonzalez, et al. 2009. Does the disturbance hypothesis explain the biomass increase in basin-wide Amazon forest plot data? Global Change Biology 15:2418–30.
Gnecco, G., and S. Mora. 1997. Late Pleistocene/early Holocene tropical forest occupations at San Isidro and Pena Roja, Colombia. Antiquity 71:683–90.
Goldammer, J. G. 2007. History of equatorial vegetation fires and fire research in Southeast Asia before the 1997–98 episode: A reconstruction of creeping environmental changes. Mitigation and Adaptation Strategies for Global Change 12:13–32.
Goldammer, J. G., and B. Seibert. 1989. Natural rain forest fires in eastern Borneo during the Pleistocene and Holocene. Naturwissenschaften 76:518–20.
Golson, J. 1991. The New Guinea highlands on the eve of agriculture. Indo-Pacific Prehistory Association Bulletin 11:82–91.
Goman, M., and R. Byrne. 1998. A 5000-year record of agriculture and tropical forest clearance in the Tuxtlas, Veracruz, Mexico. Holocene 8:83–89.
Gomes, A. C. S., and F. J. Luizão. 2012. Leaf and soil nutrients in a chronosequence of second-growth forest in Central Amazonia: Implications for restoration of abandoned lands. Restoration Ecology 20:339–45.
Gómez-Pompa, A. 1971. Posible papel de la vegetación secundaria en la evolución de la flora tropical. Biotropica 3:125–35.
———. 1987. On Maya silviculture. Mexican Studies 3:1–17.
Gómez-Pompa, A., and A. Kaus. 1999. From pre-Hispanic to future conservation alternatives: Lessons from México. Proceedings of the National Academy of Sciences 96:5982–86.
Gómez-Pompa, A., J. S. Flores, and V. Sosa. 1987. The "Pet Kot": A man-made tropical forest of the Maya. Interciencia 12:10–15.
Gómez-Pompa, A., and C. Vázquez-Yanes. 1981. Successional studies of a rain forest in México. Pages 246–66 *in* D. C. West, H. H. Shugart, and D. B. Botkin, eds. Forest succession: Concepts and applications. Springer Verlag, New York.
———. 1985. Estudios sobre la regeneración de selvas en regiones cálido-húmedas de México. Pages 1–25 *in* A. Gómez-Pompa and S. Del Amo, eds. Investigaciones sobre la regeneración de Selvas Altas en Veracruz, México. Vol. 2. Editorial Alhambra, Xalapa, Mexico.
Gonzales, R. S., N. R. Ingle, D. A. Lagunzad, and T. Nakashizuka. 2009. Seed dispersal by birds and bats in lowland Philippine forest successional area. Biotropica 41:452–58.
Goosem, S., and N. I. J. Tucker. 1995. Repairing the rainforest: Theory and practice of rainforest re-establishment in North Queensland's wet tropics. Wet Tropics Management Authority, Cairns, Australia.
Gorchov, D. L., F. Cornejo, C. Ascorra, and M. Jaramillo. 1993. The role of seed dispersal in the natural regeneration of rain forest after strip-cutting in the Peruvian Amazon. Vegetatio 107/108:339–49.

———. 1995. Dietary overlap between frugivorous birds and bats in the Peruvian Amazon. Oikos 74:235–50.

Gotelli, N. J., and R. K. Colwell. 2001. Quantifying biodiversity: Procedures and pitfalls in the measurements and comparison of species richness. Ecology Letters 4:379–91.

GPFLR. 2012. Global Partnership on Forest Landscape Restoration. Accessed July 28, 2012. http://www.ideastransformlandscapes.org/.

Grabowski, Z. J., and R. L. Chazdon. 2012. Beyond carbon: Redefining forests and people in the global ecosystem services market. SAPIENS 5. http://sapiens.revues.org/1246.

Graham, E. 2006. A Neotropical framework for terra preta. Pages 57–86 *in* W. Balée and C. L. Erickson, eds. Time and complexity in historical ecology. Columbia University Press, New York.

Grainger, A. 2010. The bigger picture—tropical forest change in context, concept and practice. Pages 15–43 in H. Nagendra and J. Southworth, eds. Reforesting landscapes: Linking pattern and process. Springer, Dordrecht, Neth.

Grau, H. R., T. M. Aide, J. K. Zimmerman, J. R. Thomlinson, E. Helmer, and X. M. Zou. 2003. The ecological consequences of socioeconomic and land-use changes in postagriculture Puerto Rico. BioScience 53:1159–68.

Grau, H. R., M. F. Arturi, A. D. Brown, and P. G. Acenolaza. 1997. Floristic and structural patterns along a chronosequence of secondary forest succession in Argentinean subtropical montane forests. Forest Ecology and Management 95:161–71.

Grayson, D. K. 2001. The archaeological record of human impacts on animal populations. Journal of World Prehistory 15:1–68.

Grimbacher, P. S., and C. P. Catterall. 2007. How much do site age, habitat structure and spatial isolation influence the restoration of rainforest beetle species assemblages? Biological Conservation 135:107–18.

Griscom, H. P., and M. S. Ashton. 2011. Restoration of dry tropical forests in Central America: A review of pattern and process. Forest Ecology and Management 261:1564–79.

Grogan, J., and J. Galvão. 2006. Factors limiting post-logging seedling regeneration by big-leaf mahogany (*Swietenia macrophylla*) in southeastern Amazonia, Brazil, and implications for sustainable management. Biotropica 38:219–28.

Grubb, P. J. 1977. The maintenance of species richness in plant communities: The importance of the regeneration niche. Biological Reviews 52:107–45.

Guadarrama, P., S. Castillo-Argüero, J. A. Ramos-Zapata, S. Camargo-Ricalde, and J. Álvarez-Sánchez. 2008. Propagules of arbuscular mycorrhizal fungi in a secondary dry forest of Oaxaca, Mexico. Revista de Biología Tropical 56:269–77.

Guariguata, M. R. 1990. Landslide disturbance and forest regeneration in the Upper Luquillo mountains of Puerto Rico. Journal of Ecology 78:814–32.

———. 1999. Early response of selected tree species to liberation thinning in a young secondary forest in northeastern Costa Rica. Forest Ecology and Management 124:255–61.

Guariguata, M. R., J. J. R. Adame, and B. Finegan. 2000. Seed removal and fate in two selectively logged lowland forests with constrasting protection levels. Conservation Biology 14:1046–54.
Guariguata, M. R., H. Arias-Le Claire, and G. Jones. 2002. Tree seed fate in a logged and fragmented forest landscape, northeastern Costa Rica. Biotropica 34:405–15.
Guariguata, M., R. Chazdon, J. Denslow, J. Dupuy, and L. Anderson. 1997. Structure and floristics of secondary and old-growth forest stands in lowland Costa Rica. Plant Ecology 132:107–20.
Guariguata, M., and J. M. Dupuy. 1997. Forest regeneration in abandoned logging roads in lowland Costa Rica. Biotropica 29:15–28.
Guariguata, M., and R. Ostertag. 2001. Neotropical secondary forest succession: Changes in structural and functional characteristics. Forest Ecology and Management 148:185–206.
Guerrero, A. C., and P. L. B. Da Rocha. 2010. Passive restoration in biodiversity hotspots: Consequences for an Atlantic rainforest lizard taxocene. Biotropica 42:379–87.
Guevara, S., J. Meave, P. Moreno-Casasola, and J. Laborde. 1992. Floristic composition and structure of vegetation under isolated trees in Neotropical pastures. Journal of Vegetation Science 3:655–64.
Guevara, S., S. E. Purata, and E. van der Maarl. 1986. The role of remnant forest trees in tropical secondary succession. Vegetatio 66:77–84.
Guix, J. C. 2009. Amazonian forests need Indians and Caboclos. Orsis 24:33–40.
Gunaratne, A., C. Gunatilleke, I. Gunatilleke, H. Madawala Weerasinghe, and D. Burslem. 2010. Barriers to tree seedling emergence on human-induced grasslands in Sri Lanka. Journal of Applied Ecology 47:157–65.
Günter, S., M. Weber, R. Erreis, and N. Aguirre. 2007. Influence of distance to forest edges on natural regeneration of abandoned pastures: A case study in the tropical mountain rain forest of Southern Ecuador. European Journal of Forest Research 126:67–75.
Guo, L., and R. Gifford. 2002. Soil carbon stocks and land use change: A meta analysis. Global Change Biology 8:345–60.
Haberle, S. G. 1996. Palaeoenvironmental changes in the eastern highlands of Papua New Guinea. Archaeology in Oceania 31:1–11.
———. 2003. The emergence of an agricultural landscape in the highlands of New Guinea. Archaeology in Oceania 38:149–58.
———. 2007. Prehistoric human impact on rainforest biodiversity in highland New Guinea. Philosophical Transactions of the Royal Society B: Biological Sciences 362:219–28.
Haberle, S. G., and B. David. 2004. Climates of change: Human dimensions of Holocene environmental change in low latitudes of the PEPII transect. Quaternary International 118–19:165–79.
Haberle, K. H., G. S. Hope, and S. van der Kaars. 2001. Biomass burning in Indonesia

and Papua New Guinea: Natural and human induced fire events in the fossil record. Palaeogeography Palaeoclimatology Palaeoecology 171:259–68.
Haberle, S. G., and M.-P. Ledru. 2001. Correlations among charcoal records of fires from the past 16,000 years in Indonesia, Papua New Guinea, and Central and South America. Quaternary Research 55:97–104.
Hacke, U. G., J. S. Sperry, W. T. Pockman, S. D. Davis, and K. A. McCulloch. 2001. Trends in wood density and structure are linked to prevention of xylem implosion by negative pressure. Oecologia 126:457–61.
Hall, A. 2012. Forests and climate change: The social dimensions of REDD in Latin America. Edward Elgar Publishing, Northampton, MA.
Hall, J., and M. Swaine. 1976. Classification and ecology of closed-canopy forest in Ghana. Journal of Ecology 64:913–51.
Hall, J. M., T. Van Holt, A. E. Daniels, V. Balthazar, and E. F. Lambin. 2012. Trade-offs between tree cover, carbon storage and floristic biodiversity in reforesting landscapes. Landscape Ecology 27:1135–47.
Halpern, C. B. 1988. Early successional pathways and the resistance and resilience of forest communities. Ecology 69:1703–15.
Hammond, D. S. 1995. Post-dispersal seed and seedling mortality of tropical dry forest trees after shifting agriculture, Chiapas, Mexico. Journal of Tropical Ecology 11:295–313.
Hammond, D. S., and H. ter Steege. 1998. Propensity for fire in Guianan rainforests. Conservation Biology 12:944–47.
Hammond, D. S., H. ter Steege, and K. van der Borg. 2006. Upland soil charcoal in the wet tropical forests of central Guyana. Biotropica 39:153–60.
Hammond, N. 1978. The myth of the milpa: Agricultural expansion in the Maya lowlands. Pages 23–34 *in* P. D. Harrison and B. L. Turner II, eds. Pre-hispanic Maya agriculture. University of New Mexico Press, Albuquerque.
Hansen, P. K., H. Sodarak, and S. Savathvong. 2007. Teak production by shifting cultivators in Northern Lao PDR. Pages 414–24 *in* M. Cairns, ed. Voices from the forest: Integrating indigenous knowledge into sustainable upland farming. Resources for the Future, Washington, DC.
Hanski, I., and Y. Cambefort. 1991. Dung beetle ecology. Princeton University Press, Princeton, NJ.
Hanson, T., S. Brunsfeld, and B. Finegan. 2006. Variation in seedling density and seed predation indicators for the emergent tree *Dipteryx panamensis* in continuous and fragmented rain forest. Biotropica 38:770–74.
Hardesty, B. D., S. P. Hubbell, and E. Bermingham. 2006. Genetic evidence of frequent long-distance recruitment in a vertebrate-dispersed tree. Ecology Letters 9:516–25.
Harms, K. E., and J. W. Dalling. 1997. Damage and herbivory tolerance through resprouting as an advantage of large seed size in tropical trees and lianas. Journal of Tropical Ecology 13:617–21.

Harrison, G. W. 1979. Stability under environmental stress: Resistance, resilience, persistence, and variability. American Naturalist 113:659–69.
Hart, T., J. Hart, M. Dechamps, M. Fournier, and M. Ataholo. 1996. Changes in forest composition over the last 4000 years in the Ituri Basin, Zaire. Pages 545–63 *in* L. J. G. van der Maesen, X. M. van der Burgt, and J. M. van Medenback de Kooy, eds. The biodiversity of African plants. Kluwer Academic Publishers, Dordrecht, Neth.
Hartshorn, G. S. 1978. Treefalls and tropical forest dynamics. Pages 616–38 *in* P. B. Tomlinson and M. H. Zimmerman, eds. Tropical trees as living systems. Cambridge University Press, Cambridge.
———. 1980. Neotropical forest dynamics. Supplement, Biotropica 12:23–30.
Hartter, J., J. Southworth, and M. Binford. 2010. Parks as a mechanism to maintain and facilitate recovery of forest cover: Examining reforestation, forest maintenance and productivity in Uganda. Pages 275–96 *in* H. Nagendra and J. Southworth, eds. Reforesting landscapes: Linking pattern and process. Springer, Dordrecht, Neth.
Harvey, C. A., O. Komar, R. Chazdon, B. G. Ferguson, B. Finegan, D. M. Griffith, M. Martinez-Ramos, et al. 2008. Integrating agricultural landscapes with biodiversity conservation in the Mesoamerican hotspot. Conservation Biology 22:8–15.
Harvey, C. A., A. Medina, D. M. Sanchez, S. Vilchez, B. Hernandez, J. C. Saenz, J. M. Maes, F. Casanoves, and F. L. Sinclair. 2006. Patterns of animal diversity in different forms of tree cover in agricultural landscapes. Ecological Applications 16:1986–99.
Hassler, S. K., B. Zimmermann, M. van Breugel, J. S. Hall, and H. Elsenbeer. 2011. Recovery of saturated hydraulic conductivity under secondary succession on former pasture in the humid tropics. Forest Ecology and Management 261:1634–42.
Haug, G., D. Gunther, L. Peterson, D. Sigman, K. Hughen, and B. Aeschlimann. 2003. Climate and the collapse of Maya civilization. Science 299:1731–35.
Haug, I., T. Wubet, M. Weib, N. Aguirre, M. Weber, S. Günter, and I. Kottke. 2010. Species-rich but distinct arbuscular mycorrhizal communities in reforestation plots on degraded pastures and in neighboring pristine tropical mountain rain forest. Tropical Ecology 51:125–48.
Haugaasen, T., J. Barlow, and C. A. Peres. 2003. Surface wildfires in central Amazonia: Short-term impact on forest structure and carbon loss. Forest Ecology and Management 179:321–31.
Hawthorne, W., D. Sheil, V. Agyeman, M. Abu Juam, and C. Marshall. 2012. Logging scars in Ghanaian high forest: Towards improved models for sustainable production. Forest Ecology and Management 271:27–36.
Hayashida, F. M. 2005. Archaeology, ecological history, and conservation. Annual Review of Anthropology 34:43–65.
Hayes, T. M. 2006. Parks, people, and forest protection: An institutional assessment of the effectiveness of protected areas. World Development 34:2064–75.

Head, L. 1989. Prehistoric aboriginal impacts on Australian vegetation: An assessment of the evidence. Australian Geographer 20:37–46.

Heartsill Scalley, T., F. N. Scatena, A. E. Lugo, S. Moya, and C. R. Estrada Ruiz. 2010. Changes in structure, composition, and nutrients during 15 yr of hurricane-induced succession in a subtropical wet forest in Puerto Rico. Biotropica 42:455–63.

Hecht, S. 2010. The new rurality: Globalization, peasants and the paradoxes of landscapes. Land Use Policy 27:161–69.

Hecht, S. B., A. B. Anderson, and P. May. 1988. The subsidy from nature: Shifting cultivation, successional palm forests, and rural development. Human Organization 47:25–35.

Hecht, S. B., S. Kandel, I. Gomes, N. Cuellar, and H. Rosa. 2006. Globalization, forest resurgence, and environmental politics in El Salvador. World Development 34:308–23.

Hecht, S. B., and S. S. Saatchi. 2007. Globalization and forest resurgence: Changes in forest cover in El Salvador. BioScience 57:663–72.

Heckenberger, M. J., A. Kuikuro, U. T. Kuikuro, J. C. Russell, and M. J. Schmidt. 2003. Amazonia 1492: Pristine forest or cultural parkland? Science 301:1710–14.

Heckenberger, M. J., and E. G. Neves. 2009. Amazonian archaeology. Annual Review of Anthropology 38:251–66.

Heckenberger, M. J., J. C. Russell, C. Fausto, J. R. Toney, and M. J. Schmidt. 2008. Pre-Columbian urbanism, anthropogenic landscapes, and the future of the Amazon. Science 321:1214–17.

Heckenberger, M. J., J. C. Russell, J. R. Toney, and M. J. Schmidt. 2007. The legacy of cultural landscapes in the Brazilian Amazon: Implications for biodiversity. Philosophical Transactions of the Royal Society B: Biological Sciences 362:197–208.

Heinimann, A., P. Messerli, D. Schmidt-Vogt, and U. Wiesmann. 2007. The dynamics of secondary forest landscapes in the lower Mekong Basin: A regional-scale analysis. Mountain Research and Development 27:232–41.

Helmer, E. H. 2000. The landscape ecology of tropical secondary forest in montane Costa Rica. Ecosystems 3:98–114.

Helmer, E. H., T. J. Brandeis, A. E. Lugo, and T. Kennaway. 2008. Factors influencing spatial pattern in tropical forest clearance and stand age: Implications for carbon storage and species diversity. Journal of Geophysical Research 113:G2, doi:10.1029/2007JG000568.

Helmer, E. H., W. B. Cohen, and S. Brown. 2000. Mapping montane tropical forest successional stage and land use with multi-date Landsat imagery. International Journal of Remote Sensing 21:2163–83.

Helmer, E. H., M. A. Lefsky, and D. A. Roberts. 2009. Biomass accumulation rates of Amazonian secondary forest and biomass of old-growth forests from Landsat time series and the Geoscience Laser Altimeter System. Journal of Applied Remote Sensing 3:1–31.

Helmer, E., O. Ramos, T. del M. López, M. Quiñones, and W. Diaz. 2002. Mapping the forest type and land cover of Puerto Rico, a component of the Caribbean biodiversity hotspot. Caribbean Journal of Science 38:165–83.

Hernández-Stefanoni, J. L., J. M. Dupuy, F. Tun-Dzul, and F. May-Pat. 2011. Influence of landscape structure and stand age on species density and biomass of a tropical dry forest across spatial scales. Landscape Ecology 26:355–70.

Herrador Valencia, D. H., M. B. I. Junca, D. V. Linde, and E. M. Riera. 2011. Tropical forest recovery and socio-economic change in El Salvador: An opportunity for the introduction of new approaches to biodiversity protection. Applied Geography 31:259–68.

Hietz, P., R. Valencia, and S. J. Wright. 2013. Strong radial variation in wood density follows a uniform pattern in two Neotropical rain forests. Functional Ecology 287:684–92.

Hilje, B., and T. M. Aide. 2012. Recovery of amphibian species richness and composition in a chronosequence of secondary forests, northeastern Costa Rica. Biological Conservation 146:170–76.

Hiratsuka, M., T. Toma, R. Diana, D. Hadriyanto, and Y. Morikawa. 2006. Biomass recovery of naturally regenerated vegetation after the 1998 forest fire in East Kalimantan, Indonesia. Japan Agricultural Research Quarterly 40:277–82.

Hjerpe, J., H. Hedenas, and T. Elmqvist. 2001. Tropical rain forest recovery from cyclone damage and fire in Samoa. Biotropica 33:249–59.

Hladik, A., and S. Miquel. 1990. Seedling types and plant establishment in an African rain forest. Pages 261–82 *in* K. S. Bawa and M. Hadley, eds. Reproductive ecology of tropical forest plants. UNESCO/Parthenon, Paris.

Hodell, D., F. Anselmetti, D. Ariztegui, M. Brenner, J. Curtis, A. Gilli, D. Grzesik, T. Guilderson, A. Müller, and M. Bush. 2008. An 85-ka record of climate change in lowland Central America. Quaternary Science Reviews 27:1152–65.

Hodgkison, R., S. T. Balding, A. Zubaid, and T. H. Kunz. 2003. Fruit bats (Chiroptera: Pteropodidae) as seed dispersers and pollinators in a lowland Malaysian rain forest. Biotropica 35:491–502.

Hoffmann, W. A., E. L. Geiger, S. G. Gotsch, D. R. Rossatto, L. C. Silva, O. L. Lau, M. Haridasan, and A. C. Franco. 2012. Ecological thresholds at the savanna-forest boundary: How plant traits, resources and fire govern the distribution of tropical biomes. Ecology Letters 15:759–68.

Höhn, A., S. Kahlheber, K. Neumann, and A. Schweizer. 2008. Settling the rain forest: The environment of farming communities in Southern Cameroon during the first millennium B.C. Pages 29–42 *in* J. Runge, ed. Dynamics of forest ecosystems in Central Africa during the Holocene: Past-present-future. Taylor & Francis, London, UK.

Holdsworth, A. R., and C. Uhl. 1997. Fire in Amazonian selectively logged rain forest and the potential for fire reduction. Ecological Applications 7:713–25.

Holl, K. D. 1999. Factors limiting tropical rain forest regeneration in abandoned

pasture: Seed rain, seed germination, microclimate, and soil. Biotropica 31:229–42.
———. 2007. Old field vegetation succession in the Neotropics. Pages 93–118 *in* V. A. Cramer and R. J. Hobbs, eds. Old fields: Dynamics and restoration of abandoned farmland. Island Press, Washington, DC.
Holl, K. D., and T. M. Aide. 2011. When and where to actively restore ecosystems? Forest Ecology and Management 261:1558–63.
Holl, K. D., M. E. Loik, E. H. V. Lin, and I. A. Samuels. 2000. Tropical montane forest restoration in Costa Rica: Overcoming barriers to dispersal and establishment. Restoration Ecology 8:339–49.
Holl, K. D., and M. W. Lulow. 1997. Effects of species, habitat, and distance from the edge on post-dispersal seed predation in a tropical rainforest. Biotropica 29:459–68.
Holl, K. D., R. A. Zahawi, R. J. Cole, R. Ostertag, and S. Cordell. 2011. Planting seedlings in tree islands versus plantations as a large-scale tropical forest restoration strategy. Restoration Ecology 19:470–79.
Holling, C. S. 1973. Resilience and stability of ecological systems. Annual Review of Ecology and Systematics 4:1–24.
Holmgren, M., and L. Poorter. 2007. Does a ruderal strategy dominate the endemic flora of the West African forests? Journal of Biogeography 34:1100–1111.
Hölscher, D., S. de A Sá, T. Bastos, M. Denich, and H. Folster. 1997. Evaporation from young secondary vegetation in eastern Amazonia. Journal of Hydrology 193:293–305.
Hölscher, D., L. Köhler, A. I. J. M. van Dijk, and L. Bruijnzeel. 2004. The importance of epiphytes to total rainfall interception by a tropical montane rain forest in Costa Rica. Journal of Hydrology 292:308–22.
Hölscher, D., C. Leuschner, K. Bohman, M. Hagemeier, J. Juhrbandt, and S. Tjitrosemito. 2006. Leaf gas exchange of trees in old-growth and young secondary forest stands in Sulawesi, Indonesia. Trees 20:278–85.
Holz, S., G. Placci, and R. D. Quintana. 2009. Effects of history of use on secondary forest regeneration in the Upper Paraná Atlantic Forest (Misiones, Argentina). Forest Ecology and Management 258:1629–42.
Hooper, E. R., P. Legendre, and R. Condit. 2004. Factors affecting community composition of forest regeneration in deforested, abandoned land in Panama. Ecology 85:3313–26.
———. 2005. Barriers to forest regeneration of deforested and abandoned land in Panama. Journal of Applied Ecology 42:1165–74.
Hope, G. 2009. Environmental change and fire in the Owen Stanley ranges, Papua New Guinea. Quaternary Science Reviews 28:2261–76.
Hope, G., A. P. Kershaw, S. van der Kaars, S. Xiangjun, P.-M. Liew, L. E. Heusser, H. Takahara, M. McGlone, N. Miyoshi, and P. T. Moss. 2004. History of vegetation and habitat change in the Austral-Asian region. Quaternary International 118–19:103–26.

Hopkins, M. S., and A. W. Graham. 1983. The species composition of soil seed banks beneath lowland tropical rainforests in North Queensland, Australia. Biotropica 15:90–99.

Hopkins, M. S., J. Ash, A. W. Graham, J. Head, and R. K. Hewett. 1993. Charcoal evidence of the spatial extent of the *Eucalyptus* woodland expansions and rain forest contractions in North Queensland during the late Pleistocene. Journal of Biogeography 20:357–72.

Hopp, P., R. Ottermanns, E. Caron, S. Meyer, and M. Rof-Nickoll. 2010. Recovery of litter inhabiting beetle assemblages during forest regeneration in the Atlantic forest of Southern Brazil. Insect Conservation and Diversity 3:103–13.

Horn, H. S. 1974. The ecology of secondary succession. Annual Review of Ecology and Systematics 5:25–37.

Horn, S. P. 2007. Lake Quaternary lake and swamp sediments: Recorders of climate and environment. Pages 4223–41 *in* J. Bundschuh, and G. E. Alvarado-Indinu, eds. Central America: Geology, resources and hazards. Vol. 1. Taylor & Francis/ Balkema, Leiden, Neth.

Horn, S. P., and R. L. J. Sanford. 1992. Holocene fires in Costa Rica. Biotropica 33:191–96.

Horvitz, C. C., J. B. Pascarella, S. McMann, A. Freedman, and R. H. Hofstetter. 1998. Functional roles of invasive non-indigenous plants in hurricane-affected subtropical hardwood forests. Ecological Applications 8:947–74.

Houghton, R. A. 1999. The annual net flux of carbon to the atmosphere from changes in land use 1850–1990. Tellus Series B: Chemical and Physical Meteorology 51:298–313.

———. 2010. How well do we know the flux of CO2 from land use change? Tellus Series B: Chemical and Physcial Meteorology 62:337–51.

Houghton, R., and C. Goodale. 2004. Effects of land-use change on the carbon balance of terrestrial ecosystems. Ecosystems and land use change. Geophysical Monograph Series 153:85–98.

Houghton, R., N. Greenglass, A. Baccini, A. Cattaneo, S. Goetz, J. Kellndorfer, N. Laporte, and W. Walker. 2010. The role of science in Reducing Emissions from Deforestation and Forest Degradation (REDD). Carbon 1:253–59.

Houghton, R., J. Hobbie, J. Melillo, B. Moore, B. Peterson, G. Shaver, and G. Woodwell. 1983. Changes in the carbon content of terrestrial biota and soils between 1860 and 1980: A net release of CO_2 to the atmosphere. Ecological Monographs 53:235–62.

Houlton, B. Z., Y. P. Wang, P. M. Vitousek, and C. B. Field. 2008. A unifying framework for dinitrogen fixation in the terrestrial biosphere. Nature 454:327–30.

Howe, H. F., and M. N. Miriti. 2004. When seed dispersal matters. BioScience 54:651–60.

Howe, H. F., and J. Smallwood. 1982. Ecology of seed dispersal. Annual Review of Ecology and Systematics 13:201–28.

Howe, H. F, Y. Urincho-Pantaleon, M. Peña-Domene, and C. Martínez-Garza.

2010. Early seed fall and seedling emergence: Precursors to tropical restoration. Oecologia 164:731–40.
Howorth, R. T., and C. A. Pendry. 2006. Post-cultivation secondary succession in a Venezuelan lower montane rain forest. Biodiversity and Conservation 15:693–715.
Huang, M., and G. P. Asner. 2010. Long-term carbon loss and recovery following selective logging in Amazon forests. Global Biogeochemical Cycles 24:GB3028.
Huante, P., E. Ceccon, A. Orozco-Segovia, M. E. Sánchez-Coronado, I. Acosta, and E. Rincon. 2012. The role of arbuscular mycorrhizal fungi on the early-stage restoration of seasonally dry tropical forest in Chamela, Mexico. Revista Árvore 36:279–89.
Hubbell, S. P. 1998. The maintenance of diversity in a Neotropical tree community: Conceptual issues, current evidence, and challenges ahead. Pages 17–44 *in* F. Dallmeier and J. A. Comiskey, eds. Forest biodiversity research, monitoring and modeling: Conceptual background and old-world case studies. Parthenon, Paris.
Hubbell, S. P., and R. B. Foster. 1991. The fate of juvenile trees in a Neotropical forest: Implications for the natural maintenance of tropical tree diversity. Pages 317–41 *in* K. S. Bawa and M. Hadley, eds. Reproductive ecology of tropical forest plants. Parthenon, Paris.
Hubbell, S. P., R. B. Foster, S. T. O'Brian, K. E. Harms, R. Condit, B. Wechsler, S. J. Wright, and S. Loo de Lao. 1999. Light gap disturbances, recruitment limitation, and tree diversity in a Neotropical forest. Science 283:554–57.
Huc, R., A. Ferhi, and J. Guehl. 1994. Pioneer and late stage tropical rainforest tree species (French Guiana) growing under common conditions differ in leaf gas exchange regulation, carbon isotope discrimination and leaf water potential. Oecologia 99:297–305.
Huettner, M. 2012. Risks and opportunities of REDD plus implementation for environmental integrity and socio-economic compatibility. Environmental Science & Policy 15:4–12.
Hughes, R. F., J. B. Kauffman, and D. L. Cummings. 2000. Fire in the Brazilian Amazon 3. Dynamics of biomass, C, and nutrient pools in regenerating forests. Oecologia 124:574–88.
———. 2002. Dynamics of aboveground and soil carbon and nitrogen stocks and cycling of available nitrogen along a land-use gradient in Rondônia, Brazil. Ecosystems 5:244–59.
Hughes, R. F., J. B. Kauffman, and V. J. Jaramillo. 1999. Biomass, carbon, and nutrient dynamics of secondary forests in a humid tropical region of Mexico. Ecology 80:1892–1907.
———. 2000. Ecosystem-scale impacts of deforestation and land use in a humid tropical region of Mexico. Ecological Applications 10:515–27.
Hunt, T. 2007. Rethinking Easter Island's ecological catastrophe. Journal of Archaeological Science 34:485–502.
Hurles, M. E., E. Marisoo-Smith, R. D. Gray, and D. Penny. 2003. Untangling oceanic settlement: The edge of the knowable. Trends in Ecology & Evolution 18:531–40.

Hviding, E., and T. Bayliss-Smith. 2000. Islands of rainforest: Agroforestry, logging and eco-tourism in Solomon Islands. Ashgate Publishing, Aldershot, UK.

Hynes, R., and A. Chase. 1982. Plants, sites and domiculture: Aboriginal influence upon plant communities in Cape York Peninsula. Archaeology in Oceania 17:38–50.

Ibarra-Manríquez, G., M. Martínez Ramos, and K. Oyama. 2001. Seedling functional types in a lowland rain forest in Mexico. American Journal of Botany 88:1801–12.

Ibrahim, M., L. Guerra, F. Casasola, and C. Neely. 2010. Importance of silvopastoral systems for mitigation of climate change and harnessing of environmental benefits. Pages 189–98 *in* M. Abberton, R. Conant, and C. Batello, eds. Grassland carbon sequestration: Management, policy and economics. FAO, Rome, Italy.

Imai, N., T. Seino, S. Aiba, M. Takyu, J. Titin, and K. Kitayama. 2012. Effects of selective logging on tree species diversity and composition of Bornean tropical rain forests at different spatial scales. Plant Ecology 213:1413–24.

Ingle, N. 2003. Seed dispersal by wind, birds, and bats between Philippine montane rainforest and successional vegetation. Oecologia 134:251–61.

Ingwell, L. L., S. Joseph Wright, K. K. Becklund, S. P. Hubbell, and S. A. Schnitzer. 2010. The impact of lianas on 10 years of tree growth and mortality on Barro Colorado Island, Panama. Journal of Ecology 98:879–87.

Iriarte, J. 2007. New perspectives on plant domestication and the development of agriculture in the New World. Pages 167–88 *in* T. P. Denham, J. Iriarte, and L. Vrydaghs, eds. Rethinking agriculture: Archaeological and ethnoarchaeological perspectives. Left Coast Press, Walnut Creek, CA.

Iriarte, J., and H. Behling. 2007. The expansion of Araucaria forest in the southern Brazilian highlands during the last 4000 years and its implications for the development of the Taquara/Itarare tradition. Environmental Archaeology 12:115–27.

Iriarte, J., M. J. Power, S. Rostain, F. E. Mayle, H. Jones, J. Watling, B. S. Whitney, and D. B. McKey. 2012. Fire-free land use in pre-1492 Amazonian savannas. Proceedings of the National Academy of Sciences 109:6473–78.

Islebe, G. A., H. Hooghiemstra, M. Brenner, J. H. Curtis, and D. A. Hodell. 1996. A Holocene vegetation history from lowland Guatemala. Holocene 6:265–71.

ITTO (International Tropical Timber Organization). 2002. ITTO guidelines for the restoration, management and rehabilitation of degraded and secondary tropical forests. ITTO Policy Development Series No. 13. ITTO, Yokohama, Japan.

———. 2005. Revised ITTO criteria and indicators for the sustainable management of tropical forests including reporting format. ITTO, Yokohama, Japan.

Jackson, S. T., and R. J. Hobbs. 2009. Ecological restoration in the light of ecological history. Science 325:567–69.

Jago, L. C. F., and W. E. Boyd. 2005. How a wet tropical rainforest copes with repeated volcanic destruction. Quaternary Research 64:399–406.

Jagoret, P., I. Michel-Dounias, and E. Malézieux. 2011. Long-term dynamics of cocoa agroforests: A case study in central Cameroon. Agroforestry Systems 81:267–78.

Jagoret, P., I. Michel-Dounias, D. Snoeck, H. T. Ngnogué, and E. Malézieux. 2012. Afforestation of savannah with cocoa agroforestry systems: A small-farmer innovation in central Cameroon. Agroforestry Systems 86:493–504.
Jankowska-Blaszczuk, M., and P. J. Grubb. 2006. Changing perspectives on the role of the soil seed bank in northern temperate deciduous forests and in tropical lowland rain forests: Parallels and contrasts. Perspectives in Plant Ecology, Evolution and Systematics 8:3–21.
Janos, D. P. 1980. Mycorrhizae influence tropical succession. Supplement, Biotropica 12:56–64.
Jansen, P. A., and P. A. Zuidema. 2001. Logging, seed dispersal by vertebrates, and natural regeneration of tropical timber trees. Pages 35–59 *in* R. Fimbel, A. Grajal, and J. G. Robinson, eds. The cutting edge: Conserving wildlife in logged tropical forest. Columbia University Press, New York.
Janzen, D. H. 1970. Herbivores and the number of tree species in tropical forests. American Naturalist 104:501–28.
———. 1988. Management of habitat fragments in a tropical dry forest—growth. Annals of the Missouri Botanical Garden 75:105–16.
———. 1990. An abandoned field is not a tree fall gap. Vida Silvestre Neotropical 2:64–67.
———. 1997. How to grow a wildland: The gardenification of nature. International Journal of Tropical Insect Science 17:269–76.
Jaramillo, V. J., R. Ahedo-Hernandez, and J. B. Kauffman. 2003. Root biomass and carbon in a tropical evergreen forest of Mexico: Changes with secondary succession and forest conversion to pasture. Journal of Tropical Ecology 19:457–64.
Jobbágy, E. G., and R. B. Jackson. 2000. The vertical distribution of soil organic carbon and its relation to climate and vegetation. Ecological Applications 10:423–36.
Johns, A. D. 1985. Selective logging and wildlife conservation in tropical rain-forest: Problems and recommendations. Biological Conservation 31:355–75.
———. 1988. Effects of "selective" timber extraction on rain forest composition and some consequences for frugivores and folivores. Biotropica 20:31–37.
Johns, R. 1986. The instability of the tropical ecosystem in New Guinea. Blumea 31:341–71.
Johnson, E. A., and K. Miyanishi. 2008. Testing the assumptions of chronosequences in succession. Ecology Letters 11:419–31.
Jones, E. W. 1955. Ecological studies of the rain forest of southern Nigeria IV. The plateau forset of the Okomu Forest Reserve, Part 1. The environment, the vegetation types of the forest, and the horizontal distribution of species. Journal of Ecology 43:564–94.
———. 1956. Ecological studies of the rain forest of southern Nigeria IV. The plateau forest of the Okomu Forest Reserve, Part 2. The reproduction and history of the forest. Journal of Ecology 44:83–117.

Jones, E., M. Wishnie, J. Deago, A. Sautu, and A. Cerezo. 2004. Facilitating natural regeneration in *Saccharum spontaneum* (L.) grasslands within the Panama Canal Watershed: Effects of tree species and tree structure on vegetation recruitment patterns. Forest Ecology and Management 191:171–83.

Jones, F., J. Chen, G. J. Weng, and S. Hubbell. 2005. A genetic evaluation of seed dispersal in the Neotropical tree *Jacaranda copaia* (Bignoniaceae). American Naturalist 166:543–55.

Jones, J. 1994. Pollen evidence for early settlement and agriculture in northern Belize. Palynology 18:205–11.

Jordan, C. F., ed. 1989. An Amazonian rain forest. UNESCO/Parthenon, Paris.

Jordan, C. F., W. Caskey, G. Escalante, R. Herrera, F. Montagnini, R. Todd, and C. Uhl. 1983. Dynamics during conversion of primary Amazonian rain forest to slash and burn agriculture. Oikos 40:131–39.

Jordan, C. F., and R. Herrera. 1981. Tropical rain forests: Are nutrients really critical? American Naturalist 117:167–80.

Jordano, P. 1987. Patterns of mutualistic interactions in pollination and seed dispersal: Connectance, dependence asymmetries, and coevolution. American Naturalist 129:657–77.

Juhrbandt, J., C. Leuschner, and D. Holscher. 2004. The relationship between maximal stomatal conductance and leaf traits in eight Southeast Asian early successional tree species. Forest Ecology and Management 202:245–56.

Junk, W. J. 1989. Flood tolerance and tree distribution in central Amazonian floodplains. Pages 47–64 *in* L. B. Holm-Nielsen, I. Nielsen, and H. Balslev, eds. Tropical forests: Botanical dynamics, speciation and diversity. Academic Press, London.

Junqueira, A. B., and C. R. Clement. 2012. Reply to Barlow et al. (2011): Towards an integrated understanding of the pre-conquest human footprint in Amazonia. Biological Conservation 152:291–92.

Junqueira, A. B., G. Shepard, and C. Clement. 2010a. Secondary forests on anthropogenic soils in Brazilian Amazonia conserve agrobiodiversity. Biodiversity and Conservation 19:1933–61.

———. 2010b. Secondary forests on anthropogenic soils of the Middle Madeira River: Valuation, local knowledge, and landscape domestication in Brazilian Amazonia. Economic Botany 20:1–15.

Kageyama, P. Y., and C. F. A. Castro. 1989. Sucessão secundária, estrutura genética e plantações de espécies arbóreas nativas. IPEF, Piracicabas 41/42:83–93.

Kaiser-Bunbury, C. N., A. Traveset, and D. M. Hansen. 2010. Conservation and restoration of plant-animal mutualisms on oceanic islands. Perspectives in Plant Ecology, Evolution and Systematics 12:131–43.

Kalacska, M., G. A. Sánchez-Azofeifa, B. Rivard, T. Caelli, H. P. White, and J. C. Calvo-Alvarado. 2007. Ecological fingerprinting of ecosystem succession: Estimating secondary tropical dry forest structure and diversity using imaging spectroscopy. Remote Sensing of Environment 108:82–96.

Kalacska, M., G. A. Sánchez-Azofeifa, B. Rivard, J. C. Calvo-Alvarado, A. R. P. Journet, J. P. Arroyo-Mora, and D. Ortiz-ortiz. 2004. Leaf area index measurements in a tropical moist forest: A case study from Costa Rica. Remote Sensing of Environment 91:134–52.

Kalame, F. B., R. Aidoo, J. Nkem, O. C. Ajayie, M. Kanninen, O. Luukkanen, and M. Idinoba. 2011. Modified taungya system in Ghana: A win-win practice for forestry and adaptation to climate change? Environmental Science & Policy 14:519–30.

Kalliola, R., J. Salo, M. Puhakka, and M. Rajasilta. 1991. New site formation and colonizing vegetation in primary succession on the western Amazon floodplains. Journal of Ecology 79:877–901.

Kammesheidt, L. 1999. Forest recovery by root suckers and above-ground sprouts after slash-and-burn agriculture, fire and logging in Paraguay and Venezuela. Journal of Tropical Ecology 15:143–57.

———. 2002. Perspectives on secondary forest management in tropical humid lowland America. AMBIO 31:243–50.

Kammu, M. 2009. Water management in Angkor: Human impacts on hydrology and sediment transportation. Journal of Environmental Management 90:1413–21.

Kang, H., and K. S. Bawa. 2003. Effects of successional status, habit, sexual systems, and pollinators on flowering patterns in tropical rain forest trees. American Journal of Botany 90:865–76.

Kanowski, J., and C. P. Catterall. 2010. Carbon stocks in aboveground biomass of monoculture plantations, mixed species plantations and environmental restoration plantings in northeast Australia. Ecological Management & Restoration 11:119–26.

Kanowski, J., C. P. Catterall, K. Freebody, A. N. D. Freeman, and D. A. Harrison. 2010. Monitoring revegetation projects in rainforest landscapes. Toolkit Version 3. Reef and Rainforest Research Cente Limited, Cairns, Australia.

Kanowski, J., C. P. Catterall, and W. Neilan. 2008. Potential value of weedy regrowth for rainforest restoration. Ecological Management & Restoration 9:88–99.

Kanowski, J., C. P. Catterall, G. Wardell-Johnson, H. Proctor, and T. Reis. 2003. Development of forest structure on cleared rainforest land in eastern Australia under different styles of reforestation. Forest Ecology and Management 183:265–80.

Kanowski, J. J., T. M. Reis, C. P. Catterall, and S. D. Piper. 2006. Factors affecting the use of reforested sites by reptiles in cleared rainforest landscapes in tropical and subtropical Australia. Restoration Ecology 14:67–76.

Kartawinata, K. 1993. A wider view of the fire hazard. Pages 261–66 *in* H. Brookfield and Y. Byron, eds. Southeast Asia's environmental future: The search for sustainability. United Nations University Press, Tokyo.

Kartawinata, K., S. Riswan, A. Gintings, and T. Puspitojati. 2001. An overview of post-extraction secondary forests in Indonesia. Journal of Tropical Forest Science 13:621–38.

Karthik, T., G. Govindhan Veeraswami, and P. Kumar Samal. 2009. Forest recovery

following shifting cultivation: An overview of existing research. Tropical Conservation Science 2:374–87.
Kasenene, J. 2001. Lost logging: Problems of tree regeneration in forest gaps in Kibale Foret, Uganda. Pages 480–90 *in* W. Weber, L. J. T. White, A. Vedder, and L. Naughton-Treves, eds. African rain forest ecology and conservation. Yale University Press, New Haven, CT.
Kassi, N. J. K., and G. Decocq. 2008. Successional patterns of plant species and community diversity in a semi-deciduous tropical forest under shifting cultivation. Journal of Vegetation Science 19:809–20.
Katovai, E., A. L. Burley, and M. M. Mayfield. 2012. Understory plant species and functional diversity in the degraded wet tropical forests of Kolombangara Island, Solomon Islands. Biological Conservation 145:214–24.
Kauffman, J. B. 1991. Survival by sprouting following fire in tropical forests of the Eastern Amazon. Biotropica 23:219–24.
Kauffman, J. B., D. L. Cummings, and D. E. Ward. 1998. Fire in the Brazilian Amazon: 2. Biomass, nutrient pools and losses in cattle pastures. Oecologia 113:415–27.
Kauffman, J. B., D. L. Cummings, D. E. Ward, and R. Babbitt. 1995. Fire in the Brazilian Amazon: 1. Biomass, nutrient pools, and losses in slashed primary forests. Oecologia 104:397–408.
Kauffman, J. B., R. F. Hughes, and C. Heider. 2009. Carbon pool and biomass dynamics associated with deforestation, land use, and agricultural abandonment in the Neotropics. Ecological Applications 19:1211–22.
Kealhofer, L. 2003. Looking into the gap: Land use and the tropical forests of Southern Thailand. Asian Perspectives 42:73–95.
Kellner, J. R., and G. P. Asner. 2009. Convergent structural responses of tropical forests to diverse disturbance regimes. Ecology Letters 12:887–97.
Kellner, J. R., D. B. Clark, and S. P. Hubbell. 2009. Pervasive canopy dynamics produce short-term stability in a tropical rain forest landscape. Ecology Letters 12:155–64.
Kennard, D. K. 2002. Secondary forest succession in a tropical dry forest: Patterns of development across a 50-year chronosequence in lowland Bolivia. Journal of Tropical Ecology 18:53–66.
Kennedy, D. N., and M. D. Swaine. 1992. Germination and growth of colonizing species in artificial gaps of different sizes in dipterocarp rain forest. Philosophical Transactions of the Royal Society B: Biological Sciences 335:357–67.
Kennedy, J., and W. Clarke. 2004. Cultivated landscapes of the Southwest Pacific. Resource Management in Asia-Pacific Program, Research School of Pacific and Asian Studies, Australian National University, Canberra, Australia.
Kennedy, L. M., and S. P. Horn. 1997. Prehistoric maize cultivation at the La Selva Biological Station, Costa Rica. Biotropica 29:368–70.
Kenzo, T., T. Ichie, D. Hattori, J. J. Kendawang, K. Sakurai, and I. Ninomiya. 2010. Changes in above- and belowground biomass in early successional tropical

secondary forests after shifting cultivation in Sarawak, Malaysia. Forest Ecology and Management 260:875–82.

Kershaw, A. P., S. C. Bretherton, and S. van der Kaars. 2007. A complete pollen record of the last 230 ka from Lynch's Crater, north-eastern Australia. Palaeogeography, Palaeoclimatology, Palaeoecology 251:23–45.

Kettle, C. J. 2010. Ecological considerations for using dipterocarps for restoration of lowland rainforest in Southeast Asia. Biodiversity and Conservation 19:1137–51.

Kettle, C. J., J. Ghazoul, P. Ashton, C. H. Cannon, L. Chong, B. Diway, E. Faridah, et al. 2011. Seeing the fruit for the trees in Borneo. Conservation Letters 4:184–91.

Kim, T. J., F. Montagnini, and D. Dent. 2008. Rehabilitating abandoned pastures in Panama: Control of the invasive exotic grass, *Saccharum spontaneum* L., using artificial shade treatments. Journal of Sustainable Forestry 26:192–203.

Kimmel, T., L. do Nascimento, D. Piechowski, E. Sampaio, M. Nogueira Rodal, and G. Gottsberger. 2010. Pollination and seed dispersal modes of woody species of 12-year-old secondary forest in the Atlantic Forest region of Pernambuco, NE Brazil. Flora-Morphology, Distribution, Functional Ecology of Plants 205:540–47.

King, D. A., S. J. Davies, and N. S. M. Noor. 2006. Growth and mortality are related to adult tree size in a Malaysian mixed dipterocarp forest. Forest Ecology and Management 223:152–58.

King, D., S. Davies, M. N. N. Supardi, and S. Tan. 2005. Tree growth is related to light interception and wood density in two mixed dipterocarp forests of Malaysia. Functional Ecology 19:445–53.

Kirch, P. V. 1989. Second millennium B.C. arboriculture in Melanesia: Archaeological evidence from the Mussau Islands. Economic Botany 43:225–40.

———. 1996. Late Holocene human-induced modifications to a central Polynesian island ecosystem. Proceedings of the National Academy of Sciences 93:5296–5300.

———. 1997. Introduction: The environmental history of Oceanic islands. Pages 1–21 *in* P. V. Kirch and T. L. Hunt, eds. Historical ecology in the Pacific Islands. Yale University Press, New Haven, CT.

———. 2005. Archaeology and global change: The Holocene record. Annual Review of Environment and Resources 30:409–40.

Kitajima, K. 1992. Relationship between photosynthesis and thickness of cotyledons for tropical tree species. Functional Ecology 6:582–89.

———. 1994. Relative importance of photosynthetic traits and allocation pattern as correlates of seedling shade tolerance of 13 tropical trees. Oecologia 98:419–28.

———. 1996. Ecophysiology of tropical tree seedlings. Pages 559–96 *in* S. S. Mulkey, R. L. Chazdon, and A. P. Smith, eds. Tropical forest plant ecophysiology. Chapman & Hall, New York.

Kitajima, K., and L. Poorter. 2008. Functional basis for resource niche partitioning by tropical trees. Pages 160–181 *in* W. P. Carson and S. A. Schnitzer, eds. Tropical forest community ecology. John Wiley & Sons, West Sussex, UK.

———. 2010. Tissue level leaf toughness, but not lamina thickness, predicts sapling

leaf lifespan and shade tolerance of tropical tree species. New Phytologist 186:708–21.
Kitching, R., L. Ashton, A. Nakamura, T. Whitaker, and C. V. Khen. 2012. Distance-driven species turnover in Bornean rainforests: Homogeneity and heterogeneity in primary and post-logging forests. Ecography 35:1–8.
Klanderud, K., H. Z. H. Mbolatiana, M. N. Vololomboahangy, M. A. Radimbison, E. Roger, O. Totland, and C. Rajeriarison. 2010. Recovery of plant species richness and composition after slash-and-burn agriculture in a tropical rainforest in Madagascar. Biodiversity and Conservation 19:187–204.
Klein, B. C. 1989. Effects of forest fragmentation on dung and carrion beetle communities in central Amazonia. Ecology 70:1715–25.
Klimes, P., C. Idigel, M. Rimandai, T. M. Fayle, M. Janda, G. D. Weiblen, and V. Novotny. 2012. Why are there more arboreal ant species in primary than in secondary tropical forests? Journal of Animal Ecology 81:1103–12.
Klooster, D. 2003. Forest transitions in Mexico: Institutions and forests in a globalized countryside. Professional Geographer 55:227–37.
Knight, D. H. 1975. A phytosociological analysis of species-rich tropical forest on Barro Colorado Island, Panama. Ecological Monographs 45:259–84.
Knogge, C., and E. W. Heymann. 2003. Seed dispersal by sympatric tamarins, *Saguinus mystax* and *Saguinus fuscicollis*: Diversity and characteristics of plant species. Folia Primatol 74:33–47.
Kochummen, K. M. 1966. Natural plant succession after farming. Malayan Forester 29:170–181.
Köhler, L., D. Hölscher, and C. Leuschner. 2006. Above-ground water and nutrient fluxes in three successional stages of Costa Rican montane oak forest with contrasting epiphyte abundance. Pages 271–98 *in* M. Kappelle, ed. Ecology and conservation of Neotropical montane oak forests. Springer, New York.
Köhler, L., C. Tobon, K. F. A. Frumau, and L. A. Bruijnzeel. 2007. Biomass and water storage dynamics of epiphytes in old-growth and secondary montane cloud forest stands in Costa Rica. Plant Ecology 193:171–84.
Kolb, M., and L. Galicia. 2012. Challenging the linear forestation narrative in the Neo-tropic: Regional patterns and processes of deforestation and regeneration in southern Mexico. Geographical Journal 178:147–61.
Körner, C. 2003. Slow in, rapid out—Carbon flux studies and Kyoto targets. Science 300:1242–43.
———. 2009. Responses of humid tropical trees to rising CO_2. Annual Review of Ecology, Evolution, and Systematics 40:61–79.
Kourampas, N., I. A. Simpson, N. Perera, S. U. Deraniyagala, and W. H. Wijeyapala. 2009. Rockshelter sedimentation in a dynamic tropical landscape: Late Pleistocene-Early Holocene archaeological deposits in Kitulgala Beli-lena, southwestern Sri Lanka. Geoarchaeology 24:677–714.
Koutika, L.-S., and H. J. Rainey. 2010. *Chromolaena odorata* in different ecosystems: Weed or fallow plant. Applied Ecology and Environmental Research 8:131–42.

Kress, W. J., and J. H. Beach. 1994. Flowering plant reproductive systems. Pages 161–82 *in* L. A. McDade, K. S. Bawa, H. A. Hespenheide, and G. S. Hartshorn, eds. La Selva: Ecology and natural history of a Neotropical rain forest. University of Chicago Press, Chicago.

Kuhlken, R. 2002. Intensive agricultural landscapes of Oceania. Journal of Cultural Geography 19:161–95.

Kull, C. A., C. K. Ibrahim, and T. C. Meredith. 2007. Tropical forest transitions and globalization: Neo-liberalism, migration, tourism, and international conservation agendas. Society & Natural Resources 20:723–37.

Kunert, N., L. Schwendenmann, C. Potvin, and D. Hölscher. 2012. Tree diversity enhances tree transpiration in a Panamanian forest plantation. Journal of Applied Ecology 49:135–44.

Kuprewicz, E. K. 2013. Mammal abundances and seed traits control the seed dispersal and predation roles of terrestrial mammals in a Costa Rican forest. Biotropica 45:333–42.

Kuusipalo, J., G. Ådjers, Y. Jafarsidik, A. Otsamo, K. Tuomela, and R. Vuokko. 1995. Restoration of natural vegetation in degraded *Imperata cylindrica* grassland: Understorey development in forest plantations. Journal of Vegetation Science 6:205–10.

Kyereh, B., M. D. Swaine, and J. Thompson. 1999. Effect of light on the germination of forest trees in Ghana. Journal of Ecology 87:772–83.

Laborde, J., S. Guevara, and G. Sánchez-Ríos. 2008. Tree and shrub seed dispersal in pastures: The importance of rainforest trees outside forest fragments. Ecoscience 15:6–16.

LaFrankie, J. V., P. S. Ashton, G. B. Chuyong, L. Co, R. Condit, S. J. Davies, R. Foster, et al. 2006. Contrasting structure and composition of the understory in species-rich tropical rain forests. Ecology 87:2298–305.

Lamb, D. 2011. Regreening the bare hills: Tropical forest restoration in the Asia-Pacific region. Springer, New York.

Lamb, D., P. D. Erskine, and J. A. Parrotta. 2005. Restoration of degraded tropical forest landscapes. Science 310:1628–32.

Lamb, D., and D. Gilmour. 2003. Rehabilitation and restoration of degraded forests. IUCN/WWF, Gland, Switzerland.

Lambin, E. F., and H. Geist. 2006. Land-use and land-cover change: Local processes and global impacts. Springer Verlag, Berlin, Ger.

Lambin, E. F., H. J. Geist, and E. Lepers. 2003. Dynamics of land-use and land-cover change in tropical regions. Annual Review of Environment and Resources 28:205–41.

Lambin, E. F., and P. Meyfroidt. 2010. Land use transitions: Socio-ecological feedback versus socio-economic change. Land Use Policy 27:108–18.

Langner, A., and F. Siegert. 2009. Spatiotemporal fire occurrence in Borneo over a period of 10 years. Global Change Biology 15:48–62.

Laporte, N. T., J. A. Stabach, R. Grosch, T. S. Lin, and S. J. Goetz. 2007. Expansion of industrial logging in Central Africa. Science 316:1451.

Larkin, C. C., C. Kwit, J. M. Wunderle Jr, E. H. Helmer, M. H. H. Stevens, M. T. Roberts, and D. N. Ewert. 2012. Disturbance type and plant successional communities in Bahamian dry forests. Biotropica 44:10–18.

Larpkern, P., S. R. Moe, and O. Totland. 2011. Bamboo dominance reduces tree regeneration in a disturbed tropical forest. Oecologia 165:161–68.

Lasco, R. D., and J. M. Pulhin. 2006. Environmental impacts of community-based forest management in the Philippines. International Journal of Environment and Sustainable Development 5:46–56.

Lasco, R., R. Visco, and J. Pulhin. 2001. Secondary forests in the Philippines: Formation and transformation in the 20th century. Journal of Tropical Forest Science 13:652–70.

Laurance, S. G. W., and M. S. Gomez. 2005. Clearing width and movements of understory rainforest Birds. Biotropica 37:149–52.

Laurance, W. F. 2006. Synergistic effects of habitat disturbance and hunting in Amazonian forest fragments. Pages 87–104 *in* W. F. Laurance and C. A. Peres, eds. Emerging threats to tropical forests. University of Chicago Press, Chicago.

———. 2007. Have we overstated the tropical biodiversity crisis? Trends in Ecology & Evolution 22:65–70.

Laurance, W. F., and T. Curran. 2008. Impacts of wind disturbance on fragmented tropical forests: A review and synthesis. Austral Ecology 33:399–408.

Laurance, W. F., A. A. Oliveira, S. G. Laurance, R. Condit, H. E. M. Nascimento, A. C. Sanchez-Thorin, T. E. Lovejoy, et al. 2004. Pervasive alteration of tree communities in undisturbed Amazonian forests. Nature 428:171–75.

Laurance, W. F., and D. Useche. 2009. Environmental synergisms and extinctions of tropical species. Conservation Biology 23:1427–37.

Lawes, M. J., and C. A. Chapman. 2006. Does the herb *Acanthus pubescens* and/or elephants suppress tree regeneration in disturbed Afrotropical forest? Forest Ecology and Management 221:278–84.

Lawrence, D. 2004. Erosion of tree diversity during 200 years of shifting cultivation in Bornean rain forest. Ecological Applications 14:1855–69.

Lawrence, D., P. D'Odorico, L. Diekmann, M. DeLonge, R. Das, and J. Eaton. 2007. Ecological feedbacks following deforestation create the potential for a catastrophic ecosystem shift in tropical dry forest. Proceedings of the National Academy of Sciences 104:20696–701.

Lawrence, D., and D. Foster. 2002. Changes in forest biomass, litter dynamics and soils following shifting cultivation in southern Mexico: An overview. Interciencia 27:400–408.

Lawrence, D., C. Radel, K. Tully, B. Schmook, and L. Schneider. 2010. Untangling a decline in tropical forest resilience: Constraints on the sustainability of shifting cultivation across the globe. Biotropica 42:21–30.

Lawrence, D., and W. H. Schlesinger. 2001. Changes in soil phosphorus during 200 years of shifting cultivation in Indonesia. Ecology 82:2769–80.

Lawton, R. O., and F. E. Putz. 1988. Natural disturbance and gap-phase regeneration in a wind-exposed tropical cloud forest. Ecology 69:764–77.
Lebrija-Trejos, E., F. Bongers, E. A. P. Garcia, and J. A. Meave. 2008. Successional change and resilience of a very dry tropical deciduous forest following shifting agriculture. Biotropica 40:422–31.
Lebrija-Trejos, E., J. A. Meave, L. Poorter, E. A. Pérez-García, and F. Bongers. 2010. Pathways, mechanisms and predictability of vegetation change during tropical dry forest succession. Perspectives in Plant Ecology, Evolution and Systematics 12:267–75.
Lebrija-Trejos, E., E. A. Pérez-García, J. A. Meave, F. Bongers, and L. Poorter. 2010. Functional traits and environmental filtering drive community assembly in a species-rich tropical system. Ecology 91:386–98.
Lebrija-Trejos, E., E. A. Pérez-García, J. A. Meave, L. Poorter, and F. Bongers. 2011. Environmental changes during secondary succession in a tropical dry forest in Mexico. Journal of Tropical Ecology 27:477–89.
Lee, J. S. H., I. Q. W. Lee, S. L. H. Lim, J. Huijbregts, and N. S. Sodhi. 2009. Changes in dung beetle communities along a gradient of tropical forest disturbance in South-East Asia. Journal of Tropical Ecology 25:677–80.
Lees, A. C., and C. A. Peres. 2009. Gap crossing movements predict species occupancy in Amazonian forest fragments. Oikos 118:280–90.
Leighton, M. 1984. The El Niño Southern Oscillation event in Southeast Asia: Effects of drought and fire in tropical forest in eastern Borneo. WWF-US, Washington, DC.
Lejju, B. J. 2009. Vegetation dynamics in western Uganda during the last 1000 years: Climate change or human induced environmental degradation? African Journal of Ecology 47:21–29.
Lejju, B. J., D. Taylor, and P. Robertshaw. 2005. Late-Holocene environmental variability at Munsa archaeological site, Uganda: A multicore, multiproxy approach. Holocene 15:1044–61.
Lentfer, C., and R. Torrence. 2007. Holocene volcanic activity, vegetation succession, and ancient human land use: Unraveling the interactions on Garua Island, Papua New Guinea. Review of Palaeobotany and Palynology 143:83–105.
Leps, J., V. Novotny, and Y. Basset. 2001. Habitat and successional status of plants in relation to the communities of their leaf-chewing herbivores in Papua New Guinea. Journal of Ecology 89:186–99.
Letcher, S. 2010. Phylogenetic structure of angiosperm communities during tropical forest succession. Proceedings of the Royal Society B: Biological Sciences 277:97–104.
Letcher, S. G., and R. L. Chazdon. 2009a. Lianas and self-supporting plants during tropical forest succession. Forest Ecology and Management 257:2150–56.
———. 2009b. Rapid recovery of biomass, species richness, and species composition in a forest chronosequence in Northeastern Costa Rica. Biotropica 41:608–17.

———. 2012. Life history traits of lianas during tropical forest succession. Biotropica 44:720–27.
Letcher, S. G., R. L. Chazdon, A. Andrade, F. Bongers, M. van Breugel, B. Finegan, S. G. Laurance, R. C. G. Mesquita, M. Martínez-Ramos, and G. B. Williamson. 2012. Phylogenetic community structure during succession: Evidence from three Neotropical forest sites. Perspectives in Plant Ecology, Evolution and Systematics 14:79–87.
Levey, D. J., and M. M. Byrne. 1993. Complex ant-plant interactions: Rain-forest ants as secondary dispersers and post-dispersal seed predators. Ecology 74:1802–12.
Levin, S. A. 1998. Ecosystems and the biosphere as complex adaptive systems. Ecosystems 1:431–36.
Levis, C., P. F. de Souza, J. Schietti, T. Emilio, J. L. P. da Veiga Pinto, C. R. Clement, and F. R. Costa. 2012. Historical human footprint on modern tree species composition in the Purus-Madeira Interfluve, Central Amazonia. PLoS One 7:e48559.
Levy Tacher, S. I., J. R. Aguirre Rivera, M. M. Martínez Romero, and A. Durán Fernández. 2002. Caracterización del uso tradicional de la flora espontánea en la comunidad Lacandona de Lacanhá, Chiapas, México. Interciencia 27:1–13.
Levy Tacher, S. I., and D. J. Golicher. 2004. How predictive is traditional ecological knowledge? The case of the Lacondon Maya fallow enrichment system. Interciencia 29:496–503.
Lewinsohn, T. M., V. Novotny, and Y. Basset. 2005. Insects on plants: Diversity of herbivore assemblages revisited. Annual Review of Ecology, Evolution, and Systematics 36:597–620.
Lewis, S. L., G. Lopez-Gonzalez, B. Sonke, K. Affum-Baffoe, T. R. Baker, L. O. Ojo, O. L. Phillips, et al. 2009. Increasing carbon storage in intact African tropical forests. Nature 457:1003–7.
Lewis, S. L., O. L. Phillips, T. R. Baker, J. Lloyd, Y. Malhi, S. Almeida, N. Higuchi, et al. 2004. Concerted changes in tropical forest structure and dynamics: Evidence from 50 South American long-term plots. Philosophical Transactions of the Royal Society B: Biological Sciences 359:421–36.
Lieberman, D., M. Lieberman, G. Hartshorn, and R. Peralta. 1985. Growth rates and age-size relationships of tropical wet forest trees in Costa Rica. Journal of Tropical Ecology 1:97–109.
Liebsch, D., M. Marques, and R. Goldenberg. 2008. How long does the Atlantic Rain Forest take to recover after a disturbance? Changes in species composition and ecological features during secondary succession. Biological Conservation 141:1717–25.
Li, Y., M. Xu, X. M. Zou, P. J. Shi, and Y. Q. Zhang. 2005. Comparing soil organic carbon dynamics in plantation and secondary forest in wet tropics in Puerto Rico. Global Change Biology 11:239–48.
Lima, H. N., C. E. R. Schaefer, J. W. V. Mello, R. J. Gilkes, and J. C. Ker. 2002. Pedogenesis and pre-Colombian land use of "Terra Preta Anthrosols "("Indian black earth") of Western Amazonia. Geoderma 110:1–17.

Lindenmayer, D. B., K. B. Hulvey, R. J. Hobbs, M. Colyvan, A. Felton, H. Possingham, W. Steffen, K. Wilson, K. Youngentob, and P. Gibbons. 2012. Avoiding bio-perversity from carbon sequestration solutions. Conservation Letters 5:28–36.

Liu, H., and J. Li. 2010. The study of the ecological problems of *Eucalyptus* plantation and sustainable development in Maoming Xiaoliang. Journal of Sustainable Development 3:197–201.

Liu, J., M. Linderman, Z. Ouyang, L. An, J. Yang, and H. Zhang. 2001. Ecological degradation in protected areas: The case of Wolong Nature Reserve for giant pandas. Science 292:98–101.

Livingstone, D. A. 1975. Late Quaternary climate change in Africa. Annual Review of Ecology & Systematics 6:249–80.

Lôbo, D., T. Leão, F. P. L. Melo, A. M. M. Santos, and M. Tabarelli. 2011. Forest fragmentation drives Atlantic Forest of northeastern Brazil to biotic homogenization. Diversity and Distributions 17:287–96.

Lobova, T. A., C. K. Geiselman, and S. A. Mori. 2009. Seed dispersal by bats in the Neotropics. New York Botanical Garden, New York.

Lohbeck, M., L. Poorter, H. Paz, L. Pla, M. van Breugel, M. Martínez-Ramos, and F. Bongers. 2012. Functional diversity changes during tropical forest succession. Perspectives in Plant Ecology, Evolution and Systematics 14:89–96.

Loiselle, B. A., and J. G. Blake. 2002. Potential consequences of extinction of frugivorous birds for shrubs of a tropical wet forest. Pages 397–406 *in* D. J. Levey, W. R. Silva, and M. Galetti, eds. Seed dispersal and frugivory: Ecology, evolution, and conservation. CABI Publishing, Wallingford, UK.

Lombardo, U., E. Canal-Beeby, S. Fehr, and H. Veit. 2011. Raised fields in the Bolivian Amazonia: A prehistoric green revolution or a flood risk mitigation strategy? Journal of Archaeological Science 38:502–12.

Lombardo, U., and H. Prümers. 2010. Pre-Columbian human occupation patterns in the eastern plains of the Llanos de Moxos, Bolivian Amazonia. Journal of Archaeological Science 37:1875–85.

Lopes, A. V., L. C. Girao, B. A. Santos, C. A. Peres, and M. Tabarelli. 2009. Long-term erosion of tree reproductive trait diversity in edge-dominated Atlantic forest fragments. Biological Conservation 142:1154–65.

López, E., G. Bocco, M. Mendoza, A. Velázquez, and J. Rogelio Aguirre-Rivera. 2006. Peasant emigration and land-use change at the watershed level: A GIS-based approach in Central Mexico. Agricultural Systems 90:62–78.

Lopez, J. E., and C. Vaughan. 2004. Observations on the role of frugivorous bats as seed dispersers in Costa Rican secondary humid forests. Acta Chiropterologica 6:111–19.

López-Ulloa, M. V., E. Veldkamp, and G. H. J. de Koning. 2005. Soil carbon stabilization in converted tropical pastures and forests depends on soil type. Soil Science Society of America Journal 69:1110–17.

Lorence, D. H., and R. W. Sussman. 1986. Species invasion into Mauritius wet forest remnants. Journal of Tropical Ecology 2:147–62.

Lozada, T., G. H. J. de Koning, R. Marche, A. M. Klein, and T. Tscharntke. 2007. Tree recovery and seed dispersal by birds: Comparing forest, agroforestry and abandoned agroforestry in coastal Ecuador. Perspectives in Plant Ecology, Evolution and Systematics 8:131–40.

Lozano-García, S., M. Caballero, B. Ortega, S. Sosa, A. Rodríguez, and P. Schaaf. 2010. Late Holocene palaeoecology of Lago Verde: Evidence of human impact and climate change in the northern limit of the Neotropics during the late formative and classic periods. Vegetation History and Archaeobotany 19:177–90.

Lucas, R. M., M. Honzak, I. D. Amaral, P. J. Curran, and G. M. Foody. 2002. Forest regeneration on abandoned clearances in central Amazonia. International Journal of Remote Sensing 23:965–88.

Lucas, R. M., M. Honzak, P. J. Curran, G. M. Foody, and D. T. Nguele. 2000. Characterizing tropical forest regeneration in Cameroon using NOAA AVHRR data. International Journal of Remote Sensing 21:2831–54.

Lugo, A. E. 1992. Comparison of tropical tree plantations with secondary forests of similar age. Ecological Monographs 62:1–41.

———. 1995. Management of tropical biodiversity. Ecological Applications 5:956–61.

———. 1997. The apparent paradox of reestablishing species richness on degraded lands with tree monocultures. Forest Ecology and Management 99:9–19.

———. 2000. Effects and outcomes of Caribbean hurricanes in a climate change scenario. Science of the Total Environment 262:243–51.

———. 2008. Visible and invisible effects of hurricanes on forest ecosystems: An international review. Austral Ecology 33:368–98.

Lugo, A. E., and S. Brown. 1992. Tropical forests as sinks of atmospheric carbon. Forest Ecology and Management 54:239–55.

Lugo, A. E., and E. Helmer. 2004. Emerging forests on abandoned land: Puerto Rico's new forests. Forest Ecology and Management 190:145–61.

Lund, H. G. 2009. What is a degraded forest? White paper prepared for FAO. Forest Information Services, Gainesville, VA. Online at http://home.comcast.net/~gyde/2009forest_degrade.doc.

Lundberg, J., and F. Moberg. 2003. Mobile link organisms and ecosystem functioning: Implications for ecosystem resilience and management. Ecosystems 6:87–98.

MacArthur, R., and E. O. Wilson. 1967. The theory of island biogeography. Princeton University Press, Princeton, NJ.

Macía, M. 2008. Woody plants diversity, floristic composition and land use history in the Amazonian rain forests of Madidi National Park, Bolivia. Biodiversity and Conservation 17:2671–90.

MacPhee, R. D. E. 2008. Insulae infortunatae: Establishing a chronology for Late Quaternary mammal extinctions in the West Indies. Pages 169–93 *in* G. Haynes, ed. American megafaunal extinctions at the end of the Pleistocene. Springer. New York.

Madeira, B., M. M. Espírito-Santo, S. D. Neto, Y. R. F. Nunes, G. A. S. Azofeifa, G. W. Fernandes, and M. Quesada. 2009. Changes in tree and liana communities along

a successional gradient in a tropical dry forest in south-eastern Brazil. Plant Ecology 201:291–304.
Madegowda, C. 2009. Traditional knowledge and conservation. Economic and Political Weekly 44:65–69.
Maginnis, S., J. Rietbergen-McCracken, and W. Jackson. 2007. Introduction. Pages 1–4 *in* J. Rietbergen-McCracken, S. Maginnis, and A. Sarre, eds. The forest landscape restoration handbook. Earthscan, London.
Magnusson, W. E., O. P. de Lima, F. Quintiliano Reis, N. Higuchi, and J. Ferreira Ramos. 1999. Logging activity and tree regeneration in an Amazonian forest. Forest Ecology and Management 113:67–74.
Mahood, S. P., A. C. Lees, and C. A. Peres. 2012. Amazonian countryside habitats provide limited avian conservation value. Biodiversity and Conservation 21:385–405.
Makana, J. R., and S. C. Thomas. 2006. Impacts of selective logging and agricultural clearing on forest structure, floristic composition and diversity, and timber tree regeneration in the Ituri Forest, Democratic Republic of Congo. Biodiversity and Conservation 15:1375–97.
Maley, J. 2002. The catastrophic destruction of African forests around 2500 years ago still exerts a major influence on present vegetation form and distribution. IDS Bulletin 33:13–30.
Maley, J., and P. Brenac. 1998. Vegetation dynamics, palaeoenvironments and climatic changes in the forests of western Cameroon during the last 28,000 years B.P. Review of Palaeobotany and Palynology 99:157–87.
Malhi, Y., L. E. Aragão, D. Galbraith, C. Huntingford, R. Fisher, P. Zelazowski, S. Sitch, C. McSweeney, and P. Meir. 2009. Exploring the likelihood and mechanism of a climate-change-induced dieback of the Amazon rainforest. Proceedings of the National Academy of Sciences 106:20610–15.
Maloney, B. K. 1999. A 10,600 year pollen record from Nong Thale Song Hong, Trang Province, South Thailand. Indo-Pacific Prehistory Association Bulletin 18:129–37.
Mann, C. C. 2005. 1491: New revelations of the Americas before Columbus. Knopf, New York.
———. 2008. Ancient earthmovers of the Amazon. Science 321:1148–52.
Mann, D., J. Edwards, J. Chase, W. Beck, R. Reanier, M. Mass, B. Finney, and J. Loret. 2008. Drought, vegetation change, and human history on Rapa Nui (Isla de Pascua, Eastern Island). Quaternary Research 69:16–28.
Mansfield, B., D. K. Munroe, and K. McSweeney. 2010. Does economic growth cause environmental recovery? Geographical explanations of forest regrowth. Geography Compass 4:416–27.
Mansourian, S., G. Davison, and J. Sayer. 2003. Bringing back the forests: By whom and for whom? Pages 27–36 *in* H. C. Sim, S. Appanah, and P. B. Durst, eds. Bringing back the forest: Policies and practices for degraded lands and forests. Food and Agriculture Organization of the United Nations, Regional Office for Asia and the Pacific, Bangkok, Thailand.

Mansourian, S., D. Vallauri, and N. Dudley. 2005. Forest restoration in landscapes: Beyond planting trees. Springer, New York.

Marchant, R., and H. Hooghiemstra. 2004. Rapid environmental change in African and South American tropics around 4000 years before present: A review. Earth-Science Reviews 66:217–60.

Marín-Spiotta, E., D. F. Cusack, R. Ostertag, and W. L. Silver. 2008. Trends in above and belowground carbon with forest regrowth after agricultural abandonment in the Neotropics. Pages 22–72 *in* R. W. Myster, ed. Post-agricultural succession in the Neotropics. Springer, New York.

Marín-Spiotta, E., R. Ostertag, and W. L. Silver. 2007. Long-term patterns in tropical reforestation: Plant community composition and aboveground biomass accumulation. Ecological Applications 17:828–39.

Marín-Spiotta, E., and S. Sharma. 2012. Carbon storage in successional and plantation forest soils: A tropical analysis. Global Ecology and Biogeography 22:105–17.

Marín-Spiotta, E., W. Silver, and R. Ostertag. 2007. Long-term patterns in tropical reforestation: Plant community composition and aboveground biomass accumulation. Ecological Applications 17:828–39.

Markesteijn, L., J. Iraipi, F. Bongers, and L. Poorter. 2011. Seasonal variation in soil and plant water potentials in a Bolivian tropical moist and dry forest. Journal of Tropical Ecology 26:497–508.

Markesteijn, L., L. Poorter, H. Paz, L. Sack, and F. Bongers. 2011. Ecological differentiation in xylem cavitation resistance is associated with stem and leaf structural traits. Plant, Cell & Environment 34:137–48.

Markewitz, D., E. Davidson, P. Moutinho, and D. Nepstad. 2004. Nutrient loss and redistribution after forest clearing on a highly weathered soil in Amazonia. Ecological Applications 14:177–99.

Marler, T. E., and R. de Moral. 2011. Primary succession along an elevation gradient 15 years after the eruption of Mount Pinatubo, Luzon, Philippines. Pacific Science 65:157–73.

Marlon, J. R., P. J. Bartlein, C. Carcaillet, D. G. Gavin, S. P. Harrison, P. E. Higuera, F. Joos, M. J. Power, and I. C. Prentice. 2008. Climate and human influences on global biomass burning over the past two millennia. Nature Geoscience 1:697–702.

Marquardt, K., R. Milestad, and L. Salomonsson. 2013. Improved fallows: A case study of an adaptive response in Amazonian swidden farming systems. Agriculture and Human Values 30:417–28.

Marquis, R. J. 1991. Herbivore fauna of *Piper* (Piperaceae) in a Costa Rican wet forest: Diversity, specificity and impact. Pages 179–208 *in* P. W. Price, T. M. Lewinsohn, G. W. Fernandes, and W. W. Benson, eds. Plant-animal interactions: Evolutionary ecology in tropical and temperate regions. John Wiley, New York.

Marshall, F., and E. Hildebrand. 2002. Cattle before crops: The beginnings of food production in Africa. Journal of World Prehistory 16:99–143.

Martin, A. R., and S. C. Thomas. 2011. A reassessment of carbon content in tropical trees. PLoS One 6:e23533.
Martin, P. H., and R. J. Fahey. 2006. Fire history along environmental gradients in the subtropical pine forests of the Cordillera Central, Dominica Republic. Journal of Tropical Ecology 22:289–302.
Martin, P. H., R. E. Sherman, and T. J. Fahey. 2004. Forty years of tropical forest recovery from agriculture: Structure and floristics of secondary and old-growth riparian forests in the Dominican Republic. Biotropica 36:297–317.
Martínez-Garza, C., A. Flores-Palacios, M. De La Peña-Domene, and H. F. Howe. 2009. Seed rain in a tropical agricultural landscape. Journal of Tropical Ecology 25:541–50.
Martínez-Garza, C., and H. F. Howe. 2003. Restoring tropical diversity: Beating the time tax on species loss. Journal of Applied Ecology 40:423–29.
Martínez-Ramos, M., E. Alvarez-Buylla, J. Sarukhan, and D. Pinero. 1988. Treefall age determination and gap dynamics in a tropical forest. Journal of Ecology 76:700–716.
Martínez-Ramos, M., and X. García-Orth. 2007. Sucesión ecológica y restauración de las selvas húmedas. Boletin De La Sociedad Botanica De Mexico. Suplemento, 80:69–84.
Mas, J. F. 2005. Assessing protected area effectiveness using surrounding (buffer) areas environmentally similar to the target area. Environmental Monitoring and Assessment 105:69–80.
Mascaro, J., G. P. Asner, H. C. Muller-Landau, M. van Breugel, J. Hall, and K. Dahlin. 2011. Controls over aboveground forest carbon density on Barro Colorado Island, Panama. Biogeosciences 8:1615–29.
Mascaro, J., I. Perfecto, O. Barros, D. H. Boucher, I. G. de la Cerda, J. Ruiz, and J. Vandermeer. 2005. Aboveground biomass accumulation in a tropical wet forest in Nicaragua following a catastrophic hurricane disturbance. Biotropica 37:600–608.
Mata Atlântica SOS. 2002. Atlas de remanescentes da Mata Atlântica. SOS Mata Atlântica/INPE, São Paulo, Brazil.
Mather, A. S. 1992. The forest transition. Area 24:367–79.
Mather, A., and C. Needle. 1999. Development, democracy and forest trends. Global Environmental Change 9:105–18.
Mather, A. S., C. Needle, and J. Fairbairn. 1999. Environmental Kuznets curves and forst trends. Geography 84:55–65.
Matricardi, E. A. T., D. L. Skole, M. A. Pedlowski, W. Chomentowski, and L. C. Fernandes. 2010. Assessment of tropical forest degradation by selective logging and fire using Landsat imagery. Remote Sensing of Environment 114:1117–29.
Matson, P. A., P. M. Vitousek, J. J. Ewel, M. J. Mazzarino, and G. P. Robertson. 1987. Nitrogen transformations following tropical forest felling and burning on a volcanic soil. Ecology 68:491–502.
Mayaux, P., P. Holmgren, F. Achard, H. Eva, H. Stibig, and A. Branthomme. 2005. Tropical forest cover change in the 1990s and options for future monitoring. Philosophical Transactions of the Royal Society B: Biological Sciences 360:373–84.

Mayfield, M. M., and J. M. Levine. 2010. Opposing effects of competitive exclusion on the phylogenetic structure of communities. Ecology Letters 13:1085–93.
Mayle, F. E., and J. Iriarte. 2013. Integrated palaeoecology and archaeology—a powerful approach for understanding Pre-Columbian Amazonia. Journal of Archaeological Science. doi:10.1016/j.jas.2012.08.038.
Mayle, F. E., R. P. Langstroth, R. A. Fisher, and P. Meir. 2007. Long-term forest-savannah dynamics in the Bolivian Amazon: Implications for conservation. Philosophical Transactions of the Royal Society B: Biological Sciences 362:291–307.
Mayle, F. E., and M. J. Power. 2008. Impact of a drier Early- Mid-Holocene climate upon Amazonian forests. Philosophical Transactions of the Royal Society B: Biological Sciences 363:1829–38.
Maza Villalobos, S., P. Balvanera, and M. Martínez Ramos. 2011. Early regeneration of tropical dry forest from abandoned pastures: Contrasting chronosequence and dynamic approaches. Biotropica 43:665–75.
Mazzei, L., P. Sist, A. Ruschel, F. E. Putz, P. Marco, W. Pena, and J. E. R. Ferreira. 2010. Above-ground biomass dynamics after reduced-impact logging in the Eastern Amazon. Forest Ecology and Management 259:367–73.
McClennan, B., and T. Garvin. 2012. Intra-regional variation in land use and livelihood change during a forest transition in Costa Rica's dry North West. Land Use Policy 29:119–30.
McConkey, K. R., and D. R. Drake. 2002. Extinct pigeons and declining bat populations: Are large seeds still being dispersed in the Tropical Pacific? Pages 381–95 *in* D. J. Levey, W. R. Silva, and M. Galetti, eds. Seed dispersal and frugivory:Ecology, evolution, and conservation. CABI Publishing, Wallingford, UK.
McConkey, K. R., D. R. Drake, J. Franklin, and F. Tonga. 2004. Effects of Cyclone Waka on flying foxes (*Pteropus tonganus*) in the Vava'u Islands of Tonga. Journal of Tropical Ecology 20:555–61.
McCulloh, K. A., F. C. Meinzer, J. S. Sperry, B. Lachenbruch, S. L. Voelker, D. R. Woodruff, and J. C. Domec. 2011 Comparative hydraulic architecture of tropical tree species representing a range of successional stages and wood density. Oecologia 167:27–37.
McElwee. 2009. Reforesting "Bare Hills" in Vietnam: Social and environmental conseqences of the 5 Million Hectare Reforestation Program. AMBIO 38:325–32.
McGrath, D. A., C. K. Smith, H. L. Gholz, and F. A. Oliveira. 2001. Effects of land-use change on soil nutrient dynamics in Amazonia. Ecosystems 4:625–45.
McKey, D., T. R. Cavagnaro, J. Cliff, and R. Gleadow. 2010. Chemical ecology in coupled human and natural systems: People, manioc, multitrophic interactions and global change. Chemoecology 20:109–33.
McMichael, C. H., M. B. Bush, D. R. Piperno, M. R. Silman, A. R. Zimmerman, and C. Anderson. 2012. Spatial and temporal scales of pre-Columbian disturbance associated with western Amazonian lakes. Holocene 22:131–41.

McMichael, C., A. Correa-Metrio, and M. Bush. 2012. Pre-Columbian fire regimes in lowland tropical rainforests of southeastern Peru. Palaeogeography, Palaeoclimatology, Palaeoecology 342–43:73–83.
McMichael, C., D. Piperno, M. Bush, M. Silman, A. Zimmerman, M. Raczka, and L. Lobato. 2012. Sparse Pre-Columbian human habitation in Western Amazonia. Science 336:1429–31.
McNamara, S., P. Erskine, D. Lamb, L. Chantalangsy, and S. Boyle. 2012. Primary tree species diversity in secondary fallow forests of Laos. Forest Ecology and Management 281:93–99.
MEA. 2005. Ecosystems and human well-being: Synthesis. Millennium Ecosystem Assessment. Island Press, Washington, DC.
Medellín, R. A., and M. Equihua. 1998. Mammal species richness and habitat use in rainforest and abandoned agricultural fields in Chiapas, Mexico. Journal of Applied Ecology 35:13–23.
Medellín, R. A., and O. Gaona. 1999. Seed dispersal by bats and birds in forest and disturbed habitats of Chiapas, México. Biotropica 31:478–85.
Meggars, B. J. 1994. Archaeological evidence for the impact of Mega Niño events on Amazonia during the past two millennia. Climate Change 28:321–38.
Meijaard, E., and D. Sheil. 2007. A logged forest in Borneo is better than none at all. Nature 446:974–74.
———. 2008. The persistence and conservation of Borneo's mammals in lowland rain forests managed for timber: Observations, overviews and opportunities. Ecological Research 23:21–34.
Meijaard, E., D. Sheil, A. J. Marshall, and R. Nasi. 2008. Phylogenetic age is positively correlated with sensitivity to timber harvest in Bornean mammals. Biotropica 40:76–85.
Meijaard, E., D. Sheil, R. Nasi, D. Augeri, B. Rosenbaum, D. Iskandar, T. Setyawati, A. Lammertink, I. Rachmatika, and A. Wong. 2005. Life after logging: Reconciling wildlife conservation and production forestry in Indonesian Borneo. CIFOR, Bogor, Indonesia.
Meinzer, F. C., D. R. Woodruff, J. C. Domec, G. Goldstein, P. I. Campanello, M. G. Gatti, and R. Villalobos-Vega. 2008. Coordination of leaf and stem water transport properties in tropical forest trees. Oecologia 156:31–41.
Mellars, P. 2006. Going east: New genetic and archaeological perspectives on the modern human colonization of Eurasia. Science 313:796–800.
Mello, M. A. R., F. M. D. Marquitti, P. R. Guimarães, E. K. V. Kalko, P. Jordano, and M. A. M. de Aguiar. 2011a. The missing part of seed dispersal networks: Structure and robustness of bat-fruit interactions. PLoS One 6:e17395.
———. 2011b. The modularity of seed dispersal: Differences in structure and robustness between bat- and bird-fruit networks. Oecologia 167:131–40.
Melo, F. P. L., V. Arroyo-Rodríguez, L. Fahrig, M. Martínez-Ramos, and M. Tabarelli. 2013. On the hope for biodiversity-friendly tropical landscapes. Trends in Ecology & Evolution 28:462–68.

Melo, F. P. L., R. Dirzo, and M. Tabarelli. 2006. Biased seed rain in forest edges: Evidence from the Brazilian Atlantic forest. Biological Conservation 132:50–60.

Melo, F. P. L., B. Rodriguez-Herrera, R. L. Chazdon, R. A. Medellín, and G. G. Ceballos. 2009. Small tent-roosting bats promote dispersal of large-seeded plants in a Neotropical forest. Biotropica 41:737–43.

Menz, M. H. M., K. W. Dixon, and R. J. Hobbs. 2013. Hurdles and opportunities for landscape-scale restoration. Science 339:526–27.

Menz, M. H. M., R. D. Phillips, R. Winfree, C. Kremen, M. A. Aizen, S. D. Johnson, and K. W. Dixon. 2011. Reconnecting plants and pollinators: Challenges in the restoration of pollination mutualisms. Trends in Plant Science 16:4–12.

Mercader, J. 2003. Under the canopy: The archaeology of tropical rain forests. Rutgers University Press, New Brunswick, NJ.

Mesquita, R. C. G., K. Ickes, G. Ganade, and G. B. Williamson. 2001. Alternative successional pathways in the Amazon Basin. Journal of Ecology 89:528–37.

Messier, C., K. J. Puettmann, and K. D. Coates. 2013. Managing forests as complex adaptive systems: Building resilience to the challenge of global change. Routledge, New York.

Mestre, L. A. M., M. A. Cochrane, and J. Barlow. 2013. Long-term changes in bird communities after wildfires in the Central Brazilian Amazon. Biotropica 45:480–88.

Metz, M. R., W. P. Sousa, and R. Valencia. 2011. Widespread density-dependent seedling mortality promotes species coexistence in a highly diverse Amazonian rainforest. Ecology 92:1723–29.

Meyfroidt, P., and E. F. Lambin. 2008. Forest transition in Vietnam and its environmental impacts. Global Change Biology 14:1319–36.

———. 2009. Forest transition in Vietnam and displacement of deforestation abroad. Proceedings of the National Academy of Sciences 106:16139–44.

———. 2010. Forest transition in Vietnam and Bhutan: Causes and environmental impacts. Pages 315–39 *in* H. Nagendra and J. Southworth, eds. Reforesting landscapes: Linking pattern and process. Springer, Dordrecht, Neth.

———. 2011. Global forest transition: Prospects for an end to deforestation. Annual Review of Environment and Resources 36:343–71.

Meyfroidt, P., T. K. Rudel, and E. F. Lambin. 2010. Forest transitions, trade, and the global displacement of land use. Proceedings of the National Academy of Sciences 107:20917–22.

Michon, G. 2005. Domesticating forests: How farmers manage forest resources. Center for International Forestry Research/World Agroforestry Center, Bogor, Indonesia.

Michon, G., H. de Foresta, P. Levant, and F. Verdeaux. 2007. Domestic forests: A new paradigm for integrating local communities' forestry into tropical forest science. Ecology and Society 12. http://www.ecologyandsociety.org/vol12/iss12/art11/.

Miles, L., A. C. Newton, R. S. DeFries, C. Ravilious, I. May, S. Blyth, V. Kapos, and J. E.

Gordon. 2006. A global overview of the conservation status of tropical dry forests. Journal of Biogeography 33:491–505.
Milne, S., and E. Niesten. 2009. Direct payments for biodiversity conservation in developing countries: Practical insights for design and implementation. Oryx 43:530–41.
Miquel, S. 1987. Morphologie fonctionelle de plantules d'especes forestiéres du Gabon. Bulletin du Museum National d'Histoire Naturelle, 9:101–21.
Moguel, P., and V. M. Toledo. 1999. Biodiversity conservation in traditional coffee systems of Mexico. Conservation Biology 13:11–21.
Moles, A. T., D. D. Ackerly, C. O. Webb, J. C. Tweddle, J. B. Dickie, and M. Westoby. 2005. A brief history of seed size. Science 307:576–80.
Molino, J.-F., and D. Sabatier. 2001. Tree diversity in tropical rain forests: A validation of the Intermediate Disturbance Hypothesis. Science 294:1702–4.
Momose, K., T. Yumoto, T. Nagamitsu, M. Kato, H. Nagamasu, S. Sakai, R. D. Harrison, T. Itioka, A. A. Hamid, and T. Inoue. 1998. Pollination biology in a lowland dipterocarp forest in Sarawak, Malaysia. I. Characteristics of the plant-pollinator community in a lowland dipterocarp forest. American Journal of Botany 85:1477–1501.
Montagnini, F., and C. Finney. 2011. Payments for environmental services in Latin America as a tool for restoration and rural development. AMBIO 40:285–97.
Montgomery, R. A., and R. L. Chazdon. 2001. Forest structure, canopy architecture and light transmittance in tropical wet forests. Ecology 82:2707–18.
Montoya, J. M., S. L. Pimm, and R. V. Solé. 2006. Ecological networks and their fragility. Nature 442:259–64.
Moran, C., C. P. Catterall, and J. Kanowski. 2009. Reduced dispersal of native plant species as a consequence of the reduced abundance of frugivore species in fragmented rainforest. Biological Conservation 142:541–52.
Moran, E. F., E. Brondizio, P. Mausel, and Y. Wu. 1994. Integrating Amazonian vegetation, land-use, and satellite data. BioScience 44:329–49.
Moran, E. F., E. Brondizio, J. M. Tucker, M. C. da Silva-Fosberg, S. McCracken, and I. Falesi. 2000. Effects of soil fertility and land-use on forest succession in Amazônia. Forest Ecology and Management 139:93–108.
Morcote-Rios, G., and R. Bernal. 2001. Remains of palms (Palmae) at archaeological sites in the New World: A review. Botanical Review 67:309–50.
Moreno, C. E., G. Castillo-Campos, and J. R. Verdú. 2009. Taxonomic diversity as complementary information to assess plant species diversity in secondary vegetation and primary tropical deciduous forest. Journal of Vegetation Science 20:935–43.
Morley, R. J. 2000. Origin and evolution of tropical rain forests. John Wiley & Sons, New York.
Morse, W. C., J. L. Schedlbauer, S. E. Sesnie, B. Finegan, C. A. Harvey, S. J. Hollenhorst, K. L. Kavanagh, D. Stoian, and J. D. Wulfhorst. 2009. Consequences

of environmental service payments for forest retention and recruitment in a Costa Rican biological corridor. Ecology and Society 14. http://www.ecologyandsociety.org/vol14/iss1/art23/.

Mostacedo, B., F. E. Putz, T. S. Fredericksen, A. Villca, and T. Palacios. 2009. Contributions of root and stump sprouts to natural regeneration of a logged tropical dry forest in Bolivia. Forest Ecology and Management 258:978–85.

Mueller, A., G. Islebe, F. Anselmetti, D. Ariztegui, M. Brenner, D. Hodell, I. Hajdas, Y. Hamann, G. Haug, and D. Kennett. 2010. Recovery of the forest ecosystem in the tropical lowlands of northern Guatemala after disintegration of Classic Maya polities. Geology 38:523–26.

Mueller, A. D., G. A. Islebe, M. B. Hillesheim, D. A. Grzesik, F. S. Anselmatti, D. Ariztegui, M. Brenner, J. H. Curtis, D. A. Hodell, and K. A. Venz. 2009. Climate drying and associated forest decline in the lowlands of northern Guatemala during the late Holocene. Quaternary Research 71:133–41.

Mueller-Dombois, D. 1992. Distributional dynamics in the Hawaiian vegetation. Pacific Science 46:221–31.

Muller-Landau, H. C. 2007. Predicting the long-term effects of hunting on plant species composition and diversity in tropical forests. Biotropica 39:372–84.

———. 2009. Carbon cycle: Sink in the African jungle. Nature 457:969–70.

Muñiz-Castro, M. A., G. Williams-Linera, and M. Martínez-Ramos. 2012. Dispersal mode, shade tolerance, and phytogeographical affinity of tree species during secondary succession in tropical montane cloud forest. Plant Ecology 213:339–53.

Murcia, C. 1997. Evaluation of Andean alder as a catalyst for the recovery of tropical cloud forests in Colombia. Forest Ecology and Management 99:163–70.

Murgueitio, E., Z. Calle, F. Uribe, A. Calle, and B. Solorio. 2011. Native trees and shrubs for the productive rehabilitation of tropical cattle ranching lands. Forest Ecology and Management 261:1654–63.

Murray, K. G., K. Willet-Murray, J. Robert, K. Horjus, W. A. Haber, W. Zuckowski, M. Kuhlmann, and T. M. Long-Robinson. 2008. The roles of disperser behavior and physical habitat structure in regeneration of post-agricultural fields. Pages 192–215 *in* R. W. Myster, ed. Post-agricultural succession in the Neotropics. Springer, New York.

Muscarella, R., and T. H. Fleming. 2007. The role of frugivorous bats in tropical forest succession. Biological Reviews 82:573–90.

Mustalahti, I., A. Bolin, E. Boyd, and J. Paavola. 2012. Can REDD+ reconcile local priorities and needs with global mitigation benefits? Lessons from Angai Forest, Tanzania. Ecology and Society 17. http://www.ecologyandsociety.org/vol17/iss1/art16/.

Muthukumar, T., L. Sha, X. Yang, M. Cao, J. Tang, and Z. Zheng. 2003. Distribution of roots and arbuscular mycorrhizal associations in tropical forest types of Xishuangbanna, southwest China. Applied Soil Ecology 22:241–53.

Mwavu, E., and E. Witkowski. 2008. Sprouting of woody species following cutting and

tree-fall in a lowland semi-deciduous tropical rainforest, North-Western Uganda. Forest Ecology and Management 255:982–92.
Myster, R. W. 2004. Post-agricultural invasion, establishment, and growth of Neotropical trees. Botanical Review 70:381–402.
———. 2007. Early successional pattern and process after sugarcane, banana, and pasture cultivation in Ecuador. New Zealand Journal of Botany 45:101–10.
———. 2008a. Introduction. Pages 3–21 *in* R. W. Myster, ed. Post-agricultural succession in the Neotropics. Springer, New York.
———. 2008b. Neotropic post-dispersal seed predation. Pages 216–20 *in* R. W. Myster, ed. Post-agricultural succession in the Neotropics. Springer, New York.
Nabe-Nielsen, J., W. Severiche, T. Fredericksen, and L. I. Nabe-Nielsen. 2007. Timber tree regeneration along abandoned logging roads in a tropical Bolivian forest. New Forests 34:31–40.
Nadkarni, N. M., and W. A. Haber. 2009. Canopy seed banks as time capsules of biodiversity in pasture-remnant tree crowns. Conservation Biology 23:1117–26.
Nagendra, H. 2007. Drivers of reforestation in human-dominated forests. Proceedings of the National Academy of Sciences 104:15218–23.
———. 2009. Drivers of regrowth in South Asia's human impacted forests. Current Science 97:1586–92.
———. 2010. Reforestation and regrowth in the human dominated landscapes of South Asia. Pages 149–74 *in* H. Nagendra and J. Southworth, eds. Reforesting landscapes: Linking pattern and process. Springer, Dordrecht, Neth.
Nagendra, H., and J. Southworth. 2010. Reforesting landscapes: Linking pattern and process. Springer, Dordrecht, Neth.
Nathan, R., and H. C. Muller-Landau. 2000. Spatial patterns of seed dispersal, their determinants and consequences for recruitment. Trends in Ecology & Evolution 15:278–85.
Nations, J., and R. Nigh. 1980. The evolutionary potential of Lacandon Maya sustained-yield tropical forest agriculture. Journal of Anthropological Research 36:1–30.
Navarrete, D., and G. Halffter. 2008. Dung beetle (Coleoptera: Scarabaeidae: Scarabaeinae) diversity in continuous forest, forest fragments and cattle pastures in a landscape of Chiapas, Mexico: The effects of anthropogenic changes. Biodiversity and Conservation 17:2869–98.
Nave, A. G., and R. R. Rodrigues. 2007. Combination of species into filling and diversity groups as forest restoration methodology. Pages 103–26 *in* R. R. Rodrigues, S. V. Martins, and S. Gandolfi, eds. High diversity forest restoration in degraded areas. Nova Science Publishers, New York.
Nayak, K. G., and P. Davidar. 2010. Pollination and breeding systems of woody plant species in tropical dry evergreen forests, southern India. Flora-Morphology, Distribution, Functional Ecology of Plants 205:745–53.
Neff, H., D. M. Pearsall, J. G. Jones, B. Arroyo de Pieters, and D. E. Freidel. 2006.

Climate change and population history in the Pacific lowlands of Southern Mesoamerica. Quaternary Research 65:390–400.

Negrón-Juárez, R. I., J. Q. Chambers, G. Guimaraes, H. Zeng, C. F. M. Raupp, D. M. Marra, G. Ribeiro, S. S. Saatchi, B. W. Nelson, and N. Higuchi. 2010. Widespread Amazon forest tree mortality from a single cross-basin squall line event. Geophysical Research Letters 37:L16701 (5 pp.).

Neill, C., and E. Davidson. 2000. Soil carbon accumulation or loss following deforestation for pasture in the Brazilian Amazon. Pages 197–211 *in* R. Lal, J. M. Kimble, and B. A. Stewart, eds. Global climate change and tropical ecosystems. CRC Press, Boca Raton, FL.

Neill, C., J. M. Melillo, P. A. Steudler, C. C. Cerri, J. F. L. de Moraes, M. C. Piccolo, and M. Brito. 1997. Soil carbon and nitrogen stocks following forest clearing for pasture in the southwestern Brazilian Amazon. Ecological Applications 7:1216–25.

Nelson, B. W. 1995. Natural disturbance and change in the Brazilian Amazon. Pages 61–65 *in* R. C. J. Somerville, W. J. Emery, and C. J. Tucker, eds. Changes in global vegetative patterns and their relationship to human activity. Aspen Global Change Institute, Aspen, CO.

Nelson, B. W., and M. N. Irmão. 1998. Fire penetration in standing Amazon forests. Proceedings of the Brazilian Remote Sensing Congress 9:3–18.

Nelson, B. W., V. Kapos, J. Adams, W. Oliveira, and O. Braun. 1994. Forest disturbance by large blowdowns in the Brazilian Amazon. Ecology 75:853–58.

Nelson, R. F., D. S. Kimes, W. A. Salas, and M. Routhier. 2000. Secondary forest age and tropical forest biomass estimation using Thematic Mapper imagery. BioScience 50:419–31.

Nelson, R. W. 1994. Natural forest disturbance and change in the Brazilian Amazon. Remote Sensing Reviews 10:105–25.

Nepstad, D. C., C. R. De Carvalho, E. A. Davidson, P. H. Jipp, P. A. Lefebvre, G. H. Negreiros, E. D. Da Silva, T. A. Stone, S. E. Trumbore, and S. Vieira. 1994. The role of deep roots in the hydrological and carbon cycles of Amazonian forests and pastures. Nature 372:666–69.

Nepstad, D. C., P. Moutinho, and D. Markewitz. 2001. The recovery of biomass, nutrient stocks and deep soil functions in secondary forests. Pages 139–55 *in* M. E. McClain, R. L. Victoria, and J. E. Richey, eds. The biogeochemistry of the Amazon Basin. Oxford University Press, New York.

Nepstad, D. C., C. M. Stickler, B. Soares, and F. Merry. 2008. Interactions among Amazon land use, forests and climate: Prospects for a near-term forest tipping point. Philosophical Transactions of the Royal Society B: Biological Sciences 363:1737–46.

Nepstad, D. C., I. M. Tohver, D. Ray, P. Moutinho, and G. Cardinot. 2007. Mortality of large trees and lianas following experimental drought in an Amazon forest. Ecology 88:2259–69.

Nepstad, D., C. Uhl, C. A. Pereira, and J. M. C. da Silva. 1996. A comparative study of

tree establishment in abandoned pasture and mature forest of eastern Amazonia. Oikos 76:25–39.
Nepstad, D. C., C. Uhl, and E. A. S. Serrão. 1991. Recuperation of a degraded Amazonian landscape: Forest recovery and agricultural restoration. AMBIO 20:248–55.
Neumann, K., K. Bostoen, A. Höhn, S. Kahlheber, A. Ngomanda, and B. Tchiengué. 2012. First farmers in the Central African rainforest: A view from southern Cameroon. Quaternary International 249:53–62.
Neumann-Cosel, L., B. Zimmermann, J. S. Hall, M. van Breugel, and H. Elsenbeer. 2011. Soil carbon dynamics under young tropical secondary forests on former pastures: A case study from Panama. Forest Ecology and Management 261:1625–33.
Neves, E. G., J. B. Petersen, R. N. Bartone, and C. A. Da Silva. 2003. Historical and socio-cultural origins of Amazonian dark earths. Pages 29–50 *in* J. Lehmann, D. Kern, B. Glaser, and W. Woods, eds. Amazonian dark earths: Origin, properties, management. Kluwer Academic Publishers, Dordrecht, Neth.
Neves, F. S., R. F. Braga, M. M. do Espírito-Santo, J. H. C. Delabie, G. W. Fernandes, and G. A. Sánchez-Azofeifa. 2010. Diversity of arboreal ants in a Brazilian tropical dry forest: Effects of seasonality and successional stage. Sociobiology 56:177–94.
Nevle, R. J., and D. K. Bird. 2008. Effects of syn-pandemic fire reduction and reforestation in the tropical Americas on atmospheric CO_2 during the European conquest. Palaeogeography, Palaeoclimatology, Palaeoecology 264:25–38.
Nevle, R. J., D. Bird, W. Ruddiman, and R. Dull. 2011. Neotropical human-landscape interactions, fire, and atmospheric CO_2 during European conquest. Holocene 21:853–64.
Newbery, D. M., I. J. Alexander, and J. A. Rother. 2000. Does proximity to conspecific adults influence the establishment of ectomycorrhizal trees in rain forest? New Phytologist 147:401–9.
Newbery, D. M., and J. Gartlan. 1996. A structural analysis of rain forest at Korup and Douala-Edea, Cameroon. Proceedings of the Royal Society of Edinburgh B: Biological Sciences 104:177–224.
Newton, A. C., R. F. del Castillo, C. Echeverría, D. Geneletti, M. González-Espinosa, L. R. Malizia, A. C. Premoli, J. M. R. Benayas, C. Smith-Ramírez, and G. Williams-Linera. 2012. Forest landscape restoration in the drylands of Latin America. Ecology and Society 17:21.
Newton, A. C., and N. Tejedor. 2011. Introduction. Pages 1–22 *in* A. C. Newton and N. Tejedor, eds. Principles and practice of forest landscape restoration: Case studies from the drylands of Latin America. IUCN, Gland, Switzerland.
Ng, F. S. P. 1978. Strategies of establishment in Malayan forest trees. Pages 129–62 *in* P. B. Tomlinson and M. H. Zimmerman, eds. Tropical trees as living systems. Cambridge University Press, Cambridge.
Ngomanda, A., A. Chepstow-Lusty, M. Makaya, C. Favier, P. Schevin, J. Maley,

M. Fontugne, R. Oslisly, and D. Jolly. 2009. Western equatorial African forest-savanna mosaics: A legacy of late Holocene climate hange? Climate of the Past 5:647–59.
Ngomanda, A., K. Neumann, A. Schweizer, and J. Maley. 2009. Seasonality change and the third millennium BP rainforest crisis in southern Cameroon (Central Africa). Quaternary Research 71:307–18.
Nichols, E., T. Larsen, S. Spector, A. Davis, F. Escobar, M. Favila, and K. Vulinec. 2007. Global dung beetle response to tropical forest modification and fragmentation: A quantitative literature review and meta-analysis. Biological Conservation 137:1–19.
Nicotra, A. B., R. L. Chazdon, and S. Iriarte. 1999. Spatial heterogeneity of light and woody seedling regeneration in tropical wet forests. Ecology 80:1908–26.
Nobre, C., and L. Borma. 2009. 'Tipping points' for the Amazon forest. Current Opinion in Environmental Sustainability 1:28–36.
Nock, C. A., D. Geihofer, M. Grabner, P. J. Baker, S. Bunyavejchewin, and P. Hietz. 2009. Wood density and its radial variation in six canopy tree species differing in shade-tolerance in western Thailand. Annals of Botany 104:297–306.
Nogueira, E. M., P. M. Fearnside, and B. W. Nelson. 2008. Normalization of wood density in biomass estimates of Amazon forests. Forest Ecology and Management 256:990–96.
Nogueira, L. R., Jr., J. L. M. Gonçalves, V. L. Engel, J. Parrotta, D. Mendes Lopes, and M. Tome. 2011. Soil dynamics and carbon stocks 10 years after restoration of degraded land using Atlantic Forest tree species. Forest Systems 20:536–45.
Norden, N., R. L. Chazdon, A. Chao, Y. H. Jiang, and B. Vilchez-Alvarado. 2009. Resilience of tropical rain forests: Tree community reassembly in secondary forests. Ecology Letters 12:385–94.
Norden, N., S. G. Letcher, V. Boukili, N. G. Swenson, and R. Chazdon. 2012. Demographic drivers of successional changes in phylogenetic structure across life-history stages in plant communities. Ecology 93:S70-S82.
Norden, N., R. C. G. Mesquita, T. V. Bentos, R. L. Chazdon, and G. B. Williamson. 2011. Contrasting community compensatory trends in alternative successional pathways in central Amazonia. Oikos 120:143–51.
Norisada, M., G. Hitsuma, K. Kuroda, T. Yamanoshita, M. Masumori, T. Tange, H. Yagi, T. Nuyim, S. Sasaki, and K. Kojima. 2005. *Acacia mangium*, a nurse tree candidate for reforestation on degraded sandy soils in the Malay Peninsula. Forest Science 51:498–510.
Normile, D. 2010. Saving forests to save biodiversity. Science 329:1278–80.
Notman, E. M., and D. L. Gorchov. 2001. Variation in post-dispersal seed predation in mature Peruvian lowland tropical forest and fallow agricultural sites. Biotropica 33:621–36.
Notman, E. M., and A. C. Villegas. 2005. Patterns of seed predation by vertebrate versus invertebrate seed predators among different plant species, seasons and spatial distributions. Pages 55–75 *in* P.-M. Forget, J. E. Lambert, P. E. Hulme, and S. B. Vander Wall, eds. Seed fate. CAB International, Cambridge, MA.

Novotny, V. 1994. Association of polyphagy in leafhoppers (Auchenorrhyncha, Hemiptera) with unpredictable environments. Oikos 70:223–32.

Nuñez-Iturri, G., O. Olsson, and H. F. Howe. 2008. Hunting reduces recruitment of primate-dispersed trees in Amazonian Peru. Biological Conservation 141:1536–46.

Nye, P. H., and D. Greenland. 1960. The soil under shifting cultivation. Commonwealth Agricultural Bureaux. Harpenden, UK.

Nykvist, N. 1996. Regrowth of secondary vegetation after the 'Borneo fire' of 1982–1983. Journal of Tropical Ecology 12:307–12.

Ochoa-Gaona, S., F. Hernández-Vázquez, B. H. J. De Jong, and F. D. Gurri-García. 2007. Loss of floristic diversity over an intensification gradient of the slash-and-burn agricultural system: A case study in the Selva Lacandona region, Chiapas, Mexico. Boletin De La Sociedad Botanica De Mexico 81:65–80.

Oliveira, L., S. Hankerson, J. Dietz, and B. Raboy. 2010. Key tree species for the golden-headed lion tamarin and implications for shade-cocoa management in southern Bahia, Brazil. Animal Conservation 13:60–70.

Oliveira, P. J. C., G. P. Asner, D. E. Knapp, A. Almeyda, R. Galván-Gildemeister, S. Keene, R. F. Raybin, and R. C. Smith. 2007. Land-use allocation protects the Peruvian Amazon. Science 317:1233–36.

Oliveira, R. R. 2007. Mata Atlântica, paleoterritorórios e história ambiental. Ambiente & Sociedade 10:11–23.

———. 2008. When the shifting agriculture is gone: Functionality of Atlantic Coastal Forest in abandoned farming sites. Boletim do Museu Paraense Emílio Goeldi. Ciências Humanas 3:213–26.

Oliveira-Filho, A. T., E. A. Vilela, M. L. Gavilanes, and D. A. Carvalho. 1994. Effect of flooding regime and understorey bamboos on the physiognomy and tree species composition of a tropical semideciduous forest in Southeastern Brazil. Vegetatio 113:99–124.

Oliver, C. D., and B. C. Larson. 1996. Forest stand dynamics. Biological resource management series. McGraw-Hill, New York.

Oliver, J. R. 2008. The archaeology of agriculture in ancient Amazonia. Pages 185–216 *in* H. Silverman and W. Isbell, eds. Handbook of South American Archaeology. Springer, New York.

Ollerton, J., R. Winfree, and S. Tarrant. 2011. How many flowering plants are pollinated by animals? Oikos 120:321–26.

Omeja, P. A., C. A. Chapman, J. Obua, J. S. Lwanga, A. L. Jacob, F. Wanyama, and R. Mugenyi. 2010. Intensive tree planting facilitates tropical forest biodiversity and biomass accumulation in Kibale National Park, Uganda. Forest Ecology and Management 261:703–9.

Omeja, P. A., J. S. Lwanga, J. Obua, and C. A. Chapman. 2011. Fire control as a simple means of promoting tropical forest restoration. Tropical Conservation Science 4:287–99.

Omeja, P. A., J. Obua, A. Rwetsiba, and C. A. Chapman. 2012. Biomass accumulation

in tropical lands with different disturbance histories: Contrasts within one landscape and across regions. Forest Ecology and Management 269:293–300.
Onguene, N., and T. Kuyper. 2002. Importance of the ectomycorrhizal network for seedling survival and ectomycorrhiza formation in rain forests of south Cameroon. Mycorrhiza 12:13–17.
Onofre, F., V. Engel, and H. Cassola. 2010. Natural regeneration of Atlantic Forest species in the understory of *Eucalyptus saligna* Smith. in a former forest production unit at the parque das Neblinas, Bertioga, SP. Scientia Forestalis 38:39–52.
Op den Camp, R. H. M., E. Polone, E. Fedorova, W. Roelofsen, A. Squartini, H. J. M. Op den Camp, T. Bisseling, and R. Geurts. 2012. Nonlegume *Parasponia andersonii* deploys a broad rhizobium host range strategy resulting in largely variable symbiotic effectiveness. Molecular Plant-Microbe Interactions 25:954–63.
Oppenheimer, S. 2009. The great arc of dispersal of modern humans: Africa to Australia. Quaternary International 202:2–13.
Oslisly, R. 2001. The history of human settlement in the Middle Ogooué Valley (Gabon). Pages 101–18 *in* W. Weber, L. J. T. White, A. Vedder, and L. Naughton-Treves, eds. African rain forest ecology and conservation: An interdisciplinary perspective. Yale University Press, New Haven, CT.
Oslisly, R., and L. White. 2003. Étude des traces de l'impact de l'homme sur l'environnement au cours de l'Holocène dans deux regions d'Afrique central forestière. Pages 77–88 *in* A. Froment and J. Guffroy, eds. Peuplements anciens et actuels de forêts tropicales: Actes de Séminaire-atelier. IRD Éditions, Paris.
———. 2007. Human impact and environmental exploitation in Gabon during the Holocene. Pages 347–60 *in* T. P. Denham, J. Iriarte, and L. Vrydaghs, eds. Rethinking agriculture: Archaeological and ethnoarchaeological perspectives. Left Coast Press, Walnut Creek, CA.
Osorio-Pérez, K., M. Barberena-Arias, and T. Aide. 2007. Changes in ant species richness and composition during plant secondary succession in Puerto Rico. Caribbean Journal of Science 43:244–53.
Ostertag, R., E. Marin-Spiotta, W. L. Silver, and J. Schulten. 2008. Litterfall and decomposition in relation to soil carbon pools along a secondary forest chronosequence in Puerto Rico. Ecosystems 11:701–14.
Ostertag, R., W. L. Silver, and A. E. Lugo. 2005. Factors affecting mortality and resistance to damage following hurricanes in a rehabilitated subtropical moist forest. Biotropica 37:16–24.
Osunkoya, O. O., T. K. Sheng, N. A. Mahmud, and N. Damit. 2007. Variation in wood density, wood water content, stem growth and mortality among twenty-seven tree species in a tropical rainforest on Borneo Island. Austral Ecology 32:191–201.
Otsamo, A., G. Ådjers, T. Hadi, J. Kuusipalo, K. Tuomela, and R. Vuokko. 1995. Effect of site preparation and initial fertilization on the establishment and growth of four plantation tree species used in reforestation of *Imperata cylindrica* (L.) Beauv. dominated grasslands. Forest Ecology and Management 73:271–77.

Otsamo, R. 2000. Secondary forest regeneration under fast-growing forest plantations on degraded *Imperata cylindrica* grasslands. New Forests 19:69–93.
Ottermanns, R., P. W. Hopp, M. Guschal, G. P. dos Santos, S. Meyer, and M. Rof-Nickoll. 2011. Causal relationship between leaf litter beetle communities and regeneration patterns of vegetation in the Atlantic rainforest of Southern Brazil (Mata Atlantica). Ecological Complexity 8:299–309.
Otterstrom, S., M. Schwartz, and I. Velázquez-Rocha. 2006. Responses to fire in selected topical dry forest trees. Biotropica 38:592–98.
Padoch, C., K. Coffey, O. Mertz, S. J. Leisz, J. Fox, and R. L. Wadley. 2007. The demise of swidden in southeast Asia? Local realities and regional ambiguities. Geografisk Tidsskrift 107:29–41.
Padoch, C., and M. Pinedo-Vásquez. 2006. Concurrent activities and invisible technologies: An example of timber management in Amazonia. Pages 172–80 *in* D. A. Posey and M. J. Balick, eds. Human impacts on Amazonia: The role of traditional ecological knowledge in conservation and development. Columbia University Press, New York.
Pagiola, S., E. Ramirez, J. Gobbi, C. de Haan, M. Ibrahim, E. Murgueitio, and J. P. Ruis. 2007. Paying for the environmental services of silvopastoral practices in Nicaragua. Ecological Economics 64:374–85.
Pagiola, S., W. Zhang, and A. Colom. 2010. Can payments for watershed services help finance biodiversity conservation? A spatial analysis of highland Guatemala. Journal of Natural Resources Policy Research 2:7–24.
Paladino, S. 2011. Tracking the fault lines of pro-poor carbon forestry. Culture, Agriculture, Food and Environment 33:117–32.
Palmeirim, J., D. Gorchoy, and S. Stoleson. 1989. Trophic structure of a Neotropical frugivore community: Is there competition between birds and bats? Oecologia 79:403–11.
Palomaki, M. B., R. L. Chazdon, J. P. Arroyo, and S. G. Letcher. 2006. Juvenile tree growth in relation to light availability in second-growth tropical rain forests. Journal of Tropical Ecology 22:223–26.
Pan, Y., R. A. Birdsey, J. Fang, R. Houghton, P. E. Kauppi, W. A. Kurz, O. L. Phillips, A. Shvidenko, S. L. Lewis, and J. G. Canadell. 2011. A large and persistent carbon sink in the world's forests. Science 333:988–93.
Pardini, R., A. de Arruda Bueno, T. A. Gardner, P. I. Prado, and J. P. Metzger. 2010. Beyond the fragmentation threshold hypothesis: Regime shifts in biodiversity across fragmented landscapes. PLoS One 5:e13666.
Pardini, R., S. M. de Souza, R. Braga-Neto, and J. P. Metzger. 2005. The role of forest structure, fragment size and corridors in maintaining small mammal abundance and diversity in an Atlantic forest landscape. Biological Conservation 124:253–66.
Parés-Ramos, I. K., W. A. Gould, and T. M. Aide. 2008. Agricultural abandonment, suburban growth, and forest expansion in Puerto Rico between 1991 and 2000. Ecology and Society 13:1. http://www.ecologyandsociety.org/vol13/iss2/art1/.
Parolin, P., O. De Simone, K. Haase, D. Waldhoff, S. Rottenberger, U. Kuhn, J.

Kesselmeier, et. al. 2004. Central Amazonian floodplain forests: Tree adaptations in a pulsing system. Botanical Review 70:357–80.
Parrotta, J. A. 2010. Restoring biodiversity and forest ecosystem services in degraded tropical landscapes. Pages 53–61 *in* T. Koizumi, K. Okabe, I. Thompson, K. Sugimura, T. Toma, and K. Fujita, eds. The role of forest biodiversity in the sustainable use of ecosystem goods and services in agro-forestry, fisheries, and forestry. Forestry and Forest Products Research Institute, Ibaraki, Japan.
Parrotta, J. A., J. K. Francis, and O. H. Knowles. 2002. Harvesting intensity affects forest structure and composition in an upland Amazonian forest. Forest Ecology and Management 169:243–55.
Parrotta, J. A., and O. H. Knowles. 1999. Restoration of tropical moist forests on bauxite-mined lands in the Brazilian Amazon. Restoration Ecology 7:103–16.
Parrotta, J., J. W. Turnbull, and N. Jones. 1997. Catalyzing native forest regeneration on degraded tropical lands. Forest Ecology and Management 99:1–7.
Parry, L., J. Barlow, and C. A. Peres. 2007. Large-vertebrate assemblages of primary and secondary forests in the Brazilian Amazon. Journal of Tropical Ecology 23:653–62.
Pärssinen, M., D. Schaan, and A. Ranzi. 2009. Pre-Columbian geometric earthworks in the upper Purús: A complex society in western Amazonia. Antiquity 83:1084–95.
Pascarella, J. B., T. M. Aide, and J. K. Zimmerman. 2004. Short-term response of secondary forests to hurricane disturbance in Puerto Rico, USA. Forest Ecology and Management 199:379–93.
Paul, G. S., and J. B. Yavitt. 2011. Tropical vine growth and the effects on forest succession: A review of the ecology and management of tropical climbing plants. Botanical Review 77:11–30.
Paul, J., A. Randle, C. Chapman, and L. Chapman. 2004. Arrested succession in logging gaps: Is tree seedling growth and survival limiting? African Journal of Ecology 42:245–51.
Paul, M., C. P. Catterall, P. C. Pollard, and J. Kanowski. 2010a. Does soil variation between rainforest, pasture and different reforestation pathways affect the early growth of rainforest pioneer species? Forest Ecology and Management 260:370–77.
———. 2010b. Recovery of soil properties and functions in different rainforest restoration pathways. Forest Ecology and Management 259:2083–92.
Paul, S., E. Veldkamp, and H. Flessa. 2008. Soil organic carbon in density fractions of tropical soils under forest—pasture—secondary forest land use changes. European Journal of Soil Science 59:359–71.
Pavelka, M. S. M., and C. A. Chapman. 2005. Population structure of black howlers (*Alouatta pigra*) in southern Belize and responses to Hurricane Iris. Pages 143–63 *in* A. Estrada, P. A. Garber, M. S. M. Pavelka, and L. Luecke, eds. New perspectives in the study of Mesoamerican primates. Springer, New York.
Paz-Rivera, C., and F. Putz. 2009. Anthropogenic soils and tree distributions in a lowland forest in Bolivia. Biotropica 41:665–75.

Pearson, T., D. Burslem, C. Mullins, and J. Dalling. 2002. Germination ecology of Neotropical pioneers: Interacting effects of environmental conditions and seed size. Ecology 83:2798–2807.
Peh, K. S. H., J. Jong, N. S. Sodhi, S. L. H. Lim, and C. A. M. Yap. 2005. Lowland rainforest avifauna and human disturbance: Persistence of primary forest birds in selectively logged forests and mixed-rural habitats of southern Peninsular Malaysia. Biological Conservation 123:489–505.
Peixoto, J., B. Nelson, and F. Wittmann. 2009. Spatial and temporal dynamics of river channel migration and vegetation in central Amazonian white-water floodplains by remote-sensing techniques. Remote Sensing of Environment 113:2258–66.
Pejchar, L., R. M. Pringle, J. Ranganathan, J. R. Zook, G. Duran, F. Oviedo, and G. C. Daily. 2008. Birds as agents of seed dispersal in a human-dominated landscape in southern Costa Rica. Biological Conservation 141:536–44.
Peña-Claros, M. 2003. Changes in forest structure and species composition during secondary forest succession in the Bolivian Amazon. Biotropica 35:450–61.
Peña-Claros, M., and H. de Boo. 2002. The effect of forest successional stage on seed removal of tropical rain forest tree species. Journal of Tropical Ecology 18:261–74.
Penot, E. 2007. From shifting cultivation to sustainable jungle rubber: A history of innovations In Indonesia. Pages 577–99 *in* M. A. Cairns, ed. Voices from the forest: Integrating indigenous knowledge into sustainable upland farming. Resources for the Future, Washington, DC.
Peres, C. A., J. Barlow, and W. F. Laurance. 2006. Detecting anthropogenic disturbance in tropical forests. Trends in Ecology & Evolution 21:227–29.
Peres, C., T. Gardner, J. Barlow, J. Zuanon, F. Michalski, A. Lees, I. Vieira, F. Moreira, and K. Feeley. 2010. Biodiversity conservation in human-modified Amazonian forest landscapes. Biological Conservation 143:2314–27.
Peres, C., and E. Palacios. 2007. Basin-wide effects of game harvest on vertebrate population densities in Amazonian forests: Implications for animal-mediated seed dispersal. Biotropica 39:304–15.
Pérez-Salicrup, D. R. 2004. Forest types and their implications. Pages 63–80 *in* B. L. I. Turner, J. Geoghegan, and D. R. Foster, eds. Integrated land-change science and tropical deforestation in the Southern Yucatan: Final Frontiers. Oxford University Press, Oxford.
Perfecto, I., R. A. Rice, R. Greenberg, and M. E. Van der Voort. 1996. Shade coffee: A disappearing refuge for biodiversity. BioScience 46:598–608.
Perfecto, I., and J. Vandermeer. 2010. The agroecological matrix as alternative to the land-sparing/agriculture intensification model. Proceedings of the National Academy of Sciences 107:5786–91.
Perfecto, I., J. Vandermeer, and A. Wright. 2009. Nature's matrix: Linking agriculture, conservation and food sovereignty. Earthscan, London.
Perz, S. G. 2007. Grand theory and context-specificity in the study of forest dynamics: Forest transition theory and other directions. Professional Geographer 59:105–14.
Perz, S., and A. Almeyda. 2010. A tri-partite framework of forest dynamics: Hierarchy,

panarchy, and heterarchy in the study of secondary growth. Pages 59–84 *in* H. Nagendra and J. Southworth, eds. Reforesting landscapes, linking pattern and process. Springer, Dordrecht, Neth.

Perz, S. G., and D. L. Skole. 2003. Secondary forest expansion in the Brazilian Amazon and the refinement of forest transition theory. Society & Natural Resources 16:277–94.

Peters, C. M. 2000. Precolumbian silviculture and indigenous management of Neotropical forests. Pages 203–23 *in* D. L. Lentz, ed. Imperfect balance: Landscape transformations in the precolumbian Americas. Columbia University Press, New York.

Peters, C. M., M. H. Balick, F. Kahn, and A. B. Anderson. 1989. Oligarchic forests of economic plants in Amazonia: Utilization and conservation of an important tropical resource. Conservation Biology 3:341–49.

Peterson, G. D., and M. Heemskerk. 2001. Deforestation and forest regeneration following small-scale gold mining in the Amazon: The case of Suriname. Environmental Conservation 28:117–26.

Pethiyagoda, R. S., and S. Nanayakkara. 2012. Invasion by *Austroeupatorium inulifolium* (Asteraceae) arrests succession following tea cultivation in the highlands of Sri Lanka. Ceylon Journal of Science (Biological Sciences) 40:175–81.

Petit, L. J., and D. R. Petit. 2003. Evaluating the importance of human-modified lands for Neotropical bird conservation. Conservation Biology 17:687–94.

Pfaff, A., and R. Walker. 2010. Regional interdependence and forest "transitions": Substitute deforestation limits the relevance of local reversals. Land Use Policy 27:119–29.

Phelps, J., D. A. Friess, and E. L. Webb. 2012. Win-win REDD+ approaches belie carbon-biodiversity trade-offs. Biological Conservation 154:53–60.

Phillips, O. L., L. E. O. C. Aragão, S. L. Lewis, J. B. Fisher, J. Lloyd, G. López-González, Y. Malhi, et al. 2009. Drought sensitivity of the Amazon rainforest. Science 323:1344–47.

Phillips, O., N. Higuchi, S. Vieira, T. Baker, K. Chao, and S. Lewis. 2009. Changes in Amazonian forest biomass, dynamics, and composition, 1980–2002. Geophysical Monograph Series 186:373–87.

Phillips, O., Y. Malhi, N. Higuchi, W. Laurance, P. Nuñez, R. Vásquez, S. Laurance, L. Ferreira, M. Stern, and S. Brown. 1998. Changes in the carbon balance of tropical forests: Evidence from long-term plots. Science 282:439–42.

Phillips, O., R. Martínez, L. Arroyo, T. Baker, T. Killeen, S. Lewis, Y. Malhi, A. Mendoza, D. Neill, and P. Vargas. 2002. Increasing dominance of large lianas in Amazonian forests. Nature 418:770–74.

Philpott, S. M., W. J. Arendt, I. Armbrecht, P. Bichier, T. V. Diestch, C. Gordon, R. Greenberg, I. Perfecto, R. Reynoso-Santos, and L. Soto-Pinto. 2008. Biodiversity loss in Latin American coffee landscapes: Review of the evidence on ants, birds, and trees. Conservation Biology 22:1093–1105.

Piacentini, V. de Q., and I. G. Varassin. 2007. Interaction network and the relationships between bromeliads and hummingbirds in an area of secondary Atlantic rain forest in southern Brazil. Journal of Tropical Ecology 23:663–72.

Pickett, S. T. A., and M. Cadenasso. 2005. Vegetation dynamics. Pages 172–98 *in* E. van der Maarel, ed. Vegetation ecology. Blackwell Publishing, Malden, MA.

Pickett, S. T. A., S. L. Collins, and J. J. Armesto. 1987. Models, mechanisms and pathways of succession. Botanical Review 53:335–71.

Pickett, S. T. A., and P. S. White, eds. 1985. The ecology of natural disturbance and patch dynamics. Academic Press, New York.

Pinard, M. A., M. G. Barker, and J. Tay. 2000. Soil disturbance and post-logging forest recovery on bulldozer paths in Sabah, Malaysia. Forest Ecology and Management 130:213–25.

Pinard, M. A., and J. Huffman. 1997. Fire resistance and bark properties of trees in a seasonally dry forest in eastern Bolivia. Journal of Tropical Ecology 13:727–40.

Pinard, M. A., and F. E. Putz. 1994. Vine infestation of large remnant trees in logged forest in Sabah, Malaysia: Biomechanical facilitation in vine succession. Journal of Tropical Forest Science 6:302–9.

———. 1996. Retaining forest biomass by reducing logging damage. Biotropica 28:278–95.

Piotto, D., D. Craven, F. Montagnini, and F. Alice. 2010. Silvicultural and economic aspects of pure and mixed native tree species plantations on degraded pasturelands in humid Costa Rica. New Forests 39:369–85.

Piotto, D., F. Montagnini, W. Thomas, M. Ashton, and C. Oliver. 2009. Forest recovery after swidden cultivation across a 40-year chronosequence in the Atlantic forest of southern Bahia, Brazil. Plant Ecology 205:261–72.

Piperno, D. R. 1994. Phytolith and charcoal evidence for prehistoric slash-and-burn agriculture in the Darién rain forest of Panama. Holocene 4:321–25.

———. 2006. Quaternary environmental history and agricultural impact on vegetation in Central America. Annals of the Missouri Botanical Garden 93:274–96.

———. 2007. Prehistoric human occupation and impacts on Neotropical forest landscapes during the Late Pleistocene and Early/Middle Holocene. Pages 193–218 *in* M. B. Bush and J. R. Flenley, eds. Tropical rainforest responses to climate change. Springer, New York.

———. 2011. Prehistoric human occupation and impacts on Neotropical forest landscapes during the Late Pleistocene and Early/Middle Holocene. Pages 185–212 *in* M. B. Bush, ed. Tropical rainforest responses to climatic change. Springer, London.

Piperno, D. R., and P. Becker. 1996. Vegetational history of a site in the central Amazon Basin derived from phytolith and charcoal records from natural soils. Quaternary Research 45:202–9.

Piperno, D. R., M. B. Bush, and P. A. Colinvaux. 1990. Paleoenvironments and human occupation in Late-Glacial Panama. Quaternary Research 33:108–16.

Piperno, D. R., and D. M. Pearsall. 1998. The origins of agriculture in the lowland Neotropics. New York, Academic Press.

Piperno, D. R., A. J. Ranere, I. Holst, J. Iriarte, and R. Duckau. 2009. Starch grain and phytolith evidence for early ninth millennium B.P. maize from the Central Balsas River Valley, Mexico. Proceedings of the National Academy of Sciences 106:5019–24.

Piperno, D. R., and K. E. Stothert. 2003. Phytolith evidence for early Holocene *Cucurbita* domestication in southwest Ecuador. Science 299:1054–57.

Pitman, N. C. A., C. E. Cerón, C. I. Reyes, M. Thurber, and J. Arellano. 2005. Catastrophic natural origin of a species-poor tree community in the world's richest forest. Journal of Tropical Ecology 21:559–68.

Pitman, N. C. A., J. W. Terborgh, M. R. Silman, P. V. Nunez, D. A. Neill, C. E. Ceron, W. E. Palacios, and M. Aulestia. 2001. Dominance and distribution of tree species in upper Amazonian terra firme forests. Ecology 82:2101–17.

Pizo, M. A., C. I. Donatti, N. M. R. Guedes, and M. Galetti. 2008. Conservation puzzle: Endangered hyacinth macaw depends on its nest predator for reproduction. Biological Conservation 141:792–96.

Pizo, M. A., L. Passos, and P. S. Oliveira. 2005. Ants as seed dispersers of fleshy diaspores in Brazilian Atlantic Forests. Pages 315–29 *in* P.-M. Forget, J. E. Lambert, P. E. Hulme, and S. B. Vander Wall, eds. Seed fate: Predation, dispersal, and seedling establishment. CABI Publishing, Cambridge, MA.

Poffenberger, M., and B. McGean, eds. 1993. Communities and forest management in East Kalimantan: Pathway to environmental stability. Southeast Asia Sustainable Forest Management Network, Center for Southeast Asia Studies, Berkeley, CA.

Pohl, M. D., and P. Bloom. 1996. Prehistoric Maya farming in the wetlands of Northern Belize: More data from Albion Island and beyond. Pages 145–64 *in* S. L. Fedick, ed. The managed mosaic: Ancient Maya agriculture and resource use. Univeristy of Utah Press, Salt Lake City.

Poorter, L., and F. Bongers. 2006. Leaf traits are good predictors of plant performance across 53 rain forest species. Ecology 87:1733–43.

Poorter, L., F. Bongers, F. J. Sterck, and H. Woll. 2005. Beyond the regeneration phase: Differentiation of height-light trajectories among tropical tree species. Journal of Ecology 93:256–67.

Poorter, L., F. Bongers, R. van Rompaey, and M. deKlerk. 1996. Regeneration of canopy tree species at five sites in West African moist forest. Forest Ecology and Management 84:61–69.

Poorter, L., M. V. de Plassche, S. Willems, and R. G. A. Boot. 2004. Leaf traits and herbivory rates of tropical tree species differing in successional status. Plant Biology 6:746–54.

Poorter, L., I. McDonald, A. Alarcon, E. Fichtler, J. C. Licona, M. Pena-Claros, F. Sterck, Z. Villegas, and U. Sass-Klaassen. 2010. The importance of wood traits and hydraulic conductance for the performance and life history strategies of 42 rainforest tree species. New Phytologist 185:481–92.

Poorter, L., S. J. Wright, H. Paz, D. D. Ackerly, R. Condit, G. Ibarra-Manriques, K. E. Harms, et al. 2008. Are functional traits good predictors of demographic rates? Evidence from five Neotropical forests. Ecology 89:1908–20.

Pope, K. O., and J. E. Terrell. 2008. Environmental setting of human migrations in the circum-Pacific region. Journal of Biogeography 35:1–21.

Porter-Bolland, L., E. A. Ellis, M. R. Guariguata, I. Ruiz-Mallón, S. Negrete-Yankelevich, and V. Reyes-García. 2012. Community managed forests and forest protected areas: An assessment of their conservation effectiveness across the tropics. Forest Ecology and Management. 268:6–17.

Portes, M., D. Damineli, R. Ribeiro, J. Monteiro, and G. Souza. 2010. Evidence of higher photosynthetic plasticity in the early successional *Guazuma ulmifolia* Lam. compared to the late successional *Hymenaea courbaril* L. grown in contrasting light environments. Brazilian Journal of Biology 70:75–83.

Posey, D. A. 1985. Indigenous management of tropical forest ecosystems: The case of the Kayapo Indians of the Brazilian Amazon. Agroforestry Systems 3:139–58.

Potvin, C., L. Mancilla, N. Buchmann, J. Monteza, T. Moore, M. Murphy, Y. Oelmann, M. Scherer-Lorenzen, B. L. Turner, and W. Wilcke. 2011. An ecosystem approach to biodiversity effects: Carbon pools in a tropical tree plantation. Forest Ecology and Management 261:1614–24.

Poulsen, J., C. Clark, G. Mavah, and P. Elkan. 2009. Bushmeat supply and consumption in a tropical logging concession in northern Congo. Conservation Biology 23:1597–1608.

Power, M., J. Marlon, N. Ortiz, P. Bartlein, S. Harrison, F. Mayle, A. Ballouche, R. Bradshaw, C. Carcaillet, and C. Cordova. 2008. Changes in fire regimes since the Last Glacial Maximum: An assessment based on a global synthesis and analysis of charcoal data. Climate Dynamics 30:887–907.

Power, M., F. Mayle, P. Bartlein, J. Marlon, R. Anderson, H. Behling, K. Brown, C. Carcaillet, D. Colombaroli, and D. Gavin. 2013. Climatic control of the biomass-burning decline in the Americas after AD 1500. Holocene 23:3–13.

Powers, J. S. 2004. Changes in soil carbon and nitrogen after contrasting land-use transitions in northeastern Costa Rica. Ecosystems 7:134–46.

Powers, J. S., and D. Pérez-Aviles. 2012. Edaphic factors are more important control on surface fine roots than stand age in secondary tropical dry forests. Biotropica 45:1–9.

Powers, J. S., J. P. Haggar, and R. F. Fisher. 1997. The effect of overstory composition on understory woody regeneration and species richness in 7-year-old plantations in Costa Rica. Forest Ecology and Management 99:43–54.

Powers, J. S., and E. Veldkamp. 2005. Regional variation in soil carbon and $\delta^{13}C$ in forests and pastures of northeastern Costa Rica. Biogeochemistry 72:315–36.

Prance, G. T. 1979. Notes on the vegetation of Amazonia III. The terminology of Amazonian forest types subject to inundation. Brittonia 31:26–38.

Prebble, M., and J. L. Dowe. 2008. The late Quaternary decline and extinction of palms on oceanic Pacific islands. Quaternary Science Reviews 27:2546–67.

Prebble, M., and J. M. Wilmshurst. 2009. Detecting the initial impact of humans and introduced species on island environments in Remote Oceania using palaeoecology. Biological Invasions 11:1529–56.

Pregill, G. K., and T. Dye. 1989. Prehistoric extinction of giant iguanas in Tonga. Copeia 2:505–8.

Premathilake, R. 2006. Relationship of environmental changes in central Sri Lanka to possible prehistoric land-use and climate changes. Palaeogeography, Palaeoclimatology, Palaeoecology 240:468–96.

Presley, S. J., M. R. Willig, J. M. Wunderle, and L. N. Saldanha. 2008. Effects of reduced-impact logging and forest physiognomy on bat populations of lowland Amazonian forest. Journal of Applied Ecology 45:14–25.

Primack, R., and R. Corlett. 2005. Tropical rain forests: An ecological and biogeographical comparison. Blackwell Publishing, Malden, MA.

Puhakka, M., R. Kalliola, M. Rajasilta, and J. Salo. 1992. River types, site evolution and successional vegetation patterns in Peruvian Amazonia. Journal of Biogeography 19:651–65.

Purata, S. E. 1986. Floristic and structural changes during old-field succession in the Mexican tropics in relation to site history and species availability. Journal of Tropical Ecology 2:257–76.

Putz, F. E., D. P. Dykstra, and R. Heinrich. 2000. Why poor logging practices persist in the tropics. Conservation Biology 14:951–56.

Putz, F. E., and N. Holbrook. 1991. Biomechanical studies of vines. Pages 73–97 *in* F. E. Putz and H. Mooney, A., eds. The biology of vines. Cambridge University Press, Cambridge.

Putz, F. E., and K. H. Redford. 2010. The importance of defining 'forest': Tropical forest degradation, deforestation, long-term phase shifts, and further transitions. Biotropica 42:10–20.

Putz, F. E., P. Sist, T. Fredericksen, and D. Dykstra. 2008. Reduced-impact logging: Challenges and opportunities. Forest Ecology and Management 256:1427–33.

Putz, F. E., P. A. Zuidema, T. Synnott, M. Peña-Claros, M. A. Pinard, D. Sheil, J. K. Vanclay, P. Sist, S. Gourlet-Fleury, and B. Griscom. 2012. Sustaining conservation values in selectively logged tropical forests: The attained and the attainable. Conservation Letters 5:296–303.

Quesada, M., G. A. Sánchez-Azofeifa, M. Alvarez-Anorve, K. E. Stoner, L. Avila-Cabadilla, J. Calvo-Alvarado, A. Castillo, et al. 2009. Succession and management of tropical dry forests in the Americas: Review and new perspectives. Forest Ecology and Management 258:1014–24.

Quintana-Ascencio, P. F., M. González-Espinosa, N. Ramirez-Marcial, G. Dominguez-Vázquez, and M. Mártinez-Icó. 1996. Soil seed banks and regeneration of tropical rain forest from milpa fields at the Selva Lacandona, Chiapas, Mexico. Biotropica 28:192–209.

Quintero, I., and G. Halffter. 2009. Temporal changes in a community of dung

beetles (Insecta: Coleoptera: Scarabeinae) resulting from the modification and fragmentation of tropical rain forest. Acta Zoolóica Mexicana (ns) 25:625–49.
Quintero, I., and T. Roslin. 2005. Rapid recovery of dung beetle communities following habitat fragmentation in Central Amazonia. Ecology 86:3303–11.
Raich, J. W., and G. W. Khoon. 1990. Effects of canopy openings on tree seed germination in a Malaysian dipterocarp forest. Journal of Tropical Ecology 6:203–17.
Ramage, B. S., D. Sheil, H. M. W. Salim, C. Fletcher, N. Z. A. Mustafa, J. C. Luruthusamay, R. D. Harrison, E. Butod, A. D. Dzulkiply, and A. R. Kassim. 2013. Pseudoreplication in tropical forests and the resulting effects on biodiversity conservation. Conservation Biology 27:364–72.
Ramakrishnan, P. S. 1998. Sustainable development, climate change and tropical rain forest landscape. Climatic Change 39:583–600.
———. 2007. Traditional forest knowledge and sustainable forestry: A north-east India perspective. Forest Ecology and Management 249:91–99.
Ramankutty, N., H. K. Gibbs, F. Achard, R. Defries, J. A. Foley, and R. Houghton. 2007. Challenges to estimating carbon emissions from tropical deforestation. Global Change Biology 13:51–66.
Randriamalala, J. R., D. Hervé, J.-C. Randriaamboavonjy, and S. M. Carrière. 2012. Effects of tillage regime, cropping duration and fallow age on diversity and structure of secondary vegetation in Madagascar. Agriculture, Ecosystems & Environment 155:182–93.
Rands, M. R., W. M. Adams, L. Bennun, S. H. Butchart, A. Clements, D. Coomes, A. Entwistle, et al. 2010. Biodiversity conservation: Challenges beyond 2010. Science 329:1298–1303.
Ranere, A. J., and R. G. Cooke. 2003. Late Glacial and early Holocene occupation of Central American tropical forest. Pages 219–48 *in* J. Mercader, ed. Under the canopy: The archaeology of tropical rain forests. Rutgers University Press, New Brunswick, NJ.
Ranere, A. J., D. R. Piperno, I. Holst, R. Dickau, and J. Iriarte. 2009. The cultural and chronological context of early Holocene maize and squash domestication in the Central Balas River Valley, Mexico. Proceedings of the National Academy of Sciences 106:5014–18.
Ranganathan, J., R. Daniels, M. Chandran, P. R. Ehrlich, and G. C. Daily. 2008. Sustaining biodiversity in ancient tropical countryside. Proceedings of the National Academy of Sciences 105:17852–54.
Ravindranath, N. H., M. Gadgil, and J. Campbell. 1996. Ecological stabilization and community needs: Manageing India's forest by objective. Pages 287–315 *in* M. Poffenberger and B. McGean, eds. Village voices, forest choices. Oxford University Press, Delhi, India.
Read, L., and D. Lawrence. 2003. Litter nutrient dynamics during succession in dry tropical forests of the Yucatan: Regional and seasonal effects. Ecosystems 6:747–61.

Redo, D. J., J. J. Bass, and A. C. Millington. 2009. Forest dynamics and the importance of place in western Honduras. Applied Geography 29:91–110.
Redo, D. J., H. R. Grau, T. M. Aide, and M. L. Clark. 2012. Asymmetric forest transition driven by the interaction of socioeconomic development and environmental heterogeneity in Central America. Proceedings of the National Academy of Sciences 109:8839–44.
Rees, M., R. Condit, M. Crawley, S. Pacala, and D. Tilman. 2001. Long-term studies of vegetation dynamics. Science 293:650–55.
Reich, P. B., C. Uhl, M. B. Walters, and D. S. Ellsworth. 1991. Leaf lifespan as a determinant of leaf structure and function among 23 Amazonian tree species. Oecologia 86:16–24.
Reid, J. L., J. B. C. Harris, and R. A. Zahawi. 2012. Avian habitat preference in tropical forest restoration in Southern Costa Rica. Biotropica 44:350–59.
Reiners, W. A., A. F. Bouwman, W. F. J. Parson, and M. Keller. 1994. Tropical rain forest conversion to pasture: Changes in vegetation and soil properties. Ecological Applications 4:363–77.
Reis, A., F. C. Bechara, and D. R. Tres. 2010. Nucleation in tropical ecological restoration. Scientia Agricola 67:244–50.
Ren, H., Z. A. Li, W. J. Shen, Z. Y. Yu, S. L. Peng, C. H. Liao, M. M. Ding, and J. G. Wu. 2007. Changes in biodiversity and ecosystem function during the restoration of a tropical forest in south China. Science in China Series C: Life Sciences 50:277–84.
Renner, S. C., M. Waltert, and M. Muhlenberg. 2006. Comparison of bird communities in primary vs. young secondary tropical montane cloud forest in Guatemala. Biodiversity and Conservation 15:1545–75.
Rerkasem, K., D. Lawrence, C. Padoch, D. Schmidt-Vogt, A. D. Ziegler, and T. B. Bruun. 2009. Consequences of swidden transitions for crop and fallow biodiversity in Southeast Asia. Human Ecology 37:347–60.
Ressel, K., F. A. G. Guilherme, I. Schiavini, and P. E. Oliveira. 2004. Ecologia morfofuncional de plântulas de espécies arbóreas da Estação Ecológica do Panga, Uberlândia, Minas Gerais. Revista Brasileira de Botánica 27:311–23.
Restrepo, C., and N. Alvarez. 2006. Landslides and their contribution to land-cover change in the mountains of Mexico and Central America. Biotropica 38:446–57.
Restrepo, C., and P. Vitousek. 2001. Landslides, alien species, and the diversity of a Hawaiian montane mesic ecosystem. Biotropica 33:409–20.
Restrepo, C., L. R. Walker, A. B. Shiels, R. Bussmann, L. Claessens, S. Fisch, P. Lozano, G. Negi, L. Paolini, and G. Poveda. 2009. Landsliding and its multiscale influence on mountainscapes. BioScience 59:685–98.
Rey Benayas, J. M., J. M. Bullock, and A. C. Newton. 2008. Creating woodland islets to reconcile ecological restoration, conservation, and agricultural land use. Frontiers in Ecology and the Environment 6:329–36.
Rey Benayas, J. M., A. Newton, A. Diaz, and J. Bullock. 2009. Enhancement of biodiversity and ecosystem services by ecological restoration: A meta-analysis. Science 325:1121–24.

Rhodes, C. C., G. E. Eckert, and D. C. Coleman. 2000. Soil carbon differences among forest, agriculture, and secondary vegetation in lower montane Ecuador. Ecological Applications 10:497–505.
Ribeiro, M. B. N., E. M. Bruna, and W. Mantovani. 2010. Influence of post clearing treatment on the recovery of herbaceous plant communities in Amazonian secondary forests. Restoration Ecology 18:50–58.
Ribeiro, M. C., J. P. Metzger, A. C. Martensen, F. J. Ponzoni, and M. M. Hirota. 2009. The Brazilian Atlantic Forest: How much is left, and how is the remaining forest distributed? Implications for conservation. Biological Conservation 142:1141–53.
Rice, D. S. 1978. Population growth and subsistence alternatives in a tropical lacustrine environment. Pages 35–61 *in* P. D. Harrison and B. L. Turner II, eds. Pre-hispanic Maya agriculture. University of New Mexico Press, Albuquerque.
Richards, P. W. 1939. Ecological studies on the rain forest of southern Nigeria. 1. The structure and floristic composition of primary forest. Journal of Ecology 27:1–61.
———. 1952. The tropical rain forest. Cambridge University Press, Cambridge.
———. 1963. Ecological notes on West African vegetation. II. Lowland forest of the Southern Bakundu Forest Reserve. Journal of Ecology 51:123–49.
———. 1971. Some problems of nature conservation in the tropics. Bulletin du Jardin Botanique National de Belgique/Bulletin van de National Plantentuin van België 41:173–87.
———. 1996. The tropical rain forest: An ecological study. 2nd ed. Cambridge University Press, Cambridge.
Rico-Gray, V., and J. G. García-Franco. 1991. The Maya and the vegetation of the Yucatan Peninsula. Journal of Ethnobiology 11:135–42.
Riswan, S., and K. Kartawinata. 1988. Regeneration after disturbance in kerangas (heath) forest in East Kalimantan, Indonesia. Pages 61–86 *in* S. Soemodihardjo, ed. Some ecological aspects of tropical forest of East Kalimantan: A collection of research papers. Indonesian National MAB Committee, MAB Indonesia Contribution No. 48. LIPI (Indonesian Institute of Sciences), Jakarta.
Riswan, S., J. B. Kentworthy, and K. Kartawinata. 1985. The estimation of temporal processes in tropical rain forest: A study of primary mixed dipterocarp forest in Indonesia. Journal of Tropical Ecology 1:171–82.
Ritter, C. D., C. B. Andretti, and B. W. Nelson. 2012. Impact of past forest fires on bird populations in flooded forests of the Cuini River in the Lowland Amazon. Biotropica 44:449–53.
Roa-Fuentes, L., C. Martínez-Garza, J. Etchevers, and J. Campo. 2013. Recovery of soil C and N in a tropical pasture: Passive and active restoration. Land Degradation & Development. doi:10.1002/ldr.2197.
Robiglio, V., and F. Sinclair. 2011. Maintaining the conservation value of shifting cultivation landscapes requires spatially explicit interventions. Environmental Management 48:289–306.
Robinson, J. G., K. H. Redford, and E. L. Bennett. 1999. Wildlife harvest in logged tropical forests. Science 284:595–96.

Robson, J. 2009. Out-migration and commons management: Social and ecological change in a high biodiversity region of Oaxaca, Mexico. International Journal of Biodiversity Science, Ecosysem Services & Management 5:21–34.
Robson, J. P., and F. Berkes. 2011. Exploring some of the myths of land use change: Can rural to urban migration drive declines in biodiversity? Global Environmental Change 21:844–54.
Rodrigues, R. R., S. Gandolfi, A. G. Nave, J. Aronson, T. E. Barreto, C. Y. Vidal, and P. H. S. Brancalion. 2011. Large-scale ecological restoration of high-diversity tropical forests in SE Brazil. Forest Ecology and Management 261:1605–13.
Rodrigues, R. R., R. A. F. Lima, S. Gandolfi, and A. G. Nave. 2009. On the restoration of high diversity forests: 30 years of experience in the Brazilian Atlantic Forest. Biological Conservation 142:1242–51.
Rodrigues, R. R., S. V. Martins, and L. H. F. Matthew. 2005. Post-fire regeneration in a semideciduous mesophytic forest, southeastern Brazil. Pages 1–19 *in* A. R. Burk, ed. New research on forest ecosystems. Nova Science Publishers, New York.
Rodrigues, R. R., R. B. Torres, L. A. F. Matthes, and A. S. Penha. 2004. Tree species sprouting from root buds in a semideciduous forest affected by fires. Brazilian Archives of Biology and Technology 47:127–33.
Roeder, M., D. Hölscher, and I. D. K. Ferraz. 2010. Liana regeneration in secondary and primary forests of central Amazonia. Plant Ecology & Diversity 3:165–74.
Roosevelt, A. C. 1998. Ancient and modern hunter-gatherers of lowland South America: An evolutionary problem. Pages 190–212 *in* W. Balee, ed. Advances in historical ecology. Columbia University Press, New York.
———. 1999a. The development of prehistoric complex societies: Amazonia, a tropical forest. Pages 13–33 *in* E. A. Bacus and L. J. Lucero, eds. Complex politics of the ancient tropical world. American Anthropological Association, Washington, DC.
———. 1999b. Twelve thousand years of human-environment interaction in the Amazon floodplain. Pages 371–92 *in* C. Padoch, J. M. Ayres, M. Pinedo-Vasquez, and A. Henderson, eds. Varzea: Diversity, development, and conservation of Amazonia's whitewater floodplains. New York Botanical Garden Press, New York.
———. 2000. The lower Amazon: A dynamic human habitat. Pages 455–91 *in* D. L. Lentz, ed. Imperfect balance: Landscape transformations in the Precolumbian Americas. Columbia University Press, New York.
Roosevelt, A. C., J. E. Douglas, A. M. Amaral, M. I. da Silveira, C. P. Barbosa, M. Barreto, W. S. da Silva, and L. J. Brown. 2009. Early hunter-gatherers in the *terra firme* rainforest: Stemmed projectile points from the Curuá goldmines. Amazônica 1:442–83.
Rös, M., F. Escobar, and G. Halffter. 2011. How dung beetles respond to a human-modified variegated landscape in Mexican cloud forest: A study of biodiversity integrating ecological and biogeographical perspectives. Diversity and Distributions 18:377–89.

Ross, N. J. 2011. Modern tree species composition reflects ancient Maya "forest gardens" in northwest Belize. Ecological Applications 21:75–84.
Rostain, S. 2013. Islands in the rainforest: Landscape management in pre-Columbian Amazonia. Left Coast Press, Walnut Creek, CA.
Rozendaal, D. M. A., C. C. Soliz-Gamboa, and P. A. Zuidema. 2011. Assessing long-term changes in tropical forest dynamics: A first test using tree-ring analysis. Trees-Structure and Function 25:115–24.
Rozendaal, D. M. A., and P. A. Zuidema. 2011. Dendroecology in the tropics: A review. Trees-Structure and Function 25:3–16.
Ruddiman, W. F. 2003. The anthropogenic greenhouse era began thousands of years ago. Climate Change 61:261–93.
———. 2005. Plows, plagues, and petroleum. Princeton University Press, Princeton, NJ.
Rudel, T. K. 2005. Tropical forests: Regional paths of destruction and regeneration in the late twentieth century. Columbia Univ Press, New York.
———. 2010. Three paths to forest expansion: A comparative historical analysis. Pages 45–57 *in* H. Nagendra and J. Southworth, eds. Reforesting landscapes: Linking pattern and process. Springer, Dordrecht, Neth.
———. 2012. The human ecology of regrowth in the tropics. Journal of Sustainable Forestry 31:340–54.
Rudel, T. K., O. T. Coomes, E. Moran, F. Achard, A. Angelsen, J. C. Xu, and E. Lambin. 2005. Forest transitions: Towards a global understanding of land use change. Global Environmental Change-Human and Policy Dimensions 15:23–31.
Rudel, T. K., R. DeFries, G. P. Asner, and W. F. Laurance. 2009. Changing drivers of deforestation and new opportunities for conservation. Conservation Biology 23:1396–1405.
Rudel, T. K., M. Perez-Lugo, and H. Zichal. 2000. When fields revert to forest: Development and spontaneous reforestation in post-war Puerto Rico. Professional Geographer 52:386–97.
Rule, S., B. W. Brook, S. G. Haberle, C. S. M. Turney, A. P. Kershaw, and C. N. Johnson. 2012. The aftermath of megafaunal extinction: Ecosystem transformation in Pleistocene Australia. Science 335:1483–86.
Runkle, J. 1982. Patterns of disturbance in some old-growth mesic forests of eastern North America. Ecology 63:1533–46.
Russell, A. E., and J. W. Raich. 2012. Rapidly growing tropical trees mobilize remarkable amounts of nitrogen, in ways that differ surprisingly among species. Proceedings of the National Academy of Sciences 109:10398–402.
Russell, A. E., J. W. Raich, O. J. Valverde-Barrantes, and R. Fisher. 2007. Tree species effects on soil properties in experimental plantations in tropical moist forest. Soil Science Society of America Journal 71:1389–97.
Russell, A. E., J. W. Raich, and P. M. Vitousek. 1998. The ecology of the climbing fern *Dicranopteris linearis* on windward Mauna Loa, Hawaii. Journal of Ecology 86:765–79.

Russell, J., S. McCoy, D. Verschuren, I. Bessems, and Y. Huang. 2009. Human impacts, climate change, and aquatic ecosystem response during the past 2000 yr at Lake Wandakara, Uganda. Quaternary Research 72:315–24.

Rutishauser, E., F. Wagner, B. Herault, E. A. Nicolini, and L. Blanc. 2010. Contrasting above-ground biomass balance in a Neotropical rain forest. Journal of Vegetation Science 21:672–82.

Saatchi, S., M. Marlier, R. Chazdon, D. Clark, and A. Russell. 2011. Impact of spatial variability of tropical forest structure on radar estimation of aboveground biomass. Remote Sensing of Environment 115:2836–49.

Sajise, P. 2003. Working with nature: Technical and social dimensions of assisted natural regeneration. Pages 5–16 in P. C. Dugan, P. B. Durst, D. J. Ganz, and P. J. McKenzie, eds. Advancing assisted natural regeneration (ANR) in Asia and the Pacific. Food and Agriculture Organization of the United Nations, Bangkok, Thailand.

Saldarriaga, J. G., and D. C. West. 1986. Holocene fires in the northern Amazon basin. Quaternary Research 26:358–66.

Saldarriaga, J. G., D. C. West, M. L. Tharp, and C. Uhl. 1988. Long-term chronosequence of forest succession in the upper Rio Negro of Colombia and Venezuela. Journal of Ecology 76:938–58.

Salo, J., R. Kalliola, I. Häkkinen, Y. Mäkinen, P. Niemalä, M. Puhakka, and P. D. Coley. 1986. River dynamics and the diversity of Amazon lowland forest. Nature 322:254–58.

Sánchez Sánchez, O., and G. A. Islebe. 1999. Hurricane Gilbert and structural changes in a tropical forest in south-eastern Mexico. Global Ecology and Biogeography 8:29–38.

Sánchez-Azofeifa, G. A., G. C. Daily, A. S. P. Pfaff, and C. Busch. 2003. Integrity and isolation of Costa Rica's national parks and biological reserves: Examining the dynamics of land-cover change. Biological Conservation 109:123–35.

Sand, C., J. Bole, and A. Ouetcho. 2006. What is archaeology for in the Pacific? History and politics in New Caledonia. Pages 321–45 *in* I. Lilley, ed. Archaeology of Oceania: Australia and the Pacific Islands. Blackwell Publishing, Oxford.

Sanford, R. L. J., and S. P. Horn. 2000. Holocene rain-forest wilderness: A Neotropical perspective on humans as an exotic, invasive species. USDA Forest Service Proceedings RMRS-P-15 3:168–73.

Sanford, R. L., J. Saldarriaga, K. E. Clark, C. Uhl, and R. Herrera. 1985. Amazon rain forest fires. Science 277:53–55.

Sang, P. M., D. Lamb, M. Bonner, and S. Schmidt. 2012. Carbon sequestration and soil fertility of tropical tree plantations and secondary forest established on degraded land. Plant and Soil 362:187–200.

Sansevero, J. B. B., P. V. Prieto, L. F. D. de Moraes, and P. J. F. P. Rodrigues. 2011. Natural regeneration in plantations of native trees in lowland Brazilian Atlantic Forest: Community structure, diversity, and dispersal syndromes. Restoration Ecology 19:379–89.

Santiago, L. S., G. Goldstein, F. C. Meinzer, J. B. Fisher, K. Machado, D. Woodruff, and T. Jones. 2004. Leaf photosynthetic traits scale with hydraulic conductivity and wood density in Panamanian forest canopy trees. Oecologia 140:543–50.
Santos, A., Jr., I. H. Ishii, N. M. R. Guedes, and F. L. R. de Almeida. 2006. Appraisal of the age of the trees used as nests by the Hyacinth Macaw in the Pantanal, Mato Grosso. Natureza & Conservacao 4:180–88.
Sardjono, M. A., and I. Samsoedin. 2001. Traditional knowledge and practice of biodiversity conservation: The Benuaq Dayak community of East Kalimantan, Indonesia. Pages 116–33 *in* C. J. P. Colfer and Y. Byron, eds. People managing forests: The links between human well-being and sustainability. Resources for the Future/Center for International Forestry Research, Bogor, Indonesia.
Sasaki, H. 2007. Innovations in swidden-based rattan cultivation by Benuaq-Dayak farmers in East Kalimantan, Indonesia. Pages 459–70 *in* M. A. Cairns, ed. Voices from the forest: Integrating indigenous knowledge into sustainable upland farming. Resources for the Future, Washington, DC.
Saulei, S. M., and M. D. Swaine. 1988. Rain forest seed dynamics during succession at Gogol, Papua New Guinea. Journal of Ecology 76:1133–52.
Sayer, J. A., U. Chokkalingam, and J. Poulsen. 2004. The restoration of forest biodiversity and ecological values. Forest Ecology and Management 201:3–11.
Sayer, J. A., C. S. Harcourt, and N. M. Collins. 1992. The conservation atlas of tropical forests: Africa. Macmillan, New York.
Sberze, M., M. Cohn Haft, and G. Ferraz. 2010. Old growth and secondary forest site occupancy by nocturnal birds in a Neotropical landscape. Animal Conservation 13:3–11.
Scariot, A., D. L. M. Vieira, A. B. Sampaio, E. Guarino, and A. Sevilha. 2008. Recruitment of dry forest tree species in Central Brazil pastures. Pages 231–44 *in* R. W. Myster, ed. Post-agricultural succession in the Neotropics. Springer, New York.
Scatena, F. N., S. Moya, C. Estrada, and J. D. Chinea. 1996. The first five years in the reorganization of aboveground biomass and nutrient use following Hurricane Hugo in the Bisley Experimental Watersheds, Luquillo Experimental Forest, Puerto Rico. Biotropica 28:424–40.
Scatena, F., W. Silver, T. Siccama, A. Johnson, and M. Sánchez. 1993. Biomass and nutrient content of the Bisley experimental watersheds, Luquillo Experimental Forest, Puerto Rico, before and after Hurricane Hugo, 1989. Biotropica 25:15–27.
Schaan, D., M. Pärssinen, S. Saunaluoma, A. Ranzi, M. Bueno, and A. Barbosa. 2012. New radiometric dates for precolumbian (2000–700 bp) earthworks in western Amazonia, Brazil. Journal of Field Archaeology 37:132–42.
Schedlbauer, J. L., and K. L. Kavanagh. 2008. Soil carbon dynamics in a chronosequence of secondary forests in northeastern Costa Rica. Forest Ecology and Management 255:1326–35.
Scheffer, M., S. R. Carpenter, T. M. Lenton, J. Bascompte, W. Brock, V. Dakos, J. van de Koppel, I. A. van de Leemput, S. A. Levin, and E. H. van Nes. 2012. Anticipating critical transitions. Science 338:344–48.

Schlawin, J., and R. A. Zahawi. 2008. 'Nucleating' succession in recovering Neotropical wet forests: The legacy of remnant trees. Journal of Vegetation Science 19:485–92.
Schleuning, M., N. Blüthgen, M. Flörchinger, J. Braun, H. M. Schaefer, and K. Böhning-Gaese. 2011. Specialization and interaction strength in a tropical plant-frugivore network differ among forest strata. Ecology 92:26–36.
Schleuning, M., N. Farwig, M. K. Peters, T. Bergsdorf, B. Bleher, R. Brandl, H. Dalitz, G. Fischer, W. Freund, and M. W. Gikungu. 2011. Forest fragmentation and selective logging have inconsistent effects on multiple animal-mediated ecosystem processes in a tropical forest. PLoS One 6:e27785.
Schmid-Vogt, D. 1997. Forests and trees in the cultural landscape of Lawa swidden farmers in northern Thailand. Pages 44–50 *in* K. Seeland, ed. Nature is culture: Indigenous knowledge and socio-cultural aspects of trees and forests in non-European cultures. Intermediate Technology Publications, London, UK.
Schmidt, M. J., and M. J. Heckenberger. 2009. Amerindian anthrosols: Amazonian dark earth formation in the Upper Xingu. Pages 163–91 *in* W. I. Woods, ed. Amazonian dark earths: Wim Sombroek's vision. Springer, New York.
Schmitt, S. F., and R. J. Whittaker. 1998. Disturbance and succession on the Krakatau Islands, Indonesia. Pages 515–43 *in* D. M. Newbery, H. N. T. Prins, and N. D. Brown, eds. Dynamics of tropical communities. Blackwell Science, Oxford, UK.
Schmook, B. 2010. Shifting maize cultivation and secondary vegetation in the Southern Yucatán: Successional forest impacts of temporal intensification. Regional Environmental Change 10:233–46.
Schmook, B., and C. Radel. 2008. International labor migration from a tropical development frontier: Globalizing households and an incipient forest transition. Human Ecology 36:891–908.
Schneider, L., and D. Fernando. 2010. An untidy cover: Invasion of bracken fern in the shifting cultivation systems of Southern Yucatán, Mexico. Biotropica 42:41–48.
Schnitzer, S. A., J. W. Dalling, and W. P. Carson. 2000. The impact of lianas on tree regeneration in tropical forest canopy gaps: Evidence for an alternative pathway of gap-phase regeneration. Journal of Ecology 88:655–66.
Schnitzer, S. A., M. P. E. Parren, and F. Bongers. 2004. Recruitment of lianas into logging gaps and the effects of pre-harvest climber cutting in a lowland forest in Cameroon. Forest Ecology and Management 190:87–98.
Schroth, G., D. Faria, M. Araujo, L. Bede, S. A. Van Bael, C. R. Cassano, L. C. Oliveira, and J. H. C. Delabie. 2011. Conservation in tropical landscape mosaics: The case of the cacao landscape of southern Bahia, Brazil. Biodiversity and Conservation 20:1635–54.
Schroth, G., and J. A. McNeely. 2011. Biodiversity conservation, ecosystem services and livelihoods in tropical landscapes: Towards a common agenda. Environmental Management 48:229–36.
Schüller, E., M. Martínez-Ramos, and P. Hietz. 2013. Radial gradients in wood specific

gravity, water and gas content in trees of a Mexican tropical rain forest. Biotropica 45:280–87.
Schulz, B., B. Becker, and E. Götsch. 1994. Indigenous knowledge in a 'modern' sustainable agroforestry system—a case study from eastern Brazil. Agroforestry Systems 25:59–69.
Schulze, M., and J. Zweede. 2006. Canopy dynamics in unlogged and logged forest stands in the eastern Amazon. Forest Ecology and Management 236:56–64.
Schwartz, D. 1992. Assechement climatique vers 3000 BP et expansion Bantu en Afrique centrale atlantique; quelques reflexions. Bulletin de la Societe Geologique de France 163:353–61.
Schwartz, G., M. Peña-Claros, J. C. A. Lopes, G. M. J. Mohren, and M. Kanashiro. 2012. Mid-term effects of reduced-impact logging on the regeneration of seven tree commercial species in the Eastern Amazon. Forest Ecology and Management 274:116–25.
Schwartzman, S. 1997. Fires in the Amazon: An analysis of NOAA-12 satellite data, 1996–1997. Environmental Defense Fund, Washington, DC.
Scriven, J. 2012. Developing REDD+ policies and measures from the bottom-up for the buffer zones of Amazonian protected areas. Environment, Development and Sustainability 14:745–65.
Secretariat of the Convention on Biological Diversity. 2002. Review of the status and trends of, and major threats to, the forest biological diversity. CBD Technical Series no. 7. Montreal, Canada.
———. 2010. Strategic Plan for Biodiversity 2011–2020 and the Aichi Targets. https://www.cbd.int/doc/strategic-plan/2011-2020/Aichi-Targets-EN.pdf.
Selaya, N. G., and N. P. R. Anten. 2008. Differences in biomass allocation, light interception and mechanical stability between lianas and trees in early secondary tropical forest. Functional Ecology 22:30–39.
———. 2010. Leaves of pioneer and later-successional trees have similar lifetime carbon gain in tropical secondary forest. Ecology 91:1102–13.
Selaya, N. G., R. J. Oomen, J. J. C. Netten, M. J. A. Werger, and N. P. R. Anten. 2008. Biomass allocation and leaf life span in relation to light interception by tropical forest plants during the first years of secondary succession. Journal of Ecology 96:1211–21.
SER (Society for Ecological Restoration). 2004. The SER International Primer on Ecological Restoration. Society for Ecological Restoration International Science & Policy Working Group, Tucson, AZ.
Servant, M., J. Maley, B. Turcq, M.-L. Absy, P. Brenac, M. Fourier, and M.-P. Ledru. 1993. Tropical forest changes during the Late Quaternary in African and South American lowlands. Global and Planetary Change 7:25–40.
Sethi, P., and H. F. Howe. 2009. Recruitment of hornbill-dispersed trees in hunted and logged forests of the Indian Eastern Himalaya. Conservation Biology 23:710–18.

Setyawan, A. D. W. I. 2010. Review: Biodiversity conservation strategy in a native perspective; case study of shifting cultivation at the Dayaks of Kalimantan. Nusantara Bioscience 2:97–108.

Sezen, U. U., R. L. Chazdon, and K. E. Holsinger. 2005. Genetic consequences of tropical second-growth forest regeneration. Science 307:891–91.

———. 2007. Multigenerational genetic analysis of tropical secondary regeneration in a canopy palm. Ecology 88:3065–75.

Sharp, J., E. Imerman, and G. Peters. 2002. Community supported agriculture (CSA): Building community among farmers and non-farmers. Journal of Extension 40:5.

———. 1999. Developing tests of successional hypotheses with size-structured populations, and an assessment using long-term data from a Ugandan rain forest. Plant Ecology 140:117–27.

Sheil, D., I. Basuki, L. German, T. W. Kuyper, G. Limberg, R. K. Puri, B. Sellato, M. van Noordwijk, and E. Wollenberg. 2012. Do Anthropogenic Dark Earths occur in the interior of Borneo? Some initial observations from East Kalimantan. Forests 3:207–29.

Sheil, D., and D. Burslem. 2003. Disturbing hypotheses in tropical forests. Trends in Ecology & Evolution 18:18–26.

Sheil, D., A. Salim, J. R. Chave, J. Vanclay, and W. D. Hawthorne. 2006. Illumination-size relationships of 109 coexisting tropical forest tree species. Journal of Ecology 94:494–507.

Shepard, G. H., and H. Ramirez. 2011. "Made in Brazil": Human dispersal of the Brazil Nut (*Bertholletia excelsa*, Lecythidaceae) in Ancient Amazonia. Economic Botany 65:44–65.

Shevliakova, E., S. W. Pacala, S. Malyshev, G. C. Hurtt, P. Milly, J. P. Caspersen, L. T. Sentman, et al. 2009. Carbon cycling under 300 years of land use change: Importance of the secondary vegetation sink. Global biogeochemical cycles 23:GB2022.

Shiels, A. B., and L. R. Walker. 2003. Bird perches increase forest seeds on Puerto Rican landslides. Restoration Ecology 11:457–65.

Shiels, A. B., C. A. West, L. Weiss, P. D. Klawinski, and L. R. Walker. 2008. Soil factors predict initial plant colonization on Puerto Rican landslides. Plant Ecology 195:165–78.

Shilton, L. A., P. J. Latch, A. McKeown, P. Pert, and D. A. Westcott. 2008. Landscape-scale redistribution of a highly mobile threatened species, *Pteropus conspicillatus* (Chiroptera, Pteropodidae), in response to Tropical Cyclone Larry. Austral Ecology 33:549–61.

Shilton, L. S., and R. H. Whittaker. 2010. The role of pteropodid bats (Megachiroptera) in re-establishing tropical forests on Krakatau. Pages 176–215 *in* T. H. Fleming and P. A. Racey, eds. Island bats: Ecology, evolution, and conservation. University of Chicago Press, Chicago.

Shlisky, A., A. A. C. Alencar, M. M. Nolasco, and L. M. Curran. 2009. Overview: Global fire regime conditions, threats, and opportunities for fire management in the tropics. Pages 65–83 *in* M. A. Cochrane, ed. Tropical fire ecology. Springer, New York.

Shono, K., E. A. Cadaweng, and P. B. Durst. 2007. Application of assisted natural regeneration to restore degraded tropical forestlands. Restoration Ecology 15:620–26.

Siddique, I., I. C. Guimaraes, S. Schmidt, D. Lamb, C. J. Reis-Carvalho, R. de Oliveira Figueredo, S. Blomberg, and E. A. Davidson. 2010. Nitrogen and phosphorus additions negatively affect tree species diversity in tropical forest regrowth trajectories. Ecology 91:2121–31.

Siegert, F., G. Ruecker, A. Hinrichs, and A. A. Hoffmann. 2001. Increased damage from fires in logged forests during droughts caused by El Nino. Nature 414:437–40.

Sierra, R., and E. Russman. 2006. On the efficiency of environmental service payments: A forest conservation assessment in the Osa Peninsula, Costa Rica. Ecological Economics 59:131–41.

Silva, J. M. C., C. Uhl, and G. Murray. 1996. Plant succession, landscape management, and the ecology of frugivorous birds in abandoned Amazonian pastures. Conservation Biology 10:491–503.

Silva, J. O., M. M. Espírito-Santo, and G. A. Melo. 2012. Herbivory on *Handroanthus ochraceus* (Bignoniaceae) along a successional gradient in a tropical dry forest. Arthropod-Plant Interactions 6:45–57.

Silva, W., P. Guimarães Jr, S. Dos Reis, and P. Guimarães. 2007. Investigating the fragility in plant-frugivore networks: A case study of the Atlantic Forest in Brazil. Pages 561–78 *in* A. J. Dennis, E. W. Schupp, R. J. Green, and D. A. Westcott, eds. Seed dispersal: Theory and its application in a changing world. CAB International, Wallingford, UK.

Silver, W., R. Ostertag, and A. Lugo. 2000. The potential for carbon sequestration through reforestation of abandoned tropical agricultural and pasture lands. Restoration Ecology 8:394–407.

Simula, M. 2009. Towards defining forest degradation: Comparative analysis of existing definitions. Forest Resources Assessment. Working Paper 154. FAO, Rome, Italy. ftp://ftp.fao.org/docrep/fao/012/k6217e/k6217e00.pdf.

Sioli, H. 1950. Das Wasser im Amazonasgebiet. Forsch. Fortschr. 26:274–80.

Sist, P., C. Sabogal, and Y. Byron, eds. 1999. Management of secondary and logged-over forests in Indonesia. Center for International Forestry Research, Bogor, Indonesia.

Sist, P., D. Sheil, K. Kartawinata, and H. Priyadi. 2003. Reduced-impact logging in Indonesian Borneo: Some results confirming the need for new silvicultural prescriptions. Forest Ecology and Management 179:415–27.

Sitzia, T., P. Semenzato, and G. Trentanovi. 2010. Natural reforestation is changing spatial patterns of rural mountain and hill landscapes: A global overview. Forest Ecology and Management 259:1354–62.

Slik, J. W. F., C. Bernard, M. Beek, F. Breman, and K. Eichhorn. 2008. Tree diversity,

composition, forest structure and aboveground biomass dynamics after single and repeated fire in a Bornean rain forest. Oecologia 158:579–88.
Slik, J. W. F., and K. A. O. Eichhorn. 2003. Fire survival of lowland tropical rain forest trees in relation to stem diameter and topographic position. Oecologia 137:446–55.
Slik, J. W. F., and S. Van Balen. 2006. Bird community changes in response to single and repeated fires in a lowland tropical rainforest of eastern Borneo. Biodiversity and Conservation 15:4425–51.
Slik, J. W. F, M. Van Beek, C. Bernard, F. Bongers, F. C. Breman, C. H. Cannon, and K. Sidiyasa. 2011. Limited edge effects along a burned unburned Bornean Forest boundary seven years after disturbance. Biotropica 43:288–98.
Slik, J. W. F., R. W. Verburg, and P. J. A. Kessler. 2002. Effects of fire and selective logging on the tree species composition of lowland dipterocarp forest in East Kalimantan, Indonesia. Biodiversity and Conservation 11:85–98.
Sloan, S. 2008. Reforestation amidst deforestation: Simultaneity and succession. Global Environmental Change-Human and Policy Dimensions 18:425–41.
Slocum, M. G., T. M. Aide, J. K. Zimmerman, and L. Navarro. 2006. A strategy for restoration of montane forest in anthropogenic fern thickets in the Dominican Republic. Restoration Ecology 14:526–36.
Slocum, M. G., and C. C. Horvitz. 2000. Seed arrival under different genera of trees in a Neotropical pasture. Plant Ecology 149:51–62.
Snook, L. K., and P. Negreros-Castillo. 2004. Regenerating mahogany (*Swietenia macrophylla* King) on clearings in Mexico's Maya forest: The effects of clearing method and cleaning on seedling survival and growth. Forest Ecology and Management 189:143–60.
Sobrado, M. 2003. Hydraulic characteristics and leaf water use efficiency in trees from tropical montane habitats. Trees-Structure and Function 17:400–406.
Soembroek, W., M. L. Ruivo, P. M. Fearnside, B. Glaser, and J. Lehmann. 2003. Amazonian dark earths as carbon stores and sinks. Pages 125–39 *in* J. Lehmann, D. Kern, B. Glaser, and W. Woods, eds. Amazonian dark earths—origin, properties, and management. Kluwer Academic Publishers, Dordrecht, Neth.
Sommer, R. F., H. Vielhauer, K. Carvalho, E. J. M. Vlek, and L. Paul. 2003. Deep soil water dynamics and depletion by secondary vegetation in the Eastern Amazon. Soil Science Society of America Journal 67:1672–86.
Song, C., and Y. Zhang. 2010. Forest cover in China from 1949–2006. Pages 341–56 *in* H. Nagendra and J. Southworth, eds. Reforesting landscapes: Linking pattern and process. Springer, Dordrecht, Neth.
Soto-Pinto, L., M. Anzueto, J. Mendoza, G. J. Ferrer, and B. de Jong. 2010. Carbon sequestration through agroforestry in indigenous communities of Chiapas, Mexico. Agroforestry Systems 78:39–51.
Southworth, J., H. Nagendra, L. A. Carlson, and C. Tucker. 2004. Assessing the impact of Celaque National Park on forest fragmentation in western Honduras. Applied Geography 24:303–22.
Southworth, J., and C. Tucker. 2001. The influence of accessibility, local institutions,

and socioeconomic factors on forest cover change in the mountains of western Honduras. Mountain Research and Development 21:276–83.
Sovu, P. Savadogo, M. Tigabu, and P. C. Odén. 2010. Restoration of former grazing lands in the highlands of Laos using direct seeding of four native tree species. Mountain Research and Development 30:232–43.
Sovu, M. Tigabu, P. Savadogo, and P. C. Odén. 2009. Recovery of secondary forests on swidden cultivation fallows in Laos. Forest Ecology and Management 258:2666–75.
Spies, T. A. 2009. Science of old growth, or a journey into wonderland. Pages 31–43 *in* T. A. Spies and S. L. Duncan, eds. Old growth in a New World. Island Press, Washington, DC.
Spriggs, M. 1997. The Island Melanesians. Blackwell, Oxford, UK.
Ssemmanda, I., D. Ryves, O. Bennike, and P. Appleby. 2005. Vegetation history in western Uganda during the last 1200 years: A sediment based reconstruction from two crater lakes. Holocene 15:119–32.
Stahl, P. 1996. Holocene biodiversity: An archaeological perspective from the Americas. Annual Review of Anthropology 25:105–26.
Stark, N. M., and C. F. Jordan. 1978. Nutrient retention by the root mat of an Amazonian rain forest. Ecology 59:434–37.
Steadman, D. W. 1995. Prehistoric extinctions of Pacific island birds: Biodiversity meets zooarchaeology. Science 267:1123–31.
———. 1997. Extinction of Polynesian birds: Reciprocal impacts of birds and people. Pages 51–79 *in* P. V. Kirch and T. L. Hunt, eds. Historical ecology in the Pacific Islands, prehistoric environmental and landscape change. Yale University Press, New Haven, CT.
Steadman, D. W., G. K. Pregill, and D. V. Burley. 2002. Rapid prehistoric extinction of iguanas and birds in Polynesia. Proceedings of the National Academy of Sciences 99:3673–77.
Steininger, M. K. 1996. Tropical secondary forest regrowth in the Amazon: Age, area and change estimation with Thematic Mapper data. International Journal of Remote Sensing 17:9–27.
Sterck, F. J., and F. Bongers. 2001. Crown development in tropical rain forest trees: Patterns with tree height and light availability. Journal of Ecology 89:1–13.
Sterck, F. J., L. Poorter, and F. Schieving. 2006. Leaf traits determine the growth-survival trade-off across rain forest tree species. American Naturalist 167:758–65.
Stiles, F. G. 1975. Ecology, flowering phenology, and hummingbird pollination of some Costa Rican *Heliconia* species. Ecology 56:285–301.
Stoner, K., P. Riba Hernández, K. Vulinec, and J. Lambert. 2007. The role of mammals in creating and modifying seed shadows in tropical forests and some possible consequences of their elimination. Biotropica 39:316–27.
Stoner, K., K. Vulinec, S. Wright, and C. Peres. 2007. Hunting and plant community dynamics in tropical forests: A synthesis and future directions. Biotropica 39:385–92.

Stothert, K. E., D. R. Piperno, and T. C. Andres. 2003. Terminal Pleistocene/Early Holocene human adaptation in coastal Ecuador: The Las Vegas evidence. Quaternary International 109–10:23–43.

Stotz, D. F., R. Bierregaard, M. Cohn-Haft, P. Petermann, J. Smith, A. Whittaker, and S. V. Wilson. 1992. The status of North American migrants in central Amazonian Brazil. Condor 94:608–21.

Stouffer, P. C., E. I. Johnson, R. O. Bierregaard, and T. E. Lovejoy. 2011. Understory bird communities in Amazonian rainforest fragments: Species turnover through 25 years post-isolation in recovering landscapes. PLoS One 6:e20543.

Strauss-Debenedetti, S., and F. A. Bazzaz. 1991. Plasticity and acclimation to light in tropical Moraceae of different successional positions. Oecologia 87:377–87.

Struhsaker, T. T., J. S. Lwanga, and J. M. Kasenene. 1996. Elephants, selective logging and forest regeneration in the Kibale Forest, Uganda. Journal of Tropical Ecology 12:45–64.

Styger, E., H. Rakotondramasy, M. Pfeffer, E. Fernandes, and D. Bates. 2007. Influence of slash-and-burn farming practices on fallow succession and land degradation in the rainforest region of Madagascar. Agriculture, Ecosystems & Environment 119:257–69.

Summerhayes, G. R., M. Leavesley, A. Fairbairn, H. Mandui, J. Field, A. Ford, and R. Fullagar. 2010. Human adaptation and plant use in highland New Guinea 49,000 to 44,000 years ago. Science 330:78–81.

Sutton, A., M. J. Mountain, K. Aplin, S. Bulmer, and T. Denham. 2009. Archaeozoological records for the Highlands of New Guinea: A review of current evidence. Australian Archaeology 69:41–58.

Swaine, M. D., and V. K. Agyeman. 2008. Enhanced tree recruitment following logging in two forest reserves in Ghana. Biotropica 40:370–74.

———. 1983. Early succession on cleared forest land in Ghana. Journal of Ecology 71:601–27.

———. 1988. The mosaic theory of forest regeneration and the determination of forest composition in Ghana. Journal of Tropical Ecology 4:253–69.

Swaine, M. D., and T. C. Whitmore. 1988. On the definition of ecological species groups in tropical rain forests. Vegetatio 75:81–86.

Swenson, N. G. 2011. Phylogenetic beta diversity metrics, trait evolution and inferring the functional beta diversity of communities. PLoS One 6:e21264.

Swenson, N. G., and B. J. Enquist. 2007. Ecological and evolutionary determinants of a key plant functional trait: Wood density and its community wide variation across latitude and elevation. American Journal of Botany 94:451–59.

Szott, L. T., C. A. Palm, and R. J. Buresh. 1999. Ecosystem fertility and fallow function in the humid and subhumid tropics. Agroforestry Systems 47:163–96.

Tabarelli, M., A. Aguiar, M. Ribeiro, J. Metzger, and C. Peres. 2010. Prospects for biodiversity conservation in the Atlantic Forest: Lessons from aging human-modified landscapes. Biological Conservation 143:2328–40.

Tabarelli, M., A. V. Lopes, and C. A. Peres. 2008. Edge-effects drive tropical forest fragments towards an early-successional system. Biotropica 40:657–61.
Tabarelli, M., and C. A. Peres. 2002. Abiotic and vertebrate seed dispersal in the Brazilian Atlantic forest: Implications for forest regeneration. Biological Conservation 106:165–76.
Tabarelli, M., B. A. Santos, V. Arroyo-Rodríguez, and F. P. L. Melo. 2012. Secondary forests as biodiversity repositories in human-modified landscapes: Insights from the Neotropics. Boletim do Museu Paraense Emílio Goeldi: Ciências Naturais, Belém 7:319–28.
Tang, H., R. Dubayah, A. Swatantran, M. Hofton, S. Sheldon, D. B. Clark, and B. Blair. 2012. Retrieval of vertical LAI profiles over tropical rain forests using waveform lidar at La Selva, Costa Rica. Remote Sensing of Environment 124:242–50.
Tanner, E. 1986. Forests of the Blue Mountains and the Port Royal Mountains of Jamaica. Pages 15–30 *in* P. K. Bretting, D. A. Thompson, and M. Humphries, eds. Forests of Jamaica. Jamaican Society of Scientists and Technologists, Kingston.
Tanner, E., and P. Bellingham. 2006. Less diverse forest is more resistant to hurricane disturbance: Evidence from montane rain forests in Jamaica. Journal of Ecology 94:1003–10.
Tanner, E., V. Kapos, S. Freskos, J. Healey, and A. Theobald. 1990. Nitrogen and phosphorus fertilization of Jamaican montane forest trees. Journal of Tropical Ecology 6:231–38.
Tansley, A. G. 1935. The use and abuse of vegetational concepts and terms. Ecology 16:284–307.
Taylor, B. W. 1957. Plant succession on recent volcanoes in Papua. Journal of Ecology 45:233–43.
Taylor, D., P. Robertshaw, and R. A. Marchant. 2000. Environmental change and political-economic upheaval in precolonial western Uganda. Holocene 10:527–36.
Taylor, Z., S. Horn, C. Mora, K. Orvis, and L. Cooper. 2010. A multi-proxy paleoecological record of late-Holocene forest expansion in lowland Bolivia. Palaeogeography, Palaeoclimatology, Palaeoecology. 293:98–107.
Teegalapalli, K., A. J. Hiremath, and D. Jathanna. 2010. Patterns of seed rain and seedling regeneration in abandoned agricultural clearings in a seasonally dry tropical forest in India. Journal of Tropical Ecology 26:25–33.
Terborgh, J. 1986. Keystone plant resources in the tropical forest. Pages 330–34 *in* M. Soulé, ed. Conservation biology: The science of scarcity and diversity. Sinauer Associates, Sunderland, MA.
Terborgh, J., R. B. Foster, and P. Nunez. 1996. Tropical tree communities: A test of the nonequilibrium hypothesis. Ecology 77:561–67.
Terborgh, J., G. Nuñez-Iturri, N. Pitman, F. Valverde, P. Alvarez, V. Swamy, E. Pringle, and C. Paine. 2008. Tree recruitment in an empty forest. Ecology 89:1757–68.
Terborgh, J., and K. Petren. 1991. Development of habitat structure through succession in an Amazonian floodplain forest. Pages 28–46 *in* S. S. Bell,

E. D. McCoy, and H. R. Mushinsky, eds. Habitat structure: The physical arrangement of objects in space. Chapman & Hall, New York.
Terrell, J. E., J. P. Hart, S. Barut, N. Cellinese, A. Curet, T. Denham, C. M. Kusimba, K. Latinis, R. Oka, and J. Palka. 2003. Domesticated landscapes: The subsistence ecology of plant and animal domestication. Journal of Archaeological Method and Theory 10:323–68.
Thomlinson, J. R., M. I. Serrano, T. M. Lopez, T. M. Aide, and J. K. Zimmerman. 1996. Land-use dynamics in a post-agricultural Puerto Rican landscape (1936–1988). Biotropica 28 (4a): 525–36.
Thornton, I. W. B., S. G. Compton, and C. N. Wilson. 1996. The role of animals in the colonization of the Krakatau Islands by fig trees (*Ficus* species). Journal of Biogeography 23:577–92.
Timm, R. M., D. Lieberman, M. Lieberman, and D. McClearn. 2009. Mammals of Cabo Blanco: History, diversity, and conservation after 45 years of regrowth of a Costa Rican dry forest. Forest Ecology and Management 258:997–1013.
Ting, T., and A. Poulsen. 2009. Understorey vegetation at two mud volcanoes in north-east Borneo. Journal of Tropical Forest Science 21:198–209.
Titiz, B., and R. L. J. Sanford. 2007. Soil charcoal in old-growth rain forests from sea level to the continental divide. Biotropica 39:673–82.
Toh, I., M. Gillespie, and D. Lamb. 1999. The role of isolated trees in facilitating tree seedling recruitment at a degraded sub-tropical rainforest site. Restoration Ecology 7:288–97.
Toledo, V. M., and P. Moguel. 2012. Coffee and sustainability: The multiple values of traditional shaded coffee. Journal of Sustainable Agriculture 36:353–77.
Toledo, V. M., B. Ortiz-Espejel, L. Cortes, P. Moguel, and M. J. Orodonez. 2003. The multiple use of tropical forests by indigenous peoples in Mexico: A case of adaptive management. Conservation Ecology 7. http://www.consecol.org/vol7/iss3/art9.
Toma, T., A. Ishida, and P. Matius. 2005. Long-term monitoring of post-fire aboveground biomass recovery in a lowland dipterocarp forest in East Kalimantan, Indonesia. Nutrient Cycling in Agroecosystems 71:63–72.
Toma, T., P. Matius, Y. Hastaniah, R. Watanabe, and Y. Okimori. 2000. Dynamics of burned lowland dipterocarp forest Sstands in Bukit Soeharto, East Kalimantan. Pages 107–20 *in* E. Guhardja, M. Fatawi, M. Sutisna, T. Mori, and S. Ohta, eds. Rainforest ecosystems of East Kalimantan: El Niño, drought, fire and human impacts. Springer, New York.
Tomimura, C., B. Singhakumara, and P. M. S. Ashton. 2012. Pattern and composition of secondary succession beneath Caribbean Pine plantations of Southwest Sri Lanka. Journal of Sustainable Forestry 31:818–34.
Townsend, A. R., C. C. Cleveland, B. Z. Houlton, C. B. Alden, and J. W. C. White. 2011. Multi-element regulation of the tropical forest carbon cycle. Frontiers in Ecology and the Environment 9:9–17.

Traveset, A., A. Robertson, and J. Rodríguez-Pérez. 2007. A review on the role of endozoochory on seed germination. Pages 78–103 *in* A. J. Dennis, E. W. Schupp, R. J. Green, and D. A. Westcott, eds. Seed dispersal: Theory and its application in a changing world. CABI Publishing, Wallingford, UK.

Tucker, C. M., D. K. Munroe, H. Nagendra, and J. Southworth. 2005. Comparative spatial analyses of forest conservation and change in Honduras and Guatemala. Conservation and Society 3:174–200.

Tucker, J., E. Brondizio, and E. Moran. 1998. Rates of forest regrowth in Eastern Amazonia: A comparison of Altamira and Bragantina regions, Pará State, Brazil. Interciencia 23:64–73.

Tunjai, P., and S. Elliott. 2012. Effects of seed traits on the success of direct seeding for restoring southern Thailand's lowland evergreen forest ecosystem. New Forests 43:319–33

Turcq, B., A. Sifeddine, L. Martin, M.-L. Absy, K. Suguio, and C. Volkmer-Ribeiro. 1998. Rainforest fires: A lacustrine record of 7000 years. Ambio 27:139–42.

Turner, B. L., II. 1978. Ancient agricultural land use in the central Maya Lowlands. Pages 163–83 *in* P. D. Harrison and B. L. Turner II, eds. Pre-hispanic Maya agriculture. University of New Mexico Press, Albuquerque.

———. 2010a. Sustainability and forest transitions in the southern Yucatán: The land architecture approach. Land Use Policy 27:170–79.

———. 2010b. Unlocking the ancient Maya and their environment: Paleo-evidence and dating resolution. Geology 38:575–76.

Turner, B. L., II, and K. W. Butzer. 1992. The Columbian encounter and land-use change. Environment 34:16–44.

Turner, I. M. 2001. The ecology of trees in the tropical rain forest. Cambridge University Press, Cambridge.

Turner, I. M., and R. T. Corlett. 1996. The conservation value of small, isolated fragments of lowland tropical rain forest. Trends in Ecology and Evolution 11:330–33.

Turner, I. M., Y. K. Wong, P. T. Chew, and A. binIbrahim. 1997. Tree species richness in primary and old secondary tropical forest in Singapore. Biodiversity and Conservation 6:537–43.

Turner, M. G., W. L. Baker, C. J. Peterson, and R. K. Peet. 1998. Factors influencing succession: Lessons from large, infrequent natural disturbances. Ecosystems 1:511–23.

Turton, S. M. 2008. Landscape-scale impacts of Cyclone Larry on the forests of northeast Australia, including comparisons with previous cyclones impacting the region between 1858 and 2006. Austral Ecology 33:409–16.

Ty, H. X. 2007. Rebuilding soil properties during the fallow: Indigenous innovations in the highlands of Vietnam. Pages 652–63 *in* M. Cairns, ed. Voices from the forest: Integrating indigenous knowledge into sustainable upland farming. Resources for the Future, Washington, DC.

Tyree, M. T., V. Velez, and J. W. Dalling. 1998. Growth dynamics of root and shoot hydraulic conductance in seedlings of five Neotropical tree species: Scaling to show possible adaptation to differing light regimes. Oecologia 114:293–98.

Uhl, C. 1987. Factors controlling succession following slash-and-burn agriculture in Amazonia. Journal of Ecology 75:377–407.

———. 1998. Perspectives on wildlife in the humid tropics. Conservation Biology 12:942–43.

Uhl, C., and R. Buschbacher. 1985. A disturbing synergism between cattle ranch burning practices and selective tree harvesting in the eastern Amazon. Biotropica 17:265–68.

Uhl, C., R. Buschbacker, and E. A. S. Serrão. 1988. Abandoned pastures in eastern Amazonia. I. Patterns of plant succession. Journal of Ecology 75:663–81.

Uhl, C., H. Clark, K. Clark, and P. Maquirino. 1982. Successional patterns associated with slash-and-burn agriculture in the upper Rio Negro region of the Amazon Basin. Biotropica 14:249–54.

Uhl, C., K. Clark, H. Clark, and P. Murphy. 1981. Early plant succession after cutting and burning in the upper Rio Negro of the Amazon Basin. Journal of Ecology 69:631–49.

Uhl, C., and C. F. Jordan. 1984. Successional and nutrient dynamics following forest cutting and burning in Amazonia. Ecology 65:1476–90.

Uhl, C., and J. B. Kauffman. 1990. Deforestation, fire susceptibility, and potential tree responses to fire in the eastern Amazon. Ecology 71:437–49.

Uhl, C., J. B. Kauffman, and D. L. Cummings. 1988. Fire in the Venezuelan Amazon 2: Environmental conditions necessary for forest fires in the evergreen rainforest of Venezuela. Oikos 53:176–84.

Uhl, C., D. Nepstad, R. Buschbacher, K. Clark, B. Kauffman, and S. Subler. 1990. Studies of ecosystem response to natural and anthropogenic disturbances provide guidelines for designing sustainable land-use systems in Amazonia. Pages 24–42 *in* A. B. Anderson, ed. Alternatives to deforestation: Steps toward sustainable use of the Amazon rain forest. Columbia University Press, New York.

Uhl, C., and I. C. G. Vieira. 1989. Ecological impacts of selective logging in the Brazilian Amazon: A case study from the Paragominas region of the state of Pará. Biotropica 21:98–106.

Unruh, J. D. 1990. Iterative increase of economic tree species in managed swidden-fallows of the Amazon. Agroforestry Systems 11:175–97.

Uriarte, M., L. W. Rivera, J. K. Zimmerman, T. M. Aide, A. G. Power, and A. S. Flecker. 2004. Effects of land use history on hurricane damage and recovery in a Neotropical forest. Plant Ecology 174:49–58.

Urquhart, G. 2009. Paleoecological record of hurricane disturbance and forest regeneration in Nicaragua. Quaternary International 195:88–97.

Urrego, D. H., M. B. Bush, M. R. Silman, B. A. Niccum, P. La Rosa, C. H. McMichael, S. Hagen, and M. Palace. 2012. Holocene fires, forest stability and human occupation in south-western Amazonia. Journal of Biogeography 40:521–33.

Valencia, R., R. Condit, H. C. Muller-Landau, C. Hernandez, and H. Navarrete. 2009. Dissecting biomass dynamics in a large Amazonian forest plot. Journal of Tropical Ecology 25:473–82.
Valladares, F., S. J. Wright, E. Lasso, K. Kitajima, and R. W. Pearcy. 2000. Plastic phenotypic response to light of 16 congeneric shrubs from a Panamanian rainforest. Ecology 81:1925–36.
Vamosi, J. C., T. M. Knight, J. A. Steets, S. J. Mazer, M. Burd, and T. L. Ashman. 2006. Pollination decays in biodiversity hotspots. Proceedings of the National Academy of Sciences 103:956–61.
Van Bloem, S., P. Murphy, and A. Lugo. 2003. Subtropical dry forest trees with no apparent damage sprout following a hurricane. Tropical Ecology 44:137–45.
———. 2007. A link between hurricane-induced tree sprouting, high stem density and short canopy in tropical dry forest. Tree Physiology 27:475–80.
van Breugel, M., F. Bongers, and M. Martinez-Ramos. 2007. Species dynamics during early secondary forest succession: Recruitment, mortality and species turnover. Biotropica 39:610–19.
van Breugel, M., M. Martinez-Ramos, and F. Bongers. 2006. Community dynamics during early secondary succession in Mexican tropical rain forests. Journal of Tropical Ecology 22:663–74.
van Breugel, M., J. Ransijn, D. Craven, F. Bongers, and J. S. Hall. 2011. Estimating carbon stock in secondary forests: Decisions and uncertainties associated with allometric biomass models. Forest Ecology and Management 262:1648–57.
van der Hammen, T. 2001. Ice age tropical South America: What was it really like? Amazoniana 16:647–52.
Vandermeer, J., A. Brenner, and I. Granzow de la Cerda. 1998. Growth rates of tree height six years after hurricane damage at four localities in eastern Nicaragua. Biotropica 30:502–9.
Vandermeer, J., D. Boucher, I. Granzow-de la Cerda, and I. Perfecto. 2001. Growth and development of the thinning canopy in a post-hurricane tropical rain forest in Nicaragua. Forest Ecology and Management 148:221–42.
Vandermeer, J., D. Boucher, I. Perfecto, and I. Granzow de la Cerda. 1996. A theory of disturbance and species diversity: Evidence from Nicaragua after Hurricane Joan. Biotropica 28:600–613.
Vandermeer, J., and I. Granzow de la Cerda. 2004. Height dynamics of the thinning canopy of a tropical rain forest: 14 years of succession in a post-hurricane forest in Nicaragua. Forest Ecology and Management 199:125–35.
Vandermeer, J., I. Granzow de la Cerda, and D. Boucher. 1997. Contrasting growth rate patterns in eighteen tree species from a post-hurricane forest in Nicaragua. Biotropica 29:151–61.
Vandermeer, J., I. Granzow de la Cerda, D. Boucher, I. Perfecto, and J. Ruiz. 2000. Hurricane disturbance and tropical tree species diversity. Science 290:788–91.
Vandermeer, J., I. Granzow de la Cerda, I. Perfecto, D. Boucher, J. Ruiz, and

A. Kaufmann. 2004. Multiple basins of attraction in a tropical forest: Evidence for nonequilibrium community structure. Ecology 85:575–79.
Vandermeer, J., M. A. Mallona, D. Boucher, K. Yih, and I. Perfecto. 1995. Three years of ingrowth following catastrophic hurricane damage on the Caribbean coast of Nicaragua: Evidence in support of the direct regeneration hypothesis. Journal of Tropical Ecology 11:465–71.
Vander Wall, S. B., and W. S. Longland. 2004. Diplochory: Are two seed dispersers better than one? Trends in Ecology & Evolution 19:155–61.
van der Werf, G., D. Morton, R. DeFries, J. Olivier, P. Kasibhatla, R. Jackson, G. Collatz, and J. Randerson. 2009. CO_2 emissions from forest loss. Nature Geoscience 2:737–38.
van Gelder, H. A., L. Poorter, and F. J. Sterck. 2006. Wood mechanics, allometry, and life-history variation in a tropical rain forest tree community. New Phytologist 171:367–78.
van Gemerden, B. S., H. Olff, M. P. E. Parren, and F. Bongers. 2003. The pristine rain forest? Remnants of historical human impacts on current tree species composition and diversity. Journal of Biogeography 30:1381–90.
van Gemerden, B. S., G. N. Shu, and H. Olff. 2003. Recovery of conservation values in Central African rain forest after logging and shifting cultivation. Biodiversity and Conservation 12:1553–70.
van Nieuwstadt, M. G. L., and D. Sheil. 2005. Drought, fire and tree survival in a Borneo rain forest, East Kalimantan, Indonesia. Journal of Ecology 93:191–201.
van Nieuwstadt, M. G. L., D. Sheil, and K. Kartawinata. 2001. The ecological consequences of logging in the burned forests of East Kalimantan, Indonesia. Conservation Biology 15:1183–86.
van Steenis, C. G. G. J. 1958. Rejuvenation as a factor for judging the status of vegetation types: The biological nomad theory. Pages 212–15 *in* Study of Tropical Vegetation: Proceedings of the Kandy Symposium. UNESCO, Paris.
Vargas, R., M. F. Allen, and E. B. Allen. 2008. Biomass and carbon accumulation in a fire chronosequence of a seasonally dry tropical forest. Global Change Biology 14:109–24.
Vasconcelos, H. L. 1999. Levels of leaf herbivory in Amazonian trees from different stages in forest regeneration. Acta Amazonica 29:615–23.
Vázquez-Yanes, C. 1976. Estudios sobre ecología de la germinación en una zona cálido-húmeda de Mexico. Pages 279–387 *in* A. Gómez-Pompa, S. del Amo, and A. Butanda, eds. Regeneración de Selvas. Editorial Continental, Mexico City.
Vázquez-Yanes, C., and A. Orozco-Segovia. 1993. Patterns of seed longevity and germination in the tropical rainforest. Annual Review of Ecology and Systematics 24:69–87.
Velázquez, E., and A. Gómez-Sal. 2007. Environmental control of early succession on a large landslide in a tropical dry ecosystem (Casita Volcano, Nicaragua). Biotropica 39:601–9.

———. 2008. Landslide early succession in a Neotropical dry forest. Plant Ecology 199:295–308.

Veldkamp, E., A. Becker, L. Schwendenmann, D. A. Clark, and H. Schulte-Bisping. 2003. Substantial labile carbon stocks and microbial activity in deeply weathered soils below a tropical wet forest. Global Change Biology 9:1171–84.

Veldman, J. W. 2008. *Guadua paniculata* (Bambusoideae) in the Bolivian Chiquitania: Fire ecology and the opportunity of a native forage. Revista Boliviana de Ecología y Conservación Ambiental 24:65–74.

Veldman, J. W., B. Mostacedo, M. Peña-Claros, and F. Putz. 2009. Selective logging and fire as drivers of alien grass invasion in a Bolivian tropical dry forest. Forest Ecology and Management 258:1643–49.

Veldman, J. W., and F. E. Putz. 2011. Grass-dominated vegetation, not species-diverse natural savanna, replaces degraded tropical forests on the southern edge of the Amazon Basin. Biological Conservation 144:1419–29.

Velho, N., and M. Krishnadas. 2011. Post-logging recovery of animal-dispersed trees in a tropical forest site in north-east India. Tropical Conservation Science 4:405–19.

Velho, N., J. Ratnam, U. Srinivasan, and M. Sankaran. 2012. Shifts in community structure of tropical trees and avian frugivores in forests recovering from past logging. Biological Conservation 153:32–40.

Vieira, D. L. M., K. D. Holl, and F. M. Peneireiro. 2009. Agro-successional restoration as a strategy to facilitate tropical forest recovery. Restoration Ecology 17:451–59.

Vieira, D. L. M., and A. Scariot. 2006. Principles of natural regeneration of tropical dry forests for restoration. Restoration Ecology 14:11–20.

Vieira, I. C. G., and J. Proctor. 2007. Mechanisms of plant regeneration during succession after shifting cultivation in eastern Amazonia. Plant Ecology 192:303–15.

Vieira, I. C. G., C. Uhl, and D. Nepstad. 1994. The role of the shrub *Cordia multispicata* Cham as a succession facilitator in an abandoned pasture, Paragominas, Amazonia. Vegetatio 115:91–99.

Vílchez Alvarado, B., R. Chazdon, and V. Milla Quesada. 2008. Dinámica de la regeneración en cuatro bosques secundarios tropicales de la región Huetar Norte, Costa Rica. Su valor para la conservación o uso comercial. Recursos Naturales y Ambiente (Costa Rica) 55:118–28.

Villegas, Z., M. Peña-Claros, B. Mostacedo, A. Alarcon, J. C. Licona, C. Leano, W. Pariona, and U. Choque. 2009. Silvicultural treatments enhance growth rates of future crop trees in a tropical dry forest. Forest Ecology and Management 258:971–77.

Vitousek, P. M. 1994. Potential nitrogen fixation during primary succession in Hawaii Volcanoes National Park. Biotropica 26:234–40.

Vitousek, P. M., L. R. Walker, L. D. Whiteaker, and P. A. Matson. 1993. Nutrient limitations to plant growth during primary succession in Hawaii Volcanoes National Park Biogeochemistry 23:197–215.

Voeks, R. A. 1996. Tropical forest healers and habitat preference. Economic Botany 50:381–400.

———. 2004. Disturbance pharmacopoeias: Medicine and myth from the humid tropics. Annals of the Association of American Geographers 94:868–88.

Vulinec, K. 2000. Dung beetles (Coleoptera: Scarabaeidae), monkeys, and conservation in Amazonia. Florida Entomologist 83:229–41.

Vulinec, K., J. E. Lambert, and D. J. Mellow. 2006. Primate and dung beetle communities in secondary growth rain forests: Implications for conservation of seed dispersal systems. International Journal of Primatology 27:855–79.

Wahl, D., R. Byrne, T. Schreiner, and R. Hansen. 2006. Holocene vegetation change in the northern Peten and its implications for Maya prehistory. Quaternary Research 65:380–89.

Waide, R. B., and A. E. Lugo. 1992. A research perspective on disturbance and recovery of a tropical forest. Pages 173–89 *in* J. G. Goldammer, ed. Tropical forests in transition. Birkhauser, Gasel, Switzerland.

Walker, L. R. 1994. Effects of fern thickets on woodland development on landslides in Puerto Rico. Journal of Vegetation Science 5:525–32.

———. 2012. The biology of disturbed habitats. Oxford University Press, New York.

Walker, L. R., and R. del Moral. 2009. Transition dynamics in succession: Implications for rates, trajectories and restoration. Pages 33–50 *in* R. J. Hobbs and K. N. Suding, eds. New models for ecosystem dynamics and restoration. Island Press, Washington, DC.

Walker, L. R., F. H. Landau, E. Velázquez, A. B. Shiels, and A. D. Sparrow. 2010. Early successional woody plants facilitate and ferns inhibit forest development on Puerto Rican landslides. Journal of Ecology 98:625–35.

Walker, L. R., J. Walker, and R. del Moral. 2007. Forging a new alliance between succession and restoration. Pages 1–18 *in* L. R. Walker, J. Walker, and R. J. Hobbs, eds. Linking restoration and ecological succession. Springer, New York.

Walker, L. R., J. Walker, and R. J. Hobbs, eds. 2007. Linking restoration and ecological succession. Springer, New York.

Walker, L. R., D. A. Wardle, R. D. Bardgett, and B. D. Clarkson. 2010. The use of chronosequences in studies of ecological succession and soil development. Journal of Ecology 98:725–36.

Walker, L. R., and M. R. Willig. 1999. An introduction to terrestrial disturbances. Pages 1–16 *in* L. R. Walker, ed. Ecosystems of disturbed ground. Elsevier, Amsterdam, Neth.

Walker, L., E. Velázquez, and A. Shiels. 2009. Applying lessons from ecological succession to the restoration of landslides. Plant and Soil 324:157–68.

Walker, L. R., D. J. Zarin, N. Fetcher, R. W. Myster, and A. H. Johnson. 1996. Ecosystem development and plant succession on landslides in the Caribbean. Biotropica 28:566–76.

Walker, R. 1993. Deforestation and economic development. Canadian Journal of Regional Science 16:481–97.
———. 2012. The scale of forest transition: Amazonia and the Atlantic forests of Brazil. Applied Geography 32:12–20.
Walsh, P. D., P. Henschel, K. A. Abernethy, C. E. G. Tutin, P. Telfer, and S. A. Lahm. 2004. Logging speeds little red fire ant invasion of Africa. Biotropica 36:637–41.
Walsh, S. J., Y. Shao, C. F. Mena, and A. L. McCleary. 2008. Integration of hyperion satellite data and a household social survey to characterize the causes and consequences of reforestation patterns in the northern Ecuadorian Amazon. Photogrammetric Engineering and Remote Sensing 74:725–35.
Waltert, M., K. S. Bobo, S. Kaupa, M. L. Montoya, M. S. Nsanyi, and H. Fermon. 2011. Assessing conservation values: Biodiversity and endemicity in tropical land use systems. PLoS One 6:e16238.
Waltert, M., A. Mardiastuti, and M. Mühlenberg. 2004. Effects of land use on bird species richness in Sulawesi, Indonesia. Conservation Biology 18:1339–46.
Wang, B. C., and T. B. Smith. 2002. Closing the seed dispersal loop. Trends in Ecology & Evolution 17:379–86.
Wang, B. C., V. Sork, M. Leong, and T. Smith. 2007. Hunting of mammals reduces seed removal and dispersal of the afrotropical tree *Antrocaryon klaineanum* (Anacardiaceae). Biotropica 39:340–47.
Wang, Z., and S. S. Young. 2003. Differences in bird diversity between two swidden agricultural sites in mountainous terrain, Xishuangbanna, Yunnan, China. Biological Conservation 110:231–43.
Wangpakapattanawong, P., N. Kavinchan, C. Vaidhayakarn, D. Schmidt-Vogt, and S. Elliott. 2010. Fallow to forest: Applying indigenous and scientific knowledge of swidden cultivation to tropical forest restoration. Forest Ecology and Management 260:1399–1406.
Ward, J., K. Tockner, D. Arscott, and C. Claret. 2002. Riverine landscape diversity. Freshwater Biology 47:517–39.
Watt, A. S. 1947. Pattern and process in the plant community. Journal of Ecology 35:1–22.
Weaver, P. L. 1986. Hurricane damage and recovery in the montane forests of the Luquillo Mountains of Puerto Rico. Caribbean Journal of Science 22:53–70.
Webb, L. J. 1958. Cyclones as an ecological factor in tropical lowland rain-forest, North Queensland. Australian Journal of Botany 6:220–28.
Weinstock, J. A. 1983. Rattan: Ecological balance in a Borneo rainforest swidden. Economic Botany 37:58–68.
Welden, C. W., S. W. Hewett, S. P. Hubbell, and R. B. Foster. 1991. Survival, growth, and recruitment of saplings in canopy gaps and forest understory on Barro Colorado Island, Panamá. Ecology 72:35–50.
Weng, C., M. B. Bush, and J. S. Athens. 2002. Holocene climate change and hydrarch succession in lowland Amazonian Ecuador. Review of Palaeobotany and Palynology 120:73–90.

Wenny, D. G., and D. J. Levey. 1998. Directed seed dispersal by bellbirds in a tropical cloud forest. Proceedings of the National Academy of Sciences 95:6204–7.

Wheelwright, N. T. 1985. Fruit size, gape width, and the diets of fruit-eating birds. Ecology 66:808–18.

Wheelwright, N. T., W. A. Haber, K. G. Murray, and C. Guindon. 1984. Tropical fruit-eating birds and their food plants: A survey of a Costa Rican lower montane forest. Biotropica 16:173–92.

White, G., D. Boshier, and W. Powell. 2002. Increased pollen flow counteracts fragmentation in a tropical dry forest: An example from *Swietenia humilis* Zuccarini. Proceedings of the National Academy of Sciences 99:2038–42.

White, J., D. Penny, L. Kealhofer, and B. Maloney. 2004. Vegetation changes from the late Pleistocene through the Holocene from three areas of archaeological significance in Thailand. Quaternary International 113:111–32.

White, L. T. J. 2001a. The African rain forest: Climate and vegetation. Pages 3–29 *in* W. Weber, L. J. T. White, A. Vedder, and L. Naughton-Treves, eds. African rain forest ecology and conservation: An interdisciplinary perspective. Yale University Press, New Haven, CT.

———. 2001b. Forest-savanna dynamics and the origins of Marantaceae. Pages 165–82 *in* W. Weber, ed. African rain forest ecology and conservation: An interdisciplinary perspective. Yale University Press, New Haven, CT.

White, L. J. T., and J. F. Oates. 1999. New data on the history of the plateau forest of Okomu, southern Nigeria: An insight into how human disturbance has shaped the African rain forest. Global Ecology and Biogeography 8:355–61.

White, P. S. 1979. Process and natural disturbance in vegetation. Botanical Review 45:229–99.

Whitfeld, T., W. Kress, D. Erickson, and G. Weiblen. 2011. Change in community phylogenetic structure during tropical forest succession: Evidence from New Guinea. Ecography 35:821–30.

Whitfeld, T., V. Novotny, S. E. Miller, J. Hrcek, P. Klimes, and G. Weiblen. 2012. Predicting tropical insect herbivore abundance from host plant traits and phylogeny. Ecology 93:S211-S222.

Whitmore, T. C. 1974. Change with time and the role of cyclones in tropical rain forest on Kolombangara, Solomon Islands. Institute Paper No. 46. Commonwealth Forestry Institute, University of Oxford. Oxford.

———. 1978. Gaps in the forest canopy. Pages 639–55 *in* P. B. Tomlinson and M. M. Zimmerman, eds. Tropical trees as living systems. Cambridge University Press, New York.

———. 1983. Secondary succession from seed in tropical rain forests. Forestry Abstracts 44:767–779.

———. 1984. Tropical rain forests of the Far East. 2nd ed. Clarendon Press, Oxford.

———. 1989a. Canopy gaps and the two major groups of forest trees. Ecology 70:536–38.

———. 1989b. Changes over twenty-one years in the Kolombangara rain forests. Journal of Ecology 77:469–83.
———. 1989c. Forty years of rain forest ecology 1948–1988 in perspective. GeoJournal 19:347–60.
———. 1990. An introduction to tropical rain forests. Oxford University Press, Oxford.
———. 1991. Tropical rain forest dynamics and its implications for management. Pages 67–89 *in* A. Gómez-Pompa, T. C. Whitmore, and M. Hadley, eds. Rain Forest Regeneration and Management. UNESCO/Parthenon, Paris.
———. 1996. A review of some aspects of tropical rain forest seedling ecology with suggestions for further enquiry. Pages 3–39 *in* M. D. Swaine, ed. The ecology of tropical forest tree seedlings. Parthenon, Paris.
———. 1998a. An introduction to tropical rain forests. 2nd ed. Oxford University Press, Oxford.
———. 1998b. A pantropical perspective on the ecology that underpins management of tropical secondary rain forests. Pages 19–34 *in* M. R. Guariguata and B. Finegan, eds. Proceedings of a conference on ecology and management of tropical secondary forest: Science, people, and policy, November 10–12, 1997. CATIE, Turrialba, Costa Rica.
———. 1998c. Potential impact of climate change on tropical rain forest seedlings and forest regeneration. Climatic Change 39:429–38.
Whitmore, T. C., N. D. Brown, M. D. Swaine, D. Kennedy, C. I. Goodwin-Bailey, and W.-K. Gong. 1993. Use of hemispherical photographs in forest ecology: Measurement of gap size and radiation totals in a Bornean tropical rain forest. Journal of Tropical Ecology 9:131–51.
Whitmore, T. C., and C. Burnham. 1975. Tropical rain forests of the Far East. Clarendon Press, Oxford.
Whitmore, T. C., and D. F. R. P. Burslem. 1988. Major disturbances in tropical rainforests. Pages 549–65 *in* D. M. Newbery, H. H. T. Prins, and N. D. Brown, eds. Dynamics of tropical communities. Blackwell Science, Oxford.
Whitmore, T. M., and B. L. Turner II. 2001. Cultivated landscapes of Middle America on the eve of conquest. Oxford University Press, Oxford.
Whittaker, R. H. 1953. A consideration of climax theory: The climax as a population and pattern. Ecological Monographs 23:41–78.
Whittaker, R. J., M. B. Bush, and K. Richards. 1989. Plant recolonization and vegetation succession on the Krakatau Islands, Indonesia. Ecological Monographs 59:59–123.
Whittaker, R. J., and S. H. Jones. 1994. The role of frugivorous bats and birds in the rebuilding of a tropical forest ecosystem, Krakatau, Indonesia. Journal of Biogeography 21:245–58.
Whittaker, R. J., T. Partomihardjo, and S. Jones. 1999. Interesting times on Krakatau: Stand dynamics in the 1990s. Philosophical Transactions of the Royal Society B: Biological Sciences 354:1857–67.

Wieland, L. M., R. C. G. Mesquita, P. E. D. Bobrowiec, T. V. Bentos, and G. B. Williamson. 2011. Seed rain and advance regeneration in secondary succession in the Brazilian Amazon. Tropical Conservation Science 4:300–316.
Wiemann, M. C., and G. B. Williamson. 1988. Extreme radial changes in wood specific gravity in some tropical pioneers. Wood and Fiber Science 20:344–49.
Wiersum, K. 1997. Indigenous exploitation and management of tropical forest resources: An evolutionary continuum in forest-people interactions. Agriculture, Ecosystems & Environment 63:1–16.
Wijdeven, S. M. J., and M. E. Kuzee. 2000. Seed availability as a limiting factor in forest recovery processes in Costa Rica. Restoration Ecology 8:414–24.
Wilcove, D. S., and L. P. Koh. 2010. Addressing the threats to biodiversity from oil-palm agriculture. Biodiversity and Conservation 19:999–1007.
Williams, M. 2008. A new look at global forest histories of land clearing. Annual Review of Environment and Resources 33:345–67.
Williams-Linera, G., C. Alvarez-Aquino, E. Hernández-Ascención, and M. Toledo. 2011. Early successional sites and the recovery of vegetation structure and tree species of the tropical dry forest in Veracruz, Mexico. New Forests 42:131–48.
Williamson, G. B., T. V. Bentos, J. B. Longworth, and R. C. Mesquita. 2012. Convergence and divergence in alternative successional pathways in Central Amazonia. Plant Ecology & Diversity. doi:10.1080/17550874.2012.735714.
Williamson, G. B., and M. C. Wiemann. 2010. Age-dependent radial increases in wood specific gravity of tropical pioneers in Costa Rica. Biotropica 42:590–97.
Willig, M. R., S. J. Presley, C. P. Bloch, C. L. Hice, S. P. Yanoviak, M. M. Díaz, L. A. Chauca, V. Pacheco, and S. C. Weaver. 2007. Phyllostomid bats of lowland Amazonia: Effects of habitat alteration on abundance. Biotropica 39:737–46.
Willis, K., M. Araujo, K. Bennett, B. Figueroa-Rangel, C. Froyd, and N. Myers. 2007. How can a knowledge of the past help to conserve the future? Biodiversity conservation and the relevance of long-term ecological studies. Philosophical Transactions of the Royal Society B: Biological Sciences 362:175–87.
Willis, K., L. Gillison, and T. Brncic. 2004. How "virgin" is virgin rainforest? Science 304:402–3.
Wirth, C., C. Messier, Y. Bergeron, D. Frank, and Fankhänel. 2009. Old-growth forest definitions: A pragmatic view. Pages 11–33 *in* C. Wirth, G. Gleixner, and M. Heimann, eds. Old-growth forests: Function, fate and value. Springer, New York.
Wiseman, F. M. 1978. Agricultural and historical ecology of the Maya lowlands. Pages 63–116 *in* P. D. Harrison and B. L. Turner, eds. Pre-Hispanic Maya agriculture. University of New Mexico Press, Albuquerque.
Wittmann, F., W. Junk, and M. Piedade. 2004. The várzea forests in Amazonia: Flooding and the highly dynamic geomorphology interact with natural forest succession. Forest Ecology and Management 196:199–212.
Wood, T. E., D. Lawrence, D. A. Clark, and R. L. Chazdon. 2009. Rain forest nutrient cycling and productivity in response to large-scale litter manipulation. Ecology 90:109–21.

Wood, T. E., D. Lawrence, and J. A. Wells. 2011. Interspecific variation in foliar nutrients and resorption of nine canopy-tree species in a secondary Neotropical rain forest. Biotropica 43:544–51.

Woods, C. L., and S. J. DeWalt. 2013. The conservation value of secondary forests for vascular epiphytes in Central Panama. Biotropica 45:119–27.

Woods, P. 1989. Effects of logging, drought, and fire on structure and composition of tropical forests in Sabah, Malaysia. Biotropica 21:290–98.

Worbes, A., R. Staschel, A. Roloff, and W. J. Junk. 2003. Tree ring analysis reveals age structure, dynamics and wood production of a natural forest stand in Cameroon. Forest Ecology and Management 173:105–23.

Worbes, M., H. Klinge, J. Revilla, and C. Martius. 1992. On the dynamics, floristic subdivision and geographical distribution of várzea forests in Central Amazonia. Journal of Vegetation Science 3:553–64.

Wright, I. J., P. B. Reich, M. Westoby, D. D. Ackerly, Z. Baruch, F. Bongers, J. Cavender-Bares, et al. 2004. The worldwide leaf economics spectrum. Nature 428:821–27.

Wright, S. J. 2005. Tropical forests in a changing environment. Trends in Ecology & Evolution 20:553–60.

———. 2010. The future of tropical forests. Annals of the New York Academy of Sciences 1195:1–27.

Wright, S. J., O. Calderón, A. Hernández, and S. Paton. 2004. Are lianas increasing in importance in tropical forests? A 17-year record from Panama. Ecology 85:484–89.

Wright, S. J., A. Hernandéz, and R. Condit. 2007. The bushmeat harvest alters seedling banks by favoring lianas, large seeds, and seeds dispersed by bats, birds, and wind. Biotropica 39:363–71.

Wright, S. J., K. Kitajima, N. J. B. Kraft, P. B. Reich, I. J. Wright, D. E. Bunker, R. Condit, et al. 2010. Functional traits and the growth-mortality trade-off in tropical trees. Ecology 91:3664–74.

Wright, S. J., and H. C. Muller-Landau. 2006. The future of tropical forest species. Biotropica 38:287–301.

Wright, S. J., H. C. Muller-Landau, R. Condit, and S. P. Hubbell. 2003. Gap-dependent recruitment, realized vital rates, and size distributions of tropical trees. Ecology 84:3174–85.

Wright, S. J., H. C. Muller-Landau, and J. Schipper. 2009. The future of tropical species on a warmer planet. Conservation Biology 23:1418–26.

Wright, S. J., and M. J. Samaniego. 2008. Historical, demographic, and economic correlates of land-use change in the Republic of Panama. Ecology and Society 13. http://www.ecologyandso ciety.org/vol13/iss2/art17/.

Wright, S. J., K. E. Stoner, N. Beckman, R. T. Corlett, R. Dirzo, H. C. Muller-Landau, G. Nunez-Iturri, C. A. Peres, and B. C. Wang. 2007. The plight of large animals in tropical forests and the consequences for plant regeneration. Biotropica 39:289–91.

Wright, S. J., J. B. Yavitt, N. Wurzburger, B. L. Turner, E. V. J. Tanner, E. J. Sayer, L. S. Santiago, M. Kaspari, L. O. Hedin, and K. E. Harms. 2011. Potassium, phosphorus,

or nitrogen limit root allocation, tree growth, or litter production in a lowland tropical forest. Ecology 92:1616–25.
Wroe, S., J. Field, and D. K. Grayson. 2006. Megafaunal extinction: Climate, humans and assumptions. Trends in Ecology & Evolution 21:61–62.
Wu, J., and O. Loucks. 1995. From balance of nature to hierarchical patch dynamics: A paradigm shift in ecology. Quarterly Review of Biology 70:439.
Wuethrich, B. 2007. Reconstructing Brazil's Atlantic rainforest. Science 315:1070–72.
Wunderle, J. M., Jr. 1997. The role of animal seed dispersal in accelerating native forest regeneration on degraded tropical lands. Forest Ecology and Management 99:223–35.
Wunderle, J. M., Jr., D. J. Lodge, and R. B. Waide. 1992. Short-term effects of Hurricane Gilbert on terrestrial bird populations on Jamaica. Auk 109:148–66.
Wyatt-Smith, J. 1954. Storm forest in Kelantan. Malayan Forester 17:5–11.
Xu, J. 2007. Rattan and tea-based intensification of shifting cultivation by Hani farmers in Southwestern China. Pages 664–72 *in* M. Cairns, ed. Voices from the forest: Integrating indigenous knowledge into sustainable upland farming. Resources for the Future, Washington, DC.
Yackulic, C. B., M. Fagan, M. Jain, A. Jina, Y. Lim, M. Marlier, R. Muscarella, et al. 2011. Biophysical and socioeconomic factors associated with forest transitions at multiple spatial and temporal scales. Ecology and Society 16:15. doi:10.5751/ES-04275-160315.
Yang, X., T. Richardson, and A. Jain. 2010. Contributions of secondary forest and nitrogen dynamics to terrestrial carbon uptake. Biogeosciences 7:3041–50.
Yarranton, G., and R. Morrison. 1974. Spatial dynamics of a primary succession: Nucleation. Journal of Ecology 62:417–28.
Yassir, I., J. van der Kamp, and P. Buurman. 2010. Secondary succession after fire in *Imperata* grasslands of East Kalimantan, Indonesia. Agriculture, Ecosystems & Environment 137:172–82.
Yen, D. E. 1974. Arboriculture in the subsistence of Santa Cruz, Solomon Islands. Economic Botany 28:247–84.
———. 1995. The development of Sahul agriculture with Australia as bystander. Antiquity 69:831–47.
———. 1996. Melanesian arboriculture: Historical perspectives with emphasis on the genus *Canarium*. Pages 36–44 *in* M. L. Stevens, R. M. Bourke, and B. R. Evans, eds. South Pacific indigenous nuts. Australian Centre for International Agricultural Research, Canberra, ACT.
Yih, K., D. H. Boucher, J. H. Vandermeer, and N. Zamora. 1991. Recovery of the rain forest of southeastern Nicaragua after destruction by Hurricane Joan. Biotropica 23:106–13.
Yu, Z. Y., Z. H. Wang, and S. Y. He. 1994. Rehabilitation of eroded tropical coastal land in Guangdong, China. Journal of Tropical Forest Science 7:28–38.
Zahawi, R. A., and C. K. Augspurger. 1999. Early plant succession in abandoned pastures in Ecuador. Biotropica 31:540–52.

———. 2006. Tropical forest restoration: Tree islands as recruitment foci in degraded lands of Honduras. Ecological Applications 16:464–78.

Zalamea, P. C., P. Heuret, C. Sarmiento, M. Rodríguez, A. Berthouly, S. Guitet, E. Nicolini, C. Delnatte, D. Barthélémy, and P. R. Stevenson. 2012. The genus *Cecropia*: A biological clock to estimate the age of recently disturbed areas in the Neotropics. PLoS One 7:e42643.

Zambrano, A. M. A., E. N. Broadbent, and W. H. Durham. 2010. Social and environmental effects of ecotourism in the Osa Peninsula of Costa Rica: The Lapa Rios case. Journal of Ecotourism 9:62–83.

Zangaro, W., R. A. Alves, L. E. Lescano, and A. P. Ansanelo. 2012. Investment in fine roots and arbuscular mycorrhizal fungi decrease during succession in three Brazilian ecosystems. Biotropica 44:141–50.

Zangaro, W., S. M. A. Nisizaki, J. C. B. Domingos, and E. M. Nakano. 2003. Mycorrhizal response and successional status in 80 woody species from south Brazil. Journal of Tropical Ecology 19:315–24.

Zanne, A. E., C. A. Chapman, and K. Kitajima. 2005. Evolutionary and ecological correlates of early seedling morphology in East African trees and shrubs. American Journal of Botany 92:972–78.

Zarin, D. J., E. A. Davidson, E. Brondizio, I. C. G. Vieira, T. Sa, T. R. Feldpausch, E. A. G. Schuur, et al. 2005. Legacy of fire slows carbon accumulation in Amazonian forest regrowth. Frontiers in Ecology and the Environment 3:365–69.

Zarin, D. J., M. J. Ducey, J. M. Tucker, and W. A. Salas. 2001. Potential biomass accumulation in Amazonian regrowth forests. Ecosystems 4:658–68.

Zarin, D. J., and A. H. Johnson. 1995. Nutrient accumulation during primary succession in a montane tropical Forest, Puerto Rico. Soil Science Society of America Journal 59:1444–52.

Zhang, Y., and C. Song. 2006. Impacts of afforestation, deforestation, and reforestation on forest cover in China from 1949 to 2003. Journal of Forestry 104:383–87.

Zhang, Z. D., R. G. Zang, and Y. D. Qi. 2008. Spatiotemporal patterns and dynamics of species richness and abundance of woody plant functional groups in a tropical forest landscape of Hainan Island, South China. Journal of Integrative Plant Biology 50:547–58.

Ziegler, A. D., J. M. Fox, E. L. Webb, C. Padoch, S. J. Leisz, R. Cramb, O. Mertz, T. B. Bruun, and T. D. Vien. 2011. Recognizing contemporary roles of swidden agriculture in transforming landscapes of Southeast Asia. Conservation Biology 25:846.

Zimmermann, B., H. Elsenbeer, and J. M. De Moraes. 2006. The influence of land-use changes on soil hydraulic properties: Implications for runoff generation. Forest Ecology and Management 222:29–38.

Zimmerman, B. L., and C. F. Kormos. 2012. Prospects for sustainable logging in tropical forests. BioScience 62:479–87.

Zimmerman, J. K., T. M. Aide, M. Rosario, M. Serrano, and L. Herrera. 1995. Effects of land management and a recent hurricane on forest structure and composition in the Luquillo Experimental Forest, Puerto Rico. Forest Ecology and Management 77:65–76.

Zimmerman, J. K., E. M. Everham, R. B. Waide, D. J. Lodge, C. M. Taylor, and N. Brokaw. 1994. Responses of tree species to hurricane winds in subtropical wet forest in Puerto Rico: Implications for tropical tree life histories. Journal of Ecology 82:911–22.

INDEX

The letter *b* following a page number denotes a box; the letter *f*, a figure; and the letter *t*, a table. References to plates are italicized.